# Handbook of Digital Innovation, Transformation, and Sustainable Development in a Post-Pandemic Era

Businesses have faced a variety of difficulties as a result of the global pandemic, and how they responded to this disruption has affected both their resilience and their ability to get through this crisis. Digital technologies have played a crucial role in addressing these issues and fostering resilience. It is, therefore, imperative to explore options for post-pandemic business transformation and rethinking sustainable development.

*Handbook of Digital Innovation, Transformation, and Sustainable Development in a Post-Pandemic Era* covers digital innovation and business transformation to build resilience for sustainable development and growth and highlights the impact of supply chain disruptions and solutions to sustain. The handbook is unique in the sense that it includes industry cases from different parts of the world. By presenting the use of big data, blockchain technology, and Industry 4.0, the handbook conveys how to work towards sustainable development and offers self-reliant and sustainable business models.

Researchers and practitioners in industrial engineering, engineering management, business management, supply chain management, and digital technologies along with businesses can apply the research and practices covered in the handbook.

## Systems Innovation Book Series

*Series Editor: Adedeji Badiru*

Systems Innovation refers to all aspects of developing and deploying new technology, methodology, techniques, and best practices in advancing industrial production and economic development. This entails such topics as product design and development, entrepreneurship, global trade, environmental consciousness, operations and logistics, introduction and management of technology, collaborative system design, and product commercialization. Industrial innovation suggests breaking away from the traditional approaches to industrial production. It encourages the marriage of systems science, management principles, and technology implementation. Particular focus will be the impact of modern technology on industrial development and industrialization approaches, particularly for developing economics. The series will also cover how emerging technologies and entrepreneurship are essential for economic development and society advancement.

**Industrial Engineering in Systems Design**
Guidelines, Practical Examples, Tools, and Techniques
*Brian Peacock and Adedeji B. Badiru*

**Leadership Matters**
An Industrial Engineering Framework for Developing and Sustaining Industry
*Adedeji B. Badiru and Melinda Tourangeau*

**Systems Engineering**
Influencing Our Planet and Reengineering Our Actions
*Adedeji B. Badiru*

**Total Productive Maintenance, Second Edition**
Strategies and Implementation Guide
*Tina Agustiady and Elizabeth A. Cudney*

**Assessing Innovation**
Metrics, Rubrics and Standards
*Adedeji B. Badiru and Melinda Tourangeau*

**Integrating Artificial and Human Intelligence through Agent Oriented Systems Design**
*Michael E. Miller and Christina F. Rusnock*

**Handbook of Digital Innovation, Transformation, and Sustainable Development in a Post-Pandemic Era**
*Edited by M. Affan Badar, Ruchika Gupta, Priyank Srivastava, Imran Ali, and Elizabeth A. Cudney*

# Handbook of Digital Innovation, Transformation, and Sustainable Development in a Post-Pandemic Era

Edited by
M. Affan Badar, Ruchika Gupta, Priyank Srivastava,
Imran Ali, and Elizabeth A. Cudney

CRC Press
Taylor & Francis Group
Boca Raton London New York

CRC Press is an imprint of the
Taylor & Francis Group, an **informa** business

Designed cover image: Shutterstock—Hurca

First edition published 2025
by CRC Press
2385 NW Executive Center Drive, Suite 320, Boca Raton FL 33431

and by CRC Press
4 Park Square, Milton Park, Abingdon, Oxon, OX14 4RN

*CRC Press is an imprint of Taylor & Francis Group, LLC*



This Open Access chapter is funded by Tallinn University of Technology.



ISBN: 978-1-032-56688-7 (hbk)
ISBN: 978-1-032-57293-2 (pbk)
ISBN: 978-1-003-43874-8 (ebk)

DOI: 10.1201/9781003438748

Typeset in Times
by Apex CoVantage LLC

# Contents

## SECTION 3 Sustainability Strategies and Management

# Preface

In the wake of global challenges posed by the COVID-19 pandemic, there has been a pronounced shift towards leveraging digital technologies not only to recover and rebuild but also to foster sustainable practices for a resilient future. This transition to digital and the pursuit of sustainability goals are generating considerable changes in organizations around the world. With the development of digital technologies such as artificial intelligence, blockchain, Internet of Things (IoT), cryptocurrency, digital twins, and others, digital transformation has a positive influence on several aspects of society. Digital transformation is seen as an efficient means for businesses to attain sustainable growth.

Moreover, the global pandemic has magnified the consequences of the expanding digital divide, thereby enhancing present inequities between developing and developed nations. It has realized and consolidated our dependence on digital technology in nearly every sphere of our daily lives, for example, online education and working outside the office, as well as healthcare, e-commerce, and market access.

Businesses have faced several problems with the global pandemic, and their adaptability as well as their resistance to this mismatch has been affected. Digital technologies have played a vital role in addressing these challenges and fostering resilience. It is, therefore, imperative to discover options for post-pandemic business transformation and for reconsidering sustainable development. The aim of *Handbook of Digital Innovation, Transformation, and Sustainable Development in a Post-Pandemic Era* is to examine how business transformation and sustainability may be effectively managed. The book has 19 chapters, grouped into four sections, with each concentrating on some significant issues where digital innovation is able to push or aid in sustaining development in the modern world.

**Section 1: Digital Innovation and Transformation for Sustainability** delves into the basic tenets of digital innovation and transformation and why they are important in achieving sustainability goals. From disruptive technologies to novel business models, this section serves as a starting point for understanding how such digital innovations can be harnessed to enable sustainable growth. This section consists of six chapters wherein Chapter 1, "A Framework for Building Resilience through Innovation and Process Optimization in AI-Powered Digital Transformation", gives a detailed outline of the basic rules which are necessary for blending machine operations into a system that is in the course of development of new AI technologies to make them operate smoothly. Chapter 2, "Driving Business Sustainability through Digital Supply Chains", highlights setting of clear sustainability goals, utilizing digital tools for real-time tracking, and enabling transparent reporting. It also explores how digital supply chains bolster resilience, ethical sourcing, traceability, and circular economy practices. Chapter 3, "Digital Transformation Framework for Logistics Service Providers: Systematic Literature Review", presents a systematic literature review that aims to develop a conceptual framework for digital transformation in logistics service providers' business models. The framework integrates the identified factors and dimensions, offering a structured approach to logistics service providers

(LSPs) and guiding logistics companies in their journey towards digitalization. Chapter 4, "Achieving Business Sustainability with Blockchain Technology", aims to provide an overarching overview of the potential of blockchain in achieving economic, environmental, and social sustainability using six case studies. Chapter 5, "Blockchain Revolution: Transforming Business Sustainability in the Era of Industry 4.0 and Financial Innovation", explores the convergence of blockchain technology and business sustainability, highlighting opportunities and challenges. Chapter 6, "Supply Chain Risk Management, Optimization, and Performance Analytics through Blockchain Integration: Bibliometric Analysis", entails a bibliometric analysis to review how blockchain technology affects optimization and risk management for better coordination and easier supply chain operations.

**Section 2: Environmental Sustainability and Digitalization** delves into the specific ways in which digital technologies are being employed to address environmental challenges. This section comprises three chapters wherein Chapter 7, "Environmental Sustainability: Digitalization and Big Data", discusses how firms may optimize their business models and develop a sustainable competitive advantage by using big data analytics to comprehend the digital customer experience. Chapter 8, "Green Logistics: A New Frontier for Economic Growth and Social Development in Türkiye and Poland", aims to assess whether Polish and Turkish logistics companies align with contemporary developments in green logistics, as documented in the existing literature. Chapter 9, "Energy-Efficient Design Strategies for Buildings: Lessons from High Performance Buildings", points out the need of energy-efficient building design for decreasing carbon emissions, energy use, climate change, and occupant quality of life.

**Section 3: Sustainability Strategies and Management** shifts focus to organizational and strategic perspectives, examining how businesses and institutions can integrate sustainability into their core operations through digital strategies. This section consists of four chapters wherein Chapter 10, "From Industry 4.0 to Industry 5.0: Transitioning to Circular Business Paradigms—A Review", aims to systematically review the extensive collaboration of Industry 5.0 with the circular economy by a model extending direct linkage between production processes and information flow. Chapter 11, "Industry 4.0 and Sustainability of Human Resources", discusses how Industry 4.0, or cyber-physical systems, could potentially help businesses eliminate or minimize factors that undermine the sustainability of human resources and optimize other factors that can improve the sustainability of human resources. Chapter 12, "Model of Dual Management at Company Scale", discusses the concept of dual management, which coordinates its activities both to achieve goals based on results, to increase its capabilities, and, in general, to improve efficiency. Chapter 13, "A Stochastic Programming Model for Managing Supply Chain Disruptions in the Post-COVID-19 Era", aims to investigate resilience strategies along with resilience metrics and to observe how a supply chain operates when faced with facility shutdowns.

**Section 4: Digital Innovation, Transformation, and Sustainability: Industry Cases** offers practical insights through case studies from various industries. These cases illustrate real-world applications of digital innovation and transformation initiatives that have contributed to sustainable outcomes, highlighting lessons learned and best practices for implementation. This section comprises six chapters, wherein Chapter, 14 "Profitability of Shipping and the Role of Fleet Ownership—North European Company Case Study", brings knowledge concerning the impact of various ownership and workforce strategies to a shipping company's profitability in different market situations. Chapter 15, "A Case Study: Lean Transformation at Sultan Qaboos University Hospital in Oman", presents a novel application of a lean methodology at the Sultan Qaboos University hospital. Chapter 16, "Tata Motors: Driving into the Future with Sustainability and Innovation", examines the various sustainability and innovative practices that helped the Tata Motors company emerge as a market leader in the Indian EV market. Chapter 17, "Blockchain in Healthcare: A Critical Evaluation", evaluates the various opportunities and challenges posed by the technology and also critically evaluates the various blockchain platforms available for healthcare sector focusing on their strengths and applicability. Chapter 18, "Digital Transformation in Advertising after COVID-19: Routing the New Landscape", examines how advertising has undergone a digital transition after COVID-19, emphasizing the expansion of online advertising platforms, including social media, search engines, and programmatic advertising. Chapter 19, "Identifying Predictors and Comparing Machine Learning Algorithms to Detect Erroneous Medicaid Adjudication in the USA", focuses on a transformational approach to verifying claim adjudication accuracy using supervised machine learning to identify payment errors.

Together, these sections provide a holistic view of the synergies between digital innovation, transformation, and sustainable development. They underscore the urgent need for collaborative efforts across sectors to harness the full potential of digital technologies in achieving global sustainability goals.

This book is intended for academicians, researchers, practitioners, and policymakers interested in understanding and advancing the nexus of digital innovation, transformation, and sustainable development. By bridging theory and practice, it aims to inspire and inform innovative approaches that can pave the way towards a more sustainable and inclusive post-pandemic era.

We extend our sincere gratitude to all the contributors whose contributions not only enrich the discourse but also provide actionable insights that can drive meaningful change in our collective journey towards a sustainable future. Additionally, we would like to thank CRC Press for their professional assistance and convey our gratitude for this opportunity.

Editors:
**M. Affan Badar**
**Ruchika Gupta**
**Priyank Srivastava**
**Imran Ali**
**Elizabeth A. Cudney**

# About the Editors

**M. Affan Badar, PhD, CPEM** is a professor, former department chair, and current director of the consortium-based PhD in Technology Management program in the Bailey College of Engineering & Technology at Indiana State University (ISU), USA. While on leave from ISU from 2016 to 2018, he was Professor and Chair of the Industrial Engineering & Engineering Management Department at University of Sharjah (UAE). Prof. Badar earned his PhD in industrial engineering (IE) from the University of Oklahoma (USA), MS in mechanical engineering (ME) from King Fahd University of Petroleum and Minerals (Saudi Arabia), and MSc in IE and BSc (Hons) in ME from Aligarh Muslim University (India). Prof. Badar has published 1 edited book, 5 book chapters, and over 90 articles in refereed journals and proceedings in the areas of lean six sigma, design, reliability, quality, engineering economy, healthcare, and supply chains. He received the ISU Theodore Dreiser Distinguished Research/Creativity Award in 2023, IEOM Fellow Award in 2022, and Albert Nelson Marquis Lifetime Achievement Award in 2017. Prof. Badar has been serving as an associate editor for the *Engineering Management Journal*, editorial board member for *Int. J. of Quality & Reliability Management* and *NED University J. of Research: Applied Sciences*, and reviewer for various journals and conferences. He was Editor-in-Chief for *Int. J. of Forensic Engr & Mgt* and Editor for *Advances in Mechanical Engineering*.

**Ruchika Gupta, PhD,** is a professor and director at Mangalmay Institute of Management & Technology, Greater Noida, INDIA. She has academic experience of over 20 years during which she has played multifaceted roles as a teacher, trainer, mentor, and administrator. She is a certified trainer and judge for "Social Innovation Generation" by HEC, Montreal, Canada, and holds a PhD with three master's degrees. Her research areas are diverse and focus on interdisciplinary research. Dr. Gupta has published 1 edited book, 1 authored book, 6 case studies, and over 80 articles in refereed journals and proceedings. She has been granted 35 copyrights and 5 patents and also serves as editor, reviewer, and advisory board member of several journals of repute. Her remarkable achievements have garnered numerous accolades, including the Women Empowerment Award and Rashtra Shakti Shiromani Samman.

**Priyank Srivastava, PhD,** is an associate professor in the Department of Operations, School of Business, Galgotias University, Greater Noida, Uttar Pradesh, India. He has vast academic and research experience of 18 years. He completed graduation in 2001, post-graduation in 2006, and PhD in 2018, from Amravati University and Madan Mohan Malviya Engineering College, Gorakhpur, and National Institute of Technology, Kurukshetra, respectively. He has published numerous articles in international journals and conferences, book chapters, and edited books. He is a reviewer of many journals of international repute. He is a certified risk management assessor from BSI. His research interests are operations management, maintenance management, risk and reliability analysis, supply chain management, business analytics, and MCDM techniques. He is a life member of ISTE.

**Imran Ali, PhD, MBA,** is a senior lecturer in the Operations and Innovation Management department at the School of Business and Law, Central Queensland University, Australia. Dr. Ali holds a PhD in business management (logistics and supply chain management) from the University of South Australia, serves as a senior editor for the *International Journal of Emerging Markets*, and sits on the review board of various journals. He has 17 years' experience of working with academic and industrial institutions within Australia and overseas. His research focuses on operations and supply chain management, covering several emerging topics such as digital transformation, risk management and resilience, sustainability, and climate change and appears in several high-quality journals. His research has received several accolades, such as Best Proceeding Paper at the Academy of Management 2019 Conference (USA); Best Track Paper Award at the Industrial Engineering and Operations Management (IEOM) Conference 2022, Sydney, Australia; Vice-Chancellor's Research Commendation Award for Outstanding Early-Career Researcher 2021; Vice-Chancellor's Award for Outstanding Early-Career Researcher 2022; and Dean's Award for Outstanding Early-Career Researcher, 2021 and 2022 (two years in a row), at CQUniversity. Owing to a diversified experience and a passion for innovative and industry-engaged education, Dr. Ali enjoys teaching at the interface of theoretical knowledge and real-world industry practices, including a range of subjects.

**Elizabeth A. Cudney, PhD, CLSSMBB,** is President of Cudney Consulting Group, LLC, and a professor of data analytics in the John E. Simon School of Business at Maryville University. She received her BS in industrial engineering from North Carolina State University, master of engineering in mechanical engineering and MBA from the University of Hartford, and doctorate in engineering management from the University of Missouri–Rolla. She received the 2023 Walter E. Masing Book Prize from the International Academy for Quality for her book on lean sustainability. Dr. Cudney received the 2022 Crosby Medal from ASQ and the 2021 Bernard R. Sarchet Award from ASEE EMD for "lifetime achievement in engineering management education". She received the 2021 Walter E. Masing Book Prize from the International Academy for Quality for her book on lean Six Sigma. In 2018, Dr. Cudney received the ASQ Crosby Medal for her book on design for Six Sigma. Dr. Cudney received the 2018 IISE Fellow Award. She also received the 2017 Yoshio Kondo Academic Research Prize from the International Academy for Quality for sustained performance in exceptional published works. In 2014, Dr. Cudney was elected as an ASEM fellow. In 2013, Dr. Cudney was elected as an ASQ fellow. In 2010, Dr. Cudney was inducted into the International Academy for Quality. She received the 2008 ASQ A.V. Feigenbaum Medal and the 2006 SME Outstanding Young Manufacturing Engineering Award. She has published 14 books and over 130 peer-reviewed journal papers. Dr. Cudney is a certified Lean Six Sigma Master Black Belt. She holds eight ASQ certifications, which include ASQ Certified Quality Engineer, Manager of Quality/Operational Excellence, and Certified Six Sigma Black Belt, among others.

# Contributors

**A. Zafer Acar**
Faculty of Applied Sciences
İstanbul Bilgi University
Istanbul, Türkiye

**Abdulaziz Aldoseri**
Engineering College
Qatar University
Doha, Qatar

**Imran Ali**
School of Business and Law
Central Queensland University
Melbourne, Victoria, Australia

**Yousuf Al Khamisi**
Quality Assurance
Sultan Qaboos University
Muscat, Oman

**Khalifa N. Al-Khalifa**
Engineering College
Qatar University
Doha, Qatar

**M. Affan Badar**
Bailey College of Engineering and Technology
Indiana State University
Terre Haute, Indiana

**Mucahit Cevik**
Industrial Engineering
Toronto Metropolitan University
Toronto, Ontario, Canada

**Atanu Chaudhuri**
Technology and Operations Management
Durham University
Durham, UK

**Gül Denktaş-Şakar**
Logistics and Operations Management
Dokuz Eylul University
Izmir, Turkey

**Svetlana De Vos**
Marketing and Entrepreneurship
Australian Institute of Business
Adelaide, South Australia, Australia

**Ganesha K. S.**
Department of Management Studies
Dayananda Sagar College of Engineering
Bengaluru, Karnataka, India

**Ritu Bala Garg**
Department of Civil Engineering
Punjabi University
Patiala, Punjab, India

**Seçil Gülmez**
Estonian Maritime Academy
Tallinn University of Technology
Tallinn, Estonia

**Ruchika Gupta**
Mangalmay Institute of Management & Technology
Greater Noida, Uttar Pradesh, India

**Abdel Magid Hamouda**
Engineering College
Qatar University
Doha, Qatar

**K. G. Hemalatha**
Department of Management Studies
Dayananda Sagar College of Engineering
Bengaluru, Karnataka, India

**Olli-Pekka Hilmola**
Estonian Maritime Academy
Tallinn University of Technology
Tallinn, Estonia
Industrial Engineering and Management
University of Gävle
Gävle, Gävleborg, Sweden

**Cyrus Hoseini**
Bailey College of Engineering and Technology
Indiana State University
Terre Haute, Indiana

**Niamat Ullah Ibne Hossain**
College of Engineering and Computer Science,
Arkansas State University
State University, Arkansas

**Devika Kabe**
Toronto Metropolitan University
Toronto, Ontario, Canada

**Mohammad Abul Kashem**
Marketing
Feni University
Trunk Road, Feni, Bangladesh

**Christopher J. Kluse**
School of Engineering
Bowling Green State University
Bowling Green, Ohio

**Vadim Lapidus**
LLC "Center Prioritet"
Nizhny Novgorod, Russia
Hadera, Israel

**Gediminas Lipnickas**
University of South Australia
Adelaide, South Australia, Australia

**Vijaya Kittu Manda**
PBMEIT
Visakhapatnam, Andhra Pradesh, India

**Farman A. Moayed**
Bailey College of Engineering and Technology
Indiana State University
Terre Haute, Indiana

**Tasnuba Nasir**
Marketing
Quincy University
Quincy, Illinois

**Ngoc Dang Khoa Nguyen**
Information Systems and Operations Management
Central Queensland University
Melbourne, Victoria, Australia

**Bora Qesja**
Marketing and Entrepreneurship
Australian Institute of Business
Adelaide, South Australia, Australia

**Ramya H. P.**
Department of Management Studies
Dayananda Sagar College of Engineering
Bengaluru, Karnataka, India

**Shiv Ranjan**
Amity Business School
Amity University
Greater Noida, Uttar Pradesh, India

**A. Mehran Shahhosseini**
Bailey College of Engineering and Technology
Indiana State University
Terre Haute, Indiana

**Abdul Rehman Shaikh**
Sukkur IBA University
Sukkur, Sindh, Pakistan

**Mohammad Shamsuddoha**
School of Management and Marketing
Western Illinois University
Macomb, Illinois

**Gurpreet Singh**
Department of Civil Engineering
Punjabi University
Patiala, Punjab, India

**Ulla Tapaninen**
Estonian Maritime Academy
Tallinn University of Technology
Tallinn, Estonia

**F. Talya Temizceri**
Faculty of Applied Sciences
İstanbul Bilgi University
Istanbul, Türkiye

**Aman Ullah**
Department of Management and Marketing
The University of Melbourne
Parkville, Victoria, Australia

**M. I. M. Wahab**
Industrial Engineering
Toronto Metropolitan University
Toronto, Ontario, Canada

**Anuradha Yadav**
Department of Management Studies
Dayananda Sagar College of Engineering
Bangalore, Karnataka, India

**Radha Pavitra Yalamarti**
UCD Michael Smurfit Graduate Business School
Blackrock, Dublin, Ireland

# *Section 1*

## *Digital Innovation and Transformation for Sustainability*

# 1 A Framework for Building Resilience through Innovation and Process Optimization in AI-Powered Digital Transformation

*Abdulaziz Aldoseri, Khalifa N. Al-Khalifa, and Abdel Magid Hamouda*

## 1.1 INTRODUCTION

In today's rapidly evolving global landscape, the COVID-19 pandemic has underscored the critical importance of resilience for organizations navigating through unprecedented challenges. As businesses grapple with volatile market conditions, supply chain disruptions, and shifting consumer behaviors, the ability to adapt and recover swiftly has become paramount. Against this backdrop, artificial intelligence (AI)–powered digital transformation emerges not only as a catalyst for enhancing operational efficiency, accuracy, and agility but also as a cornerstone for building organizational resilience.

AI-powered digital transformation transcends conventional operational improvements, embedding resilience into the fabric of organizational strategies. This resilience, characterized by the capacity to withstand disruptions and adapt to new normal, has proven indispensable during the pandemic. Organizations leveraging AI-driven solutions have demonstrated remarkable adaptability, swiftly adjusting to new operational and market conditions. From deploying AI tools for healthcare triage and diagnosis to optimizing supply chains and facilitating remote work, these solutions have enabled businesses to maintain continuity and even thrive amid crises.

The organizations are increasingly turning to artificial intelligence as a vehicle for driving digital transformation. This transformation is not only reshaping industries but also revolutionizing the way businesses operate, engage with customers, and make strategic decisions. Beyond these advantages, it plays a pivotal role in fortifying organizational resilience, equipping businesses to withstand and rapidly recover from disruptions. This resilience is particularly crucial in navigating the volatile,

DOI: 10.1201/9781003438748-2

uncertain, complex, and ambiguous (VUCA) world exacerbated by global challenges such as the COVID-19 pandemic [1].

A critical aspect of this transformation is the integration of automation and process optimization, which leverage the capabilities of AI to enhance efficiency, accuracy, and agility across various operational domains. Process automation is the use of technology to automate business processes. Generally, the transformation serves three functions: to automate processes, centralize information, and reduce the need for human input [2]. Process automation simplifies systems by eliminating human input, reducing errors, increasing delivery speed, improving quality, minimizing costs, and streamlining business processes [3].

One of the most compelling benefits of automation is its ability to offload routine and repetitive tasks from human workers to AI systems. Automation facilitates scalability, allowing businesses to seamlessly expand their operations without proportional increases in human resources [4]. In the contemporary landscape of rapid technological evolution, research on automation and process optimization within the context of AI-powered digital transformation has emerged as a vital arena of exploration [5]. This research domain delves into the synergistic relationship between automation, AI, and process optimization, uncovering novel strategies, frameworks, and insights that reshape industries and elevate organizational capabilities [6]. The surge of interest in automation and process optimization research for AI-based digital transformation is fueled by several interconnected factors:

a) Technology Advancements: Rapid advancements in AI technologies, machine learning, and automation tools have unlocked new possibilities, prompting researchers to explore innovative ways to harness their potential [7].
b) Business Imperatives: The digital age demands organizations to remain agile, efficient, and customer-centric. Researchers seek to identify how AI-powered automation can catalyze these imperatives [8].
c) Competitive Edge: As industries become increasingly competitive, research focuses on uncovering how AI-driven optimization can differentiate organizations and drive market leadership [9].
d) Operational Efficiency: Organizations are under pressure to optimize their operational processes for cost savings, reduced errors, and improved resource allocation [9].

Automation powered by AI technologies ensures a remarkable degree of accuracy and precision in performing tasks. AI systems are capable of flawlessly executing complex processes, thereby reducing the likelihood of human errors that can have far-reaching consequences. This precision is especially vital in industries where accuracy is paramount, such as healthcare, finance, and manufacturing. This fusion of AI and process automation and optimization transcends traditional operational models, offering several benefits that shape the way organizations conduct business and interact with stakeholders [6].

This research is grounded in the premise that the advent of digital transformation systems has substantially enhanced the accessibility of data related to business process execution. Coupled with advancements in AI, this evolution has paved the way for the emergence of smart systems. Within these systems, execution flows lack predetermined structures, allowing for dynamic adaptations without explicit changes to software applications. Opportunities for improvement are autonomously recognized, validated, and implemented, signifying a prospective paradigm for AI-powered digital transformation systems. Reference [10] introduces an AI-augmented business management system, offering possibilities for leveraging automation in business processes and enhancing efficiency with minimal human involvement. However, while this work presents a visionary outlook, it lacks a comprehensive managerial guideline for executing, realizing, or implementing the transformation. Partial solutions, such as robotic process automation (RPA) technology [10], identify automation opportunities strategically. A key factor in utilizing RPA involves identifying opportunities for automation, strategically implementing RPA technology in appropriate areas to maximize its potential [11]. RPA, while effective, is a partial solution, lacking the integration of AI techniques to enhance human capabilities and leverage tacit knowledge in process task execution. In [12], a goal-driven methodology for adapting processes to runtime changes using AI planning is introduced, focusing on process optimization. Paper [13] explores the feasibility of deploying an EmailBot solution within a private cloud to automate routine tasks, illustrating the practical application in installation, upgrades, guidance, and request handling. The work in [14] delves into AI's role in business process transformation, investigating implications, advantages, and commercial applications. The study examines the intricacies of business processes, shedding light on the adoption of intelligent automation and providing valuable insights for managers across diverse industries. In [15], a banking-specific AI solution is discussed, while [16] explores AI-powered processes in multi-vendor business applications, emphasizing the opportunity to optimize resources. The work in [17] applies AI-powered digital transformation to enhance business processes in the financial industry, conducting a comprehensive evaluation to understand the changes in various financial aspects. Paper [18] investigates the incorporation of AI into diverse business processes, highlighting its significant impact on organizations across industries. Furthermore, the integration of AI-powered digital transformation into supply chain management is a focal point of attention [19]. This work underscores AI's significance in optimizing inventory, demand forecasting, and logistics, offering potential benefits such as enhanced efficiency and reduced operational costs. Despite these advantages, ethical concerns and challenges associated with AI integration are acknowledged [20], emphasizing the need to address issues related to data privacy, algorithmic bias, and transparency.

However, the existing literature reveals a research gap in AI-powered digital transformation, particularly concerning the integration of automation and business process management optimization systems. While studies explore AI integration into Business Process Management (BPM), further in-depth investigations are needed to address practical challenges, implementation issues, and the actual impact on diverse business environments. Additionally, ethical considerations, risks, and factors influencing successful AI-augmented BPM adoption require more exploration. Bridging this gap is crucial for developing comprehensive frameworks, guidelines, and best practices to inform effective AI integration into BPM, ensuring successful implementation across industries.

While the adoption of AI and automation is a prevalent topic in contemporary business literature, there remains a significant gap in our understanding of how organizations can systematically integrate AI-driven automation with process optimization to achieve AI driven digital transformation. While it is acknowledged that automation and AI offer substantial benefits such as operational efficiency and enhanced customer experiences, there is limited comprehensive research that provide a structured roadmap or approach for organizations to manage this complex landscape more effectively. This paper seeks to address this knowledge gap by providing a pioneering contribution to the field. It provides a comprehensive structured roadmap of the foundational principles necessary for successfully integrating automation and optimizing processes within the context of emerging AI technologies. It does so by proposing a cohesive structure of essential pillars: data-driven insights, seamless automation, adaptive learning and continuous improvement, human-centric collaboration, ethical and responsible AI, strategic alignment, scalability, and innovation. These pillars serve as guiding principles for adoption of automation-driven initiatives in the era of AI-powered digital transformation.

## 1.2 THEORETICAL BACKGROUND

### 1.2.1 Business Process Management

Business process management serves as the orchestrator of organizational excellence, leveraging the capabilities of AI to streamline, enhance, and innovate business processes [21]. BPM focuses on improving corporate performance by managing and optimizing a company's business processes. It is a way of viewing and controlling the processes that are present in an organization and can be broken down into several components, as depicted in Figure 1.1.

BPM encompasses the identification, design, execution, monitoring, and continuous improvement of business processes. When coupled with AI-based technologies, it evolves into a dynamic force that drives transformation:

a) Process Identification: BPM begins by identifying existing processes, understanding their intricacies, and modeling them in a structured manner.
b) Process Discovery: This includes documentation of the process steps, responsibilities, and other relevant details.
c) Process Analysis: BPM focuses on optimizing processes to eliminate inefficiencies, bottlenecks, and redundancies. It often includes the use of analytical tools to simulate, analyze, and benchmark different processes. The design process often includes re-structuring existing processes or designing new ones [22]. The models can show how changes will impact performance and where potential bottlenecks might occur [23]. The process design or changes are implemented into the business operation.
d) Process Implementation: AI-powered automation seamlessly integrates into BPM. It automates routine tasks, accelerates workflows, and ensures consistency while adapting to changing requirements. This can involve changes in roles, use of new tools, and adjustments to current business rules [24].

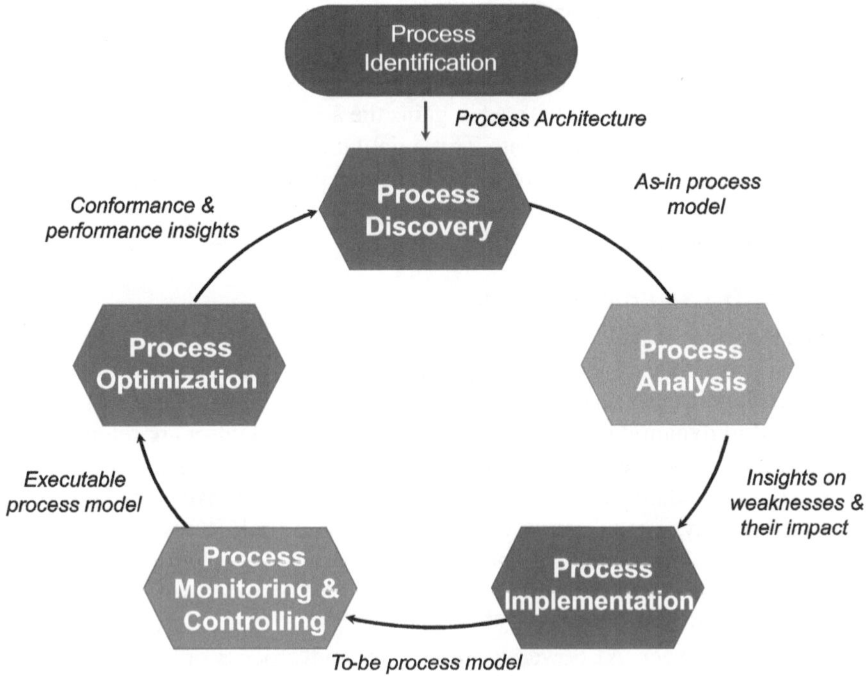


**FIGURE 1.1** Business process management lifecycle.

e) Process Monitoring and Controlling: AI technologies extract insights from data generated within processes. These insights guide optimization efforts, ensuring that decisions are based on real-time and historical data.
f) Process optimization: Once implemented, the performance of the processes is continuously monitored and controlled to ensure that they meet the desired performance metrics. This can involve using dashboards and other tools to provide real-time information on process performance [23]. This is the ongoing evaluation and refinement of processes that involves identifying areas for improvement, implementing changes, and then looping back to monitoring and management [22].

### 1.2.2 The Intersection of AI-Powered Automation and Process Optimization

At the nexus of AI and automation lies a foundation for modern business transformation. AI enhances automation with adaptability, while process optimization uses AI to refine operations. This synergy boosts performance, agility, and customer focus. Embracing this, businesses transition from minor tweaks to groundbreaking innovations, positioning themselves at the forefront of the digital transformation era [25, 26].

#### 1.2.2.1 Amplifying Automation with AI

AI empowers automation by giving it cognitive capabilities previously reserved for human decision-making. Traditional automation excels at executing predefined tasks; however, when coupled with AI, it gains the ability to learn, adapt, and evolve in response to dynamic environments. This amplification is embodied in AI-powered robotic process automation, where software robots equipped with AI can comprehend unstructured data, make context-aware decisions, and autonomously navigate intricate workflows [27].

#### 1.2.2.2 AI-Driven Process Optimization

Process optimization, fueled by AI-driven insights, takes a quantum leap from traditional analysis methods. Machine learning models ingest historical data, learn from it, and generate predictive insights that drive informed decisions. These AI-generated insights lead to dynamic process optimization, where workflows are refined in real time on the basis of changing data inputs. AI identifies bottlenecks, recommends alterations, and predicts outcomes, thus creating a continuous cycle of improvement [28]. Consequently, organizations can eliminate waste, reduce resource consumption, and achieve operational excellence.

#### 1.2.2.3 Agile, Responsive Operations

Collaboration between AI-powered automation and process optimization ushers in an era of agile, responsive operations. Automation orchestrated by AI ensures that routine tasks are executed with precision and efficiency, regardless of scale or complexity. Moreover, AI's predictive capabilities enable proactive decision making, enabling organizations to anticipate challenges, adapt to changing conditions, and make strategic pivots swiftly [29]. In the realm of process optimization, AI's real-time insights empower organizations to make dynamic adjustments to workflows as conditions evolve. This agility enables businesses to navigate market disruptions, customer fluctuations, and unexpected events with resilience. The benefits of automation and process optimization include the following, as depicted in Figure 1.2.

a) Efficiency and Productivity: Automation enhances efficiency by automating repetitive tasks, allowing human workers to focus on strategic, high-value tasks. This boosts productivity and employee satisfaction [30].
b) Cost Savings: Automation can lead to substantial cost savings. Automated systems often complete tasks more rapidly and accurately than humans, reducing the need for rework and minimizing costly errors [31].
c) Scalability: Automated processes are easily scalable. As a business grows, it can increase the capacity of its automated systems without necessarily needing to add more staff, making automation a key driver of business growth [32].
d) Consistency and Quality: Automated processes consistently deliver reliable results, ensuring that tasks are performed to the same standard each time, enhancing the quality of the business' products or services [33].

**FIGURE 1.2** Business process optimization benefits.

e) Data Analysis: Automated systems generate considerable data that can be analyzed to provide insights into business performance and identify opportunities for improvement. This is particularly relevant with AI, which can analyze large datasets and learn from the data to improve the performance of automated systems [34].
f) Competitive Advantage: Automation and process optimization can provide businesses with a significant competitive advantage. By offering services more efficiently and at a lower cost, businesses can outperform their competitors and capture a larger market share [35].
g) Simplified Operations: Implementing automation and AI helps eliminate unnecessary steps, simplifies complex tasks, and results in smoother, more efficient workflows. By automating manual, time-consuming tasks, businesses can focus on more strategic initiatives and objectives, making operations simpler and more effective [32].
h) Enhanced Security: Automation can significantly contribute to the security of business operations. Automated systems can monitor security breaches or unusual activity, instantly notifying the concerned authorities in real time. AI and machine learning can be used to predict and identify potential threats, thereby enhancing the overall security of the system [30].

## 1.2.3 Types of Automation and Process Optimization

Automation and process optimization in the context of AI-powered digital transformation encompass various types and strategies to enhance efficiency, productivity, and innovation. Here are some types of automation and process optimization that organizations can leverage:

### 1.2.3.1 Robotic Process Automation

RPA involves automating routine, rule-based tasks by deploying software robots RPA involves creating "bots" that can handle repetitive, rule-based tasks, thus freeing up human employees for more complex, value-adding activities. It can be used in various business processes, such as customer service, supply chain management, and financial operations.

The benefits of RPA are illustrated in Figure 1.3:

a) Scalability and Efficiency: Bots can work 24/7 without breaks, improving efficiency, and can be quickly scaled up or down to meet changing business demands. They also reduce the possibility of human errors.
b) Cost Savings: RPA can lead to significant cost savings by reducing the need for human labor for repetitive tasks. This frees up employees to work on higher-value tasks that require human skills such as problem-solving, creativity, and customer interaction.
c) Reduces Manual Intervention: Automation minimizes the need for manual intervention in business processes. Tasks that previously required human input can now be executed autonomously, thereby reducing the potential for errors, and enhancing overall operational efficiency. This shift allows employees to focus on strategic tasks, thus increasing the value of the business [36].

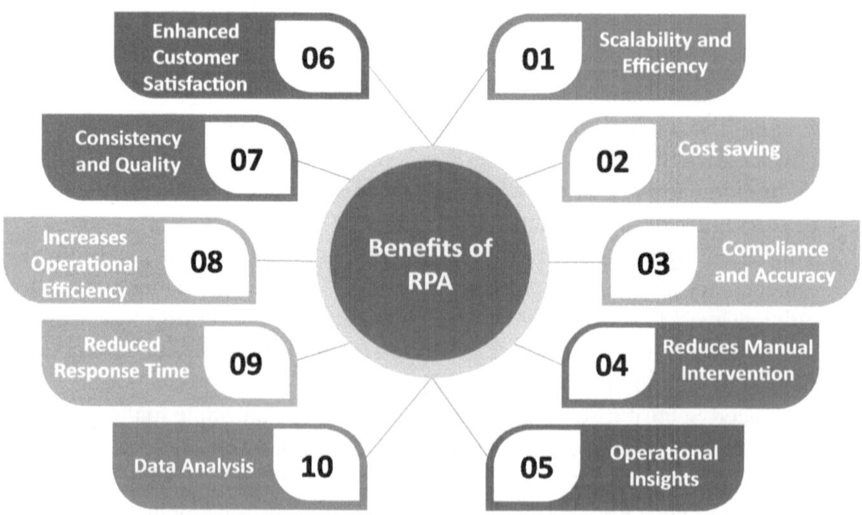


**FIGURE 1.3** Benefits of RPA.

d) Operational Insights: Data generated by automated systems provide deep insights into operational performance. Through sophisticated data analysis, companies can identify inefficiencies, bottlenecks, and opportunities for improvement, thus making operations more effective and efficient [37].
e) Enhanced Customer Satisfaction: Automation can improve customer satisfaction. Faster response times, increased accuracy, and the ability to provide personalized experiences based on data analysis contribute to a better customer experience [38].
f) Consistency and Quality: Automated processes consistently deliver reliable results, ensuring that tasks are performed to the same standard each time, enhancing the quality of the business' products or services [28].
g) Increases Operational Efficiency: Automation reduces the amount of human intervention required in processes, thus increasing operational efficiency. This allows companies to achieve their objectives with less effort and resources, leading to improved profitability and customer service [39].
h) Reduced Response Time: Automation can significantly reduce response times by streamlining workflows and eliminating bottlenecks. This results in faster service delivery and improved customer satisfaction [34].

#### 1.2.3.2 Chatbots and Virtual Assistants

Chatbots and virtual assistants are AI-powered technologies that can interact with humans in a natural, conversational manner. Figure 1.4 represents a simplified presentation of how chatbots works. The benefits of chatbots and virtual assistants are:

a) Automated Customer Service: Chatbots and virtual assistants can handle several customer service tasks, from answering frequently asked questions to guiding users through complex processes. They can provide immediate responses at any time of day, thereby enhancing customer satisfaction [40].
b) Efficiency and Scalability: These technologies can handle multiple interactions simultaneously, which would be impossible for human agents. They can also quickly access and analyze large amounts of data to provide accurate responses or recommendations [41].
c) Personalized Experiences: Through machine learning algorithms and access to customer data, chatbots and virtual assistants can provide personalized responses and recommendations. This can enhance the user experience and increase customer engagement [42].
d) Cost Savings: Implementing chatbots and virtual assistants can lead to substantial cost savings. They can handle many tasks that would otherwise require human agents, thus reducing labor costs. In addition, they can work around the clock without taking breaks or vacations [41].
e) Continuous Learning and Improvement: These systems learn from every interaction, allowing them to continually improve their responses and capabilities. Over time, they can become more effective and efficient, providing increased value to the business [43].

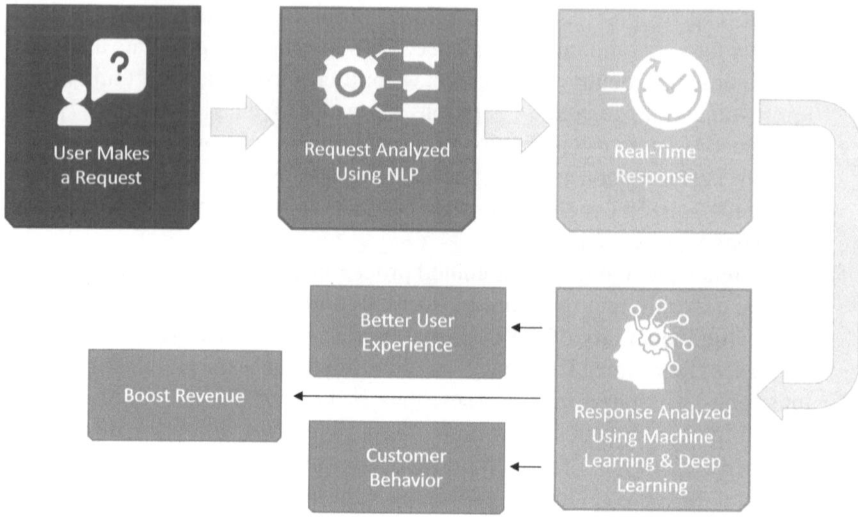


**FIGURE 1.4** How a chatbot works.

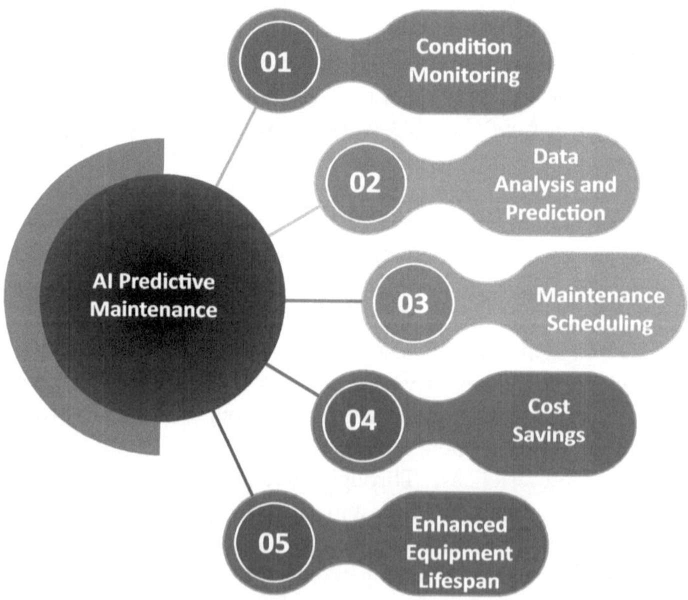


**FIGURE 1.5** AI predictive maintenance.

### 1.2.3.3 Predictive Maintenance

Predictive maintenance is an approach that leverages AI and machine learning to predict equipment failure before it occurs, based on the analysis of relevant data. AI can improve predictive maintenance, as summarized in Figure 1.5.

Condition Monitoring: Predictive maintenance involves continuous monitoring of machine conditions. Various types of data such as temperature, vibration levels, and the presence of particles in fluids are gathered and analyzed to monitor the operational status of equipment [44].

a) Data Analysis and Prediction: Data collected from sensors are processed and analyzed, often in real time, using machine learning and other advanced analytical techniques. This analysis can identify patterns and trends that may indicate potential equipment failure [45].
b) Maintenance Scheduling: By predicting when equipment is likely to fail, maintenance can be scheduled proactively to avoid unexpected downtime. This not only prevents costly disruptions in operations but also enables a more efficient use of resources [46].
c) Cost Savings: Predictive maintenance helps reduce maintenance costs by avoiding unnecessary preventive maintenance and reducing breakdowns that require reactive maintenance. Furthermore, minimizing downtime improves productivity and efficiency [43].
d) Enhanced Equipment Lifespan: By maintaining equipment in optimal condition and preventing serious damage, predictive maintenance can extend the useful life of equipment [46].

The advent of the Internet of Things (IoT) has greatly facilitated predictive maintenance by enabling the continuous collection of relevant data from various equipment and systems [47].

#### 1.2.3.4 Supply Chain and Logistics Optimization

AI can help optimize inventory management, demand forecasting, and delivery routes, leading to cost savings and improved customer service, as depicted in Figure 1.6.

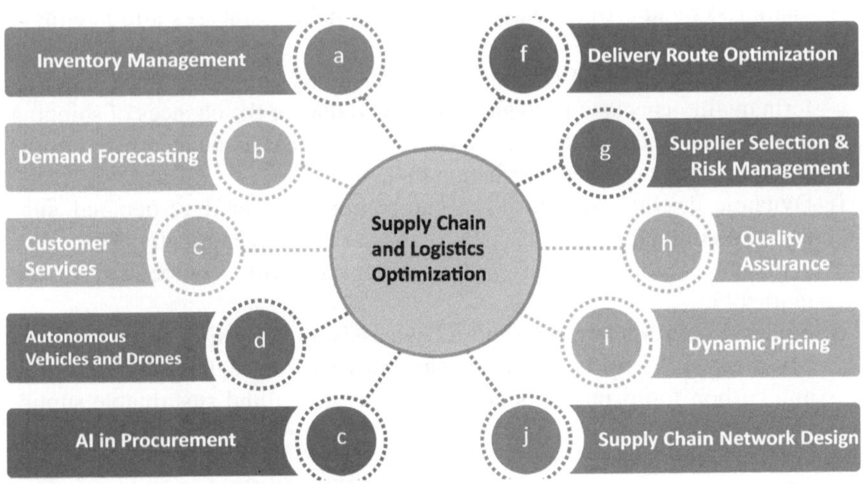


FIGURE 1.6 AI supply chain and logistic optimization.

a) Inventory Management: AI can analyze complex datasets to accurately predict inventory needs, ensuring that businesses have the right amount of stock. This can lead to significant cost savings by reducing the number of tied-up capital and storage space requirements and preventing stockouts or overstock situations [48].
b) Demand Forecasting: AI can analyze historical sales data along with external factors such as economic indicators, weather, and market trends to generate highly accurate demand forecasts. This helps businesses better plan production and logistics, reduce waste, and improve customer satisfaction [48].
c) Customer Service: AI-powered chatbots and virtual assistants can improve customer service in supply chain and logistics. They can provide real-time tracking information, answer customer queries, and handle complaints, thus increasing customer satisfaction and loyalty [49].
d) Autonomous Vehicles and Drones: AI enables the operation of autonomous vehicles and drones for transportation and delivery. This can increase efficiency, especially in last-mile delivery logistics, thereby reducing costs and time [50].
e) AI in Procurement: AI can analyze a vast amount of supplier data to help businesses make better procurement decisions. It can recommend the best suppliers based on factors such as price, delivery time, and reliability [51].
f) Delivery Route Optimization: AI algorithms can analyze factors such as traffic patterns, delivery windows, and the number and location of deliveries to determine the most efficient routes. This can reduce fuel costs, improve delivery speed, and increase customer satisfaction [52].
g) Supplier Selection and Risk Management: AI can help businesses evaluate and select suppliers based on various factors such as cost, quality, delivery performance, and risk. It can also monitor potential supply chain risks and provide early warnings to help businesses mitigate these risks [53].
h) Quality Assurance: AI technologies, particularly computer vision, can perform quality checks on products, thereby reducing the chances of shipping damaged or defective goods. This reduces the costs associated with returns and increases customer satisfaction [54].
i) Dynamic Pricing: AI algorithms can continuously analyze demand, supply, and competitor pricing data to adjust prices dynamically. This helps companies optimize their profits while remaining competitive in the market [55].
j) Supply Chain Network Design: AI can assist in designing optimal supply chain networks by factoring in variables such as delivery time, costs, and carbon footprint. This helps create resilient and sustainable supply chains [56].

The incorporation of AI into supply chain and logistics represents a significant step toward more resilient, responsive, and customer-centric operations.

#### 1.2.3.5 AI-Based Decision Systems

These systems use advanced AI algorithms to transform data into valuable insights, thereby enhancing decision-making processes. Here are some critical aspects:

a) Predictive Analysis: These systems use historical and real-time data to predict future outcomes. Machine learning models are trained on past data and can extrapolate to forecast future trends or events, thereby aiding decision-making [57]. Figure 1.7 presents the steps for the predictive analysis.
   1) Data Collection: Predictive analysis requires a large amount of high-quality data to provide accurate results. This data can come from various sources, including internal databases, customer interactions, social media, external databases, and IoT devices [58].
   2) Statistical Analysis and Data Mining: Predictive analysis involves various statistical techniques to analyze historical data. This can include regression, correlation, and cluster analysis. It also employs data mining to identify patterns and relationships within the data [59].
   3) Machine Learning: Predictive analysis often uses machine learning algorithms to learn from data and make predictions. This can involve supervised learning (where the algorithm is trained on labeled data), unsupervised learning (where the algorithm identifies patterns in unlabeled data), or reinforcement learning (where the algorithm learns by interacting with its environment) [60].

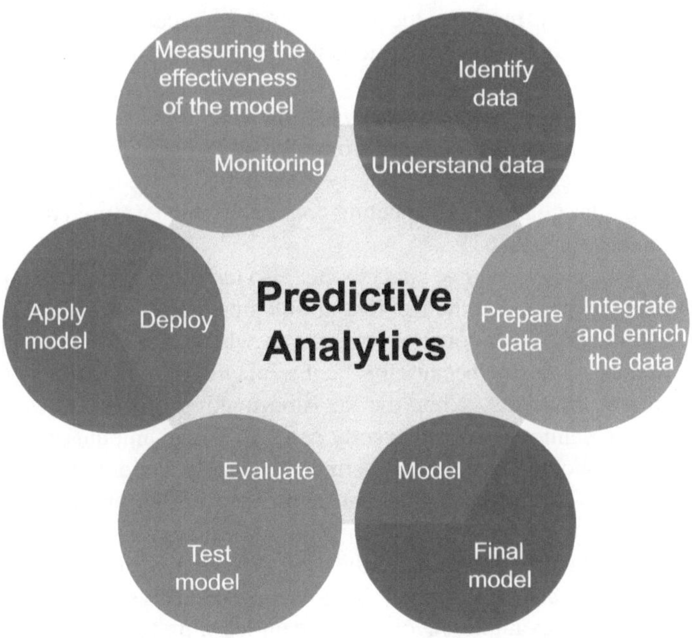


**FIGURE 1.7** Predictive analytics steps.

4) Model Building and Validation: Predictive models are created based on insights derived from the data. These models are then validated by applying them to a separate dataset to test their accuracy. The models can be continually refined and updated as new data become available [61].
5) Prediction: Once the predictive model has been validated, it can be used to make predictions about future outcomes based on current and historical data. These predictions can guide decision-making in various areas, including sales forecasting, risk management, and operational efficiency [62].
6) Decision Making: Predictive analysis not only provides insights into what might happen in the future but also proposes the recommended actions based on those predictions. This can help businesses take proactive steps to capitalize on opportunities or mitigate risks [63].

Predictive analysis can be a powerful tool for organizations of all types, providing valuable insights into future trends and guiding strategic decision-making.

b) Prescriptive Analysis: Beyond predictive capabilities, some AI systems can recommend the best course of action, given a certain business objective and set of constraints. The primary goal is to guide decision making toward the most favorable outcomes based on the predictive analysis results and a set of decision variables [64].
   1) Modeling Business Decisions: Prescriptive analysis begins with an understanding of the business scenario and the necessary decisions. This typically involves the use of mathematical and computational models that describe the relationships among various factors in a given scenario [65].
   2) Optimization: Prescriptive analysis often involves optimization, which seeks to find the best solution from all possible solutions. Optimization techniques can be used to identify the most effective actions to achieve a particular goal while respecting constraints such as limited resources, time, and budget [63].
   3) Simulation: Prescriptive analysis can also use simulation to explore different scenarios and their potential outcomes. This is especially useful when dealing with complex systems in which it is difficult to predict outcomes precisely because of the interplay of many variables [66].
   4) Heuristic Methods: When the decision-making problem is very complex and cannot be solved exactly in a reasonable amount of time, heuristic methods can be used to find reasonably good solutions. These are rule-of-thumb strategies that help speed up the process of finding satisfactory solutions [67].
   5) AI and Machine Learning: Advanced AI and machine learning techniques can also be applied in prescriptive analysis. Reinforcement learning, for example, involves an agent learning to make decisions by interacting with an environment and receiving feedback in the form of rewards or penalties [68].

c) Real-Time Decision Making: AI decision systems can process data in real time, allowing for immediate decision making. This is particularly useful in dynamic environments where circumstances can change rapidly, such as in financial trading or emergency response situations [69]. The foundation of real-time decision-making lies in the ability to process data as it arrives or is created. Technologies such as stream processing and complex event processing enable businesses to analyze and respond to incoming data in real time. AI techniques, such as online learning, can help adapt models as new data arrive [70].
   1) Predictive Analytics: AI can use real-time data to make immediate predictions about future outcomes. These predictions can help decision makers respond proactively to potential changes and seize opportunities as they arise [71].
   2) Automation: AI algorithms can automate decision-making processes, thus reducing the time required to respond to changing circumstances. This is particularly important in sectors such as financial trading, where milliseconds can make the difference between profit and loss [72].
   3) Adaptability: In real-time decision making, the ability to adapt to changing situations is crucial. AI systems can learn from new data and adjust their models and predictions accordingly. This allows them to respond to changes in the environment more effectively [73].
   4) Risk Management: Real-time decision making can help manage risks more effectively. By processing data in real time, businesses can immediately identify potential risks and take steps to mitigate them [71].

#### 1.2.3.6 Data Mining and Extraction

This process can help businesses identify patterns, trends, and relationships within the data that might not be immediately clear. AI plays a crucial role in modern data mining and extraction processes. Figure 1.8 illustrates the data mining techniques.

a) Pattern Recognition: Data mining involves pattern recognition in which algorithms search through data for repeating patterns. These patterns can then be used to predict future trends or identify anomalies that might indicate a problem [74].
b) Classification: Classification in data mining involves sorting data into categories or classes based on identified features. This is a type of supervised learning in which an AI model is trained on a dataset with pre-labeled instances and then uses that learned knowledge to categorize new, unseen data. Machine learning algorithms used for classification include logistic regression, decision trees, and support vector machines [75].
c) Clustering: Clustering is a type of unsupervised learning method used in data mining to group similar data points or objects. This method is often used in market segmentation, image segmentation, and anomaly detection. Clustering algorithms include k-means, hierarchical clustering, and DBSCAN [76].

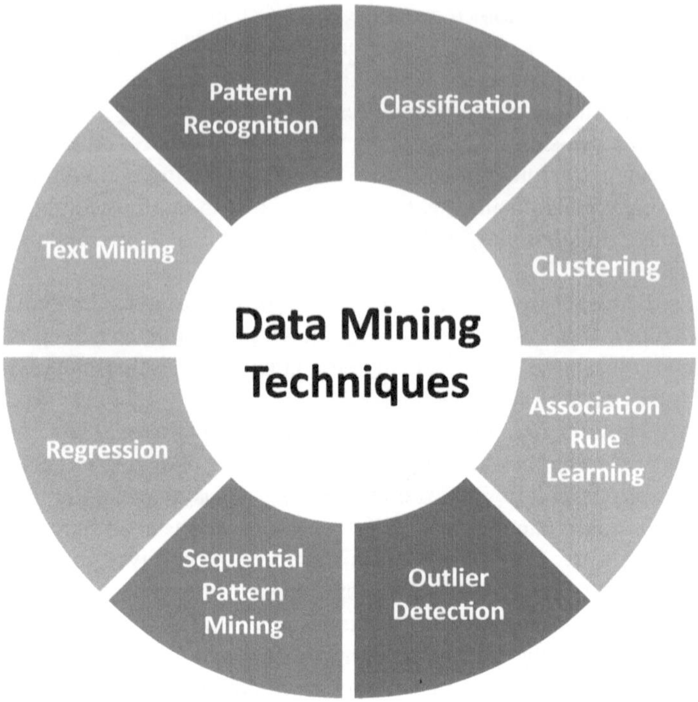


**FIGURE 1.8** Data mining techniques.

d) Association Rule Learning: This is a method for discovering relationships between variables in large databases. For example, in a supermarket dataset, association rule learning might find that people who buy bread often also buy butter [77].
e) Outlier Detection: Outlier detection is a key function of data mining, identifying data points that deviate significantly from other observations. These outliers can sometimes indicate errors, but they can also represent significant and interesting findings [78].
f) Sequential Pattern Mining: This involves finding frequent sequences or patterns in the data. This can be particularly useful for analyzing customer behavior over time [79].
g) Text Mining: Text mining is a specific application of data mining that involves extracting high-quality information from text. This can include sentiment analysis, topic modeling, and information extraction [80].

#### 1.2.3.7 Industry Paradigms and Trends

In formulating a comprehensive roadmap for the integration of automation with process optimization in the context of AI-powered digital transformation, it is imperative to delve deeply into the intricate nuances of industry paradigms and the evolving trends that underpin these transformative processes. Central to this discourse are prominent industry paradigms such as Industry 4.0, which heralds the fourth industrial revolution, and smart manufacturing, which emphasizes the synergistic convergence of advanced technologies within manufacturing ecosystems. Industry 4.0, characterized by its amalgamation of cyber-physical systems, IoT, AI, and big data analytics, fundamentally reshapes traditional manufacturing models by imbuing them with intelligence, connectivity, and autonomy. Smart manufacturing, on the other hand, underscores the strategic utilization of digital technologies to optimize manufacturing operations, enhance productivity, ensure real-time decision-making, and augment overall operational efficiency. Both paradigms provide a foundational framework upon which innovative solutions for automation and process optimization are built.

In parallel, a multitude of dynamic trends is shaping the trajectory of automation and process optimization endeavors. One such trend is the paradigm of edge computing, where data processing is executed in close proximity to the data source, mitigating latency and enabling rapid real-time analytics. This trend is particularly germane in the context of IoT devices, where instantaneous decision-making is imperative. Moreover, the integration of blockchain technology stands as a beacon of trust and transparency in the digital landscape, offering robust solutions for secure and immutable data transactions. In intricate supply chain networks, blockchain ensures the veracity of information, significantly reducing the scope for fraud and enhancing the traceability of products. Augmented reality (AR) technologies represent another pivotal trend, revolutionizing training methodologies, maintenance protocols, and collaborative endeavors within industrial settings. AR applications facilitate immersive and interactive experiences, enabling enhanced skill acquisition, streamlined maintenance procedures, and seamless remote collaboration.

Within the realm of academic research, these paradigms and trends form the crux of strategic considerations for scholars and industry practitioners alike. In-depth analyses of these paradigms provide critical insights into the theoretical underpinnings of automation and optimization processes, offering a theoretical framework for empirical studies. Likewise, an exhaustive exploration of emerging trends not only elucidates the evolving landscape of industrial technologies but also serves as a catalyst for innovation and exploration of novel research avenues. By integrating these academic explorations into the roadmap for integrating automation with process optimization for AI-powered digital transformation, scholars contribute significantly to the discourse on industry evolution, technological advancements, and the holistic transformation of contemporary manufacturing paradigms.

By intertwining the framework pillars with these industry paradigms and trends, automation and process optimization are achieved synergistically. Leveraging data-driven insights, seamless automation, adaptive learning, ethical considerations, strategic alignment, and scalability within the context of Industry 4.0 and smart manufacturing paves the way for transformative digital evolution. This integration ensures that organizations not only optimize their processes efficiently but also do so ethically and sustainably, reflecting the ethos of modern AI-powered digital transformation paradigms.

## 1.3 FRAMEWORK DEVELOPMENT

### 1.3.1 Introduction

Automation and process optimization research fuels innovation by encouraging organizations to reevaluate their existing processes and infuse AI-driven solutions. This dynamic creates novel business models and offerings. Agile organizations that incorporate AI-driven solutions into their processes can respond swiftly to market shifts, thereby gaining a competitive edge. Integration of AI-driven insights into process optimization enables data-informed decision making. BPM emerges as the linchpin that unifies AI-based digital transformation. As organizations embrace the holistic integration of BPM and AI, they embark on a journey where automation, optimization, and human ingenuity converge. This journey propels organizations toward a future where digital transformation is not just a destination but a continuous evolution of excellence. BPM aims to reduce inefficiencies, maintain quality standards, and improve the flexibility and agility of processes. It combines methodologies, technologies, and human input to achieve these results.

It is important to infer critical aspects that shape the success of automation and process optimization in the context of AI-powered digital transformation and consider them as pillars of the integration. These pillars serve as a guiding framework that offers organizations a structured approach to navigate the complex landscape of integrating automation and optimizing processes using AI. By embracing these pillars, businesses can create a holistic approach that combines technology, human collaboration, strategic vision, and ethical considerations, leading to a future characterized by efficiency, innovation, and sustainable growth in the digital age. The following are the key pillars that underpin the paradigm of automation and process optimization in the context of AI-driven digital transformation.

### 1.3.2 Data-Driven Insights

This pillar emphasizes the collection, analysis, and utilization of data to drive informed decisions and actions. Data-driven insights is fueling process optimization efforts, ensuring that improvements are based on factual information rather than assumptions [81]. This can be elaborated further on how this pillar shapes AI-driven digital transformation as follows:

a) Data Collection and Integration: Collecting relevant and accurate data from various sources is the foundation of data-driven automation and optimization. These data can originate from sensors, user interactions, transactions, and social media. The collected data must be integrated into a centralized repository to ensure its accessibility and coherence.
b) Analysis and Pattern Recognition: AI algorithms excel in analyzing vast datasets quickly and identifying patterns, trends, and correlations that might be imperceptible to human analysts. This analysis uncovers valuable insights that drive informed decision making and guide optimization efforts.

c) Predictive and Prescriptive Analytics: By leveraging historical data and machine learning techniques, AI algorithms can predict future outcomes and trends. This predictive capability empower organizations to take proactive actions, minimize risks, and seize opportunities. Prescriptive analytics goes a step further by proposing optimal courses of action based on data-driven insights [82].
d) Real-Time Responsiveness: With real-time data processing, organizations can respond swiftly to changing conditions and emerging trends. This agility enables dynamic process adjustments, leading to better resource allocation, improved customer experiences, and enhanced operational efficiency [83].
e) Identifying Inefficiencies and Bottlenecks: Data-driven insights are particularly potent in identifying bottlenecks and inefficiencies within processes. Organizations can pinpoint areas where resources are underutilized or where delays occur, allowing them to implement targeted process optimizations [84].

The data-driven pillar is the cornerstone that shapes how AI-based automation and process optimization revolutionize business operations. By harnessing the power of data and AI, organizations can make informed decisions, optimize processes, and adapt to dynamic market conditions, while ensuring that improvements are grounded in empirical evidence. This pillar embodies the transformational potential of data in the digital age.

### 1.3.3 Seamless Automation

This pillar highlights the seamless integration of AI-powered automation into existing processes. Seamless automation integration entails the harmonious fusion of technology, processes, and people [85]. By seamlessly integrating automation, organizations ensure that new technologies become an integral part of their daily operations, achieving a delicate balance between technological advancement and organizational adaptation [85]. The key aspects of this pillar are as follows:

a) User-Centered Design: This ensures that the interface between humans and automation is intuitive, user-friendly, and promotes efficient interactions. This approach minimizes resistance to change and fosters user acceptance.
b) Process Analysis and Redesign: Before integration, a thorough analysis of the existing processes is essential. This analysis identified inefficiencies, bottlenecks, and opportunities for improvement. Redesigning processes with automation in mind ensures that the technology aligns with organizational objectives [86].
c) Customization and Scalability: Automation solutions should be tailored to the specific needs of an organization. Scalability is also crucial, allowing automation to accommodate evolving requirements as the organization grows or pivots its strategies [87].
d) Interoperability: Seamless integration requires compatibility with existing systems and technologies. Automation solutions should be integrated

with other tools, databases, and software to create a unified ecosystem that enhances efficiency.

e) Real-Time Feedback and Monitoring: Continuous monitoring of automated processes is vital. Real-time feedback allows organizations to detect anomalies, rectify errors, and fine-tune processes to achieve optimal performance.
f) Agility and Adaptability: The integrated automation ecosystem should be agile enough to adapt to changing business needs. Whether responding to market shifts, regulatory changes, or emerging trends, integration should enable swift adjustments [88].

Seamless automation integration is a pivotal strategy for organizations navigating the digital age. It marries technological prowess with organizational adaptability, creating an environment in which automation becomes an integral and transformative force. By embracing this integration, businesses lay the foundation for enhanced efficiency, innovation, and sustainable growth in an increasingly automated world.

### 1.3.4 Adaptive Learning and Continuous Improvement

The pillar of adaptive learning emphasizes the dynamic nature of AI-powered automation and process optimization [89]. This pillar encourages organizations to embrace a culture of continuous improvement, where processes are iteratively refined based on AI-generated insights. The following is a deeper insight into this pillar and its implications:

a) Continuous Evolution through AI: The adaptive learning pillar underscores that AI-powered systems are not static; they have the capacity to evolve and improve over time. This evolution enables AI to adapt to changing conditions and make increasingly precise predictions and decisions.
b) Resilience and Agility: The dynamic nature of AI-powered automation and optimization cultivates resilience and agility within organizations. In a rapidly changing business landscape, the ability to swiftly adapt to new information, market shifts, and unexpected challenges is paramount. AI systems that continuously learn enable organizations to respond proactively to disruptions and seize emerging opportunities [90].
c) Insight-Driven Decision-Making: Adaptive learning fuels insight-driven decision-making. As AI systems learn from data, they unveil hidden trends, patterns, and correlations that human analysis might overlook. These insights provide organizations with a competitive advantage, allowing them to make informed, data-backed decisions that lead to better outcomes.
d) Iterative Process Improvement: The adaptive learning pillar encourages a culture of continuous improvement. Organizations leverage AI-generated insights to iteratively refine their processes. This iterative approach leads to incremental enhancements that compound over time, creating substantial efficiency gains and operational excellence.
e) Personalization and Customization: In customer-centric domains, AI's adaptive learning powers personalization. AI-driven systems analyze

individual user behavior, preferences, and interactions to tailor experiences and offerings. This personalized approach enhances customer satisfaction, engagement, and loyalty.

f) Risk Management and Predictive Capabilities: AI's ability to adapt and learn enhances risk management strategies. Organizations can develop predictive models that assess potential risks and opportunities based on changing market conditions and historical data. This aids in making informed decisions to mitigate risks and capitalize on favorable conditions [91].

The adaptive learning pillar transforms organizations into dynamic entities that learn, grow, and evolve in tandem with their environment. By leveraging AI's capacity to continuously learn and adapt, businesses foster a culture of resilience, agility, and ongoing improvement. This pillar affirms the transformative power of AI in shaping modern enterprises and their strategies for long-term success.

### 1.3.5 Human-Centric Collaboration

This pillar underscores the importance of collaboration between AI systems and human employees. This pillar emphasizes user-centered design, ensuring that AI-driven interfaces are intuitive, user-friendly, and promote effective human—machine interaction [92]. The following items explore this pillar in greater depth:

a) Human–AI Collaboration: AI systems excel at repetitive and data-intensive tasks, allowing human employees to focus on higher-order activities that require creativity, critical thinking, emotional intelligence, and complex decision-making [92].
b) Augmenting Human Abilities: AI augments human abilities by offering insights, data analysis, and support in decision making. This synergy between AI and humans empowers employees to make better-informed decisions and contribute more strategically to the organization's objectives [93].
c) Task Redistribution: By automating routine and rule-based tasks, AI creates room for human employees to engage in value-added activities. This redistribution of tasks elevates job satisfaction, increases employee morale, and leads to a more fulfilled workforce.
d) Enhancing User Experience: The user-centered design approach under this pillar ensures that AI-driven interfaces are intuitive and user-friendly. Interfaces that promote effective human—machine interaction are crucial for seamless collaboration. A well-designed interface enhances the user experience, enabling users to interact with AI systems comfortably and efficiently.
e) Creativity and Innovation: Human employees are irreplaceable regarding creativity and innovation. The time and cognitive resources saved through AI automation can be channeled into brainstorming new ideas, developing innovative solutions, and exploring uncharted territories.
f) Ethical Decision-Making: Humans bring ethical judgment, empathy, and understanding to complex decision-making scenarios that AI cannot

replicate. This is particularly crucial in sensitive areas such as healthcare, law, and customer service [94].

g) Feedback Loop and Improvement: Through interaction, humans help AI systems learn, adapt, and correct errors, leading to more accurate results over time.

The human-centric collaboration pillar underscores that the integration of AI should not diminish the role of humans; rather, it should amplify their capabilities and create a harmonious partnership. Recognizing the unique qualities that humans bring to the table and designing AI interfaces that align with human needs are pivotal in achieving the full potential of AI-driven digital transformation. This pillar ensures that technology remains a tool that enhances, rather than replaces, the intrinsic value of human skills and intellect.

### 1.3.6 Ethical and Responsible AI

This pillar emphasizes the need for clear guidelines, governance frameworks, and transparency in AI decision-making processes. Addressing ethical considerations, such as data privacy and algorithmic fairness, builds trust with stakeholders and maintains the integrity of AI-driven operations [95]. The following are core principles of ethical and responsible AI:

a) Ethical Decision-Making: Organizations must ensure that AI systems make decisions that align with moral and societal norms. Ethical frameworks guide the development and use of AI to prevent actions that might lead to harm or discrimination [96].
b) Data Privacy and Security: Ethical AI respects individuals' data privacy and ensures that personal information is handled with utmost security. Compliance with data protection regulations and obtaining informed consent are the central tenets of responsible AI deployment.
c) Algorithmic Fairness: AI systems must be designed to ensure that their decisions are fair and unbiased across different demographic groups. This pillar emphasizes the importance of avoiding discriminatory outcomes and upholding fairness in AI-driven decisions.
d) Human Oversight and Accountability: Humans should retain the ability to oversee and intervene in AI decisions, particularly in critical areas. Accountability ensures that the ultimate responsibility for AI actions rests with humans, thus preventing the abdication of ethical responsibilities.
e) Continuous Monitoring and Auditing: Organizations must continuously monitor AI systems to detect any unintended consequences or ethical violations. Regular audits ensure that AI systems align with ethical guidelines and do not deviate over time.
f) Code of Ethics and Guidelines: Developing a clear code of ethics and guidelines for AI use ensures that all stakeholders understand and adhere to the ethical principles governing AI-driven operations.
g) Ethical Culture: Promoting an ethical culture within the organization reinforces responsible AI practices. Leadership plays a crucial role in setting the tone for ethical decision making.

h) Public Engagement: Engaging the public, customers, and stakeholders in discussions about AI ethics fosters transparency, accountability, and responsiveness to concerns [97].

The ethics and responsibility pillar shapes AI-powered automation and process optimization into instruments that promote fairness, transparency, and accountability. Organizations that prioritize ethics gain the trust of stakeholders, mitigate risks, and ensure the sustained benefits of AI-driven transformation. By adhering to ethical principles, organizations not only achieve their objectives but also contribute to the betterment of society as a whole.

### 1.3.7 Strategic Alignment

The strategic alignment pillar emphasizes the integration of automation and process optimization with an organization's overarching goal and strategies [98]. This pillar underscores the paramount importance of integrating AI technologies into the fabric of an organization's overarching goals, strategies, and vision. Strategic alignment transcends the mere adoption of AI as a technological tool; it entails a profound integration that resonates across every facet of an organization's operations:

a) Identifying Value-Centric Processes: Strategic alignment begins with discerning the processes that stand to gain the most from automation and optimization. These processes, when augmented by AI, yield substantial efficiency gains, cost savings, and customer experience enhancements.
b) Clear Objectives and Milestones: Strategic alignment mandates the establishment of clear, quantifiable objectives and milestones for AI initiatives. These objectives should resonate with the organization's strategic goals, facilitating effective measurement of progress and outcomes.
c) Prioritization and Resource Allocation: In a world of finite resources, strategic alignment necessitates prioritization. Organizations must allocate resources judiciously, focusing on initiatives that align most closely with strategic imperatives and deliver the greatest impact [99].
d) Agile Adaptation to Market Dynamics: Strategic alignment empowers organizations to adapt swiftly to shifting market dynamics. AI-driven processes can be adjusted in real time to align with evolving customer demands, industry trends, and competitive pressures.
e) Cross-Functional Collaboration: Strategic alignment is a collaborative endeavor that bridges departments and functions. Cross-functional collaboration ensures that AI initiatives holistically address the organization's challenges and opportunities [100].

The strategic alignment pillar weaves AI-driven automation and process optimization into the fabric of an organization's journey toward success. This alignment transforms technology from a siloed tool into a dynamic force that propels the organization forward. When AI is harnessed with strategic clarity, it elevates an organization's competitiveness, drives market differentiation, and secures a lasting position at the forefront of the digital era.

### 1.3.8 Scalability and Innovation

The scalability and innovation pillar emphasizes AI's potential to foster transformative change. Organizations should design scalable processes, allowing AI to handle increased data and adapt to changing business needs. This pillar not only pushes for scalability but also promotes exploring innovative AI applications to discover new opportunities [101]. By prioritizing scalability and innovation, organizations can achieve efficiency improvements, market differentiation, and tap into untapped potential.

a) Designing for Growth: Scalability requires organizations to design processes, systems, and architectures that can seamlessly expand to handle larger datasets, increasing workloads, and emerging complexities. Scalable AI systems ensure that solutions remain effective as the organization's demands evolve [102].
b) Flexibility and Adaptability: Scalability mandates flexibility and adaptability. AI solutions should be capable of adapting to new data sources, diverse use cases, and changing business landscapes without significant reconfiguration.
c) Resource Efficiency: Scalable AI systems optimize resource utilization. This efficiency ensures that as operations expand, resource consumption remains manageable and cost-effective.
d) Responsive to Market Dynamics: Scalability enables organizations to react swiftly to market shifts and opportunities. As AI systems scale, they facilitate agile responses to new demands, customer preferences, and industry trends.

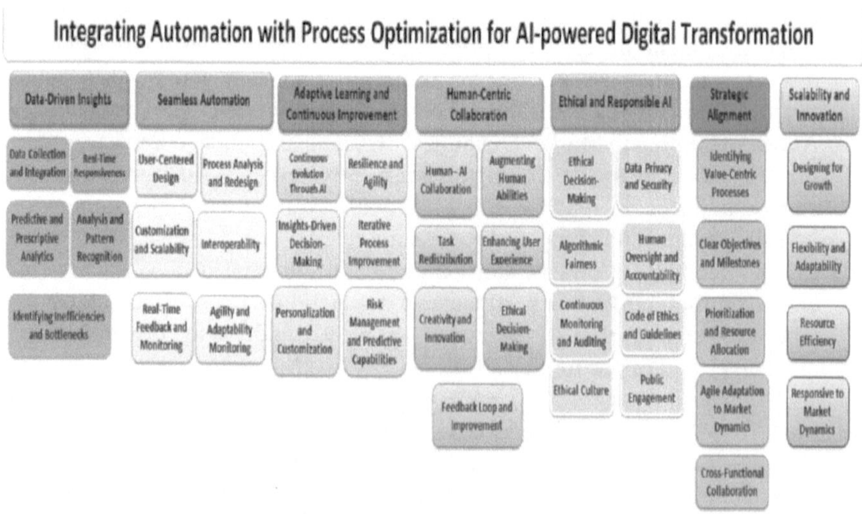


**FIGURE 1.9** Proposed framework.

The scalability and innovation pillar transcends routine optimization, propelling organizations toward visionary realms where transformational change becomes the norm. By preparing for scalability and nurturing innovation, organizations create an ecosystem where AI-driven automation not only elevates operations but also catalyzes groundbreaking shifts that redefine industries, create new norms, and forge a path toward a future marked by limitless possibilities.

The presented approaches to automation and process optimization within the context of AI-powered digital transformation leverage AI technologies to enhance efficiency, accuracy, and innovation across various industries. The proposed pillars address critical aspects that shape the success of automation and process optimization in the context of AI-powered digital transformation. These pillars serve as a guiding framework that offers organizations a structured approach to navigate the complex landscape of integrating automation and optimizing processes using AI. By embracing these pillars, businesses can create a holistic approach that combines technology, human collaboration, strategic vision, and ethical considerations, leading to a future characterized by efficiency, innovation, and sustainable growth in the digital age. The proposed framework can be displayed as follows.

## 1.4 CONCLUSION

In the realm of AI-powered digital transformation, our research has yielded a practical roadmap that identified seven fundamental pillars that are paramount for organizations aiming to leverage the transformative capabilities of AI technologies. These pillars serve as the cornerstones of successful integration and encompass a comprehensive approach to automation and process optimization. The first pillar, data-driven insights, emphasizes the strategic utilization of data, encompassing facets such as data collection, pattern recognition, and predictive analytics, ensuring that organizations make informed decisions grounded in data-driven intelligence. Seamless automation, the second pillar, underscores the importance of user-centric design and real-time adaptability, ensuring that automated processes align seamlessly with user needs while remaining agile and scalable. The third pillar, adaptive learning and continuous improvement, underscores the dynamic nature of AI technologies. It encompasses continuous evolution through AI and risk management, emphasizing the need for organizations to continuously learn, adapt, and innovate in response to evolving technological landscapes. Human-centric collaboration, the fourth pillar, recognizes the synergy between human expertise and AI capabilities. It focuses on augmenting human abilities and ethical decision-making, ensuring a harmonious collaboration that enhances efficiency and accuracy. Ethical and responsible AI, the fifth pillar, addresses the ethical considerations inherent in AI integration. It encompasses data privacy, algorithmic fairness, and public engagement, promoting responsible AI practices and ethical decision-making. Strategic alignment, the sixth pillar, emphasizes the importance of organizational alignment. It involves identifying value-centric processes and cross-functional collaboration, ensuring that AI initiatives are strategically aligned with organizational objectives. The final pillar, scalability and innovation, underscores the necessity of scalability in the face of evolving market dynamics. It includes designing for growth and responsiveness to market dynamics, enabling organizations to scale their initiatives efficiently and foster innovation.

In summary, our research provides not only a theoretical framework but also practical, actionable guidelines for organizations venturing into the realm of AI-powered digital transformation. By embracing these pillars, organizations can navigate complexities, achieve operational excellence, foster innovation, and ensure ethical, responsible, and scalable integration of AI technologies. This roadmap stands as a testament to the transformative potential of AI technologies, offering a strategic blueprint for industries and organizations worldwide.

## REFERENCES

1. Dau, Luis Alfonso, Elizabeth M. Moore, and William Newburry. "Building international business and societal resilience in an uncertain global environment." *AIB Insights* 23, no. 3 (2023): 1–5.
2. Moya, Ana, Michael Hein, and Janina Reimann. Business process automation for data teams-a practical approach at Handelsblatt Media Group. In International Conference on Subject-Oriented Business Process Management. Cham: Springer Nature, 2023, 55–60.
3. Moreira, Sílvia, Henrique S. Mamede, and Arnaldo Santos. "Process automation using RPA–a literature review." *Procedia Computer Science* 219 (2023): 244–254.
4. Asadov, Ramil. "Intelligent process automation: Streamlining operations and enhancing efficiency in management." Available at SSRN 4495188 (2023).
5. Gołąb-Andrzejak, Edyta. "AI-powered digital transformation: Tools, benefits and challenges for marketers–case study of LPP." *Procedia Computer Science* 219 (2023): 397–404.
6. Allioui, Hanane, and Youssef Mourdi. "Unleashing the potential of AI: Investigating cutting-edge technologies that are transforming businesses." *International Journal of Computer Engineering and Data Science (IJCEDS)* 3, no. 2 (2023): 1–12.
7. Masoodifar, Mahmood, İsmet Kahraman Arslan, and Aşkım Nurdan Tümbek Tekeoğlu. "Artificial intelligence in global business and its communication." *Journal of International Trade, Logistics & Law* 9, no. 1 (2023).
8. Petrillo, Antonella, Fabio De Felice, Monica Violeta Achim, and Nawazish Mirza, eds. *Digital Transformation: Towards New Frontiers and Business Opportunities.* BoD–Books on Demand, 2022.
9. Candelon, François, and Martin Reeves, eds. *The Rise of AI-powered Companies.* Walter de Gruyter GmbH & Co KG, 2022.
10. Dumas, Marlon, Fabiana Fournier, Lior Limonad, Andrea Marrella, Marco Montali, Jana-Rebecca Rehse, Rafael Accorsi, Diego Calvanese, Giuseppe De Giacomo, Dirk Fahland, Avigdor Gal, Marcello La Rosa, Hagen Völzer, and Ingo Weber. "AI-augmented business process management systems: A research manifesto." *ACM Transactions on Management Information Systems* 14, no. 1 (2023): 1–19.
11. Wil M. P. van der Aalst, Martin Bichler, and Armin Heinzl. "Robotic process automation." *Business & Information Systems Engineering* 60, no. 4 (2018): 269–272.
12. Volodymyr Leno, Artem Polyvyanyy, Marlon Dumas, Marcello La Rosa, and Fabrizio Maggi. "Robotic process mining: Vision and challenges." *Business & Information Systems Engineering* 63, no. 3 (2021): 301–314.
13. Antonio Bucchiarone, Marco Pistore, Heorhi Raik, and Raman Kazhamiakin. Adaptation of service-based business processes by context-aware replanning. In Proceedings of the 4th International Conference on Service-Oriented Computing and Applications (SOCA'11). IEEE, 2011, 1–8.

14. Al-Mekhlal, Monerah, Mohammed Al-Buraik, and Mohammad Al-Lubli. Digital transformation: AI-powered bot solutions and automation for customer services. In 2023 International Conference on Digital Applications, Transformation & Economy (ICDATE). IEEE, 2023, 1–7.
15. Prasad, Sandya, Prasanna Kulkarni, and Veronika Veronika. "Role of artificial intelligence in business process transformation." *AIP Conference Proceedings* 2736, no. 1 (2023).
16. Agarwal, Akshat, Charu Singhal, and Renny Thomas. *AI-Powered Decision Making for the Bank of the Future*. McKinsey & Company,–2021. www.mckinsey.com/industries/financial-services/our-insights/ai-powered-decision-making-for-the-bank-of-the-future
17. Girimurugan, B., S. Rajeshwari, S. P. Sreekala, and S. Revathy. "14 The smart and secured AI-powered strategies for optimizing processes in multi-vendor business applications." *Toward Artificial General Intelligence: Deep Learning, Neural Networks, Generative AI* (2023): 287.
18. Chahal, Sunil. "Navigating financial evolution: Business process optimization and digital transformation in the finance sector." *International Journal of Finance* 8, no. 5 (2023): 67–81.
19. Dash, Rupa, Mark McMurtrey, Carl Rebman, and Upendra K. Kar. "Application of artificial intelligence in automation of supply chain management." *Journal of Strategic Innovation and Sustainability* 14, no. 3 (2019): 43–53.
20. Sakala, Chibale M., and Sichone M. Bwalya. "The role of artificial intelligence in optimizing supply chain performance." *Journal of Procurement and Supply Chain Management* 2, no. 1 (2023): 1–14.
21. Pata, Kai, Merja Bauters, Petri Vesikivi, and Jaana Holvikivi. "Agile and lean methods with design thinking." *Technology Supported Active Learning: Student-Centered Approaches* (2021): 13–30.
22. Watson, Edward F. III, and Andrew H. Schwarz. "Enterprise and Business Process Automation." In *Springer Handbook of Automation*, pp. 1385–1400. Cham: Springer International Publishing, 2023.
23. Dumas, M., M. La Rosa, J. Mendling, and H. A. Reijers. *Fundamentals of Business Process Management*. Springer, 2018.
24. Harmon, P. *Business Process Change: A Business Process Management Guide for Managers and Process Professionals*. Morgan Kaufmann, 2020.
25. Fernandes, Ederson Carvalhar, and Carla Cristina Amodio Estorilio. The impact of artificial intelligence technologies in achieving better levels of maturity in business process management. In International Conference on Pattern Recognition. Cham: Springer Nature, 2022, 717–725.
26. Ughulu, Dr John. "The role of artificial intelligence (AI) in starting, automating and scaling businesses for entrepreneurs." *ScienceOpen Preprints* (2022).
27. Wirtz, Jochen, Johannes Hofmeister, Patricia Y. P. Chew, and Xin Ding. "Digital service technologies, service robots, AI, and the strategic pathways to cost-effective service excellence." *The Service Industries Journal* (2023): 1–24.
28. Martínez-Rojas, Antonio, J. Sánchez-Oliva, J. M. López-Carnicer, and Andrés Jiménez-Ramírez. Airpa: An architecture to support the execution and maintenance of AI-powered RPA robots. In International Conference on Business Process Management. Cham: Springer International Publishing, 2021, 38–48.
29. Alavian, Pooya, Yongsoon Eun, Semyon M. Meerkov, and Liang Zhang. "Smart production systems: Automating decision-making in manufacturing environment." *International Journal of Production Research* 58, no. 3 (2020): 828–845.
30. Chen, Jacky, Chee Peng Lim, Kim Hua Tan, Kannan Govindan, and Ajay Kumar. "Artificial intelligence-based human-centric decision support framework: An application to predictive maintenance in asset management under pandemic environments." *Annals of Operations Research* (2021): 1–24.

31. Makridakis, S. "The forthcoming artificial intelligence (AI) revolution: Its impact on society and firms." *Futures* 116 (2020): 102511.
32. Bughin, J., E. Hazan, S. Ramaswamy, M. Chui, T. Allas, P. Dahlström, N. Henke, and M. Trench. *Notes from the AI Frontier: Insights from Hundreds of Use Cases.* McKinsey Global Institute, 2018.
33. Tornbohm, R., and D. Miers. *Robotic Process Automation: From Hype to Reality—Now What?* Gartner Research, 2020.
34. Kaplan, A., and M. Haenlein. "Siri, Siri, in My Hand: Who's the Fairest in the land? On the interpretations, illustrations, and implications of artificial intelligence." *Business Horizons* 62 no. 1 (2019): 15–25.
35. Davenport, T. H., and R. Ronanki. "Artificial intelligence for the real world." *Harvard Business Review* 96, no. 1 (2018): 108–116.
36. Bughin, Jacques, Jeongmin Seong, James Manyika, Michael Chui, and Raoul Joshi. *Notes from the AI Frontier: Modeling the Impact of AI on the World Economy.*" McKinsey Global Institute, 2018.
37. Rogers, D., S. Medha, and U. Sivarajah. "A digital transformation and business model framework for the next generation of smart cities." *Technological Forecasting and Social Change* 153 (2020): 119927.
38. Wamba, S. F., A. Gunasekaran, S. Akter, S. J. F. Ren, R. Dubey, and S. J. Childe. "Big data and predictive analytics for supply chain and organizational performance." *Journal of Business Research* 70 (2020): 308–317.
39. Lu, Y., S. Sadiq, and V. Padmanabhan. "From business intelligence to predictive analytics." *Business & Information Systems Engineering* 62 (2020): 323–336.
40. Sharma, R., S. Mithas, and A. Kankanhalli. "Transforming decision-making processes: A research agenda for understanding the impact of business analytics on organisations." *European Journal of Information Systems* 24, no. 4 (2019): 433–441.
41. Feine, J., U. Gnewuch, S. Morana, and A. Maedche. "A taxonomy of social cues for conversational agents." *International Journal of Human-Computer Studies* 132 (2019): 138–161.
42. Xu, A., Z. Liu, J. Guerin, and R. Bults. "A comparative study of users' perceptions of a conversational agent and a social robot in a behavior change context." *Computers in Human Behavior* 105 (2020): 106214.
43. Shahin, Mohammad, F. Frank Chen, Ali Hosseinzadeh, and Neda Zand. "Using machine learning and deep learning algorithms for downtime minimization in manufacturing systems: An early failure detection diagnostic service." *The International Journal of Advanced Manufacturing Technology* 128, no. 9–10 (2023): 3857–3883.
44. Liu, C. Z., R. J. Kauffman, and D. Ma. "Competition, cooperation, and regulation: Understanding the evolution of the mobile payments technology ecosystem." *Electronic Commerce Research and Applications* 33 (2019): 100817.
45. Shukla, Khyati, Samia Nefti-Meziani, and Steve Davis. "A heuristic approach on predictive maintenance techniques: Limitations and scope." *Advances in Mechanical Engineering* 14, no. 6 (2022).
46. Wen, Yuxin, Md Fashiar Rahman, Honglun Xu, and Tzu-Liang Bill Tseng. "Recent advances and trends of predictive maintenance from data-driven machine prognostics perspective." *Measurement* 187 (2022): 110276.
47. Rojek, Izabela, Małgorzata Jasiulewicz-Kaczmarek, Mariusz Piechowski, and Dariusz Mikołajewski. "An artificial intelligence approach for improving maintenance to supervise machine failures and support their repair." *Applied Sciences* 13, no. 8 (2023): 4971.
48. Wang, Hongchao, Weiting Zhang, Dong Yang, and Yuhong Xiang. "Deep-learning-enabled predictive maintenance in industrial internet of things: Methods, applications, and challenges." *IEEE Systems Journal* 17, no. 2 (2022): 2602–2615.

49. Ivanov, D., A. Dolgui, B. Sokolov, F. Werner, and M. Ivanova. "A dynamic model and an algorithm for short-term supply chain scheduling in the smart factory Industry 4.0." *International Journal of Production Research* 57, no. 2 (2019), 383–397.
50. Dong, Y., K. Xu, and K. X. Zhu. "Blockchain technology in supply chain operations: An application in container shipping." *Journal of Cleaner Production* 277 (2020): 123201.
51. He, L., T. Yan, L. Da Xu, and D. Zeng. "An energy-efficient routing algorithm for software-defined industrial internet of things with underlaying UAVs-assisted sensor networks." *IEEE Transactions on Industrial Informatics* 14, no. 8 (2018): 337–3745.
52. Kumar, M., R. Shankar, and H. Choudhury. "A cloud-based fuzzy TOPSIS approach for supplier evaluation in sustainable supply chain." *The International Journal of Advanced Manufacturing Technology* 105, no. 1–4 (2019): 1143–1159.
53. Sar, Kubra, and Pezhman Ghadimi. "A systematic literature review of the vehicle routing problem in reverse logistics operations." *Computers & Industrial Engineering* (2023): 109011.
54. Zamani, Efpraxia D., Conn Smyth, Samrat Gupta, and Denis Dennehy. "Artificial intelligence and big data analytics for supply chain resilience: A systematic literature review." *Annals of Operations Research* 327, no. 2 (2023): 605–632.
55. Wang, Y., H. S. Ma, J. H. Yang, and K. S. Wang. "Industry 4.0: A way from mass customization to mass personalization production." *Advances in Manufacturing* 6, no. 4 (2018): 285–299.
56. Chen, Q., H. Chen, and D. Zeng. "Price competition and cooperation in a dual-channel supply chain with a direct selling option." *Journal of Cleaner Production* 215 (2019): 1481–1493.
57. Chen, Q., S. Pan, and Y. Wang. "Optimal and fair decisions in a dual-channel supply chain with risk-averse retailer and manufacturer." *European Journal of Operational Research* 264, no. 2 (2018): 534–545.
58. Punia, Sushil, and Sonali Shankar. "Predictive analytics for demand forecasting: A deep learning-based decision support system." *Knowledge-Based Systems* 258 (2022): 109956.
59. Vats, Prashant, and Siddhartha Sankar Biswas. "Big data analytics in real time for enterprise applications to produce useful intelligence." *Data Wrangling: Concepts, Applications and Tools* (2023): 187.
60. Hopkins, Emily. "Machine learning tools, algorithms, and techniques." *Journal of Self-Governance and Management Economics* 10, no. 1 (2022): 43–55.
61. Sharifani, Koosha, and Mahyar Amini. "Machine learning and deep learning: A review of methods and applications." *World Information Technology and Engineering Journal* 10, no. 7 (2023): 3897–3904.
62. McCarthy, Richard V., Mary M. McCarthy, Wendy Ceccucci, Leila Halawi, R. V. McCarthy, M. M. McCarthy, W. Ceccucci, and L. *Halawi. Applying Predictive Analytics*. Springer International Publishing, 2022.
63. W. B. Powell, and H. Topaloglu. "Stochastic programming in transportation and logistics." *Handbooks in Operations Research and Management Science* 10 (2008): 555–636.
64. Sharma, Vinod, Jeanne Poulose, and Chandan Maheshkar. "Analytics Enabled Decision Making "Tracing the Journey from Data to Decisions"." In *Analytics Enabled Decision Making*, pp. 1–22. Singapore: Springer Nature Singapore, 2023.
65. Susnjak, Teo. "A prescriptive learning analytics framework: Beyond predictive modelling and onto explainable AI with prescriptive analytics." *arXiv preprint* arXiv:2208.14582 (2022).

66. Bertsimas, D., M. V. Bjarnadóttir, M. A. Kane, J. C. Kryder, R. Pandey, S. Vempala, and G. Wang. "Algorithmic prediction of health-care costs." *Operations Research* 59, no. 4 (2021): 948–965.
67. Greasley, Andrew. *Simulating Business Processes for Descriptive, Predictive, and Prescriptive Analytics*. Walter de Gruyter GmbH & Co KG, 2019.
68. Russell, S., and P. Norvig. *Artificial Intelligence: A Modern Approach*. Malaysia: Pearson Education Limited, 2020.
69. Sutton, R. S., and A. G. Barto. *Reinforcement Learning: An Introduction*. MIT Press, 2018.
70. Liu, Y., and Y. Yang. "Real-time decision-making with recurrent neural networks." *Decision Support Systems* 130 (2020): 113215.
71. Zaharia, M., R. S. Xin, P. Wendell, T. Das, M. Armbrust, A. Dave, X. Meng, J. Rosen, S. Venkataraman, M. J. Franklin, A. Ghodsi, J. Gonzalez, S. Shenker, and I. Stoica. "Apache spark: A unified engine for big data processing." *Communications of the ACM* 59, no. 11 (2016): 56–65.
72. Bertsimas, D., and N. Kallus. "From predictive to prescriptive analytics." *Management Science* 66, no. 3 (2020): 1009–1023.
73. Chen, B., X. Dong, and C. Q. Zhao. Reinforcement learning in financial markets: A survey. Available at SSRN 3555213 (2020).
74. Arulkumaran, K., M. P. Deisenroth, M. Brundage, and A. A. Bharath. "A brief survey of deep reinforcement learning." *IEEE Signal Processing Magazine* 34, no. 6 (2019): 26–38.
75. Martínez-Plumed, F., R. Prudêncio, F. Martínez-Álvarez, and L. Torgo. "A snapshot of ongoing research on predictive modelling." *Expert Systems with Applications* 113 (2018): 368–377.
76. Zhang, Q., X. Li, H. Ning, D. Shang, and M. Gu. "Large-scale data mining using genetics-based machine learning." *IEEE Access* 7 (2019): 133231–133245.
77. Xu, R., and D. Wunsch. "Clustering algorithms in biomedical research: A review." *IEEE Reviews in Biomedical Engineering* 3 (2019): 120–154.
78. Deshpande, M., and G. Karypis. "Item-based top-N recommendation algorithms." *ACM Transactions on Information Systems (TOIS)* 22, no. 1 (2019): 143–177.
79. Souiden, Imen, Mohamed Nazih Omri, and Zaki Brahmi. "A survey of outlier detection in high dimensional data streams." *Computer Science Review* 44 (2022): 100463.
80. Nam, Kihwan. "Conversion paths of online consumers: A sequential pattern mining approach." *Expert Systems with Applications* 202 (2022): 117253.
81. Feldman, R., and J. Sanger. *The Text Mining Handbook: Advanced Approaches in Analyzing Unstructured Data*. Cambridge University Press, 2019.
82. George, A. Shaji, V. Sujatha, A. S. Hovan George, and T. Baskar. "Bringing light to dark data: A framework for unlocking hidden business value." *Partners Universal International Innovation Journal* 1, no. 4 (2023): 35–60.
83. Bharadiya, Jasmin Praful. "Machine learning and AI in business intelligence: Trends and opportunities." *International Journal of Computer (IJC)* 48, no. 1 (2023): 123–134.
84. Sattarova, Nozima. "Innovations in supply chain management: From just-in-time to agile supply chains." *International Bulletin of Young Scientist* 1, no. 1 (2023).
85. Kareska, Katerina. "Key factors for increasing the productivity of production organizations." Available at SSRN 4532554 (2023).
86. Vijayakumar, Senthilkumar, Kumar S. P. Gowtham, Navneet Nigam, and Raj Vijay Raj Singh. An novel approach in designing a security workbench with deep learning capabilities and process automation. In TENCON 2019–2019 IEEE Region 10 Conference (TENCON). IEEE, 2019, 263–268.
87. Mishra, Sumit, K. K. Sree Devi, and M. K. Badri Narayanan. "Technology dimensions of automation in business process management industry." *International Journal of Engineering and Advanced Technology* 8, no. 6 (2019): 1919–1926.

88. Dolz Ausina, G. (2023). *Evaluation of Different AI Applications for Operational Logistic Systems*. Universitat Politècnica de València. http://hdl.handle.net/10251/194044.
89. Nagy, László, Tamás Ruppert, and János Abonyi. Human-centered knowledge graph-based design concept for collaborative manufacturing. In 2022 IEEE 27th International Conference on Emerging Technologies and Factory Automation (ETFA). IEEE, 2022, 1–8.
90. Nadimpalli, Vamsi Krishna, Florian Hauser, Dominik Bittner, Lisa Grabinger, Susanne Staufer, and Jürgen Mottok. Systematic literature review for the use of AI based techniques in adaptive learning management systems. In Proceedings of the 5th European Conference on Software Engineering Education. Verdu, 2023, 83–92.
91. Rožman, Maja, Dijana Oreški, and Polona Tominc. "A multidimensional model of the new work environment in the digital age to increase a company's performance and competitiveness." *IEEE Access* 11 (2023): 26136–26151.
92. Naim, Arshi. "Role of artificial intelligence in business risk management." *American Journal of Business Management, Economics and Bank*ing 1 (2022): 55–66.
93. Pagani, Margherita, and Renaud Champion. "Creating business value through human-centric AI." *Artificial Intelligence for Sustainable Value Creation* (2021): 9–31.
94. Usmani, Usman Ahmad, Ari Happonen, and Junzo Watada. Human-centered artificial intelligence: Designing for user empowerment and ethical considerations. In 2023 5th International Congress on Human-Computer Interaction, Optimization and Robotic Applications (HORA). IEEE, 2023, 1–5.
95. Zohuri, Bahman, and Farhang Mossavar-Rahmani. "The symbiotic relationship unraveling the interplay between technology and artificial intelligence (an intelligent dynamic relationship)." *Journal of Energy and Power Engineering* 17 (2023): 63–68.
96. Gerke, Sara, Timo Minssen, and Glenn Cohen. "Ethical and Legal Challenges of Artificial Intelligence-Driven Healthcare." In *Artificial Intelligence in Healthcare*, pp. 295–336. Academic Press, 2020.
97. Pagani, Margherita, and Renaud Champion. "Creating business value through human-centric AI." *Artificial Intelligence for Sustainable Value Creation* (2021): 9–31.
98. Abdikhakimov, Islombek. Unraveling the copyright conundrum: Exploring AI-generated content and its implications for intellectual property rights. *International Conference on Legal Sciences* 1, no. 5 (2023): 18–32.
99. Shafagatova, Aygun, and Amy Van Looy. "A conceptual framework for process-oriented employee appraisals and rewards." *Knowledge and Process Management* 28, no. 1 (2021): 90–104.
100. Wang, Ke, Ziyi Ying, Shankha Shubhra Goswami, Yongsheng Yin, and Yafei Zhao. "Investigating the role of artificial intelligence technologies in the construction industry using a delphi-ANP-TOPSIS hybrid MCDM concept under a fuzzy environment." *Sustainability* 15, no. 15 (2023): 11848.
101. Brown, Solly, Darshit Gandhi, Louise Herring, and Ankur Puri. "The analytics academy: Bridging the gap between human and artificial intelligence." *McKinsey Quarterly* (2019): 1–9.
102. Czvetkó, Tímea, Alex Kummer, Tamás Ruppert, and János Abonyi. "Data-driven business process management-based development of Industry 4.0 solutions." *CIRP Journal of Manufacturing Science and Technology* 36 (2022): 117–132.

# 2 Driving Business Sustainability through Digital Supply Chains

*Abdul Rehman Shaikh and Imran Ali*

## 2.1 INTRODUCTION

In recent years, the evolution of consumer preferences has introduced a significant shift in the business landscape. The modern customer is no longer content with passive consumption; instead, they seek active participation in the supply chain and wish to learn about the origins of the products they purchase as well as their environmental and social effects on society (Wang et al., 2022). This transformation in consumer behaviour has profound implications for businesses operating within supply chains. Companies are now confronted with the imperative to adapt or risk going out of business at a time when stakeholders, from customers to investors, want not only high-quality goods and services but also ethical and transparent business practices (Asif et al., 2023).

This shift in consumer expectations has catalysed a broader paradigm change, one driven by the intersection of digital innovation and sustainability imperatives (Ratchford, 2019). As a result, businesses are compelled to explore new development opportunities that are not only commercially viable but also ecologically and socially responsible (Nayal et al., 2022). This intersection between digital innovation and sustainability isn't merely a trend but rather a fundamental reimagining of how supply chains operate. Supply chains no longer function solely as logistical conduits; instead, they have evolved into dynamic ecosystems propelled by data, technology, and a renewed dedication to reducing environmental impact (Sharma et al., 2022).

The transformation of supply chains into dynamic ecosystems is redefining how businesses engage with their partners, customers, and the world at large. This ecosystem-based approach embraces collaboration and information-sharing among supply chain participants. It involves the integration of cutting-edge technologies such as Internet-of-Things (IoT) sensors, blockchain, and artificial intelligence to track and trace products' journeys from raw materials to the end consumer (Ali et al., 2021). This transparency not only satisfies the consumer's desire for knowledge about product origins but also empowers them to make more informed and ethical purchasing decisions.

Moreover, this paradigm shift is pushing companies to re-evaluate their sourcing and production processes. They are increasingly investing in sustainable practices, from responsibly sourcing raw materials to reducing waste and carbon emissions


DOI: 10.1201/9781003438748-3

(Lerman et al., 2022). Supply chain sustainability has become a key driver of competitive advantage, as businesses recognize that their long-term viability is intricately linked to their ability to operate in an environmentally and socially responsible manner.

However, while the benefits of sustainable supply chain management initiatives are almost clear, it is unclear how different technologies contribute to developing more sustainable supply chains. The chapter also identifies potential challenges and barriers that companies may face when implementing such initiatives, including data privacy and security issues, interoperability and integration, and cultural resistance. In response, this chapter presents a critical examination of the role that digital supply chains can play in driving sustainable business practices, highlighting how key digital innovations can enable companies to optimize their operations, improve transparency and traceability, and foster collaboration with stakeholders.

To inspire and guide companies looking to invest in sustainable supply chain management initiatives, the chapter presents a variety of real-world case studies that demonstrate the successful implementation of digital supply chain technologies to drive sustainability. These case studies include companies such as Engro Foods Limited Pakistan, Americold Australia, Walmart, and Linfox Logistics Australia, all of which have used digital technologies to increase transparency, optimize operations, and reduce their carbon footprint.

The chapter concludes by emphasizing the importance of aligning digital supply chain strategies with broader business objectives thereby achieving sustainability goals. It provides a critical and insightful analysis of how digital supply chains can drive sustainable business practices and encourages companies to adopt a more sustainable approach to their supply chain management. By offering real-world case studies, the chapter motivates businesses to adopt sustainable supply chain management practices and leverage digital technologies to drive positive change across their supply chains. It encourages readers to think critically about the role that digital supply chains can play in building a more sustainable future and to take action to create a more socially and environmentally responsible business ecosystem.

### 2.1.1 The Urgency of Sustainable Business Practices

The urgency of sustainable business practices stems from the increasingly evident environmental and social challenges that businesses and society as a whole face. Organizations are recognizing that their long-term success and resilience are intertwined with their ability to operate in an environmentally responsible and socially conscious manner. Some of the major key points associated with the need for sustainable business practices have ascended from environmental degradation and resource scarcity, regulatory and legal pressures, consumer and investor demand, brand reputation and brand image, increasing effects of climate change and global responsibility, risk management and fear of pandemics, the need of resilient supply chains, long-term business viability, and increasing stakeholder pressure and engagement.

As the Earth's resources deplete and ecosystems degrade, businesses must adapt their practices to avoid resource shortages, such as water scarcity and raw material depletion. Moreover, governments worldwide are enacting stricter environmental regulations, requiring businesses to adhere to sustainable practices or face legal consequences. Likewise, consumers and investors are increasingly favouring companies that demonstrate a commitment to sustainability, creating market pressures for businesses to adopt sustainable practices.

A positive reputation for sustainability can enhance brand loyalty and attract environmentally conscious consumers, while negative practices can lead to reputational damage (Jung et al., 2020). Businesses are increasingly expected to contribute to global climate change mitigation efforts to address the severe consequences of rising temperatures and extreme weather events. Sustainable practices can enhance supply chain resilience by mitigating risks associated with disruptions, such as resource shortages, natural disasters, pandemics and social unrest. Businesses that prioritize sustainability are more likely to achieve long-term success by addressing the triple bottom line—economic, environmental, and social performance. Engaging with stakeholders and addressing their concerns has become more for maintaining a social license to operate and sustain positive relationships.

These key points underscore the urgency for businesses to adopt sustainable practices, aligning with the broader global goals of environmental conservation and social well-being. Incorporating these goals and objectives shall provide a solid foundation and help organizations achieve the benefits of integrating sustainability into digital supply chains.

### 2.1.2 The Role of Digital Supply Chains in Driving Sustainability

The role of digital supply chains in driving sustainability is multifaceted and pivotal in achieving environmentally responsible and socially conscious business practices. Digital technologies play a transformative role in enhancing supply chain operations, optimizing resource utilization, and promoting transparency, all of which contribute to sustainability goals (Yuan et al., 2023). Here are the key aspects of the role of digital supply chains in driving sustainability:

- Efficient Resource Management: Digital technologies, such as Internet of Things (IoT) sensors, data analytics, and artificial intelligence (AI), facilitate data-driven insights into resource utilization. Organizations can optimize energy consumption, water usage, and raw material usage, thereby minimizing waste and environmental footprint.
- Demand Forecasting and Inventory Management: Advanced analytics and AI algorithms enhance demand forecasting accuracy, reducing overproduction and excess inventory. This leads to reduced waste and more efficient use of resources throughout the supply chain.
- Real-Time Visibility and Transparency: Digital supply chains enable real-time monitoring and tracking of products, materials, and processes across the entire supply network. This transparency allows organizations to identify inefficiencies, waste, and areas for improvement, leading to better decision-making and reduced environmental impact.
- Sustainable Sourcing and Supplier Engagement: Digital supply chains enable traceability of raw materials and components, ensuring ethical sourcing practices and compliance with sustainability standards. Organizations

can collaborate with suppliers to promote sustainable production methods and responsible sourcing.

- Carbon Footprint Reduction: By optimizing transportation routes, modes, and scheduling, digital supply chains can minimize carbon emissions associated with logistics operations. This contributes to a reduced carbon footprint and aligns with climate change mitigation efforts.
- Circular Economy Integration: Digital supply chains facilitate the implementation of circular economy principles, such as recycling, remanufacturing, and product lifecycle extension. Reverse logistics and traceability enable the efficient return and refurbishment of products, reducing waste and promoting resource circularity.
- Risk Management and Resilience: Digital supply chain tools enhance risk identification and mitigation strategies. Organizations can respond to environmental disruptions, such as natural disasters or supply chain disruptions, more effectively, ensuring business continuity and minimizing negative social and environmental impacts.
- Stakeholder Engagement and Communication: Digital platforms and tools enable organizations to engage with stakeholders, including customers and communities, to communicate sustainability initiatives and progress. This transparency fosters trust and accountability in sustainable business practices.
- Continuous Improvement and Innovation: Digital supply chains provide real-time data for performance monitoring and evaluation. This information empowers organizations to continuously improve sustainability efforts, innovate new eco-friendly processes, and adapt to evolving environmental challenges.
- Measurement and Reporting: Digital supply chains enable accurate data collection and reporting of sustainability metrics, supporting organizations in meeting reporting requirements and demonstrating their commitment to sustainable practices.

As such, the integration of digital technologies into supply chains is a cornerstone for driving sustainability by enhancing efficiency, reducing waste, and promoting responsible business conduct. This chapter will delve deeper into each of these aspects, providing a comprehensive understanding of how digital supply chains catalyse achieving business sustainability goals.

## 2.2 DIGITAL TECHNOLOGIES AND SUPPLY CHAINS

### 2.2.1 Autonomous Supply Chain Management

An autonomous supply chain automates tasks to reduce human involvement and increase efficiency. It uses AI, machine learning, robotics, IoT, and blockchain to automate decision-making and execution. Autonomous supply chain management includes:

1. Data visibility: Autonomous supply chain management integrates data from suppliers, manufacturers, logistics providers, and customers. This connection provides accurate, real-time information.

2. Predictive analytics and demand forecasting: AI and machine learning evaluate historical data, market trends, and customer behaviour to accurately estimate demand. This optimizes inventory, production, and distribution.
3. Intelligent automation: Robotics and automation automate monotonous processes and streamline operations. Picking, packing, order fulfilment and transport management are automated. Delivery and logistics can use autonomous vehicles and drones.
4. Cognitive decision-making: AI-powered algorithms and decision support systems enable autonomous supply chain decision-making. These systems can evaluate massive amounts of data, spot trends, and make optimal decisions about inventory, order fulfilment, route optimization, and risk mitigation.
5. Blockchain technology: Blockchain improves supply chain transparency, traceability, and security. It creates a decentralized, immutable ledger to track transactions, authenticate items, and reduce fraud, counterfeiting, and supply chain disputes.
6. Continuous monitoring and adjustment: Autonomous supply chain systems monitor key performance indicators (KPIs), environmental factors, and market conditions. They can reroute shipments in the event of delays or change production schedules based on demand.

Autonomous supply chain management creates a highly efficient, flexible, and responsive supply chain network that can autonomously anticipate and adapt to market demands and disruptions. Reducing manual intervention and enabling intelligent decision-making improves operational efficiency, costs, order fulfilment, customer satisfaction, and market competitiveness. The use of drones and autonomous vehicles for distribution and logistics is a prime example. Autonomous drones are being used by companies such as United Parcel Services (UPS) and Amazon for last-mile deliveries. These drones improve routes, use less fuel, and speed up deliveries, all of which help increase economic efficiency and reduce the environmental impact of transportation by reducing carbon emissions. In addition, as workers move into positions such as drone operators and data analysts, they provide opportunities for career advancement.

### 2.2.2 Smart Warehousing

Smart warehousing uses innovative technologies and automated methods to manage and operate warehouses. It optimizes warehouse processes and increases productivity using IoT, robotics, AI, big data analytics, and cloud computing. Smart warehouses automate manual tasks for faster, more accurate operations. Smart warehouses have several key characteristics:

1. IoT sensors and devices: These devices collect real-time data on inventory, temperature, humidity, and equipment status. They improve warehouse operations through 24/7 monitoring.
2. Automated systems: Robotics and automation simplify picking, packing, and sorting. Automated guided vehicles (AGVs) and robotic arms efficiently transport and fulfil orders.

3. Inventory management: Smart warehousing uses data analytics to optimize stock levels, track items and streamline order fulfilment. This reduces out-of-stock, inventory, and errors.
4. Warehouse management systems (WMSs): WMS software manages warehouse operations. It integrates inventory management, order processing, labour management, and shipment tracking to provide real-time visibility and control of the warehouse.
5. Data analytics: Big data analytics tools examine massive amounts of sensors, devices, and other data to gain insights. This helps find patterns, optimize operations, and make data-driven process improvements.
6. Cloud computing: Cloud systems store and process warehouse data flexibly. They enable remote access, real-time collaboration, and system integration.

Smart warehousing improves operational efficiency, accuracy, speed, space utilization, labour costs, safety, and customer satisfaction. Take Alibaba's intelligent warehousing system as an example. They have improved order fulfilment efficiency by over 300% through the use of robots and automation. Reducing the need for huge, energy-intensive warehouses not only increases economic benefits but also leaves a smaller environmental footprint. In addition, reducing physically demanding tasks can improve worker safety and well-being while advancing societal goals.

### 2.2.3 Big Data and Artificial Intelligence

Big data and artificial intelligence are transforming technology, business, and society. Let's examine each:

1. Big data: Large and complex data sets that cannot be processed or evaluated using typical data processing methods. It collects, stores, and analyses massive amounts of structured and unstructured data from social media, sensors, transactions, and web interactions.

   Big data has three Vs:

   Volume: Traditional database systems cannot handle big data. It requires scalable storage and processing.

   Velocity: Big data is generated rapidly and requires real-time or near–real-time processing. Social media and sensor data need to be analysed quickly.

   Variety: Big data includes text, photos, audio, and video. It includes relational databases and unstructured data such as social media posts and emails.

   Big data analytics extracts insights, patterns, and trends from massive data sets to inform decision-making. Data mining, machine learning, and statistical analysis reveal patterns, correlations, and predictive models. Big data analytics helps businesses streamline processes, improve customer experiences, and compete.

2. Artificial intelligence: Computer systems that can learn, reason, solve problems, and make decisions are called AI. AI includes machine learning, natural language processing (NLP), computer vision, robotics, and expert systems. Machine learning (ML) is a subset of AI that uses algorithms and models to teach computers to predict and behave without being programmed. ML algorithms improve with experience, using patterns and statistics. AI systems can analyse massive amounts of data to gain insights and make intelligent decisions. AI can evaluate customer behaviour, automate tedious tasks, streamline supply chain processes, make personalized suggestions, detect fraud, and analyse advanced data.

Big data provides the raw material, while AI algorithms extract useful insights and patterns. AI algorithms can sift through massive amounts of data faster, improving decision-making, automation, and understanding of complicated phenomena. Big data and AI can help companies use data and intelligent systems to innovate and improve in industries such as healthcare, finance, manufacturing, marketing, transportation, and others. The healthcare industry is a prime example. IBM's Watson for Oncology uses AI and big data to analyse medical literature and patient information to provide personalized cancer therapies. By improving health outcomes and reducing the costs associated with poor treatment, this has a positive economic impact. It is also socially beneficial by improving access to state-of-the-art healthcare, especially in underprivileged regions, and environmentally friendly by minimizing the need for trial-and-error therapies.

### 2.2.4 Blockchain

Blockchain is a decentralized and distributed digital ledger that enables secure and transparent transactions and record-keeping without intermediaries. It secures and verifies data.

Blockchain elements and concepts include:

1. Decentralized: Blockchain data is decentralized, unlike centralized systems. Each node in the network stores the entire blockchain. Decentralization improves transparency, security, and single points of failure.
2. Blocks and chain: Blockchain stores transactions in blocks. Cryptographically linked blocks form a chain. This connectivity provides the integrity and immutability of blockchain data.
3. Cryptographic security: Blockchain uses cryptography to secure data. Digital signatures verify each transaction or record. Cryptographic hash algorithms generate unique block identifiers (hashes), making data tampering nearly impossible.
4. Consensus mechanisms: These ensure the integrity and agreement of the blockchain. These technologies allow network nodes to agree on the validity of transactions and the order of the blockchain. Proof of Work (PoW), Proof of Stake (PoS), and practical Byzantine fault tolerance (PBFT) are popular consensus algorithms.

5. Smart contracts: Blockchain technologies provide self-executing contracts with pre-defined rules and conditions. Smart contracts automate and program business processes. They are executed without intermediaries under certain conditions.
6. Transparency and auditability: All network participants can verify blockchain data. Transactions and records are transparent, which builds trust. The immutability of the blockchain provides a trusted audit trail for traceability and accountability.

Beyond Bitcoin, blockchain technology has attracted attention. Supply chain management, financial services, healthcare, voting systems, and intellectual property protection have all been explored. Blockchain's security, transparency, and decentralized trust could transform industries and enable new digital interactions. By reducing fraud and increasing transparency in financial transactions and supply networks, blockchain can increase economic efficiency. By providing traceability and verification of sustainable practices, it can also help achieve environmental goals, and consumers have access to verify and check the practices of the organization. In terms of social empowerment, blockchain can give people more control over their data and financial assets, including labour practices, and product traceability. Blockchain is used by the diamond industry to trace the origin of stones. For example, Everledger uses blockchain technology to track a diamond from the mine to the market. By improving the working conditions of miners and reducing the funding of armed conflict, it ensures ethical sourcing, reduces the trade in conflict diamonds, and advances societal goals. Economically, it increases consumer confidence, which can lead to increased demand for diamonds, while environmentally, it promotes ethical mining practices.

### 2.2.5 Internet of Things

The Internet of Things is a network of embedded physical devices, vehicles, appliances, and other objects with sensors, software, and connectivity capabilities. These devices collect and exchange data over the Internet, enabling them to communicate, interact, and perform tasks autonomously.

Important aspects of the Internet of Things are:

1. Connectivity: IoT devices are equipped with Wi-Fi, Bluetooth, cellular networks, or low-power wide-area networks (LPWANs) to connect and communicate with each other or with central systems.
2. Sensors and actuators: IoT devices are equipped with various sensors (e.g. temperature, humidity, motion, proximity) to monitor and collect data from the environment. Actuators allow devices to perform tangible actions in response to the data received, such as turning on/off appliances, adjusting settings, and controlling machinery.
3. Internet of Things devices collect and generate large amounts of data from their environment. This data can be processed locally on the device or sent to cloud-based platforms for storage and further analysis. Using advanced

analytics techniques, such as machine learning, it is possible to gain insights from the data collected.
4. Interoperability and standards: For interoperability and seamless integration, IoT devices and systems must adhere to common standards. Standardization efforts enable different devices and platforms to communicate and exchange data effectively.
5. Applications and use cases: The applications of the Internet of Things are vast and diverse, spanning industries such as healthcare, transportation, agriculture, smart homes, and industrial automation, among others. IoT enables a variety of use cases such as remote monitoring, predictive maintenance, asset tracking, energy management, and enhanced automation.

Smart thermostats that adjust temperature based on occupancy and weather conditions, wearable fitness trackers that monitor health and activity levels, connected cars that provide real-time traffic information, and industrial sensors that optimize manufacturing are examples of IoT in action. The expansion of the IoT has the potential to revolutionize industries, increase productivity, improve decision-making, and enable new services and experiences. However, it also raises concerns about privacy, security, data governance, and the need for a robust infrastructure to accommodate the growing number of connected devices.

IoT increases productivity through automation and data-driven decision-making, which helps drive economic development. Environmentally, it can support more effective resource management and monitoring, reducing waste. From a social perspective, IoT can improve safety and quality of life by providing real-time monitoring and support, especially for the elderly and people with disabilities. IoT sensors in agriculture can monitor weather and soil conditions to optimize irrigation and fertilization. This boosts the economy by improving food yields and reducing resource waste. For the environment, it saves water and uses fewer chemicals. For society, it promotes ethical farming practices and ensures food security.

### 2.2.6 Cloud Enterprise Resource Planning

Cloud ERP, also known as enterprise resource planning, refers to the use of an ERP system that is accessed and delivered via a cloud computing infrastructure. The process involves the use of remote servers to host the ERP software and databases, allowing users to access the system via the Internet, as opposed to operating it locally on-site. The following are some of the key features and benefits associated with cloud-based enterprise resource planning systems:

1. Scalability and flexibility: An advantage of cloud ERP is its ability to provide companies with scalability and flexibility, allowing them to tailor their resource allocation to their specific needs. By using cloud servers, organizations can easily change their user base, storage capacity, and computing resources in response to changing needs.
2. Accessibility and remote collaboration: The use of cloud ERP allows users to access the system at any time and from any location, facilitating remote

working and fostering collaboration between people in different geographical locations. This enables seamless, real-time data sharing; improves decision making; and promotes effective collaboration between team members.

3. Reduce information technology (IT) infrastructure and maintenance costs: By adopting cloud ERP, companies can eliminate the need for physical servers and infrastructure on their premises. They are no longer responsible for hardware maintenance, upgrades, and system administration. This can lead to financial savings and a reduction in IT administration costs.
4. Faster implementation and easy upgrades: Cloud-based ERP solutions generally have shorter implementation times than on-premises systems. The infrastructure setup is managed by the cloud provider, allowing users to get up and running quickly. The provider is also responsible for managing upgrades and system updates, ensuring that users always have access to the latest features and security enhancements.
5. Data security and backup: Data security and backup are important considerations when using cloud-based enterprise resource planning systems. Cloud ERP providers typically implement comprehensive security measures and use effective disaster recovery mechanisms to protect data. Various measures are implemented to ensure data integrity and availability, including the use of encryption, access controls, regular backups, and data redundancy.
6. Integration and connectivity: Cloud-based enterprise resource planning systems can integrate with various other cloud-based applications and services, facilitating the smooth and uninterrupted flow of data and integration across various business functions. This enables improved operational efficiency, reduces the occurrence of redundant data, and provides a consolidated view of the organization's activities.
7. Vendor support and expertise: Cloud-based enterprise resource planning systems are typically supported by the vendor or service provider, ensuring access to technical support and expertise. The company provides technical support, regular system maintenance, and ongoing software updates. Users can benefit from the vendor's specialized knowledge, allowing them to focus on using the system effectively rather than being burdened with the responsibility of monitoring and maintaining the underlying IT infrastructure.

Cloud-based enterprise resource planning systems are a viable option for businesses of all sizes, including small and medium-sized enterprises (SMEs), which may be limited in their ability to invest in on-premises infrastructure. The software offers organizations the benefits of flexibility, scalability, cost-effectiveness, and access to advanced ERP functionality. These benefits enable organizations to streamline operations, improve decision-making, and respond effectively to evolving business needs. ERP solutions in the cloud increase economic efficiency by reducing IT costs and simplifying business processes. They can have a positive impact on the environment by reducing the need for on-premises data centres and therefore energy consumption. From a social perspective, cloud ERP supports inclusion and work-life balance by enabling remote collaboration and work. Cloud-based ERP solutions are used by companies like Salesforce to effectively manage their customer data. Improving customer relationship management not only increases economic efficiency but also reduces the need for on-premises data centres, which saves energy and has less of a negative impact on the environment. From a social perspective, it enables companies to offer more flexible working conditions that support work-life balance.

**CASE STUDY—ENGRO FOODS LIMITED**

*In the face of evolving market dynamics, Engro Foods Limited (EFL) embarked on a transformative journey with Project iEFL, a comprehensive systems, applications and products (SAP) ERP implementation aimed at bolstering its supply chain management. EFL's previous challenges, including disjointed systems, inefficient planning, and limited visibility, were tackled head-on. Through the deployment of SAP ERP modules across production scheduling, sales and distribution, finance, and human resource management, EFL achieved a holistic integration that redefined its operational landscape.*

*The implementation process was accompanied by rigorous training programs and a strategic change management approach, ensuring a seamless transition for employees and minimizing resistance to change. This initiative bore fruit as EFL reaped a plethora of benefits. Real-time data access ushered in heightened visibility, enabling accurate demand forecasting, efficient inventory management, and reduced stockouts. Automated production scheduling not only streamlined operations but also optimized resource allocation, resulting in enhanced production efficiency.*

*The ERP system's impact extended to the decision-making process, as timely and precise data empowered informed decisions across departments. This, in turn, translated to customer satisfaction, with faster order processing and improved delivery tracking fostering loyalty. Notably, the newfound efficiency led to tangible cost savings, as resource allocation became more targeted, production delays diminished, and processes were streamlined.*

*Engro Foods Limited's successful SAP ERP implementation epitomizes the transformative potential of digital technologies in revolutionizing supply chain management. Project iEFL underscores the pivotal role of seamless integration across diverse departments, offering profound insights for businesses seeking to navigate the dynamic landscape of modern business challenges with precision and ingenuity.*

## 2.3 LEVERAGING DIGITAL SUPPLY CHAINS FOR SUSTAINABLE BUSINESS PRACTICES

The digital technologies discussed in Section 2 help organizations achieve their economic, environmental, and social goals. A mix of these technologies may be leveraged by firms to achieve various benefits, including optimizing operations with real-time data, enhanced transparency and traceability, smarter resource allocation, and setting up sustainability goals and metrics.

### 2.3.1 Optimizing Operations with Real-Time Data

Optimizing operations and supply chain with real-time data involves leveraging digital technologies to collect, analyse, and act upon up-to-the-minute information from various points within the supply chain. This approach enables organizations to make informed decisions, enhance efficiency, and respond swiftly to changing conditions. For optimum and integrated real-time data, today most organizations are relying on enterprise resource planning software (e.g. SAP, Oracle, Microsoft Dynamics). With the help of ERP software, all the departments are integrated and have visibility to cross-departmental information. This ensures timely adjustments to plans and organizations can quickly respond to any disruptions in no time. Here's a detailed explanation of the process:

Real-time data optimization mainly consists of three steps, data collection, data integration, and continuous monitoring. All these steps are crucial and sequential; for instance, data cannot be integrated before collection, and likewise, monitoring cannot take place before integration. Real-time data can be collected through various sources, such as IoT sensors, global positioning system (GPS) trackers, radio frequency identification (RFID) tags, and connected devices. These devices gather information on factors like production rates, inventory levels, transportation status, and environmental conditions. In the next step, data is integrated into a central platform or system, often using cloud-based solutions. This is where the role of ERP software is crucial. All the departments have access to ERP, and all the employees can view cross-departmental information for their decision making. This consolidation allows for a holistic view of the supply chain, making it easier to identify bottlenecks, inefficiencies, and areas for improvement. For instance, the warehouse can plan their storage by observing the purchase order logs; likewise, production can plan their production schedule by observing the real-time sales data. Real-time data is continuously monitored, providing insights into current operations. This monitoring allows organizations to detect anomalies or deviations from expected performance, enabling proactive interventions. Such optimization of operations with real-time data can help achieve the following benefits:

1. Predictive Analytics: Advanced analytics and AI algorithms analyse real-time data to predict future trends, demand fluctuations, and potential disruptions. This foresight helps organizations make proactive decisions and plan for contingencies.
2. Demand Forecasting: Real-time data aids in accurate demand forecasting, enabling organizations to align production and inventory levels with actual market needs. This minimizes overproduction and reduces waste.
3. Inventory Management: Real-time visibility into inventory levels helps optimize stock levels and prevent stockouts or excess inventory. This ensures efficient resource allocation and reduces carrying costs.

4. Agile Manufacturing and Production: Real-time data allows for agile manufacturing and production adjustments based on changing demand or supply conditions. This flexibility enhances responsiveness and reduces lead times.
5. Supplier Collaboration: Real-time data sharing with suppliers fosters collaboration, enabling them to align their production schedules and deliveries with actual requirements. This reduces delays and minimizes disruptions.
6. Route Optimization: For logistics and transportation, real-time data on traffic, weather, and road conditions helps optimize routes, reducing transit times, fuel consumption, and emissions.
7. Quality Control: Real-time data monitoring can help identify quality issues in real time, enabling immediate corrective actions. This reduces defects and potential recalls.
8. Regulatory Compliance: Real-time data ensures that processes and products remain compliant with relevant regulations and standards, avoiding costly penalties.
9. Continuous Improvement: Analysis of real-time data provides insights into ongoing operational performance. Organizations can identify trends and areas for improvement, facilitating a culture of continuous enhancement.
10. Risk Mitigation: Real-time data aids in identifying potential supply chain disruptions or risks, allowing organizations to take swift corrective actions to minimize negative impacts.
11. Sustainability Impact: Real-time data can be used to track and monitor sustainability-related metrics, such as carbon emissions, energy consumption, and waste generation. This aids in achieving sustainability goals and reporting progress.

By optimizing operations and supply chains with real-time data, organizations can achieve greater efficiency, reduced costs, improved customer satisfaction, and enhanced sustainability practices. The integration of digital technologies and real-time data analytics provides a powerful toolset for organizations to navigate the complexities of modern supply chain management. In summary, using real-time data to optimize operations and supply chains is a powerful tactic for achieving sustainability goals. Real-time data gives companies the flexibility and insight they need to eliminate waste, conserve resources, reduce carbon emissions, and meet sustainability goals. As a result, real-time data helps companies remain competitive and resilient in a rapidly changing global economy. The following case study further illustrates how digitalization supports sustainability at Americold LLC.

### CASE STUDY: HOW DIGITALIZATION SUPPORTS SUSTAINABILITY AT AMERICOLD LLC

*Americold LLC, a leader in temperature-controlled warehousing and logistics, has made substantial strides in sustainability by integrating digital technologies into its operations.*

*On the environmental front, Americold has prioritized energy efficiency through the implementation of advanced energy management systems. These systems rely on sensors and data analytics to optimize temperature control within their warehouses. By maintaining precise storage conditions, the company reduces energy consumption and minimizes greenhouse gas emissions. This dual benefit not only aligns with environmental sustainability goals but also yields cost savings, contributing to economic sustainability. Moreover, Americold has embraced smart refrigeration technologies that continuously monitor and adjust temperature settings. This not only minimizes energy wastage but also enhances food safety by reducing the risk of spoilage. The integration of renewable energy sources, such as solar panels and wind turbines, further demonstrates their commitment to environmental sustainability and reduces reliance on fossil fuels.*

*In terms of social sustainability, Americold employs digital tools for workforce management. These include workforce management software and mobile apps that facilitate efficient scheduling, monitoring, and communication with their employees. This not only enhances operational efficiency but also supports employee well-being by ensuring fair scheduling practices. Additionally, the company utilizes digital platforms for comprehensive training programs focused on safety and sustainability. This investment in employee development contributes to a safer work environment and overall social well-being. Furthermore, Americold engages with local communities through digital platforms and social media. They use these channels to share information about their sustainability initiatives, community involvement, and job opportunities. This engagement fosters positive relationships with the areas where they operate, reinforcing their commitment to social sustainability.*

*Economically, Americold leverages digital technologies to optimize its supply chain. Advanced analytics and predictive modelling enable them to streamline the flow of goods, reduce delays, and cut costs. This not only benefits their bottom line but also enhances economic sustainability for both Americold and their clients. Digital inventory management systems are another economic sustainability driver. By efficiently tracking inventory levels, Americold reduces excess stock and minimizes waste, aligning with environmental sustainability goals by decreasing food waste. Moreover, the company employs digital platforms for improved customer engagement. Through real-time information sharing on inventory, temperature monitoring, and logistics, Americold enhances customer satisfaction and nurtures long-term business relationships.*

*Last, Americold's commitment to data-driven decision-making underpins its approach to sustainability. By harnessing data analytics, they can make informed choices about their operations, identify cost-saving opportunities, and enhance overall economic sustainability.*

### 2.3.2 Enhancing Transparency and Traceability across the Supply Chain

Enhancing transparency and traceability across the supply chain using digital technologies involves implementing tools and systems that provide real-time visibility into the movement, origin, and processes of goods and materials. This traceability in some organizations goes beyond the focal organization, when the whole supply chain is connected, and data of one organization is shared with other supply chain members in real time. This also helps analyse the sales trends of distributors and quickly react to any increasing customer demand. When other supply chain actors are connected with the focal organization, the focal organization can monitor the distribution of their products and can get notifications if the shelf stock is reduced or goes to minimum stock levels. This not only promotes ethical and sustainable practices but also builds trust among stakeholders, especially the customers. Following digital technologies can help organizations achieve transparency and traceability across the supply chain network:

1. Digital Tracking and Identification: Implement technologies like RFID tags, barcodes, and quick response (QR) codes to uniquely identify and track products, materials, and components. These identifiers can be scanned at various points in the supply chain to record and transmit data.
2. IoT Sensors: Integrate Internet of Things sensors into products or packaging to collect data on factors like temperature, humidity, location, and handling conditions. These sensors transmit real-time information that enhances visibility and ensures compliance with quality standards.
3. Blockchain Technology: Utilize blockchain to create an immutable and transparent ledger of transactions. This technology ensures that every step in the supply chain is securely recorded, providing an auditable and tamper-proof record.
4. Cloud-Based Platforms: Leverage cloud-based platforms to centralize and share supply chain data among stakeholders. This enables real-time collaboration, ensuring that all parties have access to accurate and up-to-date information.
5. Supplier Integration: Require suppliers to adopt digital tracking and reporting systems, enabling seamless data exchange and visibility into their processes. This integration enhances transparency across the entire supply chain network.

Apart from various other advantages, some of the obvious advantages of traceability and transparency across the supply chain network are presented in the following:

1. Data Analytics and Visualization: As organizations have access to all the information in real time, it is much easier to achieve data analytics for quick decision-making. Visualization of data can be useful in understanding big data, which may otherwise be difficult to comprehend. Visualization techniques such as dashboards and interactive reports help stakeholders understand complex supply chain information easily.

2. Geographic Information Systems (GIS): By integrating GIS with real-time data, organizations can geographically map and visualize supply chain activities. This aids in understanding the physical flow of goods across distributors, distribution centres, and even retailers and identifying potential bottlenecks.
3. Supplier Audits and Compliance: With real-time data and access to suppliers' networks and databases, it has become much easier to conduct supplier audits and compliance checks. Regularly monitoring supplier activities can help ensure adherence to sustainability and ethical standards.
4. Consumer Engagement: Digital technologies like mobile apps and QR codes can help consumers trace the origin and journey of products. This empowers consumers with information and promotes brand transparency ultimately leading to brand trust and brand equity.
5. Real-Time Alerts and Notifications: With automated alerts and notifications to inform stakeholders about critical supply chain events, such as delays, disruptions, or quality issues, managers can coordinate and make informed decision-making to handle the disruptions.
6. Environmental Impact Tracking: Integrate environmental metrics into the traceability system to monitor the carbon footprint, water usage, and other environmental impacts associated with the supply chain.
7. Cross-Functional Collaboration: Using digital platforms different departments can collaborate within an organization and outside the organization, enabling a holistic approach to transparency and traceability.
8. Supplier Code of Conduct: Require suppliers to adhere to a digital supplier code of conduct, outlining expectations for transparency, ethical practices, and sustainability performance.
9. Data Security and Privacy: Despite having access to supply chain network data, there are always provisions for access rights and roles (like a user, power user, administrator, etc.). This ensures that data security and privacy measures are in place to protect sensitive information while maintaining transparency, with the right information for the right people.

### 2.3.3 Leveraging Artificial Intelligence for Smarter Resource Allocation

Leveraging AI for smarter resource allocation in the supply chain involves using advanced algorithms and machine learning techniques to optimize the allocation of resources such as inventory, production capacity, transportation, and labour. AI-driven solutions can analyse big data, identify patterns, and make data-driven decisions to ensure efficient resource utilization. AI-driven technology can help organizations meet all the standards of an efficient supply chain, including but not limited to demand planning, production scheduling, inventory optimization, route planning and scheduling, real-time decision-making, maintenance scheduling, resource allocation and optimization, quality control, and environmental impact management.

AI algorithms can analyse historical data, market trends, seasonality, and external factors to generate accurate **demand planning** and **demand forecasts** for future periods. This helps in aligning resources with expected consumer demand, minimizing stockouts, and avoiding the excessive costs associated with overstocking. At the same time, AI can generate optimum **production schedules** by considering the production capacity of individual machines, resource availability, order priorities, and required setup times. This reduces production bottlenecks improves throughput and helps in inventory optimization, ultimately benefiting production-related KPIs. AI can **optimize inventory** levels by considering factors like suppliers' lead times, demand variability, storage costs, and production times. This leads to reduced carrying costs while ensuring sufficient stock availability to meet demand forecasts. AI algorithms can also help in transport **route planning**. By integrating geographic information systems with sales order systems, AI can generate ideal transportation routes based on factors like truck loads, vehicle capacities, route conditions, traffic congestions, and drop points. This reduces transportation costs and enhances on-time deliveries, ultimately benefiting consumer-related supply chain KPIs.

AI systems can process real-time data from various sources to make **informed decisions** on resource allocation, considering factors like demand spikes, supply disruptions, and changing market conditions. With this real-time data, organizations can plan **predictive maintenance** schedules. AI systems also help monitor equipment sensor data and predict maintenance schedules proactively. This prevents unplanned downtime and optimizes equipment usage, leading to increased overall equipment efficiency. With planned maintenance schedules, organizations can achieve better resource allocation and plan better human workload factors. This reduces human resource (HR) management costs while achieving **optimum resource utilization** at the same time. Besides, AI can analyse production and operational data to optimize the utilization of manufacturing and processing capacities. This minimizes under-utilization and ensures efficient use of resources. With optimum resource utilization, organizations can achieve **efficient energy management**. AI algorithms can also help optimize energy usage in manufacturing and distribution facilities by analysing energy consumption patterns and recommending strategies for reducing energy waste. By learning from the past and adapting the future process and controls, AI can help achieve **quality control** over time to reduce the deficiencies and chances of error. Based on the qualities of continuous learning, AI leads to zero-error systems, on the principles of continuous improvement and quality assurance.

Some other benefits of AI include dynamic pricing, supplier collaboration, cognitive procurement, scenario analysis, and simulations to model difference scenarios. By leveraging AI for smarter resource allocation, organizations can achieve greater efficiency, reduced costs, improved customer satisfaction, and enhanced sustainability practices within their supply chains.

**CASE STUDY—WALMART USA**

*One of the world's largest retailers, Walmart, is aggressively using artificial intelligence and other cutting-edge technologies to improve its digital supply chain and promote ethical business practices. Walmart has used AI algorithms to improve the accuracy of demand forecasting and to manage inventory levels.*

*Walmart has made it a priority to reduce overstocking and understocking to reduce waste, increase sustainability, and save costs. AI-powered solutions are used to optimize delivery times, load sizes, and transportation routes. This optimization improves supply chain efficiency and reduces emissions and fuel consumption. To monitor and manage energy use in its stores and distribution centres, Walmart is pioneering the use of AI-based energy management systems. These technologies support sustainability goals and reduce operating costs by identifying opportunities for energy savings. Walmart is using AI and big data to analyse the sustainability performance of its suppliers and service partners, further improving its supply chain. This allows it to make smart choices, ensure ethical sourcing, and keep an eye on suppliers' labour policies.*

*Walmart uses blockchain technology to improve traceability in its food supply chain, often using AI for data analysis. This promotes ethical sourcing and provides transparency, allowing buyers to track the origin of specific items. AI is also being used to streamline the circulation of inventory within retailers, ensuring effective product placement and minimizing waste. AI systems used in stores help maintain appropriate temperatures and lighting levels, significantly reducing energy consumption.*

### 2.3.4 Setting Clear Sustainability Goals and Metrics

Setting clear sustainability goals and metrics using digital supply chains involves a systematic approach that aligns environmental and social objectives with digital technology capabilities. Organizations need to identify the specific sustainability areas that are most relevant to industry, markets, and other relevant stakeholders. Such goals and targets might be related to circular manufacturing, carbon emissions reduction, waste reduction, sustainable sourcing, energy efficiency, or social responsibility. These sustainability goals must be aligned with the overall business strategy and organizational values. Sustainability objectives should complement and reinforce your organization's mission and long-term vision. Once the goals have been established, organizations need to develop metrics for measuring the performance in achieving those goals. A metrics or a dashboard shall be helpful to keep an eye on organizational performance and ensure the direction of employees.

These metrics or dashboards shall require baseline data related to chosen sustainability focus areas. This data will serve as a starting point for measuring progress and assessing the impact of organizational digital supply chain initiatives. While measuring the metrics and performance of the organization, organizations need to choose measurable and relevant metrics that reflect sustainability goals. For example, if the goal is to reduce carbon emissions, metrics could include $CO_2$ emissions per unit of production or transportation. Organizations also need to identify digital tools and technologies that can collect, analyse, and report data related to their chosen sustainability metrics. This could involve IoT sensors, data analytics platforms, blockchain for traceability, and cloud-based reporting systems. The goals need to be specific, measurable, achievable, relevant, and time-bound (SMART), allowing for clear tracking and evaluation. It is a preferred practice to break down larger sustainability goals into smaller, actionable targets. This makes it easier to track progress and identify areas that require attention. The focal organization needs to leverage digital supply chain technologies to enable real-time monitoring and reporting of sustainability metrics by involving key stakeholders, including suppliers, customers, and partners. Digital platforms can provide up-to-date insights that facilitate timely decision-making. Last, organizations need to set up performance benchmarks and compare their sustainable goals with those of the industry; this helps to identify the opportunities for development and continuous improvement. A regular review of set goals and objectives is also required to reflect changing market conditions and technological advancements along with transparent reporting. Digital platforms can be used to enhance data visualization, accessibility, and communication with relevant stakeholders.

By leveraging digital supply chains to set clear sustainability goals and metrics, organizations can effectively measure their progress, drive continuous improvement, and demonstrate their commitment to sustainable business practices to stakeholders.

### CASE STUDY—LINFOX LOGISTICS AUSTRALIA

*The integration of big data analytics emerges as a powerful tool for businesses like Linfox Logistics. By analysing historical trends, companies can make informed decisions that enhance sustainability in both current operations and future planning. Another technology, blockchain, takes centre stage for enhancing product traceability and visibility, addressing concerns related to fraud, tampering, and waste. The integration of these digital technologies culminates in heightened operational visibility, which contributes to a safer work environment for employees.*

*Linfox Logistics serves as a tangible illustration of these concepts. The company deploys advanced systems such as the BevChain warehouse management system and customer enterprise resource planning systems to automate warehouse operations. Incorporating smart transport technology enables real-time optimization of transport networks through data analysis, thereby boosting efficiency. Embracing cloud-based solutions like Snowflake facilitates data sharing and real-time shipment tracking, exemplifying Linfox's commitment to enhancing customer interaction and visibility.*

*The implementation of digital technologies at Linfox sheds light on how the amalgamation of digital technologies and sustainability principles can revolutionize logistics practices. With its embrace of IoT, blockchain, big data analytics, and cloud computing, Linfox exemplifies the potential of digitalization to transform logistics into a realm of efficient, transparent, and sustainable operations. This real-world example offers valuable insights into the profound benefits that arise from harnessing digital technologies in the pursuit of a sustainable supply chain management perspective.*

## 2.4 OVERCOMING CHALLENGES AND OBSTACLES IN DIGITAL SUPPLY CHAIN SUSTAINABILITY

### 2.4.1 Ensuring Data Privacy and Security

A comprehensive strategy that addresses a variety of factors is essential to achieving data privacy and security within the context of a sustainable digital supply chain. Protecting sensitive information must be a top priority for organizations throughout their lifecycle. This requires the implementation of effective authentication and authorization systems, data encryption strategies, and secure communication channels to protect data both in transit and at rest. A clear data classification and access control system must be developed to limit data access to authorized personnel. It is important to conduct regular security audits and assessments so that vulnerabilities can be identified and remediated as quickly as possible. Working with technology vendors should include a comprehensive review of their security protocols to ensure that their products meet relevant industry standards. Employee training and awareness initiatives are critical to fostering a culture of data protection. In the event of a breach, a well-defined incident response plan should be implemented to guide the containment, investigation, communication, and recovery processes. Compliance with applicable data protection and privacy regulations further strengthens data security measures. By implementing these safeguards, companies can build trust with their stakeholders, protect sensitive information, and provide a foundation of confidence for their digital supply chain sustainability projects.

Ensuring data privacy and security in digital supply chain sustainability is of paramount importance due to the sensitive nature of the information being exchanged. Protecting data safeguards not only your organization's operations but also the trust of the stakeholders involved. For data privacy and security, it is recommended that data be categorized as sensitive and non-sensitive or important and not important, and access levels should be defined accordingly with robust authorization and authentication protocols to ensure only authorized personnel have access to the information they need. Access roles and responsibilities should be developed to avoid unauthorized access to data. Furthermore, organizations can use various encryption standards and technologies and use secure communication channels to avoid data theft and hacker attacks. Such protocols and communications channels should be regularly audited by IT experts to identify vulnerabilities and proactive measures should be taken, for example, by installing regular security updates and patches. In case of a data breach, organizations should have a comprehensive incident response plan for securing further breaches and recovery of lost data. Relevant stakeholders also need to be informed in case of data breaches.

By implementing these measures, organizations can create a robust framework for data privacy and security in digital supply chain sustainability efforts. This framework ensures that sensitive information remains protected while enabling the benefits of digital technologies and sustainable practices.

### 2.4.2 Addressing Interoperability and Integration Challenges

Interoperability and integration challenges in the context of digital supply chain sustainability can hinder the seamless exchange of data and information across different systems, processes, and stakeholders. Overcoming these challenges requires careful planning, technological solutions, and collaborative efforts. The first and foremost important thing is to have clearly defined and communicated goals and objectives. This ensures that data needs to flow between systems and stakeholders. Also, all the data sharing formats and reporting formats should be agreed upon between the stakeholders to ensure a common data format and protocols. This reduces the complexity of data interpretation and yields the same results for everyone involved.

Integration technologies that best fit the organization's needs, considering factors like data volume, frequency of updates, and system compatibility are chosen. This could include application programming interfaces (APIs); middleware; or extract, transform, load (ETL) tools. APIs provide a structured way for different systems to communicate. Data mapping and transformation are required when different systems are used by different stakeholders. Implementing data mapping and transformation processes to convert data from one format to another is ensured in such a way that data retains its integrity and meaning during the transformation. Different supply chain networks also rely on enterprise integration platforms (EIPs) which offer tools for connecting and coordinating data flow between various systems. These platforms can centralize integration efforts and provide scalability. Organizations also adopt collaborative data governance to work with both internal and external stakeholders and establish cross-function teams involving IT, supply chain, sustainability, and other relevant departments to develop design integration solutions that meet technical and business requirements.

Another interoperability and integration challenge is vendor collaboration while maintaining data security and compliance. It is essential to collaborate closely with software vendors and technology partners to ensure that their solutions can integrate effectively with organizations' existing systems. The focal organization has to ensure that data security and compliance measures are in place during integration and protect sensitive information as it moves between systems to avoid data breaches.

By addressing interoperability and integration challenges, organizations can create a cohesive digital supply chain ecosystem that enhances sustainability practices, promotes data-driven decision-making, and facilitates collaboration among stakeholders.

### 2.4.3 Overcoming Skilled Labor Shortages and Cultural Resistance

Overcoming challenges related to skilled labour shortages and cultural resistance in digital supply chain sustainability requires a combination of strategies that focus on talent development, change management, and fostering a culture of innovation. Organizations can tackle these challenges effectively by implementing the following solutions.

#### 2.4.3.1 Skilled Labor Shortages

Several measures can be taken to address the issue of skill shortages. First, a proactive strategy involves investing in training and development. This includes offering comprehensive training programs aimed at improving the skills of current employees, particularly in the areas of digital technology and supply chain sustainability practices. These programs offer a variety of options, such as certificates, workshops, and online courses, and successfully address the mismatch between individuals' existing skill levels and the needs of the industry.

Partnering with educational institutions is another important avenue. By establishing collaborative alliances with universities, colleges, and vocational institutions, it becomes possible to develop customized educational programs aimed at producing graduates who possess the necessary skills and competencies required for employment in the digital supply chain industry. Such collaborative efforts facilitate a direct congruence between industry needs and educational curriculum content, thereby cultivating a skilled workforce capable of seamlessly integrating into the dynamic supply chain landscape.

In addition, the integration of apprenticeships and internships serves as a practical approach. By implementing such structured programs, companies can attract young and ambitious individuals who are looking for opportunities for growth and development. These initiatives provide participants with great experiential learning opportunities and mentorship, facilitating the acquisition of practical skills and abilities that are highly relevant to digital supply chain positions. By engaging in immersive learning experiences, it becomes possible to effectively bridge the gap that exists between theoretical knowledge and its practical application in the real world.

The introduction of cross-functional training is another strategic strategy. Promoting cross-functional learning opportunities among employees from different departments within the supply chain enhances their understanding of operational processes and sustainability goals, thereby cultivating a comprehensive perspective. This integrated approach not only fosters the development of a workforce that fully understands the interrelated aspects of supply chain functions but also enables them to make meaningful contributions to the company's overarching goals. To improve the distribution of expertise and knowledge, organizations can choose to develop internal platforms specifically designed for knowledge sharing. These platforms facilitate the creation of forums where employees can discuss best practices, insights, and lessons learned related to digital supply chain operations. Collaborative workplaces can enhance the collective skill set and cultivate a culture that emphasizes continuous learning and development.

In conclusion, building a robust talent pipeline requires active participation in industry associations, networking groups, and communities. Through active engagement in these areas, companies have the opportunity to identify and cultivate individuals who have a strong passion for pursuing a career in the digital supply chain space. These strategic alliances enable the acquisition of individuals who are already in sync with the changing needs of the industry, making the process of integrating talent more efficient.

#### 2.4.3.2 Cultural Resistance

Mitigating the cultural resistance often encountered when integrating digital supply chain sustainability requires the implementation of several strategic activities. Securing executive commitment is paramount. The support and buy-in of senior leadership not only serve as a guiding influence but also set a precedent for the entire organization to follow. Leaders who actively promote the concept of digital supply chain sustainability serve as compelling role models, effectively allaying concerns and inspiring a shared sense of purpose.

The importance of effective communication is critical to overcoming resistance. Effective and persistent communication of the benefits of digital supply chain sustainability is essential. This includes not only explaining the benefits but also addressing concerns, providing rationale, and presenting empirical evidence of successful implementations. By employing effective communication strategies, companies can bridge gaps in understanding and foster a more positive view of the transformation process.

To effectively manage cultural resistance, it is valuable to identify change agents and champions within the workforce. These individuals, who have a strong propensity to accept change and demonstrate an unwavering commitment to the transformation process, have the potential to act as influential advocates. Empowered to lead by example and influence their peers, these individuals play a critical role in generating widespread support that effectively counteracts any resistance.

The use of collaborative decision-making is a powerful method. Companies cultivate a sense of ownership and commitment by involving employees in the decision-making processes related to digital supply chain initiatives. By actively seeking information, considering a range of perspectives, and incorporating employee insights, it is possible to transform resistance into a collective sense of responsibility. This transformation occurs when employees are empowered to participate as collaborators in the pursuit of sustainable goals.

Recognizing the efforts and commitment of employees is paramount. Providing incentives, prizes, or recognition for individuals who actively engage in sustainable practices within the digital supply chain can be an effective way to build enthusiasm and mitigate potential resistance. This strategic approach aims to align individual interests with the goals of the organization, fostering a climate in which acceptance of change is perceived as beneficial to all parties.

The implementation of pilot projects, albeit on a smaller scale, provides a means of demonstrating the benefits of digital supply chain methods. These projects serve as prime examples of the tangible benefits and positive results, fostering a sense of security and enthusiasm among the workforce. Presenting concrete evidence in support of an idea serves to reduce doubt and resistance, cultivating an atmosphere that is more open and accepting.

Establishing a culture that encourages continuous learning and innovation plays a critical role in effectively managing resistance. Encouraging employees to explore new technologies, share insights, and experiment with digital tools cultivates an adaptive mindset. The prevailing culture of continuous growth serves to mitigate resistance by fostering a workforce that is adaptable and receptive to change.

Establishing forums to facilitate open communication allows employees to voice concerns, seek clarification, and offer comments on efforts to promote sustainability within the digital supply chain. These platforms serve to increase transparency, debunk myths, and promote mutual understanding. By proactively addressing concerns and fostering open dialogue, companies can mitigate resistance and cultivate a culture of collaborative engagement.

Establishing a precise and clear plan for digital transformation is paramount. This roadmap provides employees with a concrete representation of the future path. Organizations create a clear sense of direction and a unified framework for collective efforts by delineating the stages and milestones of the transformation process. This practice helps reduce ambiguity and provides a structured approach to navigating the transformation journey.

Finally, it is critical to commemorate accomplishments and significant milestones in the effort to achieve digital supply chain sustainability. The act of recognizing achievements, whether in terms of impact on an organization or broader contributions to society, serves to underscore the importance of these efforts. Organizations foster a culture of achievement by highlighting the positive consequences, thereby mitigating resistance and underscoring the importance of change.

By addressing skilled labour shortages and cultural resistance through these strategies, organizations can create a workforce that is equipped with the necessary skills and motivated to embrace digital supply chain sustainability practices, driving positive change and long-term success.

### 2.4.4 Aligning Digital Supply Chain Strategies with Broader Business Goals

Successfully aligning digital supply chain plans with broader business goals requires a systematic and comprehensive approach. This is a significant undertaking that must be completed. To achieve this alignment, companies must first clearly define their overarching business objectives. These objectives may be related to increasing revenue, reducing costs, expanding markets, or achieving specific sustainability goals. It is imperative to identify specific supply chain priorities that directly support the achievement of these goals. This requires collaboration among key stakeholders from various departments, such as supply chain, operations, marketing, finance, and sustainability, to collectively determine how digital supply chain strategies can best serve the goals of each department while also advancing the organization's broader mission.

Data-driven insights play an essential role in identifying opportunities for alignment. When companies conduct data analysis on their supply chains, they can identify bottlenecks, inefficiencies, and places where digital solutions have the potential to have the greatest impact on achieving business goals. When analysing digital technologies, it is critical to consider the ability of these tools to address not only specific supply chain issues but also broader business objectives. Solutions must be tailored to the unique needs of the business. This will ensure that the technologies selected provide real value that goes beyond simple trends.

To facilitate successful alignment, companies should make it a point to verify that the KPIs for the digital supply chain are aligned with the metrics that pertain to the business as a whole. The metrics monitored within the supply chain will, as a result of this synchronization, directly contribute to the performance of the organization as a whole. Evaluating performance against these aligned metrics regularly allows you to measure progress and adjust plans as needed. Communication and transparency play an important role in establishing a shared understanding of the alignment process among stakeholders. In addition, strong support from senior leadership is essential for success. It is important to foster a culture of continuous improvement within an organization, as this will enable companies to routinely assess, refine, and perfect their digital supply chain strategies in line with their ever-evolving business goals. Ultimately, successful alignment ensures that digital supply chain projects directly contribute to the growth, innovation, and long-term success of the organization.

## 2.5 CONCLUSION

In conclusion, this chapter presents a compelling argument for the transformative power of digital technologies in shaping the future of sustainable supply chain management. By seamlessly integrating these technologies, businesses can not only achieve their sustainability objectives but also reduce their environmental footprint while upholding ethical standards. The chapter underscores the importance of real-time data and transparency in enhancing operational efficiency, allowing for agile decision-making, and optimizing resource allocation. Through vivid case studies and examples, it becomes evident that data-driven decision-making, empowered by analytics and AI-driven insights, is the linchpin for informed choices, ultimately leading to resource optimization, waste reduction, and enhanced sustainability practices. Collaboration among stakeholders—suppliers, partners, customers, and even competitors—takes centre stage, facilitated by digital tools that enable seamless data sharing and the co-creation of sustainable initiatives.

Moreover, the chapter highlights the necessity of setting clear sustainability metrics and goals, with digital tools serving as instrumental trackers of progress and transparent reporting mechanisms. It delves into how digital supply chains bolster resilience and risk management, ensuring business continuity even in the face of disruptions. Ethical sourcing, traceability, and circular economy practices are explored in the context of digital supply chains, demonstrating how technology can be harnessed to promote responsible sourcing and minimize environmental impact. Innovation, exemplified through AI, IoT, and other digital solutions, emerges as a driving force behind continuous improvement in sustainable supply chain practices. Furthermore, the chapter emphasizes the critical role of stakeholder engagement and transparent communication, with digital tools serving as facilitators in building trust and fostering accountability. It underscores the global relevance of sustainable digital supply chains in addressing pressing environmental challenges and advancing broader sustainability goals.

In essence, this chapter equips students, scholars and practitioners with a persuasive narrative: that embracing digital technologies is not merely a choice but a strategic imperative in shaping a more sustainable, responsible, and resilient future in supply chain management. It calls upon us to lead the charge for positive change, champion innovation, and contribute to a world where business success goes hand in hand with environmental stewardship.

## REFERENCES

Ali, I., Arslan, A., Khan, Z., & Tarba, S. Y. (2021). The role of Industry 4.0 technologies in mitigating supply chain disruption: Empirical evidence from the Australian food processing industry. *IEEE Transactions on Engineering Management*. https://doi.org/10.1109/TEM.2021.3088518

Asif, M., Searcy, C., & Castka, P. (2023). ESG and Industry 5.0: The role of technologies in enhancing ESG disclosure. *Technological Forecasting and Social Change*, *195*, 122806. https://doi.org/10.1016/J.TECHFORE.2023.122806

Jung, J., Kim, S. J., & Kim, K. H. (2020). Sustainable marketing activities of traditional fashion market and brand loyalty. *Journal of Business Research*, *120*, 294–301. https://doi.org/10.1016/J.JBUSRES.2020.04.019

Lerman, L. V., Benitez, G. B., Müller, J. M., de Sousa, P. R., & Frank, A. G. (2022). Smart green supply chain management: A configurational approach to enhance green performance through digital transformation. *Supply Chain Management*, *27*(7), 147–176. https://doi.org/10.1108/SCM-02-2022-0059/FULL/PDF

Nayal, K., Raut, R. D., Yadav, V. S., Priyadarshinee, P., & Narkhede, B. E. (2022). The impact of sustainable development strategy on sustainable supply chain firm performance in the digital transformation era. *Business Strategy and the Environment*, *31*(3), 845–859. https://doi.org/10.1002/BSE.2921

Ratchford, B. T. (2019). The impact of digital innovations on marketing and consumers. *Review of Marketing Research*, *16*, 35–61. https://doi.org/10.1108/S1548-643520190000016005/FULL/XML

Sharma, M., Kumar, A., Luthra, S., Joshi, S., & Upadhyay, A. (2022). The impact of environmental dynamism on low-carbon practices and digital supply chain networks to enhance sustainable performance: An empirical analysis. *Business Strategy and the Environment*, *31*(4), 1776–1788. https://doi.org/10.1002/BSE.2983

Wang, M., Kumar, V., Ruan, X., Saad, M., Garza-Reyes, J. A., & Kumar, A. (2022). Sustainability concerns on consumers' attitude towards short food supply chains: An empirical investigation. *Operations Management Research*, *15*(1–2), 76–92. https://doi.org/10.1007/S12063-021-00188-X/TABLES/5

Yuan, Y., Tan, H., & Liu, L. (2023). The effects of digital transformation on supply chain resilience: A moderated and mediated model. *Journal of Enterprise Information Management*. https://doi.org/10.1108/JEIM-09-2022-0333/FULL/XML

# 3 Digital Transformation Framework for Logistics Service Providers

## *A Systematic Literature Review*

*Seçil Gülmez and Gül Denktaş-Şakar*

## 3.1 INTRODUCTION

Due to globalization and the competitive business environment, businesses need to produce goods and provide services more efficiently. To do this, it is necessary to combine emerging technologies with novel business practices and models. The Fourth Industrial Revolution (Industry 4.0) provided a suitable climate for business to find paths for differentiation as well as efficiency in their business processes by offering brand new technologies such as augmented reality, cloud computing, robotics, Internet of Things (IoT) and big data technologies (Ghaffari et al., 2020). Although the impacts of such technologies vary across industries, it is evident that the logistics industry is one which is exposed to dramatic changes both in the digitalized business environment and logistics business landscape itself. Considering the rapid changes in customer requirements and demand, customized service expectations of customers, inadequate management of goods in warehouses and possible risk of damage and loss in transportation processes, logistics service providers may face uncertainties as well as complicated information processing issues (Borgstrom et al., 2021). Apart from existing challenges resulting from the business environment itself, the COVID-19 crisis and sustainability targets of countries and firms have required business to transform its methods. The recent pandemic as one of the most impactful disruptions has changed the structure of supply chains. The logistical cost related to transportation has seen a significant surge, affecting companies' performance and profitability (Spieske and Birkel, 2021). Furthermore, customer demand, which is already unpredictable in nature, has risen and become even more complicated given the changes in consumer behavior and new e-commerce approaches (Al Mashalah et al., 2022). This drive has pushed businesses to build more resilient business systems to survive in challenging conditions. In such a complex business environment, a key component that emerges as a bounding mechanism between the logistics service providers and their customers is digitalization.


DOI: 10.1201/9781003438748-4

Within a logistics context, digitalization represents technologically enhanced interconnected operations between companies and customers. The establishment of technological hardware and software tools within the concept of digitalization requires a thorough application of a proper business model within logistics service providers' organizations. The business model explains how an organization creates, delivers and captures value. Understanding customers' perceptions regarding value is a primary step in the creation of value propositions. This is followed by finding out the proper ways of delivering value to the market. In value capture, organization's pricing techniques as well as revenue sources play an important role. Digitalization of these stages in the overall business models of logistics service providers is widely accepted as a critical need in the industry (Xu et al., 2023; Meier et al., 2023; Bigliardi et al., 2022; Jović et al., 2022). An updated review of the existing business models and necessary changes in such models is expected to provide additional revenue from new sources (Gong and Ribiere, 2021), entry into new markets where logistics service providers were not available before (Jović et al., 2022), launch of new services (El Hilali et al., 2020; Chinoracky and Corejova, 2019), introduction of new sales channels (Jović et al., 2022) and new ways of charging for services (Pucihar et al., 2019). Dynamic and flexible business models enriched with improved digital tools are expected to provide increased customer engagement, efficient utilization of resources and better anticipation of customer demand, as well as changes in prices, tax and depreciation rates (Straková et al., 2022). Such dramatic changes in the logistics industry with novel business models trigger the development of a business climate fostering customer-centric innovation by involving customers in various stages not only in logistics processes but also in supply chain networks, real-time data sharing among the actors and business practices with flexible and agile solutions in order to allow a quick response to changes in the logistics industry. Based on these expected benefits, logistics services providers have started to implement new digitalized business models in order to remain competitive.

There has been increasing interest in the digital applications of logistics service providers in the literature. While some specific components have been discussed through the lens of digital transformation, such as government support and customer collaboration (Zhou et al., 2023), some studies have focused on the sustainability perspective of digital business models in the logistics industry (Xu et al., 2023; Piscicelli, 2023, Romagnoli et al., 2023; Liu et al., 2022). Some systematic literature reviews have also been conducted by focusing on specific supply chains (Ding, 2018; Rejeb et al., 2022; Rad et al., 2022). There have also been studies focusing on the maritime transport segment of the supply chain by discussing the maritime transport industry's response to digitalization (Balci, 2021; Poulis et al., 2020; Lambrou et al., 2019). Although the current practices of logistics service providers in the context of digitalization and emerging business models are widely implemented, academic research concentrating on the digital transformation processes in the logistics industry is rather limited. In addition, it was stated that uncertainty exists around how digitalization will impact logistics service providers' business strategies (Bjorkdahl, 2020; Kohtamaki et al., 2020). With this view, the main aim of this chapter is to present a framework for digital transformation in logistics service providers' business models based on a systematic literature review. The main research questions of this study are as follows:

- What are the main components of a digital business model for the logistics service industry?
- Which dynamics are required to digitally transform the business models of logistics service providers (LSPs)?
- How can LSPs digitally redesign their business models?

The chapter is organized as follows. The first section explains methodology employed in the study by presenting the main stages. Then, the findings and discussion section provides the findings of the systematic literature review by highlighting the main components in the business models and how such components may be adopted in the digitalized business landscape of logistics service providers. The chapter ends with a conclusion.

## 3.2 METHODOLOGY

In this study, a systematic literature review was adopted to identify the topics in a digital business model specific to the logistics industry and develop a digital business model for logistics service providers. Systematic literature reviews represent a methodological, scientific and a clear approach to delineating a specific area of study. They allow readers insights into the systematic process researchers follow in reaching their conclusions (Tranfield et al., 2003).

A systematic literature review strategy was followed based on Tranfield et al. (2003). The procedure involves three sequential stages: planning, execution and reporting. The planning stage refers to developing a review protocol which ensures objectivity through ensuring explicit illustration of the steps to be taken. The execution process starts with the identification of the research, selecting studies based on specific criteria, quality assessment of data to minimize biases and extracting and synthesis of the data, and finally, reporting implies synthesizing the findings through the predefined protocol (Tranfield et al., 2003). Figure 3.1 schematically illustrates the review process. In this study, the reporting part of systematic literature review is explained in the findings and discussion section.

### 3.2.1 PLANNING

The research questions and the objectives of the study mainly directed the review process. The analysis engaged in defining the concept of a digital business model and the framework which can facilitate the implementation of designing a new digital business model and managing the digital transformation process of existing business models of logistics service providers. In pursuit of that objective, the ("digitalization" OR "digitalisation" OR "digitization" OR "digitisation" OR "digital transformation") AND "business model" AND ("logistics" OR "shipping" OR "transport" OR "logistics service providers") search terms were defined for the query. After screening the outputs from the first query, the authors decided to reorganize the keywords in the research string to find more studies related to the field. In the second query, one additional research string was generated by excluding the "business model" keyword

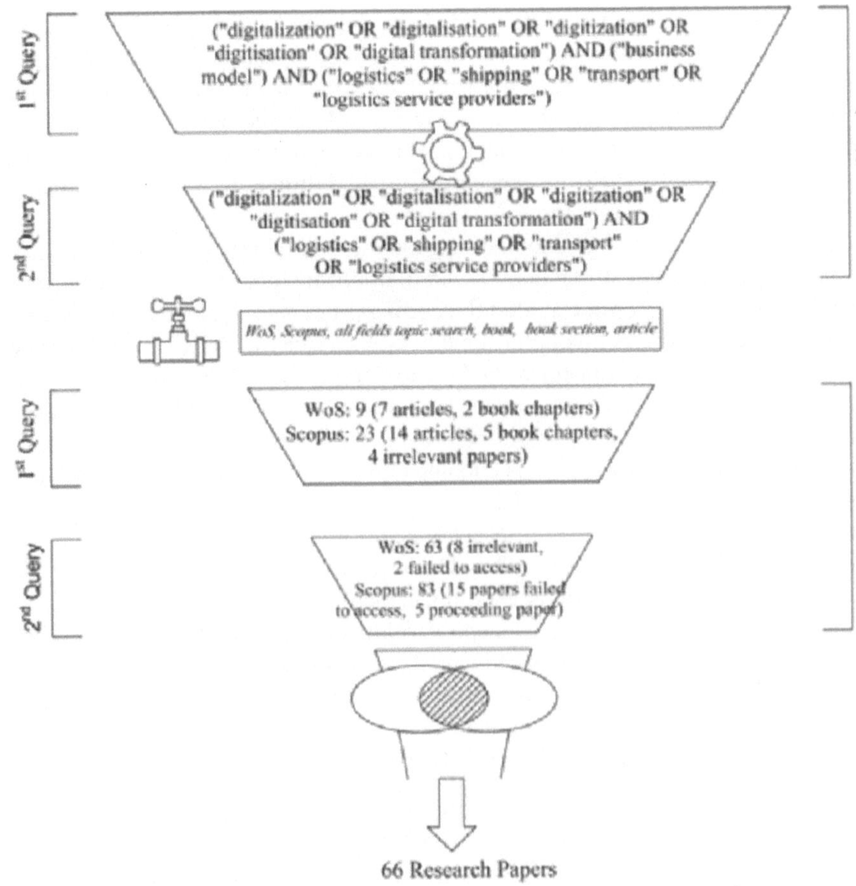


**FIGURE 3.1** Search inquiry and process.

and then unified with the first research string. Thus, two different research streams were combined into a single compact research string. With the two-stage research query, those keywords were also reaffirmed during the coding process. To ensure the quality of the papers, the ISI Web of Science (WoS) and Scopus databases were chosen without setting any time constraint and in all search topics. A topic search was performed in both databases, and inclusion criteria were set regarding the text language (non-English text was excluded) and type of research (book, book section and article were determined for the query. Proceeding papers were excluded due to concerns about not being included in a comprehensive peer-review process).

### 3.2.2 Execution

In the first step of the two-stage research query, WoS yielded 9 research papers (7 articles and 2 book chapters), while Scopus yielded 23 papers (14 articles, 5 book chapters, 2 irrelevant articles, 2 irrelevant book chapters). Because of obtaining limited articles in the field, the second query was performed with reorganized keywords.

The WoS query at the second stage yielded 63 research papers (8 irrelevant papers and 2 papers for which access failed), while the Scopus search provided 83 papers (30 book chapters/books, 53 articles, 4 papers for which access failed, 5 proceeding papers and 15 book chapters with limited access). After the outputs of each query were merged into a single file, duplicated papers were eliminated and the contents of all paper re-assessed by their relevance of the field. Ultimately, a total of 66 papers were identified for coding purposes.

A deductive approach was adopted to generate an archetype for a digital business model for LSPs and strategic points in digital transformation of their business model. The coding framework guided the coding process to identify the main components of digital business model for the logistics service industry, which dynamics are required for digitally transform the business models of LSPs and how they can digitally redesign their business models. The process was inspired by the business logic triangle (see Figure 3.1), as outlined by Osterwalder and Pigneur (2010) and Rachinger et al. (2019). The analysis encompassed strategic, architectural and technological levels of the business model.

At the architectural level, the initial set of codes was rooted in the business model (BM) value architecture, as described by Osterwalder and Pigneur (2002) and Teece (2010). While various frameworks exist for depicting a business model's structure, they tend to converge toward the concept of value architecture (Morris et al., 2005; Teece, 2010; Saebi et al., 2017). Consequently, this study revolved around key aspects such as value creation, value proposition, value delivery and value capture.

Nvivo 11 Plus supported the content analysis and coding the activities. Regarding the coding framework depicted in Figure 3.2, open codes were determined deductively

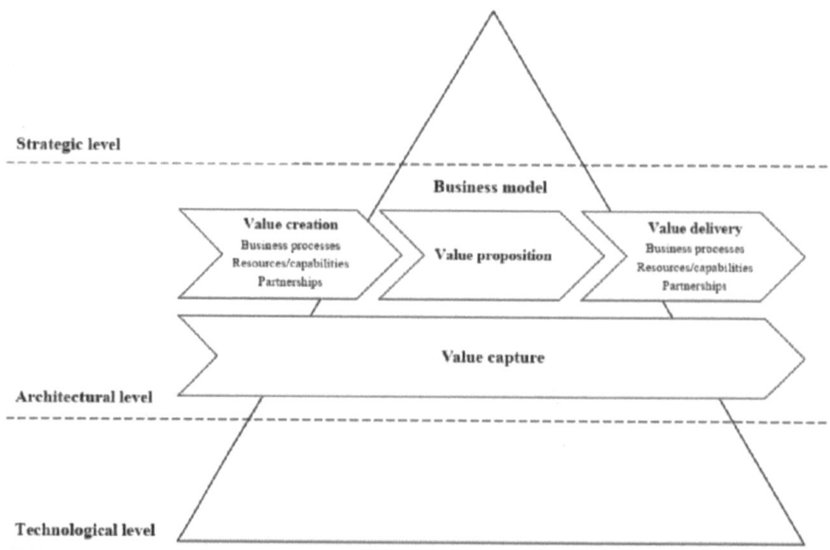


**FIGURE 3.2** Coding schema for digital business model of LSP.

**Source: Rachinger et al. (2019); Osterwalder and Pigneur (2002).**

**TABLE 3.1**
**Business Model–Associated Coding Terms**

| Business Model–Based Value Segment | BM Components | Terms Associated with BM Components |
|---|---|---|
| Value Proposition *What value is suggested and for whom is it intended?* | Offer | Product/service type, value creation and delivery, offerings (Morris et al., 2005; Bocken et al., 2014) |
| | Value Proposition | Customer segment, offering, customer needs, service, product, experience, channel pricing, offering, generic strategy to attract customer, relationship, operations, quality, availability, innovativeness (Linder and Cantrell, 2000; Osterwalder et al., 2005; Richardson, 2008; Clauss, 2016; Bocken et al., 2014) |
| | Customer Segments | Target customer, distribution channel, relationship (Bocken et al., 2014; Morris et al., 2005) |
| | Customer/ Partner Relationship | Customer, partner, supplier relationship (Morris et al., 2005; Bocken et al., 2014) |
| Value Creation *How is value created?* Value Delivery *How is value delivered?* | Key Activities | Activities/activity system (Richardson, 2008) |
| | Key Resources/ Capabilities | Execution, distinct capabilities, technology, equipment, operating system, supply chain management, networking, resource leveraging, (Osterwalder et al., 2005; Clauss, 2016; Bocken et al., 2014; Richardson, 2008; Morris et al., 2005) |
| | Key Partners | Partnership, suppliers, customers, partners (Richardson, 2008; Bocken et al., 2014) |
| | Channels | Process, value chain, business process (Clauss, 2016; Richardson, 2008) |
| Value Capture *How is value captured?* | Cost Revenue Flows | Cost structure, revenue sources, economics of the business, revenue model, financial structure (Richardson, 2008; Bocken et al., 2014; Clauss, 2016; Osterwalder et al., 2005; Linder and Cantrell, 2000) |

*Source:* Compiled through the categorization of Ranta et al. (2018) and Nußholz (2018)

based on the value segments specified in the coding framework, and papers were coded deductively. BM-related open codes, which are illustrated in Table 3.1, were used in coding the collected papers.

While coding the papers, BM components were defined as free nodes and associated terms coded as tree nodes. In the articles, related words were queried; then the paragraphs involving the word groups were coded to the nodes. Although using software tools provides greater advantages for analyzing the bulk of qualitative data, it may not always address the analysis requirement. Because of exploring the main components and the dynamics of a digital business model specific to the logistics service industry, the coded data were deeply reanalyzed and re-coded to according to the business logic triangle (Figure 3.2).

## 3.3 FINDINGS AND DISCUSSION

Figure 3.3 outlines the findings of the systematic literature review in developing a business model framework for digital transformation of logistics service providers. The results were discussed by business model triangle aspects.

### 3.3.1 Value Proposition

As Figure 3.3 shows, the findings underscore the multifaceted value proposition stemming from the digital transformation of LSP business models. Customer engagement emerges as a cornerstone, as involving clients in the design process not only empowers them but also fosters a sense of ownership, thereby creating a more profound bond between customers and services. This trend continues as businesses further integrate digital tools to enhance customer experience. The focus remains on empowering customers and fostering loyalty through interactive and personalized digital experiences.

The pivotal role of strategic marketing efforts is illuminated, elucidating how efforts aimed at building consumer trust have a cascading effect on brand credibility and long-term customer loyalty, highlighting the symbiotic relationship between customer perception and business success.

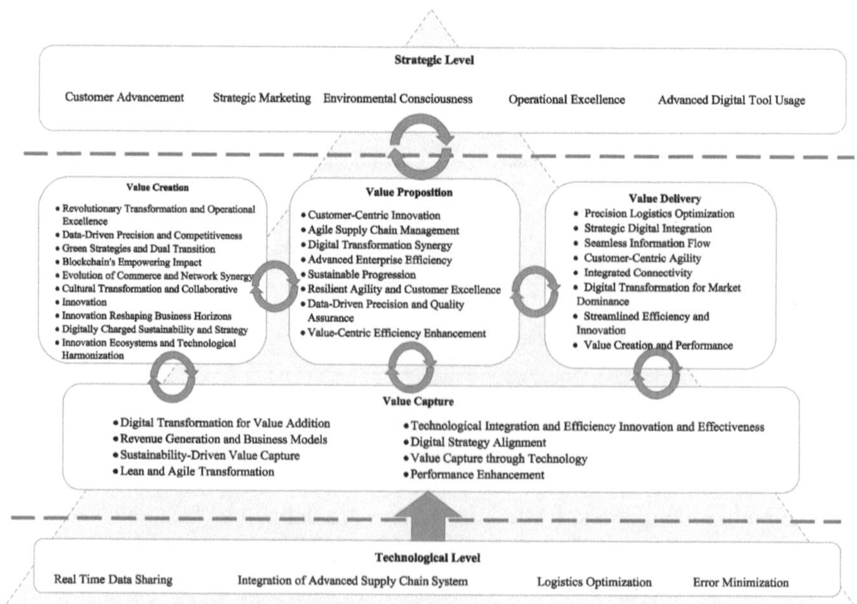


**FIGURE 3.3** Business model approach to digitalization of logistics service providers.

Furthermore, the findings emphasize the strategic imperative of addressing the burgeoning environmental consciousness of consumers. The provision of sustainable products and services emerges as a strategic alignment that reinforces the brand's significance in an era characterized by heightened environmental awareness (Iglesias-Pradas et al., 2022). This resonance with societal trends underscores the importance of staying adaptable and relevant in a market driven by ecological considerations. LSPs can also accelerate the adoption of digital solutions to provide sustainable services, ensuring their operations align with ecological considerations and customer expectations.

The integration of advanced digital software for supply chain optimization takes center stage as a transformative force in logistics operations. This digital integration, which prioritizes factors such as cost, time, quality, resilience and environmental impact, is shown to be essential in achieving operational excellence. Additionally, the role of digital platforms as proactive sentinels, alerting companies to global risks and enabling strategic responses, highlights the real-time responsiveness that digital tools can offer to the dynamic and resilient logistics landscape (Varriale et al., 2021).

The significance of real-time data sharing among logistics partners is commonly emphasized, accentuating how this practice not only enhances operational efficiency but also promotes collaborative decision-making, ultimately resulting in cost-effective and streamlined operations. Especially in disruptions, real-time data sharing and collaboration remain critical. Enhancing the usage of related platforms facilitates the continuous collaboration and information exchange, driving more efficient and agile logistics operations (Belmoukari et al., 2023).

Digital technologies extend their transformative reach to waste management, optimizing routes for material collection. This integration within scalable cloud environments is shown to enhance accessibility and data flow, streamlining waste management processes and broadening the operational spectrum of logistics service providers (Sorkun, 2020). Innovative business models driven by digital tools reshape value propositions, enabling the expansion of market reach and catering to evolving customer demands, thereby fostering a competitive edge in the industry landscape (Korucuk et al., 2022).

Among software technologies, the strategic role of enterprise resource planning (ERP) systems as organizational orchestrators should not be neglected in terms of ensuring agile and responsive schedules across the supply chain. Moreover, the integration of radio frequency identification (RFID), Internet of Things and automation technologies enhances precision and minimizes errors in various logistical processes, further attesting to the role of digital tools in operational optimization (Varriale et al., 2021).

In the realm of sustainability, the findings through the literature review highlight the alignment of digital transformation with eco-consciousness, stressing the significance of businesses fostering environmental awareness and offering eco-friendly products. Strategic agility emerges as a key factor in anticipating disruptions, fostering innovation ecosystems and providing businesses with essential digital capabilities to enhance resilience in the face of market uncertainties (Wohlleber et al., 2022). LSPs can invest in enhancing digital capabilities to stay ahead on a market that continues to face uncertainties and disruptions.

### 3.3.2 Value Creation

The insights gathered through the analysis highlight a fundamental shift in the industry, driven by the integration of new technologies, resulting in operational excellence and increased value for customers. The incorporation of novel technologies signifies a paradigm shift in the management of the service life cycle. This transition not only facilitates operational excellence but also generates unprecedented value for discerning customers. The flexibility achieved through organizational changes empowers businesses to evolve seamlessly, leading to personalized experiences that redefine benchmarks for value. This customer-centric approach disrupts traditional models, effectively enhancing customer value.

An evidence-based approach amplifies precision and competitiveness as Industry 4.0 technologies converge to empower stakeholders with improved services and operational capabilities (Rad et al., 2022). The notion of servitization unlocks substantial value by offering comprehensive, impactful solutions to customers. The acceleration of operational speed not only heightens customer satisfaction but also reinforces the dedication to value creation (Jabłoński and Jabłoński, 2019). The integration of software solutions and simulation models mitigates risks and confers competitive advantages, enhancing value across the spectrum.

Sustainability initiatives in information and communication technology (ICT) play a pivotal role in driving sustainability, catering to environmentally conscious consumers and enhancing societal well-being (Belmoukari et al., 2023). The dual transition—green and digital—creates a synergistic effect, promoting sustainability while catalyzing technological advancement. The fusion of green strategies and digital integration extends market dominance, providing a competitive edge and augmenting customer value (Cavalli et al., 2021). By combining circular economy principles with digital solutions, businesses redefine their paradigms, achieving a balance between commercial growth and ecological sustainability.

The transformative impact of blockchain is evident. Smart contracting systems redefine efficiency, precision and automation, strengthening logistics entities with superior capabilities and enhancing client value. The convergence of digital tools and self-executing contracts ensures transparency, trust and unparalleled excellence (Silkina et al., 2021). Blockchain empowers stakeholders to become data-enabled decision-makers, guiding processes and strategies toward optimal outcomes.

The evolution of commerce, particularly the shift towards network commerce, revolutionizes value delivery by connecting marketplaces and providing consumers with unmatched convenience. Shared supply management platforms revolutionize stock control, leading to operational efficiency and elevated stakeholder satisfaction (Kassou et al., 2021). The dynamics of network competition reshape the market landscape, with empowered corporate ecosystems influencing market direction and delivering paramount value.

One another achievement factor in digital transformation is cultural transformation, which enables organizations to extract enduring value from change (Lomakina et al., 2021). Engagement and knowledge investments enhance employees' expertise, serving clients with elevated value. Collaborative data exchanges expedite service optimization and excellence, resulting in significant cost savings and heightened stakeholder satisfaction.

The synergy between digitally charged sustainability and strategic alignment is evident. Organizational transformation and knowledge exchange expedite value enhancement, enriching both business prosperity and client satisfaction (Awan et al., 2022; Wohlleber et al., 2022). The crafting of digital strategies not only aligns with the digital economy's rhythm but also fosters value amplification across stakeholder landscapes. Environmental stewardship through green technologies exemplifies the holistic essence of value, uniting commercial advancement with global betterment. Process optimization through emerging technologies extends value horizons by promoting sustainability, financial prudence and operational excellence (Silkina et al., 2021).

The innovation topic, in this realm, emerges as a driving force reshaping business horizons, invigorating service quality and introducing novel value dimensions. Open innovation, coupled with technology partnerships, crystallizes value by propelling businesses towards greater creativity and efficacy. The innovation ecosystems nurtured through this transformation cultivate entrepreneurial potential, magnifying industry worth by synergizing collective ingenuity. Technological integration propels streamlined processes, creating opportunities for value amplification and delivering enhanced services.

### 3.3.3 Value Delivery

Value delivery in digitally logistics service organizations highlights the strategic integration, information flow, customer centric agility, integrated connectivity, efficiency and innovation and value creation and performance topics.

Precision logistics optimization emerges as a key driver, leveraging smart transportation to ensure the timely delivery of materials while reducing both delays and costs. By harnessing data-driven insights, LSPs engage in industrial symbiosis, effectively minimizing waste and optimizing resource efficiency, thereby maximizing the value they deliver (Alhawari et al., 2021).

Strategic digital integration becomes a cornerstone in this transformation journey, where the implementation of software solutions enhances network optimization, leading to efficient resource allocation (Wei et al., 2019). Risk-analyzing platforms enable proactive risk management, empowering decision-making and minimizing risks. Automation-driven tour planning and freight portals redefine routes, ultimately saving valuable time and resources (Meier et al., 2023).

Seamless information flow, facilitated by blockchain technology, redefines transparency and collaboration. The pandemic especially highlighted issues like lack of transparency and collaboration in supply chains. Blockchain technology gained attention for its potential to address these issues (Kamble et al., 2020). Through smart contracts, manual errors are significantly reduced, ensuring accuracy across processes. The real-time data sharing capabilities of blockchain bridge gaps in supply and demand, driving operational efficiency and value delivery.

Customer-centric agility stands out as a pivotal factor, as logistics services tailored to customer needs redefine the value proposition. Real-time collaboration with customers guarantees rapid responses to their preferences and concerns, fostering meaningful engagement (Raza et al., 2023). Co-creating value with customers leads to personalized solutions, elevating customer satisfaction.

Integrated connectivity, enabled by cloud computing, establishes a scalable IT infrastructure that bolsters efficient operations (Heilig et al., 2017). Integration of ICT systems, along with stakeholder engagement, further amplifies information flow and collaboration. The exchange of real-time data enhances transparency, thereby elevating overall operational efficiency (Varriale et al., 2021).

Digital transformation for market dominance becomes a driving force, as digital capabilities redefine operational excellence, yielding cost savings and improved services. By harnessing digital technologies, LSPs achieve revenue growth and penetrate new markets. Collaborative innovation through networks solidifies their competitive advantage, ensuring sustained success.

Streamlined efficiency and innovation emerge as direct outcomes of process optimization through digital technologies. The integration of blockchain, smart contracts and the Internet of Things augments operational performance and accuracy (Varriale et al., 2021). With transparency and collaboration flourishing through blockchain, profound impacts are observed across core business processes.

Value creation and performance are intrinsically linked to digitalization. Operational performance reaches new heights, resulting in the delivery of higher-quality products and services. Digital technologies act as catalysts for revenue growth and innovation, reshaping entire industries. Open collaboration and digital networks reframe value propositions, ultimately amplifying overall success.

### 3.3.4 Value Capture

The research findings illustrated several value capture perspectives within the context of the digital transformation of logistics service providers' business models. The integrated framework presents a structured approach that aligns identified factors and dimensions, guiding LSPs in their journey towards digitalization while maximizing value creation. In the pursuit of value addition, LSPs are directed towards upgrading their digital infrastructure to minimize waste and augment process value. This also led to alignment with sustainability goals as well. LSPs can foster nuanced human–machine interactions for heightened process efficiency, enabling real-time data visibility, transparency and liberation from spatial and temporal constraints to cope with changing market demands and maintain resilience against future disruptions. Increasing omnichannel capabilities in service provision can provide greater resilience in LSPs' business model and lead to capture value through enhanced stakeholder engagement.

Within the realm of revenue generation and business models, systematic analysis of the extant research underscores the significance of charging fees for software-driven network optimization solutions. Implementation of risk management platforms and the monetization of freight portals and logistics optimization services have the potential for revenue generation. Especially capturing value through digital transformation extends to efficiency and risk reduction offerings, further accentuated through the implementation of fee-based scheduling and logistics management software (Kwilinski et al., 2022; Wei et al., 2019).

One other perspective to value capture in digitalization of LSPs is sustainability-driven value capture, which is identified as a strategic imperative, as LSPs leverage green strategies and digital systems to achieve enhanced market share and cost advantages. Strategic policies are recommended to promote sustainable approaches in logistics, and special financing programs are highlighted as tools to foster the adoption of green technologies, thereby capturing value through sustainable practices.

Lean and agile transformation emerges as another pivotal dimension in the digitalization of LSPs' activities. It advocates for the adoption of lean principles to establish standardized efficiency, while agile practices facilitate rapid adaptation and alignment with dynamic market conditions (Cichosz et al., 2020). Furthermore, the integration of external partnerships is recommended to empower LSPs with cutting-edge technology and innovation. However, this integration gives rise to another concern, particularly regarding technological integration and the effectiveness of technological usage.

This concern becomes critical for operational enhancement, as it encourages LSPs to effectively plan and execute using advanced systems like MRP, MRP 2, ERP and distribution resource planning (DRP). Additionally, further techniques such as the shared supply management approach can be integrated into freight flow to optimize distribution and stock management (Kassou et al., 2021). This integration ultimately leads to increased efficiency and value capture.

Innovation and effectiveness are championed, prompting LSPs to renovate outdated processes, strategically assess competitive potential and transform business models to match the pace of innovation in the digital age. The alignment of digital and business strategies is crucial, requiring preparedness for cultural shifts and the proactive customization of technology implementation to maximize its tailored impact (Raza et al., 2023).

Existing studies generally accentuate value capture through technology, advocating for the proactive adoption of advanced technologies to drive digitalization efforts. Strengthening information processing capabilities is identified as a critical step, while the harnessing of government policies for logistical enhancements further amplifies value capture. The creation of novel digital products, services and business models is recognized as a conduit for expanding value capture potential.

### 3.3.5 Strategic Level of Business Model Triangle

As strategic level, customer engagement, strategic marketing efforts and environmental consciousness have a critical role in shaping the value proposition. The integration of advanced digital software for logistics process optimization is properly portrayed as a transformative force that aligns with the strategic objectives of operational excellence. The emphasis on real-time data sharing, particularly through blockchain technology, demonstrates strategic agility in enhancing collaboration and operational efficiency. The focus on customer-centric agility and sustainable practices indicates a strategic shift towards personalized solutions and market differentiation. The strategic significance of integrating green strategies and digital systems is well articulated, showcasing a thoughtful approach to value creation aligned with evolving societal trends.

### 3.3.6 Technological Level of Business Model Triangle

At the technological level, integration of various digital tools and technologies within LSPs' business models should not be neglected. The results of the thorough analysis of the literature review imply the integration of precision logistics optimization, where smart transportation and data-driven insights play pivotal roles in enhancing delivery efficiency and minimizing costs. The mention of risk management platforms and automation-driven planning of freight movement illustrate the application of advanced technologies in risk reduction and route optimization. The emphasis on blockchain technology and real-time data sharing point up technological advancement in ensuring transparency, accuracy and collaboration. The integration of advanced systems like MRP, MRP 2, ERP and DRP highlights the technological underpinning of efficient planning and execution. The reference to RFID, IoT and automation technologies further validates the incorporation of technology in enhancing precision and minimizing errors. Moreover, the fusion of green strategies with digital integration showcases how technological advancements contribute to sustainability initiatives.

## 3.4 CONCLUSION

The motivation for this chapter stems from the lack of existing studies concentrating on digital transformation in the logistics industry with a specific lens of emerging business models. Through an extensive systematic literature review, this chapter provided a framework for the digital transformation components in the business models of logistics service providers. The study highlighted some components within the perspective of value proposition, value creation, value delivery and value capture by considering the impact of digitalization. With regard to value proposition, the main components that emerged in the systematic literature are customer-centric innovation, agile supply chain management, digital transformation synergy, advanced enterprise efficiency, sustainable progression, resilient agility and customer excellence, data-driven precision and quality assurance and value-centric efficiency. Digitalization offers new ways to communicate with the current and potential customers, and this is also reflected in logistics service providers' information processing speed. In addition, by having historical transaction records in order to analyze customer preferences and provide customized services, logistics service providers can meet the expectations of their customers in a more efficient way. In the case of advanced enterprise efficiency, the use of robots, hybrid and unmanned vehicles and self-driving cars has huge potential in the value propositions of service providers.

In the value creation process, servitization, use of software solutions and simulation models, adoption of circular economy principles and the need for cultural transformation are highlighted. In servitization literature, digitalization is considered a critical driver of business model, value creation and value capture (Grubic, 2014; Parida et al., 2019). Additional benefits of digital servitization regarding environment and society can also be listed as reduction in energy consumption and increases in resource efficiency and product life span (Bressanelli et al., 2018). All these efforts of logistics service providers must also be supported by the organization itself through cultural transformation which includes cultural readiness and collaborative actions among the stakeholders in order to achieve flexibility and transparency.

The value delivery process includes logistics optimization, strategic digital integration, seamless information flow, customer-centric agility, integrated connectivity, digital transformation for market dominance, streamlined efficiency and innovation, value creation and performance. Facilitation of storage and sharing of data can be achieved by strategic digital integration, integrated connectivity and seamless information flow between logistics services providers and their customers as well as their relevant stakeholders.

Value capture is reflected through specific components like digital transformation for value addition, revenue generation, sustainability-driven value capture, lean and agile transformation, technology integration and efficiency, innovation and effectiveness, digital strategy alignment, value capture through technology and performance enhancement. Logistics service providers may benefit from income streams in order to achieve more effective management of their fleets. As Ehret and Wirtz (2017) and Parida et al. (2019) stated, a critical approach to reducing operational expenses is by implementing process optimization and monitoring techniques to facilitate cost-effective resource allocation and improve overall capacity.

Considering such findings originated from a systematic literature review, the study provided a framework by focusing on the digital transformation concept in the business models of logistics service providers. The study systematically provided an evaluation of value proposition, value creation, value delivery and value capture components of business models through the specific investigation of logistics service providers' practices. The study has several limitations. The systematic literature review was limited to the analysis of two databases, ISI Web of Science and Scopus. For future research, three main suggestions can be made. First, other academic databases may also be considered for further studies. Also, some keywords, as explained in the methodology section, were used in the search query, and these can be increased or narrowed based on the aims of the future research. Last, the framework discussed in the current study can be evaluated and discussed by panel of experts. This can be conducted via in-depth interviews or focus groups. A panel of experts may specifically include practitioners from the logistics industry to get industry-specific feedback. In addition, a combination of various stakeholders, including logistics professionals, digital service providers and customers, may be considered for detailed discussion of the components in the framework.

## REFERENCES

Alhawari, O., Awan, U., Bhutta, M. K. S., & Ali Ülkü, M. (2021). Insights from circular economy literature: A review of extant definitions and unravelling paths to future research. *Sustainability (Switzerland)*, *13*(2), 1–22. https://doi.org/10.3390/su13020859

Al Mashalah, H., Hassini, E., Gunasekaran, A., & Bhatt, D. (2022). The impact of digital transformation on supply chains through e-commerce: Literature review and a conceptual framework. *Transportation Research Part E: Logistics and Transportation Review*, *165*, 102837.

Awan, U., Sroufe, R., & Bozan, K. (2022). Designing value chains for Industry 4.0 and a circular economy: A review of the literature. *Sustainability (Switzerland)*, *14*(12). https://doi.org/10.3390/su14127084

Balci, G. (2021). Digitalization in container shipping services: Critical resources for competitive advantage. *Journal of ETA Maritime Science*, *9*(1), 1–3. 3–1

Belmoukari, B., Audy, J. F., & Forget, P. (2023). Smart port: A systematic literature review. *European Transport Research Review*, *15*(1). https://doi.org/10.1186/s12544-023-00581-6

Bigliardi, B., Filippelli, S., Petroni, A., & Tagliente, L. (2022). The digitalization of supply chain: A review. *Procedia Computer Science*, *200*, 1806–1815.

Bjorkdahl, J. (2020). Strategies for digitalization in manufacturing firms. *California Management Review*, *62*(4), 17–36.

Borgstrom, B., Hertz, S., & Jensen, L. M. (2021). Strategic development of third-party logistics providers (tpls): 'Going under the floor' or 'raising the roof'? *Industrial Marketing Management*, *97*, 183–192.

Bocken, N. M. P., Short, S. W., Rana, P., & Evans, S. (2014). A literature and practice review to develop sustainable business model archetypes. *Journal of Cleaner Production*, *65*, 42e56. https://doi.org/10.1016/j.jclepro.2013.11.039

Bressanelli, G., Adrodegari, F., Perona, M., & Saccani, N. (2018). Exploring how usage focused business models enable circular economy through digital technologies. *Sustainability*, *10*(3), 639.

Cavalli, L., Lizzi, G., Guerrieri, L., Querci, A., De Bari, F., Barbieri, G., Ferrini, S., Di Meglio, R., Cardone, R., Tardo, A., Pagano, P., Tesei, A., & Lattuca, D. (2021). Addressing efficiency and sustainability in the port of the future with 5g: The experience of the livorno port. A methodological insight to measure innovation technologies' benefits on port operations. *Sustainability (Switzerland)*, *13*(21). https://doi.org/10.3390/su132112146

Chinoracky, R., & Corejova, T. (2019). Impact of digital technologies on labor market and the transport sector. *Transportation Research Procedia*, *40*, 994–1001.

Cichosz, M., Wallenburg, C. M., & Knemeyer, A. M. (2020). Digital transformation at logistics service providers: Barriers, success factors and leading practices. *International Journal of Logistics Management*, *31*(2), 209–238. https://doi.org/10.1108/IJLM-08-2019-0229

Clauss, T. (2017). Measuring business model innovation: Conceptualization, scale development and proof of performance. *R&D Management*, *47*, 385–403. https://doi.org/10.1111/radm.12186

Ding, B. (2018). Pharma Industry 4.0: Literature review and research opportunities in sustainable pharmaceutical supply chains. *Process Safety and Environmental Protection*, *119*, 115–130.

Ehret, M., & Wirtz, J. (2017). Unlocking value from machines: Business models and the industrial internet of things. *Journal of Marketing Management*, *33*, 111–130.

El Hilali, W., El Manouar, A., & Janati Idrissi, M. A. (2020). Reaching sustainability during a digital transformation: A PLS approach. *International Journal of Innovation Science*, *12*, 52–79.

Ghaffari, K., Lagzian, M., Kazemi, M., & Malekzadeh, G. (2020). A comprehensive framework for Internet of Things development: A grounded theory study of requirements. *Journal of Enterprise Information Management*, *33*(1), 23–50.

Gong, C., & Ribiere, V. (2021). Developing a unified definition of digital transformation. *Technovation*, *102*, 102217.

Grubic, T. (2014). Servitization and remote monitoring technology: A literature review and research agenda. *Journal of Manufacturing Technology Management*, *25*(1), 100–124.

Heilig, L., Lalla-Ruiz, E., & Voß, S. (2017). Digital transformation in maritime ports: Analysis and a game theoretic framework. *NETNOMICS: Economic Research and Electronic Networking*. https://doi.org/10.1007/s11066-017-9122-x

Iglesias-Pradas, S., Acquila-Natale, E., & Del-Río-Carazo, L. (2022). Omnichannel retailing: A tale of three sectors. *Economic Research-Ekonomska Istrazivanja*, *35*(1), 3305–3336. https://doi.org/10.1080/1331677X.2021.1991825

Jabłoński, M., & Jabłoński, A. (2019). Social factors as a basic driver of the digitalization of the business models of railway companies. *Sustainability (Switzerland)*, *11*(12), 1–29. https://doi.org/10.3390/su10023367

Jović, M., Tijan, E., Vidmar, D., & Pucihar, A. (2022). Factors of digital transformation in the maritime transport sector. *Sustainability*, *14*, 9776.

Kamble, S. S., Gunasekaran, A., & Sharma, R. (2020). Modeling the blockchain enabled traceability in agriculture supply chain. *International Journal of Information Management*, *52*, 101967.

Kassou, M., Bourekkadi, S., El Imrani, O., Boulaksili, A., Aharouay, S., Ourdi, A., Chikri, H., & Khoulji, S. (2021). Digital transformation in flow planning: The case of container terminals at a smart port. *Journal of Theoretical and Applied Information Technology*, *99*(9), 1966–1976.

Korucuk, S., Aytekin, A., Ecer, F., Karamaşa, Ç., & Zavadskas, E. K. (2022). Assessing green approaches and digital marketing strategies for twin transition via fermatean fuzzy SWARA-COPRAS. *Axioms*, *11*(12), 1–25. https://doi.org/10.3390/axioms11120709

Kohtamaki, M., Parida, V., Patel, P. C., & Gebauer, H. (2020). The relationship between digitalization and servitization: The role of servitization in capturing the financial potential of digitalization. *Technological Forecasting and Social Change*, *151*, 119804.

Kwilinski, A., Hnatyshyn, L., Prokopyshyn, O., & Trushkina, N. (2022). Managing the logistic activities of agricultural enterprises under conditions of digital economy. *Virtual Economics*, *5*(2), 43–70. https://doi.org/10.34021/ve.2022.05.02(3)

Lambrou, M., Watanabe, D., & Lida, J. (2019). Shipping digitalization management: Conceptualization, typology and antecedents. *Journal of Shipping & Trade*, *4*, 1–17.

Linder, J., & Cantrell, S. (2000). Changing business models: Surveying the landscape. *Accenture Institute for Strategic Change*, 1e15. https://doi.org/10.4018/978-1-59904-939-7.ch249

Liu, J., Quddoos, M. U., Akhtar, M. H., Amin, M. S., Tariq, M., & Lamar, A. (2022). Digital technologies and circular economy in supply chain management: In the era of COVID-19 pandemic. *Operations Management Research*, *15*, 326–341.

Lomakina, O., Kookueva, V., & Makarenko, A. (2021). Redistribution of economic resources in the digital society. *Business and Society Review*, *126*(1), 25–35. https://doi.org/10.1111/basr.12220

Meier, T., Makyšová, H., & Pauliková, A. (2023). Evaluation of the economic, ecological and ethical potential of big data solutions for a digital Utopia in logistics. *Sustainability*, *15*(6), 5088. https://doi.org/10.3390/su15065088

Morris, M., Schindehutte, M., & Allen, J. (2005). The entrepreneur's business model: Toward a unified perspective. *Journal of Business Research*, *58*(6), 726–735. https://doi.org/10.1016/j.jbusres.2003.11.001

Nußholz, J. L. K. (2018). A circular business model mapping tool for creating value from prolonged product lifetime and closed material loops. *Journal of Cleaner Production*, *197*, 185–194. https://doi.org/10.1016/j.jclepro.2018.06.112.

Parida, V., Sjödin, D. R., & Reim, W. (2019). Reviewing literature on digitalization, business model innovation, and sustainable industry: Past achievements and future promises. *Sustainability (Switzerland)*, *11*(2).

Piscicelli, L. (2023). The sustainability impact of a digital circular economy. *Current Opinion in Environmental Sustainability*, *61*, 101251.

Poulis, K., Galanakis, G. C., Triantafillou, G. T., & Poulis, E. (2020). *Value migration: Digitalization of Shipping as a Mechanism of Industry Dethronement. Journal of Shipping and Trade*, *5*(1), 1–18.

Pucihar, A., Lenart, G., Borstnar, M. K., Vidmar, D., & Marolt, M. (2019). Drivers and outcomes of business model innovation—micro, small and medium-sized enterprises perspective. *Sustainability*, *11*, 344.

Rachinger, M., Rauter, R., Müller, C., Vorraber, W., & Schirgi, E. (2019). Digitalization and its influence on business model innovation. *Journal of Manufacturing Technology Management*, *30*(8), 1143–1160.

Rad, F. F., Oghazi, P., Palmié, M., Chirumalla, K., Pashkevich, N., Patel, P. C., & Sattari, S. (2022). Industry 4.0 and supply chain performance: A systematic literature review of the benefits, challenges, and critical success factors of 11 core technologies. *Industrial Marketing Management*, *105*(June), 268–293.https://doi.org/10.1016/j.indmarman.2022.06.009

Ranta, V., Aarikka-Stenroos, L., & Mäkinen, S. J. (2018). Creating value in the circular economy: A structured multiple-case analysis of business models. *Journal of Cleaner Production*, *201*, 988–1000. https://doi.org/10.1016/j.jclepro.2018.08.072

Raza, Z., Woxenius, J., Vural, C. A., & Lind, M. (2023). Digital transformation of maritime logistics: Exploring trends in the liner shipping segment. *Computers in Industry*, *145*(November 2022), 103811. https://doi.org/10.1016/j.compind.2022.103811

Rejeb, A., Rejeb, K., Abdollahi, A., Zailani, S., Iranmanesh, M., & Ghobakhloo, M. (2022). Digitalization in food supply chains: A bibliometric review and key-route main path analysis. *Sustainability*, *14*, 83.

Richardson, J. (2008). The business model: An integrative framework for strategy execution. *Strategic Change*, *17*, 133e144. https://doi.org/10.1002/jsc.821

Romagnoli, S., Tarabu', C., & Maleki Vishkaei, B., & De Giovanni, P. (2023). The impact of digital technologies and sustainable practices on circular supply chain management. *Logistics*, *7*, 1.

Saebi, T., Lien, L., Foss, N. J., 2017. What drives business model adaptation? The impact of opportunities, threats and strategic orientation. *Long Range Planning*, *50*, 567e581. https://doi.org/10.1016/j.lrp.2016.06.006

Silkina, G. Y., Shevchenko, S., & Sharapaev, P. (2021). Digital innovation in process management. *Academy of Strategic Management Journal*, *20*(Special Issue 2), 1–25.

Sorkun, M. F. (2020). Digitalization in Logistics Operations and Industry 4.0: Understanding the Linkages with Buzzwords. In *Contributions to Management Science* (pp. 177–199). https://doi.org/10.1007/978-3-030-29739-8_9

Spieske, A., & Birkel, H. (2021). Improving supply chain resilience through industry 4.0: A systematic literature review under the impressions of the COVID-19 pandemic. *Computers and Industrial Engineering*, *158*, 107452. https://doi.org/10.1016/J.CIE.2021.107452.

Straková, J., Talíř, M., & Váchal, J. (2022). Opportunities and threats of digital transformation of business models in SMEs. *Economics and Sociology*, *15*(3), 159–171

Teece, D. J. (2010). Business models, business strategy and innovation. *Long Range Planning*, *43*, 172e194. https://doi.org/10.1016/j.lrp.2009.07.003

Tranfield, D., Denyer, D., & Smart, P. (2003). Towards a methodology for developing evidence-informed management knowledge by means of systematic review. *British Journal of Management*, *14*(3), 207–222

Osterwalder, A., & Pigneur, Y. (2002). *An eBusiness model ontology for modeling eBusiness*. BLED 2002 Proceedings. 2. http://aisel.aisnet.org/bled2002/2

Osterwalder, A., & Pigneur, Y. (2010). *Business model generation: A handbook for visionaries, game changers, and challengers*. Hoboken, NJ: Wiley.

Osterwalder, A., Pigneur, Y., & Tucci, C. L. (2005). Clarifying business models: Origins, present, and future of the concept. *Communications of the Association for Information Systems*, *15*, 1e43. https://doi.org/10.1.1.83.7452

Xu, A., Qian, F., Ding, H., & Zhang, X. (2023). Digitalization of logistics for transition to a resource-efficient and circular economy. *Resources Policy*, *83*, 103616.

Varriale, V., Cammarano, A., Michelino, F., & Caputo, M. (2021). Sustainable supply chains with blockchain, IoT and RFID: A simulation on order management. *Sustainability (Switzerland)*, *13*(11). https://doi.org/10.3390/su13116372

Wei, F., Alias, C., & Noche, B. (2019). Applications of digital technologies in sustainable logistics and supply chain management. *Innovative Logistics Services and Sustainable Lifestyles: Interdependencies, Transformation Strategies and Decision Making*, 235–263. https://doi.org/10.1007/978-3-319-98467-4_11

Wohlleber, A. J., Bock, M., Birkel, H., & Hartmann, E. (2022). Implementing vital dynamic capabilities to succeed in digital transformation: A multiple-case study in maritime container shipping. *IEEE Transactions on Engineering Management*, 1–19. https://doi.org/10.1109/TEM.2022.3201770

Zhou, H., Wang, Q., Wang, L., Zhao, X., & Feng, G. (2023). Digitalization and third-party logistics performance: Exploring the roles of customer collaboration and government support. *International Journal of Physical Distribution & Logistics Management*, *53*(4), 467–488. https://doi.org/10.1108/IJPDLM-12-2021-0532

# 4 Achieving Business Sustainability with Blockchain Technology

*Ngoc Dang Khoa Nguyen, Atanu Chaudhuri, and Aman Ullah*

## 4.1 INTRODUCTION

In today's rapidly changing world, contemporary organisations are seeking innovative solutions to alleviate their ecological footprint, improve supply chain transparency and promote inclusivity (Ali et al., 2022; Mishra et al., 2023; Zheng et al., 2022). Simultaneously, emerging technologies such as blockchain technology offer unprecedented opportunities to drive sustainability efforts forward (Ali et al., 2023; D. Biswas et al., 2023; Chaudhuri et al., 2021; Dal Mas et al., 2023).

Within this landscape, blockchain emerges as a promising catalyst for advancing sustainability (Chaudhuri et al., 2022; Nazam et al., 2022; OECD, 2022). It offers a transparent, decentralised, and immutable platform for recording and verifying transactions, revolutionising the way businesses operate and interact with their stakeholders in diverse sectors including supply chain management, healthcare, finance, and energy (Choi & Siqin, 2022; Friedman & Ormiston, 2022; Oracle, 2023). Specifically, by leveraging blockchain's inherent features including transparency, security, and immutability, organisations can establish a secure and tamper-proof record of every transaction within the supply chain (Arunmozhi et al., 2022; Mercuri et al., 2021; Patil et al., 2023). This allows for enhanced visibility, accountability, and trust, facilitating responsible sourcing, eliminating environmental impact, and addressing issues such as deforestation and forced labour (Chaudhuri et al., 2022; Di Vaio & Varriale, 2020; Feng et al., 2020).

To exemplify the importance of blockchain in fostering sustainability, we illustrate six compelling case studies from various regions around the globe. Such case studies showcase how businesses have successfully harnessed blockchain to drive sustainable operations, addressing challenges and reaping the rewards of sustainable practices. Through these real-world examples, we aim to inspire businesses across industries to adopt a more sustainable approach to their operations and supply chains.

The findings presented in this chapter contribute to the growing body of knowledge on the role of blockchain in accomplishing sustainable development goals. Our exploration goes beyond the technological aspects of blockchain implementation and delves into the socio-economic and environmental elements. Through the analysis


DOI: 10.1201/9781003438748-5

of six global case studies, we offer practical insights, lessons learned, and best practices that can serve as a blueprint for organisations seeking to embrace blockchain for sustainability.

This chapter aims to establish a strong rationale for businesses to embrace blockchain as a catalyst for sustainable practices. By investigating the interplay between blockchain and sustainability across economic, environmental, and social domains, we aim to equip readers with the necessary knowledge and inspiration to harness the full potential of blockchain and drive positive change towards a more sustainable future.

## 4.2 BLOCKCHAIN TECHNOLOGY

Blockchain technology has undeniably garnered immense hype and attention in the past decade (Dal Mas et al., 2023; Witt & Schoop, 2023). Its inception can be traced back to 2008 when Satoshi Nakamoto developed it to address the issue of double spending in digital currencies (Nakamoto, 2008). Nakamoto conceptualised and framed Bitcoin as a peer-to-peer electronic cash system, utilising digital signatures to timestamp transactions. The underlying technology supporting Bitcoin is known as blockchain, which allows for secure online transactions without the need for intermediaries such as financial institutions (Dal Mas et al., 2023; Komulainen & Nätti, 2023). Nevertheless, it is important to differentiate between the cryptocurrency Bitcoin and the broader concept of blockchain (Ahluwalia et al., 2020; Tiscini et al., 2020).

Blockchain is a shared, open, and distributed ledger that records transactions across peer-to-peer network (Feng et al., 2020; Kuzior & Sira, 2022). The name "blockchain" itself reflects its functioning: information is stored in individual blocks that are linked together to form a chain of validated data (Dutta et al., 2020; Kumar Singh et al., 2023). Several authors (Choi & Siqin, 2022; Di Vaio & Varriale, 2020; Kumar Singh et al., 2023) outline three key steps in the operation of blockchain: (i) participants exchange data encrypted by blockchain, (ii) such transactions are verified by blockchain network and grouped into a locked data "block", and (iii) such blocks are linked to previous ones using a hash key, thereby forming a chain based on a validation principle call a blockchain. This blockchain exhibits three basic features: (i) the storage of permanent, auditable, and unchangeable digital records ensuring data security; (ii) real-time exchange of digital assets enhancing transparency; and (iii) the execution of smart contracts contributing to efficiency. These three features are underpinned by the decentralisation in blockchain (Arunmozhi et al., 2022; Choi et al., 2020; Tiscini et al., 2020). Blockchain therefore is universally acknowledged in various disciplines due to its conspicuous features (see Figure 4.1) such as security, transparency, decentralisation, and efficiency.

*Data security*: The consensus mechanism in the system ensures that each transaction is verified by all participants before it can be added as a block (Park & Li, 2021). Further, once a public ledger is approved by all nodes, it cannot be deleted, providing the feature of immutability that safeguards data against tampering (Chaudhuri et al., 2022; Cole et al., 2019). Another

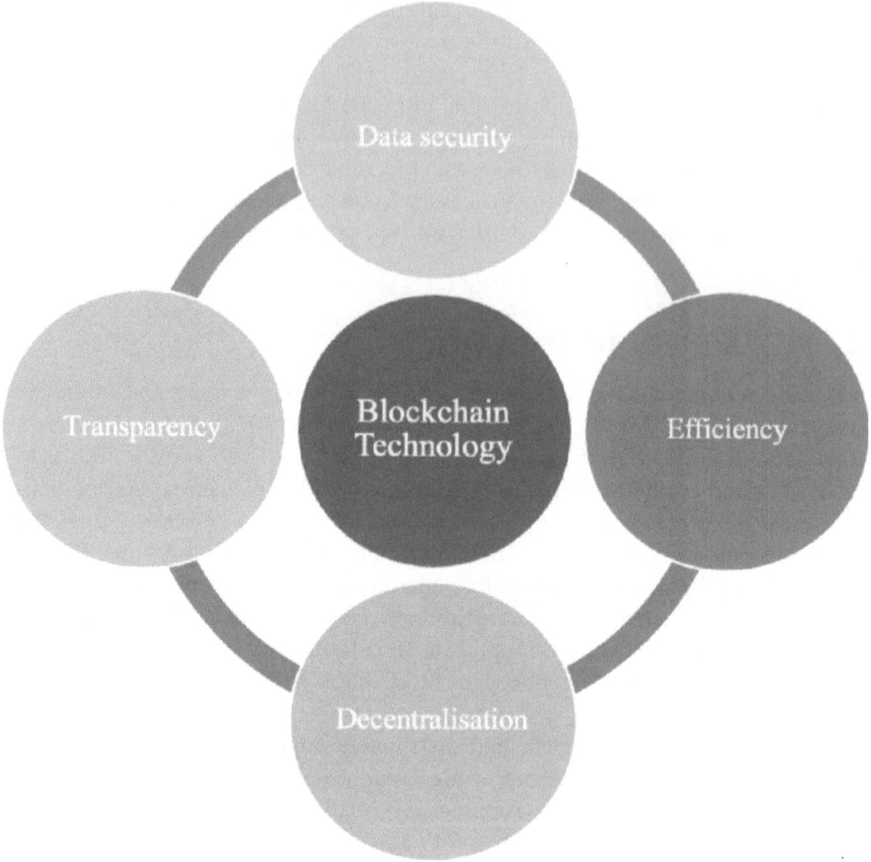


**FIGURE 4.1** Benefits of blockchain technology.

**(Source: Authors' elaboration).**

feature of security is achieved through public key cryptography, where authorised users are assigned their unique key (digital code) to access the data (Chittipaka et al., 2022; Liu et al., 2020). Before a peer-to-peer exchange can occur, algorithms validate the digital identities of the transaction partners (Feng et al., 2020; Liu et al., 2020).

*Transparency*: Blockchain combines secure access for authorised entities with public visibility, leading to enhanced transparency (Gazzola et al., 2023; Kiu et al., 2022). While only authorised peers can access the blockchain, all network members can view the complete record of transactions. The consensus process ensures that the history of records is reliable, as all participants in the blockchain network have to verity each block (Alqarni et al., 2023; Khan et al., 2022). Digital signatures further contribute to transparency by easily proving the identity of participants and preventing identity fraud (Hallioui et al., 2022; Pandey et al., 2022).

*Decentralisation*: The decentralised nature of blockchain fosters a peer-to-peer environment where all members collectively contribute to building the network (Cai et al., 2023; Li et al., 2022). Each member possesses their own copy of the ledger, distributing data across the network and eliminating the reliance on a centralised database (Centobelli et al., 2022; Mukherjee et al., 2021). Not only does this decentralised structure enhance security by making the system more robust against external disruptions but also establishes trust between distrustful parties, facilitating safe transactions and fair compensation for honest participants (Hallioui et al., 2022; Park & Li, 2021). With no single party having control over the entire database, the network becomes more transparent and enables trust among its members (Chittipaka et al., 2022; Yontar, 2023).

*Efficiency*: Efficiency is a fundamental attribute of blockchain due to its capability to digitally and instantaneously share data (Han & Rani, 2022). This efficiency paves the way for the execution of smart contracts, which are automated and digital agreements that reduce transaction costs and increase security (Chang et al., 2019; J. Li et al., 2023). Smart contracts offer a tamper-proof and conflict-free method of exchanging value, providing advantages over traditional paper-based contracts in the digital realm (Alqarni et al., 2023; Yousefi & Mohamadpour Tosarkani, 2022). By maintaining records of asset provenance and ensuring instant outcomes without the involvement of third parties, smart contracts optimise the contracting process (Mukherjee et al., 2021; Wang et al., 2021). The absence of intermediaries reduces data errors, lowers transaction costs, and mitigates reliance on individual servers (Arunmozhi et al., 2022; Kuzior & Sira, 2022).

## 4.3 CHALLENGES IN BLOCKCHAIN

While blockchain holds immense promise, its successful implementation faces various challenges and obstacles as follows:

*Technological limitations:* As blockchain networks expand, limitations surrounding transaction processing speed and capacity become major focal points (Köhler et al., 2022; Zheng et al., 2022), thereby impeding the widespread adoption and utilisation in high-volume business environments (Baysal et al., 2023; Chittipaka et al., 2022). Likewise, the absence of interoperability between different blockchain platforms poses difficulties in integrating existing systems with blockchain solutions (Li et al., 2022; Nagariya et al., 2022).

*Legal and regulatory issues:* The legal and regulatory landscape surrounding blockchain is still evolving, which gives rise to uncertainty and inconsistencies in regulations across jurisdictions (Chittipaka et al., 2022; Cole et al., 2019; Li et al., 2022). Issues pertaining to data privacy, intellectual property rights, and compliance with existing regulations present obstacles that need to be addressed (Agi & Jha, 2022; Yontar, 2023). Collaborative efforts between industry stakeholders and regulatory bodies are necessary to establish clear guidelines and frameworks that facilitate the integration of blockchain while ensuring compliance and protection for all parties involved (Nagariya et al., 2022; Tan & Salo, 2023).

*Adoption barriers and education gap:* The adoption of blockchain necessitates a significant shift in mindset and organisational processes (Alimohammadlou & Alinejad, 2023; Pandey et al., 2022). Lack of awareness and understanding of blockchain among business leaders and stakeholders can breed resistance to its adoption (Komulainen & Nätti, 2023; Xu et al., 2022). Similarly, the complexity of blockchain and its potential impact on existing systems can deter business from embracing it (Xu et al., 2022; Yadlapalli et al., 2022).

As such, we can create a more sustainable future where blockchain plays a central role in fostering data security, decentralisation, transparency, and efficiency by addressing these hurdles.

## 4.4 SUSTAINABILITY

Sustainability is a multifaceted concept, firmly rooted in three core pillars: economic, environmental, and social sustainability.

1. *Economic sustainability* relates to the promotion of economic growth and development. It often emphasises cost reduction and efficiency enhancement to foster an economically sound future (Bhubalan et al., 2022; Choi, 2022). Specifically, economic growth remains a foremost priority for most organisations, often facilitated by an efficient interorganisational information system that bolsters supply chain capabilities (Bunge et al., 2022; Wong et al., 2023). The relentless demands of stakeholders necessitate organisations to keep pace with technological evolution and align with the dynamic shifts in the business landscape (Akinradewo et al., 2023; Joshi et al., 2023). Likewise, advancements in technology and transformations in the business ecosystem precipitates the requirement for procedural adaptations through industrial revolutions, in turn, potentially amplifying profit margins (Fatorachian & Kazemi, 2018; Kashem et al., 2023). Yaqub and Alsabban (2023) delineate numerous enablers, particularly pertinent to Industry 4.0, which can serve as resources to achieve sustainable economic performance. Research by Polas et al. (2022) and Niu et al. (2021) elucidates initiatives that strong economic performance not only scaffolds the operations of business process ecosystems but also offers a framework to meet and exceed stakeholder expectations.
2. *Environmental sustainability*, on the other hand, is the conscious preservation of our planet's resources. It mandates the protection of raw materials sources and the implementation of strategies to significantly curtail waste production (N. Li et al., 2023; Rejeb et al., 2023). Specifically,

as businesses extend their reach across international borders, they are obliged to align with globally recognised norms (Ronaghi & Mosakhani, 2022; Tang et al., 2022). With revisions in governmental environmental policies and a rising emphasis on ecological preservation amongst stakeholders, industries and businesses are increasingly gravitating towards environmentally friendly practices (Mohamed et al., 2023; Stuit et al., 2022). The notion of the green supply chain, introduced and explored by academic scholars, suggests that the adoption of eco-friendly practices can lead to improved economic performance (Kholaif et al., 2023; Wang et al., 2023). Environmental sustainability and green initiatives now occupy a prominent place on managerial agendas (González-Benito & González-Benito, 2006; Khan et al., 2021). Organisations are thus encouraged to proactively implement environmental management systems. By minimising by-products and emissions and leveraging lean production processes, businesses can not only enhance environmental productivity but also boost their economic performance (Oliveira et al., 2019; Parmentola et al., 2022).

3. *Social sustainability* underscores the importance of promoting social development and well-being. It encapsulates practices like advocating for social equity, responsibility, health equity, justice, fairness, ethics, and cultural competence (Parmentola et al., 2022; Ronaghi & Mosakhani, 2022). From their inception, organisations have prioritised social initiatives to cultivate a positive reputation among stakeholders beyond their immediate sphere (Liu et al., 2023; Mangla et al., 2021). As businesses extend their global footprint, the understanding and role of social performance have evolved (Chaudhuri et al., 2021; Hervani et al., 2022). Both external pressures and governmental regulations act as catalysts, stimulating organisations to strive for superior social sustainability (Nobanee et al., 2021; Ronaghi & Mosakhani, 2022). Contemporary organisations, in their pursuit of sustainable performance, ensure robust social sustainability by endorsing ethical practices, fostering open communication, and fulfilling societal obligations (Joshi et al., 2023; Nandi et al., 2023). While social sustainability may not exert a direct influence on the organisation, it can serve as a stabilising force during times of disruption (Hervani et al., 2022; Quayson et al., 2021). As such, social sustainability is deemed as an integral facet of sustainable organisational performance.

By synergistically integrating these three foundational pillars—economic, environmental, and social—we can devise and actualise sustainable strategies that pay the way for a balanced and harmonious future.

## 4.5 BLOCKCHAIN AND BUSINESS SUSTAINABILITY

The intersection of blockchain and sustainability including economic, environmental, and social holds immense promise in revolutionising the way organisations approach sustainable development (see Figure 4.2).

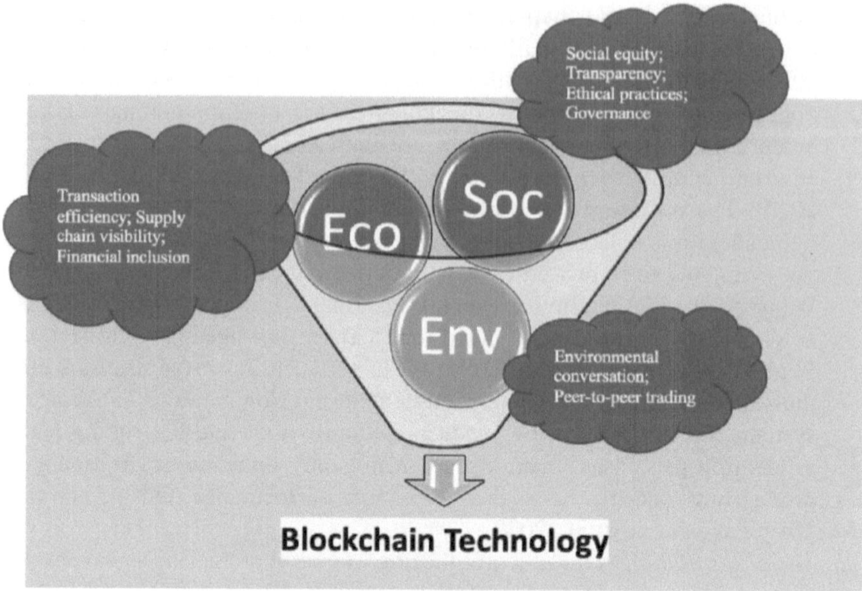


**FIGURE 4.2** Blockchain and sustainability triple bottom line.

**(Source: Authors' elaboration).**

### 4.5.1 Blockchain and Economic Sustainability

Blockchain has swiftly emerged as a promising solution for various sectors due to its unique capabilities. Its potential to foster economic sustainability has drawn significant scholarly attention (Ada et al., 2021; Choi, 2022; Esmaeilian et al., 2020).

*Transaction efficiency:* Blockchain, as a decentralised and secure ledger system, sparks widespread investigation into the economic implications of this technology (Bunge et al., 2022; Di Vaio & Varriale, 2020). Several authors (e.g., Varavallo et al., 2022; Wang et al., 2020; Zhao, 2023) delve into the subject, detailing how the intrinsic transparency and immutability of blockchain could reshape business transactions and, in turn, stimulate sustainable economic growth. Each transaction or event is securely recorded on the blockchain, establishing an auditable trail that enables stakeholders to track the origin, production processes, and movement of goods (Centobelli et al., 2022; Philsoophian et al., 2022).

*Supply chain visibility:* Building upon this discourse, blockchain's transparent and immutable nature can encourage ethical business practices, thereby amplifying traceability across supply chains (de Boissieu et al., 2021; Manzoor et al., 2022). By enabling a distributed ledger that records every transaction and movement of goods, blockchain can enhance visibility across the supply chain, thereby promoting accountability and reducing fraud (Centobelli et al., 2022; Philsoophian et al., 2022). It could potentially address common supply chain management issues such as counterfeiting, product diversion, and poor product quality (Sahoo et al., 2022; Sharma et al., 2023).

*Financial inclusion:* The most-discussed potential of blockchain pertains to its capacity for fostering economic inclusion. The decentralisation and democratisation of financial systems are well articulated by Chung et al. (2023) and Paglietti and Rabitti (2022). They posit that blockchain can spur economic sustainability by enabling the unbanked population to access financial services. Several authors (García-Monleón et al., 2023; Hoffmann, 2021) further this argument by highlighting that blockchain could lower the costs of remittances, making them more accessible to low-income populations, thereby facilitating efficient and cost-effective financial transactions.

#### 4.5.1.1 Comprehensive Cases

##### *4.5.1.1.1 Case 1*

**IBM Food Trust** in partnership with Walmart utilises blockchain to enhance transparency and traceability in the food supply chain (IBM, 2019; Vitasek et al., 2022). By leveraging the decentralised and immutable nature of blockchain, this platform ensures transparency, reduces fraud, and improves collaboration and efficiency (IBM, 2020, 2023; Park & Li, 2021) through:

*Traceability and transparency:* Every transaction and movement of food products are recorded and verified on an auditable and transparent ledger. Each participant in the network contributes to the shared ledger, recording crucial data such as product origins, certifications, and handling practices. This level of traceability empowers businesses and consumers to track the journey of food products, identify potential issues, and make well-informed decisions with respect to sustainability, ethical sourcing, and responsible consumption.

*Collaboration and efficiency:* The IBM Food Trust fosters collaboration among supply chain participants by providing a secure and efficient platform for sharing information. By streamlining communication and data exchange, the platform reduces inefficiencies, eliminates redundant processes, and optimises the allocation of resources. This collaborative approach enhances supply chain efficiency, reduces waste, and promotes sustainable practices, in turn contributing to business sustainability and resilience.

*Consumer empowerment and trust:* One of the key advantages of the IBM Food Trust is its capability to empower consumers by providing them with access to detailed information about the products they consume. Through QR codes and mobile applications, consumers can obtain information about the food's origin, production methods, certifications, and sustainability practices. This transparency fosters consumer trust, promotes responsible purchasing decisions, and incentivises businesses to adopt sustainable and ethical practices to meet evolving consumer demands.

By digitising the tracking of food products, from farm to store shelves, this blockchain-based solution enables collaboration and efficiency, reduces waste, and ensures responsible sourcing. It allows consumers and retailers to access detailed information about the origin, quality, and journey of products, fostering trust and supporting sustainable practices.

##### 4.5.1.1.2 Case 2

**ViettelPay** is a Vietnamese mobile payment blockchain platform offering numerous digital financial services, including mobile banking, remittances, payments, and microfinance (Donkin, 2020; RMIT Vietnam, 2021; Viet Nam News, 2018; Viettel, 2018, 2023b; Viettel Group, 2020a). By leveraging blockchain's decentralised nature, ViettelPay aims to bridge the financial inclusion gap and foster sustainable economic growth (Henderson, 2019; Moises, 2022; Tech, 2022; Tri, 2023; Viettel, 2023a; Viettel Group, 2020b, 2022), including:

*Financial inclusion:* ViettelPay's blockchain-based platform plays an imperative role in advancing financial inclusion by providing access to financial services for unbanked and underbanked individuals. These individuals can create digital wallets and perform numerous financial transactions, including payments, transfers, and savings. This empowers previously excluded populations, namely low-income individuals and rural communities, to join the formal financial system, allowing them to save, transact, and build financial resilience.

*Remittances and cross-border transactions:* The ViettelPay platform facilitates cost-effective and efficient remittances and cross-border transactions. Users can securely transfer funds internationally, eliminating the need for intermediaries and reducing transaction costs. This platform has significant implications for overseas users, enabling faster and more affordable remittance services and improving economic opportunities for individuals and communities.

*Microfinance and credit access:* The blockchain-based platform provides microfinance services to small businesses and individuals that lack access to traditional credit channels. Through smart contracts and decentralised lending platforms, ViettelPay can offer microloans and credit facilities to underserved entrepreneurs, promoting entrepreneurship, job creation, and economic development. Given the blockchain's transparency and traceability, ViettelPay ensures fair and secure lending practices, reducing the risk of fraud and fostering responsible financial behaviour.

Therefore, ViettelPay's adoption of blockchain for financial services enables individuals and businesses to participate in the digital economy, fostering economic growth and sustainability.

### 4.5.2 Blockchain and Environmental Sustainability

Blockchain's potential implications for environmental sustainability represent a rapidly evolving research area (Javed et al., 2022; Sriyono, 2020). While initially conceptualised for the financial sector, blockchain's secure, decentralised, and immutable features have begun to interest environmental scholars and practitioners (Mohamed et al., 2023; Wong et al., 2023).

*Environmental conversation:* The literature consistently underscores blockchain's capability to bolster transparency in environmental conservation efforts. Gazzola et al. (2023) and Liu et al. (2020) argue that incorporating blockchain in natural resource governance could augment transparency and traceability. For instance, applying blockchain to supply chain management empowers stakeholders to authenticate goods' provenance and confirm adherence to sustainable and ethical practices (Patil et al., 2023; Wan et al., 2020; Wang et al., 2020). Such applications could motivate consumers to make eco-friendlier decisions and enable companies to demonstrate their commitment to sustainability.

*Peer-to-peer energy trading:* The promise of blockchain in facilitating decentralised peer-to-peer (P2P) energy trading represents another crucial focal point in the literature (Javed et al., 2022; X. Li et al., 2023). Blockchain can optimise resource management by tracking and monitoring resource usage, such as energy, water, and raw materials, in real time (Dutta et al., 2020; Kumar et al., 2023; Teodorescu & Korchagina, 2021). Through smart contracts, blockchain can automate processes, ensuring efficient utilisation and reducing waste (Baysal et al., 2023; Centobelli et al., 2022). Several authors (e.g., Ada et al., 2021; Tan & Salo, 2023) explore how blockchain can catalyse P2P energy networks, allowing households to trade surplus renewable energy. Such democratisation of the energy sector could facilitate a more sustainable energy mix, thus advancing environmental sustainability.

### 4.5.2.1 Comprehensive Cases

#### *4.5.2.1.1 Case 3*

**Australian Dairy Farmers (ADF)** partnered with blockchain technology providers to implement a blockchain-based platform for its dairy supply chain. This platform incorporated features such as smart contracts, immutable record-keeping, and real-time data sharing (Australian Dairy Farmers, 2021; Colquhoun, 2020; Daniel, 2022). Key stakeholders, involving farmers, processors, distributors, and retailers were onboarded into the blockchain network, enabling end-to-end traceability and transparency (Australian Dairy Farmers, 2020, 2022a, 2022b; Coles, 2020) through:

*Quality assurance and compliance:* Blockchain enabled ADF to implement robust quality assurance measures. Smart contracts and sensors integrated into the supply chain facilitated real-time monitoring of critical parameters, including temperature, humidity, and storage conditions. Any deviations from the predefined standards triggered automatic alerts, enabling prompt corrective actions. This proactive approach improved product quality, reduced waste, and ensured compliance with industry regulations.

*Supply chain transparency and traceability:* Every stage of dairy production process, from farm to consumer, was recorded on the blockchain, generating an immutable and auditable trail. This transparency allowed consumers to access detailed information about the origin, farming practices, and quality of dairy products, promoting trust and supporting sustainable purchasing decisions.

*Collaboration and trust-building:* Blockchain facilitated collaboration and trust-building across dairy supply chain. By sharing data on a distributed ledger, ADF and its partners could access accurate and reliable information, leading to improved communication, coordination, and decision-making. Collaborative efforts to address sustainability challenges, including mitigating water usage or carbon emissions, were more effective through shared data and insights.

By leveraging blockchain's capabilities, ADF improved supply chain transparency, promoted sustainable farming practices, and built trust with consumers.

##### *4.5.2.1.2 Case 4*

**Powerledger** is an Australian blockchain platform that facilitates peer-to-peer energy trading and supports renewable energy adoption (Australian Trade and Investment Commission, 2021; Government House Western Australia, 2020; Patrick, 2023). Through blockchain, Powerledger allows individuals and businesses to trade excess renewable energy directly, promoting a decentralised and sustainable energy ecosystem (Arunmozhi et al., 2022; Energy Matters, 2022; Foti & Vavalis, 2021; Powerledger, 2019, 2023), including:

*Transparency and traceability:* Each transaction is recorded on the blockchain, generating an immutable and auditable record of energy generation, consumption, and trading. This transparency enables consumers to verify the origin of the energy they purchase, ensuring that it comes from renewable sources. It also promotes trust and accountability among energy market participants, driving the adoption of sustainable energy practices and enabling consumers to support renewable energy producers directly.

*Energy access and financial inclusion:* Powerledger's platform fosters energy access and financial inclusion by enabling individuals and communities to join the energy market as both producers and consumers. Through blockchain-enabled energy trading, this platform empowers individuals to generate income from their renewable energy assets and reduce their dependence on centralised energy providers, thereby leading to better energy independence and resilience.

*Grid flexibility and resilience:* During peak energy demand periods, excess energy from distributed renewable sources can be efficiently distributed to areas with higher demand, reducing strain on the grid and minimising the need for additional fossil-based generation. This flexibility promotes the integration of intermittent renewable energy sources and helps balance supply and demand, contributing to a more stable and reliable energy system.

As a result, this system encourages the use of renewable sources, reduces reliance on traditional energy grids, and empowers consumers to make environmentally conscious energy choices.

#### *4.5.2.1.3 Case 5*

**Circulor** is a pioneering company in the United Kingdom, offering sustainable supply chain solutions to businesses across diverse industries (Circulor, 2022b, 2023a, 2023c, 2023d). By harnessing the power of blockchain, Circulor aims to improve transparency, boost accountability, and drive positive environmental impact within supply chains (Circulor, 2022a, 2023b, 2023e) through:

*Traceability and responsible sourcing*: Circulor utilises blockchain to establish a transparent and immutable record of every step in the supply chain, enabling businesses to trace the origin and lifecycle of raw materials. By integrating blockchain with IoT devices and other data sources, Circulor effectively tracks the movement of goods and verifies responsible sourcing practices, including conflict-free material supply chains or sustainable forestry practices.

*Carbon footprint monitoring and offset programs*: Blockchain empowers Circulor to accurately monitor and record carbon emissions at each stage of the supply chain. Through smart contracts and decentralised applications, Circulor automates the calculation and reporting of carbon footprints, allowing businesses to make informed decisions to mitigate their environmental impact. Furthermore, blockchain facilitates the integration of carbon offset programs, fostering a more sustainable and carbon-neutral business ecosystem.

*Ethical labour practices and human rights*: Circulor employs blockchain to ensure compliance with ethical labour standards and human rights practices. Through immutable records of labour conditions, certifications, and audits, Circulor's blockchain platform enables businesses to verify and monitor the working conditions of their suppliers, promoting fair treatment, worker safety, and ethical labour practices.

Thus, by leveraging blockchain's inherent features including decentralisation, transparency, and immutability, Circulor seeks to address critical challenges relevant to trust, traceability, and accountability in the supply chain.

### 4.5.3 Blockchain and Social Sustainability

As blockchain continues to evolve beyond its original application in cryptocurrencies, scholars have begun to investigate its potential to foster social sustainability (Mangla et al., 2021; Massaro et al., 2020). The decentralised and secure nature of blockchain holds potential for profound societal impacts, especially in the realms of equity, inclusion, and transparency (Chaudhuri et al., 2021; Yontar, 2023).

*Social equity:* The potential of blockchain to democratise access to resources is a recurring theme in its contribution to social sustainability. Ahluwalia et al. (2020) and Treiblmaier (2018) posit that blockchain could revolutionise financial systems by enabling secure, peer-to-peer transactions that bypass traditional intermediaries and reduce transaction costs. These implications for financial inclusivity for social sustainability are profound, as it could engender economic opportunities for underprivileged populations (Agi & Jha, 2022; Karakas et al., 2021).

*Transparency and ethical practices:* Transparency, another critical area where blockchain can contribute to social sustainability, is well-documented (Hervani et al., 2022; Ronaghi & Mosakhani, 2022). As stated by Sadeghi et al. (2023) and Quayson et al. (2021), blockchain's inherent public and decentralised ledger can instil greater accountability across a myriad of sectors. For instance, supply chains often involve complex networks of stakeholders, raising concerns about accountability and ethical practices (Boutkhoum et al., 2021; Mangla et al., 2021). Blockchain's decentralised and immutable ledger offers the potential to enhance supply chain transparency, ensuring fair labour practices, sustainable sourcing, and reduced environmental impact (D. Biswas et al., 2023; Mercuri et al., 2021).

*Enhanced public services and governance:* Blockchain has emerged as a promising avenue for transforming public services and governance (Massaro et al., 2020; Wong et al., 2022). By leveraging its decentralised, transparent, and tamper-resistant properties, blockchain holds the potential to revolutionise how governments interact with citizens and manage public resources (Havidz et al., 2023; Wei et al., 2023). One of the key areas where blockchain has shown promise is in identity management (M. Biswas et al., 2023; Sun et al., 2023). Through blockchain-based digital identity systems, individuals can have more control over their personal information and securely access public services (Ietto et al., 2023; Tan et al., 2023). Likewise, by providing an immutable record of transactions and ensuring data integrity, blockchain can reduce bureaucratic inefficiencies and enhance trust between citizens and governments (Quayson et al., 2021; Rim, 2023).

#### 4.5.3.1 Comprehensive Cases

##### *4.5.3.1.1 Case 6*

**CropData** is an Indian blockchain platform, addressing agricultural challenges by providing farmers with a transparent and secure system to manage their land records, crop history, and financial transactions (Agrawal, 2020; Naik, 2020). By leveraging blockchain's immutability and decentralised nature, CropData improves data integrity, eliminates fraud, and facilitates access to credit for farmers (CIO, 2020; CropData, 2020; Raghavan et al., 2021), through:

*Supply chain traceability:* Blockchain allows for the recording and verification of every transaction and movement of crops throughout the supply chain. From seed sourcing and cultivation to harvesting and distribution, each step is documented on the blockchain, ensuring traceability and accountability.

This level of transparency enables stakeholders to track the origin, quality, and sustainability attributes of crops, empowering consumers to make informed choices and encouraging sustainable agricultural practices.

*Certifications and standards compliance:* CropData facilitates the integration of certifications and standards related to sustainable farming practices into the blockchain. Farmers can record and provide evidence of compliance with certifications such as organic, fair trade, or environmentally friendly practices. This enables buyers and consumers to verify the authenticity and credibility of sustainability claims, fostering trust and promoting the adoption of sustainable agricultural practices.

*Efficiency and resource optimisation:* Through smart contracts, CropData enables real-time monitoring and data collection on key parameters, consisting of water usage, fertiliser application, and pest management. This data is securely stored on the blockchain, allowing farmers to analyse and optimise their resource allocation, reduce waste, and minimise the environmental impact of agricultural activities. CropData supports sustainable farming practices and contributes to overall business sustainability through efficiency and resource optimisation.

As such, this technology promotes sustainable agriculture by empowering farmers and enhancing their productivity and well-being.

In a nutshell, these six case studies exemplify the diverse applications of blockchain in promoting business sustainability across different industries and regions (see Table 4.1). By harnessing the power of blockchain, these organisations have demonstrated the potential for increased transparency, efficiency, and trust in sustainable business practices and thereby shedding light on the potential impact of this innovative solution on business sustainability.

**TABLE 4.1**
**Summary of Case Studies of Blockchain Applications for Sustainability**

| Organisation | Industry and Context | Key Features | Sustainable Benefits to Stakeholders |
|---|---|---|---|
| **IBM Food Trust** | US Retail Food Industry | *Traceability*<br>*Transparency*<br>*Collaboration*<br>*Efficiency*<br>*Consumer empowerment*<br>*Trust* | Allows consumers and retailers to access detailed information about the origin, quality, and journey of products, fostering trust and supporting sustainable practices. |
| **ViettelPay** | Vietnamese Financial Sector | *Financial inclusion*<br>*Remittances*<br>*Cross-border transaction*<br>*Microfinance*<br>*Credit access* | Enables individuals and businesses to participate in the digital economy, facilitating economic growth and financial sustainability. |

(*Continued*)

**TABLE 4.1**
**(Continued)**

| Organisation | Industry and Context | Key Features | Sustainable Benefits to Stakeholders |
|---|---|---|---|
| **ADF** | Australian Dairy Industry | *Traceability*<br>*Transparency*<br>*Quality assurance*<br>*Compliance*<br>*Collaboration*<br>*Trust-building* | Incorporated features such as smart contracts, immutable record-keeping, blockchain networks, and real-time data sharing for key stakeholders, involving farmers, processors, distributors, and retailers. |
| **Powerledge** | Australian Energy Sector | *Traceability*<br>*Transparency*<br>*Energy access*<br>*Financial inclusion*<br>*Grid flexibility*<br>*Resilience* | Encourages the use of renewable sources, reduces reliance on traditional energy grids, and empowers consumers to make environmentally conscious energy choices. |
| **Circulor** | UK Supply Chain Sector | *Traceability*<br>*Responsible sourcing*<br>*Carbon footprints monitoring and offset programs*<br>*Ethical labour practices*<br>*Human rights practices* | Allows businesses to verify and monitor the working conditions of their suppliers, promoting fair treatment, worker safety, protecting human rights, and ethical labour practices. |
| **CropData** | Indian Agriculture Sector | *Traceability*<br>*Certifications*<br>*Standards compliance*<br>*Efficiency*<br>*Resource optimisation* | Promotes sustainable agriculture by empowering farmers and enhancing their productivity and financial well-being.<br>Also improves data integrity, eliminates fraud, and facilitates access to credit for farmers. |

*Source:* Authors' elaboration.

## 4.6 CONCLUSION

The convergence of blockchain and sustainability has stimulated remarkable possibilities across various sectors worldwide. Our rigorous analysis, informed by six unique case studies—Circulor, ViettelPay, Powerledger, Australian Dairy Farmers, CropData, and IBM Food Trust—accentuates the significant promise blockchain harbours in advancing sustainable initiatives.

From Circulor's trailblazing efforts to enhance supply chain transparency in the mining industry to ViettelPay's innovative use of blockchain in advancing financial inclusion in Vietnam, each case study contributes to a deeper understanding of blockchain's transformative power. Powerledger's ground-breaking initiative in democratising renewable energy trading in Australia, Australian Dairy Farmers' novel application of blockchain for ensuring traceability of dairy products, CropData's empowering project assisting smallholder farmers in India, and IBM Food Trust's ambitious endeavour to ameliorate global food safety and reduce waste using blockchain, all illustrate the extensive potential of this technology.

In each scenario, the intrinsic features of blockchain, including decentralisation, transparency, and traceability, are recognised as crucial benefits supporting sustainable systems. Blockchain's decentralised architecture reduces the necessity for intermediaries, facilitating efficient resource allocation. Its inherent transparency cultivates an environment of accountability, crucial for optimising efficiency and minimising waste. Furthermore, blockchain's capability to ensure traceability fortifies supply chain integrity, fostering ethical and sustainable practices. However, it's essential to address the challenges associated with blockchain adoption, namely scalability issues and regulatory complexities to fully unlock its potential for sustainable development.

The future of blockchain hinges on the ability of both practitioners and researchers to judiciously harness this technology. This will necessitate ongoing innovation and regulatory adaptability with the aim of maximising blockchain's benefits while mitigating its limitations. Future research should focus on uncovering the full potential of blockchain in emerging areas such as renewable energy trading, sustainable finance, and carbon accounting. Moreover, it's crucial to examine the social and economic implications of blockchain adoption, including its impact on job creation, economic disparity, and power distribution. Technical challenges, including scalability, energy efficiency, and privacy in blockchain systems, also warrant further research. The development of standardised protocols, governance models, and best practices will enhance interoperability, security, and scalability, facilitating the seamless integration of blockchain into sustainable business practices.

In conclusion, our meticulous exploration of six case studies of various industries across globe from reaffirms blockchain's transformative power as an indispensable enabler for sustainability, thereby providing significant impetus towards a more sustainable and equitable world.

## 4.7 IMPLICATIONS, LIMITATIONS, AND FUTURE RESEARCH DIRECTIONS

Our study has tremendous theoretical implications, as it identifies several key elements encompassing all aspects of blockchain and its application to business sustainability across various global sectors. These pillars, including economic, social, and environmental, offer a foundation for further theorisation and empirical testing of blockchain in diverse contexts and different countries.

We enrich the existing literature on sustainable development by highlighting the significance of interconnected factors such as sustainable finance, renewable energy trading and carbon accounting. The dynamics of these factors could differ across disciplines, paving the way for future theorisation and exploration in diverse contexts.

From a managerial standpoint, a key takeaway from our study is the importance of blockchain for achieving business sustainability. It serves as a critical tool for socio-economic and environmental elements, becoming an increasingly essential requirement for a range of business leaders. Non-compliance and lack of standardised protocols could lead to reputational challenges. Furthermore, a holistic approach that incorporates sustainable development as part of blockchain implementation offers strategic insights into sustainability. This enables C-level executives in different industries to view sustainability from an organisational standpoint, rather than a functional, short-term viewpoint.

However, like any research, out study is not without limitations. It is confined to blockchain and sustainability within specific industries worldwide, and therefore may not fully account for the unique dynamics of other sectors. In addition, our case studies rely on publicly available reports, news articles, and internet resources. As such, the lack of primary data collection may be seen as a limitation, but it also opens the door for future empirical research.

Despite these limitations, our study represents a pioneering exploration of blockchain's importance for business sustainability in different industries. It underscores its crucial role in reducing carbon footprint, addressing economic disparity, and distributing power more equitably. Future studies can build upon our work to delve into both technical and human dynamics of blockchain implementation in different countries and across different disciplines.

Finally, future research should incorporate blockchain-specific theorisation within the broader context of sustainability in specific countries and industries. This approach would provide in-depth insights into this specific phenomenon and further solidify its importance within the sustainability discourse. As blockchain continues to evolve, ongoing research and analysis will be indispensable to unlock its full potentials in advancing business sustainability across diverse sectors and regions.

## REFERENCES

Ada, N., Ethirajan, M., Kumar, A., Vimal, K. E. K., Nadeem, S. P., Kazancoglu, Y., & Kandasamy, J. (2021). Blockchain technology for enhancing traceability and efficiency in automobile supply chain—a case study. *Sustainability (Switzerland), 13*(24), Article 13667.

Agi, M. A. N., & Jha, A. K. (2022). Blockchain technology in the supply chain: An integrated theoretical perspective of organizational adoption. *International Journal of Production Economics, 247*, Article 108458.

Agrawal, R. (2020). How CropData is enabling small farmers in india dream big. *Microsoft*. Retrieved 3 July from https://news.microsoft.com/en-in/features/how-cropdata-is-enabling-small-farmers-in-india-dream-big/

Ahluwalia, S., Mahto, R. V., & Guerrero, M. (2020). Blockchain technology and startup financing: A transaction cost economics perspective. *Technological Forecasting and Social Change, 151*, Article 119854.

Akinradewo, O., Aigbavboa, C., Oke, A., & Edwards, D. (2023). A roadmap for present focus and future trends of blockchain technology in the built environment. *African Journal of Science, Technology, Innovation and Development, 15*(2), 153–165. https://doi.org/10.1080/20421338.2022.2046249

Ali, I., Arslan, A., Chowdhury, M., Khan, Z., & Tarba, S. Y. (2022). Reimagining global food value chains through effective resilience to COVID-19 shocks and similar future events: A dynamic capability perspective. *Journal of Business Research, 141*, 1–12. https://doi.org/10.1016/j.jbusres.2021.12.006

Ali, I., Nguyen, N. D. K., & Gupta, S. (2023). A multi-disciplinary review of enablers and barriers to Cloud ERP implementation and innovation outcomes. *Journal of Enterprise Information Management, ahead-of-print* (ahead-of-print). https://doi.org/10.1108/JEIM-08-2022-0273

Alimohammadlou, M., & Alinejad, S. (2023). Challenges of blockchain implementation in SMEs' supply chains: An integrated IT2F-BWM and IT2F-DEMATEL method. *Electronic Commerce Research*. https://doi.org/10.1007/s10660-023-09696-3

Alqarni, M. A., Alkatheiri, M. S., Chauhdary, S. H., & Saleem, S. (2023). Use of blockchain-based smart contracts in logistics and supply chains [Article]. *Electronics (Switzerland), 12*(6), Article 1340.

Arunmozhi, M., Venkatesh, V. G., Arisian, S., Shi, Y., & Raja Sreedharan, V. (2022). Application of blockchain and smart contracts in autonomous vehicle supply chains: An experimental design. *Transportation Research Part E: Logistics and Transportation Review, 165*, Article 102864. https://doi.org/10.1016/j.tre.2022.102864

Australian Dairy Farmers. (2020). Blockchain technology to give aussie dairy competitive edge. *Australian Dairy Farmers*. Retrieved 4 July from https://australiandairyfarmers.com.au/mediareleasespost/blockchain-technology-to-give-aussie-dairy-competitive-edge/

Australian Dairy Farmers. (2021). Australian dairy traceability: Implementation guideline v1.0. *Australian Dairy Farmers*. Retrieved 4 July from https://australiandairyfarmers.com.au/wp-content/uploads/2021/09/Dairy-Traceability-Implementation-Guideline-Sept-2021_v01.pdf

Australian Dairy Farmers. (2022a). Annual report 21/22: Delivering for dairy farmers. *Australian Dairy Farmers*. Retrieved 4 July from https://australiandairyfarmers.com.au/media-corner/annual-reports/

Australian Dairy Farmers. (2022b). Blockchain and real time payment system. *Australian Dairy Farmers*. Retrieved 4 July from https://australiandairyfarmers.com.au/blockchain-and-real-time-payment-system/

Australian Trade and Investment Commission. (2021). Powerledger: The world's first 'New Energy' trading platform. *Australian Trade and Investment Commission*. Retrieved 5 July from www.austrade.gov.au/news/success-stories/powerledger-the-world-s-first-new-energy-trading-platform

Baysal, M. V., Özcan-Top, Ö., & Betin-Can, A. (2023). Blockchain technology applications in the health domain: A multivocal literature review. *Journal of Supercomputing, 79*(3), 3112–3156.

Bhubalan, K., Tamothran, A. M., Kee, S. H., Foong, S. Y., Lam, S. S., Ganeson, K., Vigneswari, S., Amirul, A.-A., & Ramakrishna, S. (2022). Leveraging blockchain concepts as water-markers of plastics for sustainable waste management in progressing circular economy. *Environmental Research, 213*, 113631. https://doi.org/https://doi.org/10.1016/j.envres.2022.113631

Biswas, D., Jalali, H., Ansaripoor, A. H., & De Giovanni, P. (2023). Traceability vs. sustainability in supply chains: The implications of blockchain. *European Journal of Operational Research, 305*(1), 128–147. https://doi.org/10.1016/j.ejor.2022.05.034

Biswas, M., Das, D., Banerjee, S., Mukherjee, A., Al-Numay, W., Biswas, U., & Zhang, Y. (2023). Blockchain-enabled communication framework for secure and trustworthy internet of vehicles. *Sustainability (Switzerland), 15*(12), Article 9399. https://doi.org/10.3390/su15129399

Boutkhoum, O., Hanine, M., Nabil, M., Barakaz, F. E. L., Lee, E., Rustam, F., & Ashraf, I. (2021). Analysis and evaluation of barriers influencing blockchain implementation in moroccan sustainable supply chain management: An integrated IFAHP-DEMATEL framework. *Mathematics, 9*(14), Article 1601.

Bunge, A. C., Wood, A., Halloran, A., & Gordon, L. J. (2022). A systematic scoping review of the sustainability of vertical farming, plant-based alternatives, food delivery services and blockchain in food systems. *Nature Food*, *3*(11), 933–941. https://doi.org/10.1038/s43016-022-00622-8

Cai, C., Hao, X., Wang, K., & Dong, X. (2023). The impact of perceived benefits on blockchain adoption in supply chain management. *Sustainability (Switzerland)*, *15*(8), Article 6634.

Centobelli, P., Cerchione, R., Vecchio, P. D., Oropallo, E., & Secundo, G. (2022). Blockchain technology for bridging trust, traceability and transparency in circular supply chain. *Information and Management*, *59*(7), Article 103508.

Chang, S. E., Chen, Y. C., & Lu, M. F. (2019). Supply chain re-engineering using blockchain technology: A case of smart contract based tracking process. *Technological Forecasting and Social Change*, *144*, 1–11.

Chaudhuri, A., Bhatia, M. S., Kayikci, Y., Fernandes, K. J., & Fosso-Wamba, S. (2021). Improving social sustainability and reducing supply chain risks through blockchain implementation: Role of outcome and behavioural mechanisms. *Annals of Operations Research*. https://doi.org/10.1007/s10479-021-04307-6

Chaudhuri, A., Bhatia, M. S., Subramanian, N., Kayikci, Y., & Dora, M. (2022). Socio-technical capabilities for blockchain implementation by service providers: Multiple case study of projects with transaction time reduction and quality improvement objectives. *Production Planning and Control*. https://doi.org/10.1080/09537287.2022.2128865

Chittipaka, V., Kumar, S., Sivarajah, U., Bowden, J. L. H., & Baral, M. M. (2022). Blockchain technology for supply chains operating in emerging markets: An empirical examination of technology-organization-environment (TOE) framework. *Annals of Operations Research*, *327*(1), 465–492.

Choi, D., Chung, C. Y., Seyha, T., & Young, J. (2020). Factors affecting organizations' resistance to the adoption of blockchain technology in supply networks. *Sustainability (Switzerland)*, *12*(21), 1–37, Article 8882.

Choi, T.-M., & Siqin, T. (2022). Blockchain in logistics and production from Blockchain 1.0 to Blockchain 5.0: An intra-inter-organizational framework. *Transportation Research Part E: Logistics and Transportation Review*, *160*, 102653. https://doi.org/https://doi.org/10.1016/j.tre.2022.102653

Choi, T. M. (2022). Achieving economic sustainability: Operations research for risk analysis and optimization problems in the blockchain era. *Annals of Operations Research*. https://doi.org/10.1007/s10479-021-04394-5

Chung, S., Kim, K., Lee, C. H., & Oh, W. (2023). Interdependence between online peer-to-peer lending and cryptocurrency markets and its effects on financial inclusion. *Production and Operations Management*, *32*(6), 1939–1957. https://doi.org/10.1111/poms.13950

CIO. (2020). With Microsoft Cloud, CropData helping Indian farmers earn better. *CIO*. Retrieved 3 July from https://cio.economictimes.indiatimes.com/news/cloud-computing/with-microsoft-cloud-cropdata-helping-indian-farmers-earn-better/78783358

Circulor. (2022a). Economies are demanding supply chain transparency. The good news—it's possible today. *Circulor*. Retrieved 6 July from www.circulor.com/transparency-is-possible-today

Circulor. (2022b). Making mining more environmentally and socially responsible with traceability. *Circulor*. Retrieved 7 July from www.circulor.com/environmentally-socially-responsible-mining

Circulor. (2023a). About us. *Circulor*. Retrieved 6 July from www.circulor.com/about-us

Circulor. (2023b). The hidden cost of gemstone mining. *Circulor*. Retrieved 7 July from www.circulor.com/gemstones

Circulor. (2023c). Traceability is the bedrock to transparent, sustainable and secure supply chains. *Circulor*. Retrieved 7 July from www.circulor.com/traceability-apathtosustainability

Circulor. (2023d). Transparency ensures the ESG and growth of Europe's critical raw materials economy. *Circulor*. Retrieved 7 July from www.circulor.com/transparency-ensures-the-esg-and-growth-of-europe-critical-raw-materials-economy

Circulor. (2023e). Why start now? In your material traceability journey. *Circulor*. Retrieved 6 July from www.circulor.com/why-start-now-in-your-material-traceability-journey

Cole, R., Stevenson, M., & Aitken, J. (2019). Blockchain technology: Implications for operations and supply chain management. *Supply Chain Management*, *24*(4), 469–483.

Coles, B. (2020). Blockchain technology to project Australia's dairy supply chain. *MHD Supply Chain*. Retrieved 4 July from https://mhdsupplychain.com.au/2020/09/01/blockchain-technology-to-protect-australias-dairy-supply-chain/

Colquhoun, L. (2020). Milking with blockchain. *CDO Trends*. Retrieved 4 July from www.cdotrends.com/story/14975/milking-blockchain

CropData. (2020). Agriota—Connecting buyers directly with farmlands. *CropData*. Retrieved 3 July from https://cropdata.in/agriota.html

Dal Mas, F., Massaro, M., Ndou, V., & Raguseo, E. (2023). Blockchain technologies for sustainability in the agrifood sector: A literature review of academic research and business perspectives. *Technological Forecasting and Social Change*, *187*, Article 122155. https://doi.org/10.1016/j.techfore.2022.122155

Daniel, P. (2022). Australian dairy farmers adding milk to the blockchain. *CoinCulture*. Retrieved 4 July from https://coinculture.com/au/business/aussie-dairy-farmers-adding-milk-blockchain/

de Boissieu, E., Kondrateva, G., Baudier, P., & Ammi, C. (2021). The use of blockchain in the luxury industry: Supply chains and the traceability of goods. *Journal of Enterprise Information Management*, *34*(5), 1318–1338.

Di Vaio, A., & Varriale, L. (2020). Blockchain technology in supply chain management for sustainable performance: Evidence from the airport industry. *International Journal of Information Management*, *52*, 102014. https://doi.org/https://doi.org/10.1016/j.ijinfomgt.2019.09.010

Donkin, C. (2020). Vietnam operators poised for mobile money move. *Mobile World Live*. Retrieved 7 July from www.mobileworldlive.com/featured-content/money-home-banner/vietnam-operators-poised-for-mobile-money-move/

Dutta, P., Choi, T. M., Somani, S., & Butala, R. (2020). Blockchain technology in supply chain operations: Applications, challenges and research opportunities. *Transportation Research Part E: Logistics and Transportation Review*, *142*, Article 102067.

Energy Matters. (2022). Powerledger: User control over consumption and solar generation. *Energy Matters*. Retrieved 5 July from www.energymatters.com.au/renewable-news/powerledger-user-control-over-consumption-and-solar-generation/

Esmaeilian, B., Sarkis, J., Lewis, K., & Behdad, S. (2020). Blockchain for the future of sustainable supply chain management in Industry 4.0. *Resources, Conservation and Recycling*, *163*, Article 105064. https://doi.org/10.1016/j.resconrec.2020.105064

Fatorachian, H., & Kazemi, H. (2018). A critical investigation of Industry 4.0 in manufacturing: Theoretical operationalisation framework. *Production Planning & Control*, *29*(8), 633–644.

Feng, H., Wang, X., Duan, Y., Zhang, J., & Zhang, X. (2020). Applying blockchain technology to improve agri-food traceability: A review of development methods, benefits and challenges. *Journal of Cleaner Production*, *260*, 121031. https://doi.org/https://doi.org/10.1016/j.jclepro.2020.121031

Foti, M., & Vavalis, M. (2021). What blockchain can do for power grids? *Blockchain: Research and Applications*, *2*(1), 100008. https://doi.org/https://doi.org/10.1016/j.bcra.2021.100008

Friedman, N., & Ormiston, J. (2022). Blockchain as a sustainability-oriented innovation?: Opportunities for and resistance to blockchain technology as a driver of sustainability in global food supply chains. *Technological Forecasting and Social Change*, *175*, 121403. https://doi.org/10.1016/j.techfore.2021.121403

García-Monleón, F., Erdmann, A., & Arilla, R. (2023). A value-based approach to the adoption of cryptocurrencies. *Journal of Innovation and Knowledge*, *8*(2), Article 100342. https://doi.org/10.1016/j.jik.2023.100342

Gazzola, P., Pavione, E., Barge, A., & Fassio, F. (2023). Using the transparency of supply chain powered by blockchain to improve sustainability relationships with stakeholders in the food sector: The case study of Lavazza. *Sustainability (Switzerland)*, *15*(10), Article 7884.

González-Benito, J., & González-Benito, Ó. (2006). The role of stakeholder pressure and managerial values in the implementation of environmental logistics practices. *International Journal of Production Research*, *44*(7), 1353–1373. https://doi.org/10.1080/00207540500435199

Government House Western Australia. (2020). Governor visits WA company Power Ledger—Developers of world-first blockchain-enabled energy trading platform. *Government House Western Australia*. Retrieved 5 July from https://govhouse.wa.gov.au/2020/07/governor-visits-wa-company-power-ledger-developers-of-world-first-blockchain-enabled-energy-trading-platform/

Hallioui, A., Herrou, B., Santos, R. S., Katina, P. F., & Egbue, O. (2022). Systems-based approach to contemporary business management: An enabler of business sustainability in a context of Industry 4.0, circular economy, competitiveness and diverse stakeholders. *Journal of Cleaner Production*, *373*, Article 133819. https://doi.org/10.1016/j.jclepro.2022.133819

Han, X., & Rani, P. (2022). Evaluate the barriers of blockchain technology adoption in sustainable supply chain management in the manufacturing sector using a novel Pythagorean fuzzy-CRITIC-CoCoSo approach. *Operations Management Research*, *15*(3–4), 725–742.

Havidz, S. A. H., Calvilus, I. C., & Angelita, Z. (2023). COVID-19 full vaccination and blockchain size: Empirical evidence from the cryptocurrency market. *EuroMed Journal of Business*, *18*(2), 229–247. https://doi.org/10.1108/EMJB-12-2021-0200

Henderson, J. (2019). Viettel creates digital subsidiary in Vietnam. *ChannelAsia*. Retrieved 7 July from www.channelasia.tech/article/664176/viettel-creates-digital-subsidiary-vietnam/

Hervani, A. A., Nandi, S., Helms, M. M., & Sarkis, J. (2022). A performance measurement framework for socially sustainable and resilient supply chains using environmental goods valuation methods. *Sustainable Production and Consumption*, *30*, 31–52. https://doi.org/10.1016/j.spc.2021.11.026

Hoffmann, C. H. (2021). Blockchain use cases revisited: Micro-lending solutions for retail banking and financial inclusion. *Journal of Systems Science and Information*, *9*(1), 1–15. https://doi.org/10.21078/JSSI-2021-001-15

IBM. (2019). About IBM food trust. *IBM*. Retrieved 1 July from www.ibm.com/downloads/cas/8QABQBDR

IBM. (2020). Focus on food safety. *IBM*. Retrieved 2 July from www.ibm.com/downloads/cas/KWNWD3ZM

IBM. (2023). IBM supply chain intelligence suite: Food trust. *IBM*. Retrieved 2 July from www.ibm.com/products/supply-chain-intelligence-suite/food-trust

Ietto, B., Rabe, J., Muth, R., & Pascucci, F. (2023). Blockchain for citizens' participation in urban planning: The case of the city of Berlin. A value sensitive design approach. *Cities*, *140*, Article 104382. https://doi.org/10.1016/j.cities.2023.104382

Javed, H., Irfan, M., Shehzad, M., Abdul Muqeet, H., Akhter, J., Dagar, V., & Guerrero, J. M. (2022). Recent trends, challenges, and future aspects of P2P energy trading platforms in electrical-based networks considering blockchain T\technology: A roadmap toward environmental sustainability. *Frontiers in Energy Research*, *10*, Article 810395. https://doi.org/10.3389/fenrg.2022.810395

Joshi, P., Tewari, V., Kumar, S., & Singh, A. (2023). Blockchain technology for sustainable development: A systematic literature review. *Journal of Global Operations and Strategic Sourcing*. https://doi.org/10.1108/JGOSS-06-2022-0054

Karakas, S., Acar, A. Z., & Kucukaltan, B. (2021). Blockchain adoption in logistics and supply chain: A literature review and research agenda. *International Journal of Production Research*. https://doi.org/10.1080/00207543.2021.2012613

Kashem, M. A., Shamsuddoha, M., Nasir, T., & Chowdhury, A. A. (2023). The role of artificial intelligence and blockchain technologies in sustainable tourism in the Middle East. *Worldwide Hospitality and Tourism Themes*, *15*(2), 178–191. https://doi.org/10.1108/WHATT-10-2022-0116

Khan, S. A., Mubarik, M. S., Kusi-Sarpong, S., Gupta, H., Zaman, S. I., & Mubarik, M. (2022). Blockchain technologies as enablers of supply chain mapping for sustainable supply chains. *Business Strategy and the Environment*, *31*(8), 3742–3756.

Khan, S. A. R., Razzaq, A., Yu, Z., & Miller, S. (2021). Industry 4.0 and circular economy practices: A new era business strategies for environmental sustainability. *Business Strategy and the Environment*, *30*(8), 4001–4014. https://doi.org/10.1002/bse.2853

Kholaif, M. M. N. H. K., Xiao, M., & Hamdy, A. (2023). COVID-19's effect on green supply chains and environmental sustainability; innovative technologies moderation. *Journal of Cleaner Production*, *406*, Article 137083. https://doi.org/10.1016/j.jclepro.2023.137083

Kiu, M. S., Chia, F. C., & Wong, P. F. (2022). Exploring the potentials of blockchain application in construction industry: A systematic review. *International Journal of Construction Management*, *22*(15), 2931–2940.

Köhler, S., Bager, S., & Pizzol, M. (2022). Sustainability standards and blockchain in agro-food supply chains: Synergies and conflicts. *Technological Forecasting and Social Change*, *185*, Article 122094.

Komulainen, R., & Nätti, S. (2023). Barriers to blockchain adoption: Empirical observations from securities services value network. *Journal of Business Research*, *159*, Article 113714. https://doi.org/10.1016/j.jbusres.2023.113714

Kumar, M., Choubey, V. K., Raut, R. D., & Jagtap, S. (2023). Enablers to achieve zero hunger through IoT and blockchain technology and transform the green food supply chain systems. *Journal of Cleaner Production*, *405*, Article 136894.

Kumar Singh, R., Mishra, R., Gupta, S., & Mukherjee, A. A. (2023). Blockchain applications for secured and resilient supply chains: A systematic literature review and future research agenda. *Computers and Industrial Engineering*, *175*, Article 108854.

Kuzior, A., & Sira, M. (2022). A bibliometric analysis of blockchain technology research using VOSviewer. *Sustainability (Switzerland)*, *14*(13), Article 8206.

Li, H., Gao, P., Zhan, Y., & Tan, M. (2022). Blockchain technology empowers telecom network operation. *China Communications*, *19*(1), 274–283.

Li, J., Han, D., Wu, Z., Wang, J., Li, K. C., & Castiglione, A. (2023). A novel system for medical equipment supply chain traceability based on alliance chain and attribute and role access control. *Future Generation Computer Systems*, *142*, 195–211.

Li, N., Gu, Z., Albasher, G., Alsultan, N., & Fatemah, A. (2023). Nexus of financial management, blockchain, and natural resources: Comparing the impact on environmental sustainability and resource productivity. *Resources Policy*, *83*, Article 103730. https://doi.org/10.1016/j.resourpol.2023.103730

Li, X., Wang, W., & Gu, H. (2023). Redesign of sharing charging system for electric vehicles using blockchain technology. *Journal of Cleaner Production*, *415*, Article 137775. https://doi.org/10.1016/j.jclepro.2023.137775

Liu, H., Crespo, R. G., & Martínez, O. S. (2020). Enhancing privacy and data security across healthcare applications using blockchain and distributed ledger concepts. *Healthcare (Switzerland)*, *8*(3), Article 243.

Liu, X., Yang, Y., Jiang, Y., Fu, Y., Zhong, R. Y., Li, M., & Huang, G. Q. (2023). Data-driven ESG assessment for blockchain services: A comparative study in textiles and apparel industry. *Resources, Conservation and Recycling*, *190*, Article 106837. https://doi.org/10.1016/j.resconrec.2022.106837

Mangla, S. K., Kazancoglu, Y., Ekinci, E., Liu, M., Özbiltekin, M., & Sezer, M. D. (2021). Using system dynamics to analyze the societal impacts of blockchain technology in milk supply chainsrefer. *Transportation Research Part E: Logistics and Transportation Review*, *149*, Article 102289. https://doi.org/10.1016/j.tre.2021.102289

Manzoor, R., Sahay, B. S., & Singh, S. K. (2022). Blockchain technology in supply chain management: An organizational theoretic overview and research agenda. *Annals of Operations Research*, 1–48.

Massaro, M., Dal Mas, F., Chiappetta Jabbour, C. J., & Bagnoli, C. (2020). Crypto-economy and new sustainable business models: Reflections and projections using a case study analysis. *Corporate Social Responsibility and Environmental Management*, *27*(5), 2150–2160. https://doi.org/10.1002/csr.1954

Mercuri, F., della Corte, G., & Ricci, F. (2021). Blockchain technology and sustainable business models: A case study of devoleum. *Sustainability (Switzerland)*, *13*(10), Article 5619.

Mishra, R., Singh, R., Kumar, S., Mangla, S., & Kumar, V. (2023). Critical success factors of Blockchain technology adoption for sustainable and resilient operations in the banking industry during an uncertain business environment. *Electronic Commerce Research*, 1–35. https://doi.org/10.1007/s10660-023-09707-3

Mohamed, S. K., Haddad, S., Barakat, M., & Rosi, B. (2023). Blockchain technology adoption for improved environmental supply chain performance: The mediation effect of supply chain resilience, customer integration, and green customer information sharing. *Sustainability (Switzerland)*, *15*(10), Article 7909. https://doi.org/10.3390/su15107909

Moises, H. (2022). HSBC Vietnam and Viettel teams up for sustainable data center project. *W. Media*. Retrieved 7 July from https://w.media/hsbc-vietnam-and-viettel-teams-up-for-sustainable-data-center-project/

Mukherjee, A. A., Singh, R. K., Mishra, R., & Bag, S. (2021). Application of blockchain technology for sustainability development in agricultural supply chain: Justification framework. *Operations Management Research*, *15*(1), 46–61.

Nagariya, R., Mukherjee, S., Baral, M. M., Patel, B. S., & Venkataiah, C. (2022). The Challenges of blockchain technology adoption in the agro-based industries. *International Journal of Mathematical, Engineering and Management Sciences*, *7*(6), 949–963.

Naik, A. R. (2020). Can this agritech startup's cross-border blockchain marketplace fix India's farm income drought? *Inc42*. Retrieved 3 July from https://inc42.com/startups/can-cropdatas-blockchain-marketplace-fix-indias-farm-income-drought/

Nakamoto, S. (2008). Bitcoin: A peer-to-peer electronic cash system. Retrieved 2 July from https://bitcoin.org/bitcoin.pdf

Nandi, S., Hervani, A. A., Helms, M. M., & Sarkis, J. (2023). Conceptualising circular economy performance with non-traditional valuation methods: Lessons for a post-pandemic recovery. *International Journal of Logistics Research and Applications*, *26*(6), 662–682. https://doi.org/10.1080/13675567.2021.1974365

Nazam, M., Hashim, M., Nută, F. M., Yao, L., Zia, M. A., Malik, M. Y., Usman, M., & Dimen, L. (2022). Devising a mechanism for analyzing the barriers of blockchain adoption in the textile supply chain: A sustainable business perspective. *Sustainability (Switzerland)*, *14*(23), Article 16159.

Niu, B., Shen, Z., & Xie, F. (2021). The value of blockchain and agricultural supply chain parties' participation confronting random bacteria pollution. *Journal of Cleaner Production*, *319*, Article 128579. https://doi.org/10.1016/j.jclepro.2021.128579

Nobanee, H., Hamadi, F. Y. A., Abdulaziz, F. A., Abukarsh, L. S., Alqahtani, A. F., Alsubaey, S. K., Alqahtani, S. M., & Almansoori, H. A. (2021). A bibliometric analysis of sustainability and risk management. *Sustainability (Switzerland)*, *13*(6), Article 3277. https://doi.org/10.3390/su13063277

OECD. (2022). Environmental impact of digital assets: Crypto-asset mining and DLT consensus mechanism. *OECD Business and Finance Policy Papers*. Retrieved 3 July from www.oecd-ilibrary.org/deliver/8d834684-en.pdf

Oliveira, T., Martins, R., Sarker, S., Thomas, M., & Popovič, A. (2019). Understanding SaaS adoption: The moderating impact of the environment context. *International Journal of Information Management*, *49*, 1–12.

Oracle. (2023). Blockchain platform service. *Oracle*. Retrieved 2023 from www.oracle.com/au/blockchain/cloud-platform/

Paglietti, M. C., & Rabitti, M. (2022). A matter of time. Digital-financial consumers' vulnerability in the retail payments market. *European Business Law Review*, *33*(4), 581–606.

Pandey, V., Pant, M., & Snasel, V. (2022). Blockchain technology in food supply chains: Review and bibliometric analysis. *Technology in Society*, *69*, Article 101954.

Park, A., & Li, H. (2021). The effect of blockchain technology on supply chain sustainability performances. *Sustainability*, *13*(4), 1726.

Parmentola, A., Petrillo, A., Tutore, I., & De Felice, F. (2022). Is blockchain able to enhance environmental sustainability? A systematic review and research agenda from the perspective of sustainable development goals (SDGs). *Business Strategy and the Environment*, *31*(1), 194–217. https://doi.org/10.1002/bse.2882

Patil, K., Ojha, D., Struckell, E. M., & Patel, P. C. (2023). Behavioral drivers of blockchain assimilation in supply chains—A social network theory perspective. *Technological Forecasting and Social Change*, *192*, 122578. https://doi.org/10.1016/j.techfore.2023.122578

Patrick, A. (2023). Power ledger-linked crypto exchange files for bankruptcy. *Financial Review*. Retrieved 5 July from www.afr.com/technology/power-ledger-linked-crypto-exchange-files-for-bankruptcy-20230509-p5d6yj

Philsoophian, M., Akhavan, P., & Namvar, M. (2022). The mediating role of blockchain technology in improvement of knowledge sharing for supply chain management. *Management Decision*, *60*(3), 784–805.

Polas, M. R. H., Kabir, A. I., Sohel-Uz-Zaman, A. S. M., Karim, R., & Tabash, M. I. (2022). Blockchain technology as a game changer for green innovation: Green entrepreneurship as a roadmap to green economic sustainability in Peru. *Journal of Open Innovation: Technology, Market, and Complexity*, *8*(2), Article 62. https://doi.org/10.3390/joitmc8020062

Powerledger. (2019). Powerledger: About. *Powerledger*. Retrieved 5 July from www.powerledger.io/company/about

Powerledger. (2023). Global Blockchain Business Council—2022 Annual Report. *Powerledger*. Retrieved 6 July from www.powerledger.io/media/global-blockchain-business-council-2022-annual-report

Quayson, M., Bai, C., & Sarkis, J. (2021). Technology for social good foundations: A perspective from the smallholder farmer in sustainable supply chains. *IEEE Transactions on Engineering Management*, *68*(3), 894–898, Article 9113288. https://doi.org/10.1109/TEM.2020.2996003

Raghavan, K. S., Singh, R., Kolluri, S., Pradeep, S., & Pandey, M. (2021). Building a fortress of blockchain to protect farmers and livelihoods. *United Nations Development Programme*. Retrieved 3 July from www.undp.org/india/blog/building-fortress-blockchain-protect-farmers-and-livelihoods

Rejeb, A., Appolloni, A., Rejeb, K., Treiblmaier, H., Iranmanesh, M., & Keogh, J. G. (2023). The role of blockchain technology in the transition toward the circular economy: Findings from a systematic literature review. *Resources, Conservation and Recycling Advances*, *17*, Article 200126. https://doi.org/10.1016/j.rcradv.2022.200126

Rim, H. (2023). Decentralized identity (DID): New technology adoption and diffusion in South Korea. *Transforming Government: People, Process and Policy*, *17*(2), 251–270. https://doi.org/10.1108/TG-11-2021-0189

RMIT Vietnam. (2021). Mobile money: The newcomer in financial services. *RMIT Vietnam*. Retrieved 7 July from www.rmit.edu.vn/news/all-news/2021/apr/mobile-money-the-newcomer-in-financial-services

Ronaghi, M. H., & Mosakhani, M. (2022). The effects of blockchain technology adoption on business ethics and social sustainability: Evidence from the Middle East. *Environment, Development and Sustainability*, *24*(5), 6834–6859. https://doi.org/10.1007/s10668-021-01729-x

Sadeghi, M., Mahmoudi, A., Deng, X., & Luo, X. (2023). Prioritizing requirements for implementing blockchain technology in construction supply chain based on circular economy: Fuzzy ordinal priority approach. *International Journal of Environmental Science and Technology*, *20*(5), 4991–5012. https://doi.org/10.1007/s13762-022-04298-2

Sahoo, S., Kumar, S., Sivarajah, U., Lim, W. M., Westland, J. C., & Kumar, A. (2022). Blockchain for sustainable supply chain management: Trends and ways forward. *Electronic Commerce Research*, 1–56.

Sharma, A., Sharma, A., Singh, R. K., & Bhatia, T. (2023). Blockchain adoption in agri-food supply chain management: An empirical study of the main drivers using extended UTAUT. *Business Process Management Journal*, *29*(3), 737–756.

Sriyono, E. (2020). Digitizing water management: Toward the innovative use of blockchain technologies to address sustainability. *Cogent Engineering*, *7*(1), 1769366. https://doi.org/10.1080/23311916.2020.1769366

Stuit, A., Brockington, D., & Corbera, E. (2022). Smart, commodified and encoded: Blockchain technology for environmental sustainability and nature conservation. *Conservation and Society*, *20*(1), 12–23. https://doi.org/10.4103/cs.cs_41_21

Sun, N., Zhang, Y., & Liu, Y. (2023). A universal privacy-preserving multi-blockchain aggregated identity scheme. *Applied Sciences (Switzerland)*, *13*(6), Article 3806. https://doi.org/10.3390/app13063806

Tan, T. M., & Salo, J. (2023). Ethical marketing in the blockchain-based sharing economy: Theoretical integration and guiding insights. *Journal of Business Ethics*, *183*(4), 1113–1140. https://doi.org/10.1007/s10551-021-05015-8

Tan, Y., Huang, X., & Li, W. (2023). Does blockchain-based traceability system guarantee information authenticity? An evolutionary game approach. *International Journal of Production Economics*, *264*, Article 108974. https://doi.org/10.1016/j.ijpe.2023.108974

Tang, Y. M., Chau, K. Y., Fatima, A., & Waqas, M. (2022). Industry 4.0 technology and circular economy practices: Business management strategies for environmental sustainability. *Environmental Science and Pollution Research*, *29*(33), 49752–49769. https://doi.org/10.1007/s11356-022-19081-6

Tech, V. H. (2022). Viettel won big reward at Vietnam Digital Awards 2021. *Viettel High Tech*. Retrieved 7 July from https://viettelhightech.com.vn/en/viettel-won-big-reward-at-vietnam-digital-awards-2021

Teodorescu, M., & Korchagina, E. (2021). Applying blockchain in the modern supply chain management: Its implication on open innovation. *Journal of Open Innovation: Technology, Market, and Complexity*, *7*(1), Article 80.

Tiscini, R., Testarmata, S., Ciaburri, M., & Ferrari, E. (2020). The blockchain as a sustainable business model innovation. *Management Decision*, *58*(8), 1621–1642. https://doi.org/10.1108/MD-09-2019-1281

Treiblmaier, H. (2018). The impact of the blockchain on the supply chain: A theory-based research framework and a call for action. *Supply Chain Management*, *23*(6), 545–559.

Tri, D. (2023). Viettel, Intel to cooperate in digital services, 5G, cloud. *The Investor*. Retrieved 7 July from https://theinvestor.vn/viettel-intel-to-cooperate-in-digital-services-5g-cloud-d3881.html

Varavallo, G., Caragnano, G., Bertone, F., Vernetti-Prot, L., & Terzo, O. (2022). Traceability platform based on green blockchain: An application case study in dairy supply chain. *Sustainability (Switzerland)*, *14*(6), Article 3321.

Viet Nam News. (2018). Viettel successfully enters blockchain arena. *Viet Nam News*. Retrieved 7 July from https://vietnamnews.vn/economy/468587/viettel-successfully-enters-blockchain-arena.html

Viettel. (2018). Viettel join remittance research team of global telecom blockchain alliance. *Viettel*. Retrieved 7 July from https://international.viettel.vn/news-detail/viettel-join-remittance-research-team-of-global-telecom-blockchain-alliance

Viettel. (2023a). Sustainability perceptions index 2023: Viettel is the only Vietnamese name. *Viettel*. Retrieved 8 July from https://viettel.com.vn/en/news-events/news/sustainability-perceptions-index-2023-viettel-is-the-only-vietnamese-name/

Viettel. (2023b). Viettel continues to push for accessible technology for everyone at MWC 2023. *Viettel*. Retrieved 8 July from https://viettel.com.vn/en/news-events/event/viettel-continues-to-push-for-accessible-technology-for-everyone-at-mwc-2023/

Viettel Group. (2020a). Viettel group helps millions of customers in developing country to stay online through the pandemic. *Viettel Group*. Retrieved 7 July from www.prnewswire.com/in/news-releases/viettel-group-helps-millions-of-customers-in-developing-country-to-stay-online-through-the-pandemic-892734424.html

Viettel Group. (2020b). ViettelPay—the fastest growing FinTech app in Vietnam—expects triple regular users in 2020. *Viettel Group*. Retrieved 7 July from www.prnewswire.com/news-releases/viettelpay—the-fastest-growing-fintech-app-in-vietnam—expects-triple-regular-users-in-2020–301012196.html

Viettel Group. (2022). Viettel's financial ecosystem named winner of 2022 excellence awards. *Viettel Group*. Retrieved 7 July from https://en.prnasia.com/releases/apac/viettel-s-financial-ecosystem-named-winner-of-2022-excellence-awards-377891.shtml

Vitasek, K., Bayliss, J., Owen, L., & Srivastava, N. (2022). How Walmart Canada uses blockchain to solve supply-chain challenges. *Harvard Business Review*. Retrieved 3 July from https://hbr.org/2022/01/how-walmart-canada-uses-blockchain-to-solve-supply-chain-challenges

Wan, P. K., Huang, L., & Holtskog, H. (2020). Blockchain-enabled information sharing within a supply chain: A systematic literature review. *IEEE Access*, *8*, 49645–49656, Article 9032112.

Wang, B., Lin, Z., Wang, M., Wang, F., Xiangli, P., & Li, Z. (2023). Applying blockchain technology to ensure compliance with sustainability standards in the PPE multi-tier supply chain. *International Journal of Production Research*, *61*(14), 4934–4950. https://doi.org/10.1080/00207543.2022.2025944

Wang, M., Wang, B., & Abareshi, A. (2020). Blockchain technology and its role in enhancing supply chain integration capability and reducing carbon emission: A conceptual framework. *Sustainability (Switzerland)*, *12*(24), 1–17, Article 10550.

Wang, M., Wu, Y., Chen, B., & Evans, M. (2021). Blockchain and supply chain management: A new paradigm for supply chain integration and collaboration. *Operations and Supply Chain Management*, *14*(1), 111–122.

Wei, J., Yi, X., Yang, X., & Liu, Y. (2023). Blockchain-based design of a government incentive mechanism for manufacturing supply chain data governance. *Sustainability (Switzerland)*, *15*(8), Article 6968. https://doi.org/10.3390/su15086968

Witt, J., & Schoop, M. (2023). Blockchain technology in e-business value chains. *Electronic Markets*, *33*(1), Article 15.

Wong, P. F., Chia, F. C., Kiu, M. S., & Lou, E. C. W. (2022). Potential integration of blockchain technology into smart sustainable city (SSC) developments: A systematic review. *Smart and Sustainable Built Environment*, *11*(3), 559–574. https://doi.org/10.1108/SASBE-09-2020-0140

Wong, S., Yeung, J. K. W., Lau, Y. Y., & Kawasaki, T. (2023). A case study of how maersk adopts cloud-based blockchain integrated with machine learning for sustainable practices. *Sustainability (Switzerland), 15*(9), Article 7305. https://doi.org/10.3390/su15097305

Xu, X., Tatge, L., Xu, X., & Liu, Y. (2022). Blockchain applications in the supply chain management in German automotive industry. *Production Planning and Control, 35*(9), 917–931.

Yadlapalli, A., Rahman, S., & Gopal, P. (2022). Blockchain technology implementation challenges in supply chains—evidence from the case studies of multi-stakeholders. *International Journal of Logistics Management, 33*(5), 278–305.

Yaqub, M. Z., & Alsabban, A. (2023). Industry-4.0-enabled digital transformation: Prospects, instruments, challenges, and implications for business strategies. *Sustainability (Switzerland), 15*(11), Article 8553. https://doi.org/10.3390/su15118553

Yontar, E. (2023). The role of blockchain technology in the sustainability of supply chain management: Grey based dematel implementation. *Cleaner Logistics and Supply Chain, 8*, Article 100113.

Yousefi, S., & Mohamadpour Tosarkani, B. (2022). An analytical approach for evaluating the impact of blockchain technology on sustainable supply chain performance. *International Journal of Production Economics, 246*, 108429. https://doi.org/10.1016/j.ijpe.2022.108429

Zhao, L. (2023). Blockchain adoption and contract coordination driven by supplier encroachment and retail service: An analysis from consumers' information traceability awareness. *Technology in Society, 73*, Article 102237.

Zheng, K., Zheng, L. J., Gauthier, J., Zhou, L., Xu, Y., Behl, A., & Zhang, J. Z. (2022). Blockchain technology for enterprise credit information sharing in supply chain finance. *Journal of Innovation & Knowledge, 7*(4), 100256. https://doi.org/https://doi.org/10.1016/j.jik.2022.100256

# 5 Blockchain Revolution

## *Transforming Business Sustainability in the Era of Industry 4.0 and Financial Innovation*

*Ramya H. P. and K. G. Hemalatha*

## 5.1 INTRODUCTION

The digital revolution is an ongoing transformation that has brought substantial changes in the way we network and do business with companies and clients. As a result, traditional business methods and financial services are undergoing significant changes, with new technologies emerging as key drivers of this transformation. The global financial crisis of 1929 and 2007–2008 created extensive awareness among financial service users, leading regulators throughout the world to impose complex regulations on all financial activities. However, these regulations are often seen as impeding economic growth, particularly in developed economies where they can stifle innovation and competition. Blockchain is a groundbreaking invention that has had an impact similar to that of the internet. The innovative blockchain technology has rapidly gained popularity and has brought about significant transformations in a wide range of industries. From financial services to healthcare, supply chain management to education and energy consumption, blockchain has revolutionized numerous business functions. This cutting-edge technology provides secure, immutable, decentralized, and transparent data services, which ensure that all transactions carried out among stakeholders in the business network are swift and low cost. By leveraging blockchain, businesses can significantly increase trust levels and boost overall efficiency. According to Yan and Bellavitis (2020), blockchain has numerous benefits, particularly regarding secure data transactions. The implementation of this technology results in an assurance of data integrity, which renders any unauthorized attempts to manipulate or alter the data futile. This has huge implications for businesses that rely on trust and transparency, as it enables them to build trust with stakeholders in a way that was not possible before.

Over the past three centuries, there have been three industrial revolutions that have had a significant impact on the progress of human civilization. According to economists, currently we are witnessing the new industrial revolution, also known

DOI: 10.1201/9781003438748-6

as Industry 4.0. Each of these revolutions has been characterized by technological advancements that have profoundly affected the course of economic and human development. The financial and banking industry currently experiencing a revolution in digital finance, commonly referred to as the FinTech Revolution. This transformative shift is driven by innovative solutions developed by both start-ups and major players in the technology industry. Given the crucial importance of financial services in economic transactions and sustainability, the rise of FinTech has led to a substantial impact on the financial industry. As noted in the Global FinTech Report (PwC 2019), FinTech significantly disrupts traditional financial services, influencing the strategies of market players. In exploring the convergence of sustainability with emerging fields like FinTech, InsurTech, and blockchain, it is essential to scrutinize the diverse approaches being employed. Many businesses are leveraging these technologies to instill more sustainable practices, such as using blockchain for transparent supply chain tracking in eco-friendly product production. Additionally, InsurTech is being employed to incentivize sustainability, offering lower premiums to those engaging in eco-friendly practices. The integration of these emerging fields with sustainability signals a promising trajectory toward a more sustainable future.

To maintain the reliability of financial systems and prevent money laundering, certain measures need to be implemented to validate and verify financial big data for credibility. With technology advancing rapidly, businesses face a need to adopt innovative solutions to enhance operational efficiency and also contribute to achieving the Sustainability Development Goals (SDGs) globally. Blockchain technology has garnered significant attention as one such solution. Central banks, retail banking and financial institutions, and technological organizations are keenly interested in its vast potential, anticipating transformative effects on the financial industry's structure and operations. Blockchain technology has the power to revolutionize the way we handle transactions and contractual information. With its real-time recording capabilities and shared ledger system, it offers an effective solution to the challenges we face in traditional methods. By adopting this innovative technology, we can streamline processes and enhance transparency, leading to a more secure and efficient way of doing business. Recent advancements in digital technology, particularly the rise of blockchain through the virtual currency Bitcoin, offer innovative solutions for financial reliability. Studies suggest that blockchain not only transforms the essence of banking systems but also has the potential to reshape the entire economic landscape. Blockchain technology can help improve the financial sector. By using security encryption and smart contracts, it can enhance transparency and privacy. Studies show that blockchain technology creates a detailed digital accounting system that reduces the risk of illegal interference.

Blockchain technology has evolved through four stages since 2008, starting with the emergence of cryptocurrencies, notably Bitcoin. The second stage marked a significant shift in financial processes with the introduction of monetary transactions and smart contracts, streamlining operations like mortgages and loans. The third stage witnessed increased integration into digital finance, by improving the capabilities of monetary transactions and smart contracts. The fourth era involves the implementation of industrial decentralized ledger systems across various sectors such as public sector units, tourism, healthcare, logistics and supply chain management, renewable energy, and finance (Trivedi et al. 2021). From its origins as a platform for digital currencies, blockchain technology is now utilized for creating decentralized applications (DApps), providing rapid and decentralized storage with ease of communication across various sectors. The potential of blockchain technology, in conjunction with Industry 4.0 infrastructure, is enormous, contributing to streamlined processes and enhanced efficiency.

The Fourth Industrial Revolution, also commonly referred to as Industry 4.0, represents a significant milestone in the evolution of modern manufacturing and production processes. It was first introduced by the German federal government in 2011 and has since been adopted globally as a game-changing movement. This revolution focuses on improving industrial capabilities through technological applications and digitalization. Industry 4.0 utilizes smart systems that incorporate information technology (IT), allowing for real-time data exchange and updates that are interconnected. It aims to merge digital, physical, and biological technologies using software and programming expertise. One of the key goals of Industry 4.0 is to enable the connection of machines, digital devices, transactions, and stakeholders such as suppliers, buyers, and customers in real time. The Internet of Things (IoT) facilitates this communication to some extent. Industry 4.0 involves the digitization, optimization, and customization of production, automation and adaptation, human–machine interaction (HMI), value-added services and businesses, and automatic data exchange and communication. However, blockchain technology is a new technology that utilizes cryptographic algorithms to ensure the integrity and security of each transaction.

Blockchain technology has been designed to ensure that transaction records can be authenticated reliably. These records are also irreversible, which adds to their credibility. Moreover, blockchain is a decentralized system that does not require unified control over financial transactions and access to record-keeping. This makes it possible to bring together transactions, stakeholders, and machines, which is a unique feature of Industry 4.0. The Internet of Things is one of the important components of Industry 4.0, which involves interconnectivity between electronic and smart devices using unique codes. Blockchain has the potential to secure this aspiration of IoT and bring automation to financial transactions in an eternal and verifiable manner. This can help save time and resources by avoiding inefficiencies (Iansiti & Lakhani 2017).

***Research Question:*** "*How does the integration of blockchain technology contribute to the convergence of business sustainability and Industry 4.0, particularly in addressing environmental challenges and reshaping financial systems?*"

**Research Gap:** While existing literature has extensively explored the importance of blockchain technology in financial services and its potential impact on sustainability, there is a notable research gap in understanding the specific synergies between blockchain, business sustainability, and the Fourth Industrial Revolution. The evolving stages of blockchain evolution, especially its transformative potential in industrial decentralized ledger systems, have not been comprehensively explored in the context of Industry 4.0 and its inferences for sustainable practices.

**Objectives:**

1. To examine the transformative opportunities presented by the evolution of distributed ledger technologies (DLT), focusing on blockchain, in the post-COVID-19 economic landscape.
2. To analyze the relationship between blockchain technology and business sustainability, emphasizing the intrinsic qualities of the blockchain (transparency, immutability, and durability) and their transformative force in fostering sustainability.
3. To investigate how blockchain enhances business sustainability through secure and immutable records, fostering trust and accountability in supply chains, and reducing reliance on intermediaries for cost savings and increased efficiency.
4. To explore the role of blockchain-powered smart contracts in automating and ensuring transparency in sustainable practices, such as renewable energy trading and tracking carbon emissions.
5. To examine the influence of blockchain technology on financial systems, including its role in transforming the structure of financial services, reinventing the banking industry, and addressing modern distributed industry demands.

## 5.2 UNDERSTANDING THE FUNDAMENTALS OF BLOCKCHAIN TECHNOLOGY

In November 2008, a revolutionary electronic cash system was introduced that completely transformed the way financial transactions were conducted. This pioneering technology is based on a distributed ledger system that supports Bitcoin transactions, and it is considered the foundation of today's blockchain technology. Interestingly, the identity of the programmer or group behind the creation of this technology remains a mystery to this day, adding to its enigmatic allure (Nakamoto 2008). Blockchain, a type of distributed ledger technology, facilitates the recording and secure storage of monetary transactions in a transparent and secure manner among multiple parties. Operating on a decentralized network, obviates the need for intermediaries, thereby fostering trust among participants. Each transaction, or "block", is sequentially linked to the preceding one, forming an immutable record of events. The distributed digital ledger system uses technology that records transactions as an unalterable list of blocks shared within a peer-to-peer network. This eliminates reliance on a single server and reduces the risk of failure at a single point. Information within each block is encrypted, and a unique identifier generated through the hashing algorithm links each block to its predecessor (Malhotra et al. 2021).

When a new block of data is added to the ledger, it must be approved by the majority of nodes to be considered valid. This process is called mining, which requires high-performance hardware to verify transactions and add them to the blockchain. When multiple participants (nodes) need to agree on whether a block is valid or not, they use a consensus algorithm. These components work together to reinforce the attributes of blockchain technology, such as security, trust, privacy, and anonymity. Nodes use consensus algorithms to determine the validity of a block. Common algorithms include PoS, PoW, PoET, DPoS, and PBFT. Each block secures transactions through Hash256. This reinforces the security, trust, privacy, and anonymity of blockchain technology (Malhotra et al. 2021).

Mining in cryptocurrency is the process of including new blocks in the chain by solving a complex mathematical problem through large combinations of different computations until the target value is reached. Think of it as trying different combinations to open a safe. Miners on the network compete to find blocks by solving the problem the fastest and adding them to the blockchain. As a reward, they receive new virtual currency. Every block must include a verified proof-of-work by other nodes on the network. This ensures the chain's security and resistance to tampering. The system is designed to limit the amount of cryptocurrency in circulation. Blockchain is built on the principle of simultaneous cooperation and competition within the network, making it a secure and reliable technology.

## 5.3 CHARACTERISTICS OF BLOCKCHAIN

Blockchain technology is an extremely secure and decentralized system that places trust at the forefront of transactions. Its unparalleled transparency and dependability set it apart from other technologies. The profound influence it has on multiple industries is truly exciting to observe (Singh & Kim 2019). The characteristics are summarized in the following:

1. Decentralization: It is used in the blockchain to avoid the issues presented by a centralized architecture. This approach employs the processing capabilities of all participating users in the blockchain network, which reduces latency and eliminates the single point of failure.
2. Immutability: The blockchain has become an essential feature in technology due to its ability to ensure the integrity of transactions by creating immutable ledgers. It's a stark contrast to traditional centralized architectures where databases can easily be altered and trust needs to be established with a third party to guarantee information integrity. With blockchain, the blocks are permanently saved and never changed as long as the participating user continues to maintain the network. This is possible because each block in the distributed ledger relates to the previous block, constituting a chain of blocks that are unchangeable.
3. Transparency: When using a blockchain, all involved parties have access to transaction details, ensuring transparency and eliminating the need for third-party intermediaries.
4. Better security: As we continue to explore emerging technologies, safeguarding sensitive information remains our utmost concern. It's worth noting that leveraging blockchain technology provides the ultimate solution to security concerns. This is because of its public key infrastructure, which ensures the protection of data against malicious attacks. Participants in the blockchain network have complete faith in the integrity and security

features of the consensus mechanism. Moreover, the blockchain effectively eliminates any potential single point of failure, which could wreak havoc on the entire system.

5. Efficiency: Blockchain technology offers a revolutionary way of storing database records among various users involved in the blockchain network. By distributing transactions, it makes it more transparent to verify all the records stored in the database. Compared to the classical centralized architecture, a blockchain is more cost effective, is faster in settlement speed, and provides better risk management. This innovative technology has the potential to transform the way we store and manage data and improve the overall efficiency of various industries (Singh & Kim 2019).

## 5.4 TYPES OF BLOCKCHAIN TECHNOLOGY

Blockchain can be classified into four types.

1. Public Blockchain: Blockchain technology has revolutionized financial transactions by making them accessible and transparent. Public blockchains like Bitcoin and Ethereum provide the means for anyone to verify and participate in the process of reaching consensus. It's an unequivocally democratic system that empowers everyone to be part of the financial ecosystem.
2. Consortium Blockchain: When it comes to blockchain, it is crucial to consider the authority-holding node. This authority is often pre-selected and may include partnerships, such as business-to-business. The data on a blockchain can be either public or private, and the level of decentralization can vary; consortium blockchains, such as Hyperledger and Ethereum, fall under this category.
3. Private Blockchain: Private blockchain restricts nodes from participating and has strict data access management.
4. Hybrid Blockchain: It merges features of private and public blockchain. What sets it apart is its capability to maintain the secrecy of data within the network while enabling verification. Even if a private entity owns the hybrid blockchain, it cannot tamper with transactions. This makes it a safe and reliable solution for organizations and institutions that require security for sensitive information yet want to verify transactions.

The global financial system, tasked with serving billions while managing trillions, faces a myriad of challenges inherent to its ambitious scope. Over time, the sector has grappled with issues ranging from the complexities of multiple stakeholders to persistent delays, excessive paperwork, and the looming threat of data breaches, resulting in substantial annual losses for businesses. Despite significant technological advancements that have addressed some of these challenges, the emergence of FinTech solutions has also introduced new complexities, leaving financial service providers inundated with choices. In the quest for a comprehensive solution capable of tackling these persistent issues, attention has turned to blockchain technology.

Blockchain technology presents a promising avenue for addressing the entrenched problems of the financial sector. The current financial landscape is burdened by the presence of various entities such as regulators, intermediaries, and stock exchanges, all of which contribute to increased costs. However, adopting a decentralized approach facilitated by blockchain technology can streamline processes and mitigate the need for external regulators. Through the use of smart contracts, blockchain offers a means of conducting investor–company interactions securely and efficiently, eliminating the need for intermediaries and reducing associated costs. The allure of blockchain technology lies in its potential to revolutionize financial services by offering a decentralized and secure alternative to traditional systems. Centralization within the financial industry necessitates substantial investments in accounting, database maintenance, security protocols, and intermediary commissions. These recurring expenses contribute to the overall cost burden shouldered by financial service providers. By leveraging blockchain technology, financial institutions stand to streamline operations, reduce costs, and enhance efficiency across the board. As such, blockchain represents a vital tool in addressing the financial challenges of today and paving the way for a more resilient and cost-effective financial ecosystem.

## 5.5 NEED FOR BLOCKCHAIN TECHNOLOGY IN THE FINANCIAL SERVICE SECTOR

Interest in blockchain technology within the financial sector has been steadily increasing, driven by its potential to revolutionize various aspects of finance. Its utilization spans a wide array of applications, including secure money transfers, cross-border payments, identity verification, contractual agreements, trade finance, insurance, smart contracts, auctions, and currency trading. Consequently, the adoption of blockchain in finance has seen remarkable growth.

The 2018 global blockchain survey conducted by Deloitte revealed a significant endorsement of blockchain technology among executives from seven countries. An astounding 74% of respondents acknowledged the presence of a "compelling business case" for implementing blockchain solutions. This endorsement underscores the growing recognition of the value that blockchain can bring as a platform integrating operational processes like supply chain management, customer/channel operations, and service alongside financial processes.

One of the driving forces behind this interest is the potential for blockchain to revolutionize transaction processing. By streamlining operations and reducing the need for multiple handoffs and laborious data entry and reconciliations, blockchain technology offers the promise of replacing conventional transaction processing methods. This transformative potential has captured the imagination of financial industry stakeholders, fueling the rapid adoption and exploration of blockchain solutions across various financial domains.

Blockchain technology, based on a distributed ledger system, ensures the secure and tamper-proof logging of transactions on a global network. It plays a crucial role in maintaining the balance between technology, user data, and privacy amidst the digital revolution. With an emphasis on confidentiality and improved data management, blockchain facilitates transparent and expedited audit processes. This technology collaborates seamlessly with artificial intelligence, particularly in financial services, enabling efficient communication and eliminating the need for recordkeeping in various financial processes. Smart contracts, enabled by blockchain, create legally binding agreements executed with a guarantee once conditions are met, enhancing responsibility and transparency. Additionally, blockchain technology facilitates real-time cash transfers between financial institutions, reducing friction and expediting settlement while offering automation potential for tracking transactions and monitoring payments and deliverables.

In the realm of blockchain technology for financial services, a diverse array of tools and strategies has emerged, each catering to specific needs and challenges encountered within the industry. These tools and methods have evolved to address real-time financial issues efficiently and intelligently. Notably, soft tools such as parity, geth, solc, mtyhx, truffle, infura, and metamask have garnered attention for their effectiveness in enhancing blockchain applications within financial services. Blockchain technology, fundamentally a ledger of financial transactions, has captured the imagination of the financial services industry in recent years. It operates by distributing, publishing, and storing transaction records across multiple locations. Each transaction is recorded in every copy of the ledger through the creation of blocks, ensuring accuracy and transparency. The distributed nature of blockchain renders it virtually immutable and highly secure; altering or falsifying any part of the ledger would necessitate simultaneous changes to every copy, a task exceedingly difficult for hackers to accomplish. Consequently, blockchain instills confidence among commercial partners and facilitates secure, streamlined transactions. Blockchain technology enables the creation and deployment of deterministic smart contracts, which are tamper-proof programs automating business logic. These contracts not only enhance efficiency but also foster trust among parties by ensuring adherence to predefined terms and conditions. Additionally, blockchain offers leading-edge technologies for granular data privacy at every layer of the software stack, enabling selective data sharing within corporate networks. This capability enhances security while promoting collaboration and transparency within financial services domains (Javaid et al. 2022).

## 5.6 INTERRELATIONSHIP BETWEEN BLOCKCHAIN TECHNOLOGY AND SUSTAINABILITY

Blockchain technology offers several essential competencies that make it a promising tool for driving sustainable infrastructure. These include the ability to provide transparency, auditability of data, privacy, value transfer, and process efficiency and automation. Utilizing these competencies, blockchain technology can create more sustainable and efficient infrastructure, as demonstrated by (Navas et al., 2021) in their study comparing the sustainability, transparency, and trustworthiness of traditional and blockchain ecolabels among Generation X and Y consumers. Blockchain technology is based on the principles of decentralized trust and immutable records. Decentralized trust means that transactions are validated by a network of users rather than a centralized authority. This eliminates the need for intermediaries, such as banks, to verify transactions, which can reduce costs and increase efficiency.

Immutable records refer to the fact that once a transaction is recorded on the blockchain, it cannot be altered or deleted. This ensures that the ledger is tamper-proof and provides a high level of security and transparency.

One of the biggest advantages of blockchain technology is that it enables the real transfer of ownership. In the past, it was only possible to copy data via the internet, but blockchain has accelerated the move to an internet of value. This means that intangible or tangible assets such as currencies, shares, infrastructure securities, data, or obligations like contracts can now be exchanged without the need for intermediaries, thanks to the trusted ledger. The elimination of intermediaries can help reduce costs associated with transactions, as well as increasing the speed of transactions. Additionally, blockchain technology can help reduce fraud and corruption, as all transactions are recorded on a public ledger that is immutable and transparent.

When we think about blockchain, the concept of carbon neutrality is not the first thing that comes to mind. We tend to associate blockchain technology with Bitcoin, which is known for being an environmental polluter. The process of validating transactions and maintaining the network consumes vast amounts of energy and emits significant amounts of $CO_2$. However, this is not the case for all applications of blockchain. Depending on the network's architecture and protocols used, blockchain can be deployed in more energy-efficient ways. Private blockchain that uses algorithms like proof-of-authority (PoA) can be set up in a way that does not consume more energy than traditional database solutions.

The potential of blockchain technology in transforming financing for infrastructure projects cannot be overstated. It can significantly lower the cost of capital, enhance transparency, and widen access to finance. Through its ability to track data and information on infrastructure projects, stakeholders can be confident of alignment with sustainability goals. By standardizing data, blockchain platforms can also evaluate asset performance and increase compliance with sustainability or ESG standards. Furthermore, the technology can act as a transaction-enabling infrastructure for new market models that incentivize institutions and consumers to contribute to long-term sustainability. In essence, blockchain technology is a game-changer that can drive incredibly positive changes within industries while simultaneously adapting to the shifting demands of consumers. The role of blockchain in the framework of sustainable infrastructure is envisioned to be far beyond enabling efficient data collection, monitoring, reporting, and steering services. The technology can potentially address the key opportunities in supporting mitigation and adaptation-related activities, particularly in the energy, transport, and agriculture industries.

## 5.7 BLOCKCHAIN TECHNOLOGY IN BUSINESS SUSTAINABILITY: CHALLENGES AND CONSIDERATIONS

Blockchain technology presents a number of opportunities for businesses to enhance sustainability, but it also brings forth several challenges. The implementation of blockchain solutions requires substantial investments in terms of time, resources, and technical expertise. This means that organizations need to allocate sufficient resources to develop and integrate blockchain systems effectively. Another significant challenge revolves around scalability and energy consumption. Some blockchain networks struggle to handle a large number of transactions quickly, which restricts their ability to operate on a larger scale. Additionally, certain blockchain protocols consume a significant amount of energy, raising concerns about their environmental impact. It is important for blockchain developers and users to find solutions to address these scalability and energy consumption issues. This way, they can ensure the viability and widespread adoption of blockchain technology.

Moreover, the successful incorporation of blockchain technology into existing business processes requires the development of appropriate regulatory and legal frameworks. These frameworks should outline guidelines and standards to facilitate the integration and use of blockchain in a secure and compliant manner. By establishing robust regulations and legal frameworks, businesses can navigate potential risks and uncertainties associated with blockchain implementation while ensuring that their operations remain within legal boundaries. The utilization of decentralized ledger technology alongside blockchain automation has fostered the development of a technological blueprint for a payment infrastructure. DLT is regarded as by low transaction costs and a high level of transparency. These features are expected to have a significant impact on the global financial market, reshaping traditional financial systems and revolutionizing the way transactions are conducted worldwide.

Blockchain promotes smart contracts, which in turn helps to increase the efficacy of transactions. Smart contracts are computer programs designed to autonomously execute predetermined actions upon meeting certain conditions. They are built on blockchain technology, which is a decentralized and transparent ledger that ensures trust and immutability. The usage of smart contracts and blockchain technology has the potential to significantly mitigate financial fraud with features such as:

- **Transparency**—Blockchain provides a transparent and immutable record of transactions. This transparency helps prevent fraudulent activities, as every transaction is visible to all participants in the network. Any attempt at tampering with or altering the data becomes highly improbable due to the consensus mechanisms employed by blockchain.
- **Decentralization**—It removes the need for intermediaries and central authorities, which in turn reduces the risk of fraud. Decentralization also promotes trust and reduces the reliance on a single point of failure. By distributing the transaction history across multiple nodes, blockchain ensures that no single individual or entity can manipulate the data for fraudulent purposes.

Additionally, the use of smart contracts automates the execution of predefined actions, which leads to reducing the potential for human error or malicious intent and also eliminating the need for intermediaries. Overall, the combination of transparency, decentralization, and automation provided by blockchain and smart contracts can help mitigate financial fraud and promote a more secure and efficient financial ecosystem. The usage of smart contracts in the stock market can revolutionize the way transactions are conducted. Some potential use cases include:

1. **Efficient Settlements**: It streamlines and automates the settlement process, reducing the time and cost involved in clearing and settling trades.
2. **Improved Transparency**: By recording all transactions on a blockchain, smart contracts can enhance transparency and enable stakeholders to verify and audit transactions easily.
3. **Automated Dividend Distribution**: Smart contracts can automate the distribution of dividends to shareholders based on predefined rules, eliminating the need for manual intervention.
4. **Tokenization of Assets**: Smart contracts can be used to tokenize real-world assets like stocks, allowing for increased liquidity, fractional ownership, and simplified trading.

While blockchain technology is popular for its security features, no system is entirely immune to vulnerabilities. There is a risk of potential security breaches, hacking attempts, and malware attacks that could compromise the integrity of blockchain systems in banking and stock market operations. Blockchain technology operates in a decentralized manner, which can present challenges with regulatory compliance in the banking and stock market sectors. The regulatory frameworks are still evolving and may not adequately address all the unique aspects of blockchain technology. Although blockchain provides transparency and immutability, it can also raise privacy concerns.

Numerous real-world applications of blockchain technology offer opportunities for fostering sustainable development. By enhancing the transparency and traceability of supply chains, blockchain can ensure products adhere to sustainability standards. It can help with tracking the origin of goods, verifying their authenticity, and monitoring their environmental impact. Blockchain can provide the peer-to-peer trading of renewable energy, allowing individuals and businesses to directly buy and sell excess energy. This decentralized approach can optimize energy distribution, reduce reliance on fossil fuels, and promote the growth of clean energy sources.

As we delve deeper into the potential applications of blockchain technology in supply chain management, we must also address the barriers that currently exist. These barriers fall into four main categories: inter-organizational, intra-organizational, technical, and external. While progress has been made towards adopting blockchain technology for business and supply chains, it is still in its early stages, and there is much work to be done. Saberi et al. (2019) suggest future research propositions and directions that can help us overcome these barriers and unlock the full potential of blockchain technology for supply chain management.

## 5.8 CHALLENGES IN THE BANKING DIGITIZATION PROCESS

The banking industry is going through a significant transformation in the digital era. As part of this process, blockchain technology has developed as a potential solution to many of the challenges faced by banks. However, implementing blockchain in the banking digitization process is not without its hurdles.

One major challenge is the regulatory framework surrounding blockchain technology. As blockchain operates in a decentralized manner, it disrupts the traditional centralized control that regulators have over financial transactions. This creates uncertainty and a need for new regulations and standards to ensure compliance and security in the use of blockchain in banking. Additionally, the global nature of blockchain introduces the challenge of harmonizing regulations across different jurisdictions.

### 5.8.1 Applications of Blockchain Technology in Financial Activities

Blockchain can streamline the payment and settlement process by enabling faster, more secure, and transparent transactions. It eliminates the need for intermediaries, reducing costs and increasing efficiency. Blockchain can facilitate cross-border payments by providing a decentralized network that allows for faster, cheaper, and more secure transactions across different currencies. Blockchain enables the creation and execution of self-executing smart contracts, which can automate specific financial agreements, such as loan disbursements, insurance claims, or supply chain logistics.

As the prevalence of blockchain smart contracts increases, discerning their legality is becoming progressively challenging (Upadhyay 2020). Addressing this challenge necessitates the establishment of rules, regulations, and standards to govern blockchain and smart contracts. This endeavor requires a delicate balance between economic and environmental considerations. Through this approach, businesses and investors can cultivate circular economy models that yield benefits for both the economy and the environment (Ghisellini et al. 2016). Additionally, effective governance and enforcement mechanisms can mitigate illegal activities associated with blockchain nodes and crypto-jacking, thereby safeguarding the public (Dierksmeier & Seele 2019).

### 5.8.2 Application of Blockchain Technology in Enhancing Supply Chain Transparency

The most important contributions of blockchain technology to business sustainability lie in its ability to enhance supply chain transparency. Companies often struggle to trace the origins of their products and materials, leading to issues related to unethical sourcing, environmental damage, and poor labor practices. Blockchain provides a solution by enabling the recording of every step in the supply chain, from raw material extraction to final product delivery. This transparency empowers consumers to make informed choices and encourages companies to adopt responsible practices, ultimately leading to reduced environmental impact and improved working conditions.

### 5.8.3 Application of Blockchain Technology in Promoting Fair Labor Practices

Blockchain can also have a pivotal role in ensuring fair labor practices within businesses. By creating an immutable record of employment contracts, wages, and working conditions, the technology helps prevent the exploitation and abuse of workers. Smart contracts are known for self-executing agreements with predefined conditions, which will automate payment processes, ensuring that workers are compensated fairly and promptly. This not only safeguards human rights but also enhances the overall reputation of the business.

### 5.8.4 Application of Blockchain Technology in Renewable Energy and Carbon Credits

Blockchain's potential in driving renewable energy adoption and carbon reduction is substantial. The technology can facilitate peer-to-peer energy trading within microgrids, allowing individuals and businesses to buy and sell excess renewable energy directly. This democratization of energy distribution supports the transition to cleaner energy sources and reduces dependence on fossil fuels. Additionally, blockchain can verify and tokenize carbon credits, creating a transparent market for companies to offset their carbon emissions. This fosters accountability and encourages enterprises to actively participate in mitigating climate change. Likewise, according to Tang and Tang (2019), the application of blockchain technology enables the establishment of a distributed carbon ledger, linking carbon asset management with emissions trading schemes. This innovation can serve as a potent instrument for carbon accounting and mitigating greenhouse gas emissions. Furthermore, it's noteworthy that blockchain technology is being advocated as a tool for constructing intelligent, environmentally sustainable cities with reduced carbon footprints (Sun & Zhang 2020).

### 5.8.5 Application of Blockchain Technology in Leveraging Decentralization for Data Security

Data security is a critical aspect of business sustainability, and blockchain's decentralized nature offers robust protection against cyber threats. Traditional centralized systems are susceptible to single points of failure and data breaches. Blockchain's distributed architecture makes it exceedingly difficult for malicious actors to compromise the entire network. Furthermore, data deposited on the blockchain is immutable, reducing the risk of unauthorized alterations. This heightened security not only safeguards sensitive business information but also protects customer data, promoting trust and loyalty.

## 5.9 IMPLICATIONS OF THE CHAPTER

This research significantly contributes to academic scholarship by advancing our understanding of the transformative interplay between blockchain technology, business sustainability, and Industry 4.0. Synthesizing concepts from blockchain evolution stages, Industry 4.0, and sustainability, the chapter establishes a comprehensive framework for future research endeavors. This interdisciplinary approach enhances theoretical foundations and offers educators valuable content to integrate into academic curricula, providing students with insights into the practical applications of blockchain technology in addressing global challenges. From an industry perspective, it guides strategic decision-making for businesses looking to leverage blockchain's potential. The chapter emphasizes how blockchain enhances transparency, reduces costs, and fosters sustainability, encouraging industry leaders to innovate and integrate blockchain into their operations. Businesses in finance, healthcare, supply chains, and energy can use blockchain to track and verify sustainable practices, meeting consumer demands for eco-friendly products and services. Policymakers can adapt regulations to support blockchain innovation, while investors may find opportunities in blockchain-based solutions. The chapter also underscores the importance of global collaboration in establishing industry standards and fostering interoperability.

## 5.10 CONCLUSION

In conclusion, the introduction of new digital technologies has significantly disrupted the traditional role of banks as intermediaries within the financial landscape. Blockchain technology, renowned as the cornerstone of trust, fosters collaboration and the cultivation of shared values among diverse stakeholders, transcending traditional boundaries. This transformative potential heralds a paradigm shift where businesses are compelled to pivot towards adaptability and cooperation over rigid competition. The synergy among banks, technology firms, market players, regulatory bodies, and customers holds immense promise for enhancing consumer experiences, fortifying the resilience of the financial system, and bolstering economic prosperity. As we stand at the precipice of the digital revolution, it is imperative for financial service providers, business organizations, and other entities to embrace this transformative wave and earnestly contemplate the integration of blockchain technology into their financial operations, thereby charting a course towards sustainable development and enduring relevance in the era of Industry 4.0 and financial innovation.

The detailed discussion presented in this chapter clearly demonstrates the undeniable link and vast potential of blockchain technology to make a positive contribution to the business as well as the economy through the sustainability and social responsibility agenda. It is essential to note that blockchain is already being utilized in various sectors due to its remarkable ability to enhance, track, and trace operations. The integration of blockchain into different areas of business perfectly aligns with sustainability and social responsibility concerns. Throughout this chapter, it is evident that blockchain decentralization, distribution, and high-security features contribute significantly to the economy. Significantly, blockchain technology has the potential to foster the advancement of businesses and economies through the reduction of transaction costs, enhancement of supply chain performance and communication, safeguarding of human rights, bolstering healthcare patient confidentiality and well-being, and lowering carbon emissions. However, it's crucial to recognize the obstacles associated with blockchain adoption, such as trust issues, illicit activities, susceptibility to hacking, and the necessity for appropriate legislative and policy frameworks to tackle these challenges. Despite the potential upfront costs involved, it is worth noting that the benefits of adopting blockchain technology will undoubtedly outweigh the challenges.

## REFERENCES


Dierksmeier, C., & Seele, P. (2019). Blockchain and business ethics. *Business Ethics: A European Review, 29*. https://doi.org/10.1111/beer.12259

Ghisellini, P., Cialani, C., & Ulgiati, S. (2016). A review on circular economy: The expected transition to a balanced interplay of environmental and economic systems. *Journal of Cleaner Production, 114*, 11e32. https://doi.org/10.1016/j.jclepro.2015.09.007

Iansiti, M., & Lakhani, K. R. (2017). *The truth about blockchain*. Harvard Business Review.

Javaid, M., Haleem, A., Singh, R. P., Suman, R., & Khan, S. (2022). A review of blockchain technology applications for financial services. *Bench Council Transactions on Benchmarks, Standards and Evaluations*, *2*(3), 100073, ISSN 2772–4859, https://doi.org/10.1016/j.tbench.2022.100073.

Malhotra, D., Saini, P., & Singh, A. K. (2021). How blockchain can automate KYC: Systematic review. *Wireless Personal Communications*, *122*, 1987–2021.

Nakamoto, S. (2008). Bitcoin: A peer-to peer electronic cash system. Retrieved from https://bitcoin.org/bitcoin.pdf

Navas, R., Chang, H. J., Khan, S., & Chong, J. W. (2021). Sustainability transparency and trustworthiness of traditional and blockchain ecolabels: A comparison of generations X and Y consumers. *Sustainability*, *13*(15), 8469. https://doi.org/10.3390/su13158469

Pwc (2019). Global fntech report 2019. Retrieved from www.pwc.com/gx/en/industries/fnancial-services/assets/pwc-global fntech-report-2019.pdf.

Saberi, S., Kouhizadeh, M., Sarkis, J., & Shen, L. (2019). Blockchain technology and its relationships to sustainable supply chain management. *International Journal of Production Research*, *57*(7), 2117–2135. https://doi.org/10.1080/00207543.2018.1533261.

Singh, M., & Kim, S. (2019). Chapter four—blockchain technology for decentralized autonomous organizations. *Advances in Computers*, *115*, 115–140. https://doi.org/10.1016/bs.adcom.2019.06.00

Sun, M., & Zhang, J. (2020). Research on the application of block chain big data platform in the construction of new smart city for low carbon emission and green environment. *Computer Communications*, *149*, 332–342. https://doi.org/10.1016/j.com

Tang, Q., & Tang, L. M. (2019). Toward a distributed carbon ledger for carbon emissions trading and accounting for corporate carbon management. *Journal of Emerging Technology*. https://doi.org/10.2308/jeta-52409

Trivedi, S., Mehta, K., & Sharma, R. (2021). Systematic literature review on application of blockchain technology in E-finance and financial services. *Journal of Technology Management & Innovation*, *16*(3), 89–102. https://doi.org/10.4067/S0718-27242021 00030008

Upadhyay, N. (2020). Demystifying blockchain: A critical analysis of challenges, applications and opportunities. *International Journal of Information Management*. https://doi.org/10.1016/j.ijinfomgt.2020.102120

Yan, C., & Bellavitis, C. (2020). Blockchain disruption and decentralized finance: The rise of decentralized business models. *Journal of Business Venturing Insights*, *13*, e00151. https://doi.org/10.1016/j.jbvi.2019.e00151

# 6 Supply Chain Risk Management, Optimization, and Performance Analytics through Blockchain Integration

## *A Bibliometric Analysis*

*Mohammad Abul Kashem,*
*Mohammad Shamsuddoha, Tasnuba Nasir,*
*and Niamat Ullah Ibne Hossain*

## 6.1 INTRODUCTION

Effective risk management is of the utmost significance in supply chains, particularly in today's dynamic and linked corporate environment (Kalbouneh et al., 2023; Hong, 2022; Cole et al., 2019; Wang et al., 2019a). This management strategy aims to enhance supply chain efficiency and mitigate disruptions, particularly those caused by technological factors (Hu and Ghadimi, 2022). Cutting-edge technology mitigates potential hazards and improves resilience, ensuring uninterrupted operational continuity (Kalbouneh et al., 2023). However, effective communication channels, supplier diversification, and contingency plans are risk mitigation methods used to prevent disruptions and enhance the efficacy of the supply chain (Petratos and Faccia, 2023). Still, effective risk management ensures customer satisfaction expeditiously, ensuring continuous business operations and maintaining the company's reputation in the market (Hong, 2022; Pu et al., 2022). Therefore, these efforts assign resources toward technology and data analytics to improve traceability and visibility.

To achieve responsiveness and efficiency, distinctive strategies and coordination across stakeholders and activities are crucial to maintain (Rahman et al., 2023; Hastig and Sodhi, 2020; Dutta et al., 2020; Li et al., 2019). Promptness related to these tasks also includes delivery, order accuracy, inventory levels, and costs associated with

 DOI: 10.1201/9781003438748-7

transportation (Schmidt and Wagner, 2019). Relating to these, supply chains often generate large amounts of data from many sources, which makes it challenging to collect, integrate, and refine data for proper analysis (Esmaeilian et al., 2020), complicating the process (Korpela et al., 2017). However, conventional data availability and quality result in constraints and limitations for the entire supply chain operations (Perboli et al., 2018; Singh et al., 2023). Thus, organizations try to update data and anticipate potential challenges through predictive analytics (Kashem et al., 2023a, 2023b; Saberi et al., 2019).

On the other hand, the reliance on past data fails to accurately represent current or future market conditions where it is necessary to achieve precise measurements by fostering collaboration among many stakeholders. This ambiguity results in faulty forecasting and planning, ultimately affecting supply chain performance (Rasi et al., 2021). Seemingly, the complex nature and interdependence of global supply chains create difficulties in acquiring and examining data across different levels and geographic regions (Julsvoll and Sickenberger, 2022; Wang et al., 2019a). So, the absence of openness and trust in traditional methods (Tian, 2017) and information imbalances complicate the task of monitoring and verifying product origin, authenticity, and movement (Hastig and Sodhi, 2020). As a result, efficiency in supply chain management (SCM) operations has been poorly achieved (Hossain et al., 2022; Khan et al., 2023; Fu and Zhu, 2019), leading to the risk of delays, errors, and hazards (Kshetri, 2018). Despite these difficulties, integrating contemporary technology into the supply chain might offer an extensive opportunity for growth.

Fortunately, blockchain technology maintains decentralization and immutability as innovative and unique answers to these problems (Hackius and Petersen, 2017; Queiroz and Wamba, 2019; Queiroz et al., 2020). In addition, it ensures the creation of a transparent and auditable record for all supply chain transactions (Kamble et al., 2019). For instance, this allows stakeholders to track the product's origin and follow their journey to the final consumer, thereby improving accountability and addressing issues such as counterfeiting, fraud, and product recalls (Wang et al., 2019b). For order fulfillment, payment, and compliance verification, the smart contract mechanism of blockchain leads to automation of supply chain activities by reducing manual intervention. This mechanism reduces delays and increases operational efficiency (Julsvoll and Sickenberger, 2022). Moreover, smart contracts guarantee compliance with quality standards, fair trade practices, and environmental regulations (Wang et al., 2020; Kamilaris et al., 2019). Despite being in a state of constant development, forward-thinking companies see the new potential of this technology and are actively looking for ways to use blockchain in their SCM processes.

SCM optimization is crucial for balancing consumer demand and reducing inventory carrying costs (Hastig and Sodhi, 2020). These strategies utilize data analysis, forecasting, and strategic decision-making to guarantee efficient inventory management with just-in-time precision (Nethravathi et al., 2022). On the other hand, optimizing safety stock and aligning production and procurement with real consumer demand offer a competitive edge in the market (Wang et al., 2019a). Organizations may boost operational efficiency, save expenses, and improve customer satisfaction by managing inventory levels (Nathani and Singh, 2020). Insufficiently trained staff and a lack of coordination within the company in evaluating and applying analytical findings result in business difficulties (Rasi et al., 2021). To address these obstacles and limits, it is crucial to develop a thorough data strategy, provide resources to enhance technological infrastructure, and foster a firm commitment inside the company to utilize data-driven decision-making in the SCM field (Esmaeilian et al., 2020). Thus, SCM with blockchain enables firms to promptly address issues efficiently, resulting in enhanced customer satisfaction, loyalty, and, eventually, more revenue and profitability.

## 6.2 SUPPLY CHAIN DISRUPTIONS AND BLOCKCHAIN

For efficient risk management, SCM operations, along with BT, resolve many problems afflicting conventional supply chain systems, creating novel prospects for improved risk reduction (Kadadevaramth et al., 2020) due to its decentralized and irreversible nature (Kshetri, 2018). Various studies (Caro et al., 2018; Francisco and Swanson, 2018; Abeyratne and Monfared, 2016; Hackius and Petersen, 2017; Kamilaris et al., 2019) have explored the benefits of blockchain technology with enhanced traceability and verification of the origin and history of goods in the SCM operations. For illustration, blockchain integration in sectors such as food and pharmaceuticals might effectively detect and minimize risks associated with contamination, counterfeit goods, and interruptions in their SCM operations (Abidi et al., 2021; Guan, 2021; Mylrea and Gourisetti, 2018; Salah et al., 2019). Hence, implementing this approach enhances the ability to manage risks effectively and promotes customer confidence and the brand's reputation (Kashem et al., 2023a; Treiblmaier, 2018) through the use of smart contracts (Chang et al., 2019; Wang et al., 2019b; Schmidt and Wagner, 2019). Per Azzi et al. (2019), smart contracts can execute themselves and have pre-established terms and conditions stored on the blockchain. Furthermore, smart contracts permit transactions to be automatically verified, executed, and enforced, eliminating the need for intermediaries and lowering the possibility of manipulation and human mistakes (Schmidt and Wagner, 2019). Last, blockchain technology improves risk management skills by reducing the likelihood of fraud, counterfeiting, and tampering across the supply chain.

## 6.3 BLOCKCHAIN IN PRACTICE

Implementing blockchain technology in Walmart's supply chain improved traceability, recall speed, efficiency, waste reduction, and consumer confidence, leading to its success and inspiring other companies in the food industry to explore similar solutions. However, with regard to the Maersk and TradeLens cases, the blockchain-based platform streamlined procedures and removed paper-based documentation, which decreased administrative costs, minimized errors, improved supply chain visibility, allowed for faster logistics, and attracted the participation of other significant shipping companies and port authorities for improved cooperation and efficiency. These examples are listed in Table 6.1.

**TABLE 6.1**
**Blockchain Industry Integration**

| Industry | Impacts |
|---|---|
| **IBM Food Trust and Walmart (Wang et al., 2019a)** | Walmart and IBM Food Trust launched their partnership. In order to improve SCM transparency and tackle the issues of food-borne diseases and product recalls, Walmart and IBM collaborated in 2018 to build a blockchain-based solution. Using BT, the system allowed instantaneous tracing and tracking of goods from the farm to the shop shelves. Every stage of SCM operations, including farmers, suppliers, distributors, and vendors, required access to a shared, unchangeable ledger containing vital data from the product's origin. The system maintains quality certificates, temperature information, and handling protocols in this process. With an open and validated approach, stakeholders could promptly locate possible sources of contamination and take the appropriate precautions to reduce risks. |
| **Maersk and TradeLens (Schmidt and Wagner, 2019)** | Usually, the maritime shipping industry is highly complex, involving numerous stakeholders and a vast amount of documentation, making it prone to inefficiencies and delays. Maersk and TradeLens streamline processes and increase transparency in the shipping industry by blockchain, allowing participants in the supply chain, including shippers, ports, customs authorities, and freight forwarders, to share securely and access real-time shipping data. Additionally, participants could check container conditions, follow the flow of products, and expedite customs procedures thanks to the blockchain's openness, which made supply chain operations faster and more effective. |

Wang et al. (2019a) found that Walmart's supply chain created substantial advantages by implementing blockchain technology. This technology facilitated faster retrieval of food products in the case of safety issues by rapidly tracking their origins, reducing the time from days to seconds. Ensuring the freshness and authenticity of the commodities also enhanced efficiency, reduced wastage, and bolstered client trust. Thanks to the successful integration of blockchain, other food-related organizations are also considering similar solutions to improve their supply chain management. Nevertheless, the blockchain-based platform eliminates the need for paper-based documentation, reduces administrative expenses, and minimizes the risk of errors by serving as a reliable and authoritative source of information for all stakeholders, ensuring data quality and consistency (Julsvoll and Sickenberger, 2022). It enhanced the transparency of the supply chain, minimized administrative tasks, and facilitated expedited and dependable logistical operations. This endeavor's triumph has enticed more prominent shipping corporations and port authorities to partake in the TradeLens platform, hence augmenting collaboration and efficacy throughout the worldwide shipping ecosystem.

According to the literature, achieving more transparency, traceability, and confidence among all parties involved necessitates a specific investigation demonstrating blockchain technology's need to analyze performance and manage risks in supply chains. Information asymmetries, a lack of real-time visibility, and inefficient data transmission frequently hinder traditional supply chains. These issues result in delays, mistakes, and increased risk. Nevertheless, blockchain technology presents a practical remedy to these problems through its decentralized and immutable ledger. However, the individuals involved in the supply chain will securely store and share data using blockchain technology. This will allow for the creation of a transparent and verified system, leading to enhanced performance analysis and risk management.

## 6.4 METHODOLOGY

This chapter presents a qualitative approach integrating BT, SCM optimization, and associated risk supervision. The approach involves the use of bibliometric and network analysis techniques. Bibliometric analysis is a quantitative research technique used to examine academic and scientific publications. Statistical and mathematical methodologies are utilized to discern patterns, trends, and correlations within a specific set of publications or a particular field of research. Bibliometric analysis aims to understand the structure and impact of scientific knowledge and the productivity and influence of authors, journals, institutions, and countries (Hossain et al., 2020; Munim et al., 2023). To analyze relevant literature and organize the chapter systematically, this study concentrates on how supply chain acceleration is facilitated by blockchain integration by following optimization and performance analysis techniques against risks and disruption in the supply chain (see Table 6.2).

These research gaps highlight areas that require more study and development to use blockchain technology for SCM, optimization, and performance analytics. This will advance the field and offer valuable insights for industry practitioners. The research questions corresponding to these gaps and their potential contributions also highlight them. The authors find pertinent material using databases and keyword analysis, highlighting highly referenced and excellent works. The study increases prominence by citing often-referenced literature. A distinct search strategy, a trustworthy repository, and precise inclusion and exclusion principles are followed in the study for an exhaustive analysis of SCM and BT (Table 6.3).

**TABLE 6.2**
**Research Gap, Question, and Contribution**

| Research Gap | Research Question | Contribution |
|---|---|---|
| **Limited inquiry on blockchain-based optimization techniques in supply chain management** | How can blockchain technology make supply chain operations more efficient and effective? | Propose a blockchain-enabled optimization model to improve supply chain efficiency by considering several variables, including demand forecasts, inventory control, and transportation logistics. |

## TABLE 6.3
## Inclusion and Exclusion Parameters

| Inclusion Parameters | Exclusion Parameters |
|---|---|
| **1. Research was published in conference proceedings or peer-reviewed publications between 2016 and 2023 (using English).** | 1. Studies that are not presented in conference proceedings or peer-reviewed publications. |
| **2. Examine how supply chain performance statistics, metrics, optimization, and risk management strategies are impacted by blockchain technology.** | 2. Research articles written in languages other than English. |
| **3. Examine the pros and cons of integrating blockchain technology into SCM processes such as demand forecasting and planning, inventory control, and so on.** | 3. Studies that don't address how blockchain affects supply chains. |
| **4. Studies on how supply chain operations are made more effective and efficient by state-of-the-art technology such as blockchain.** | 4. Research examining further facets of blockchain, excluding the context of this study, like performance analytics and route optimization. |
| **5. Research on current trends and emerging practices demonstrates empirically how blockchain affects the supply chain, both in practical and theoretical phases, leading to increased solutions and efficiency.** | 5. Old research doesn't accurately represent the latest developments and trends. |

## TABLE 6.4
## Keyword Searching and Paper Selection

| Search String/Keywords | No. of Papers |
|---|---|
| **Blockchain Supply Chain** | 3710 |
| **Blockchain Risk Management** | 101 |
| **Blockchain Optimization** | 408 |
| **Blockchain Performance Analytics** | 4 |
| **Supply Chain Risk Management** | 2760 |
| **Supply Chain Optimization** | 4760 |
| **Supply Chain AND Performance Analytics** | 131 |
| **Blockchain AND Supply Chain AND Risk Management** | 21 |
| **Blockchain AND Supply Chain AND Optimization** | 28 |
| **Blockchain AND Supply Chain AND Performance Analytics** | 1 |
| **Total Papers Selected For Bibliometric Analysis** | 1,180 |

The quantity of articles about blockchain and supply chains is apparent in Table 6.4. Several search engines provide the data, including Scopus, Google Scholar, and PubMed. Yet, searching for terms like "blockchain" or "supply chain" on Google Scholar produces 3,710 publications with citations but no explicit inclusion or exclusion criteria. Different articles were sorted for relevance based on the keywords. Nonetheless, 1,180 of them satisfied the inclusion criteria, which were to be accessible and to include the keywords in the title.

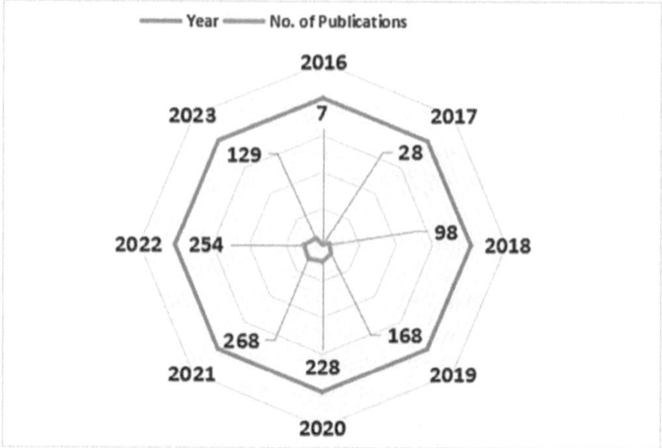


**FIGURE 6.1** Year-wise publications.

## TABLE 6.5
## Citation Metrics

| Papers | Citations | Years | Cite Year | Cite Paper | Cites Author | h_index | g_index | hc_index | hl_index |
|---|---|---|---|---|---|---|---|---|---|
| **1180** | 90537 | 2016–2023 | 12933.86 | 76.73 | 55920.7 | 153 | 260 | 191 | 70.94 |

Table 6.5 presents citation metrics for the publications between 2016 and 2023. It includes bibliometric data, such as the number of publications (also in Figure 6.1) and citations earned during that period. A total of 1,180 publications received 90537 citations, with an average of 12,933 citations per year. The values of the h-index and g-index are 260 and 153, correspondingly.

Wamba and Mishra's (2017) guidelines are followed for bibliometric analysis, and VOSviewer aids in graph visualization and author cooperation network analysis. Figures 6.2(a & b) demonstrate network studies on author connections where the minimum connection for authors is 3, and they found that 155 meet the criteria with 15 clusters.

The low degree of compatibility among the authors' cluster nodes implies weak connections and a significant distance among clusters. Citation analysis has reduced the likelihood that "similarity breeds connection" because of the many opinions on emerging technology. A thorough analysis of the industry's transition to these cutting-edge domains is essential to maximizing the potential of BT in the early stages of SCM operations (Etemadi et al., 2021). These contributions highlight the qualitative improvement in potential risk management policy and optimization improvements. To address the issues raised by the digital revolution of the SCM, the state and society must be able to communicate electronically through new means of route optimization, inventory management, data-driven decision-making, and performance analytics.

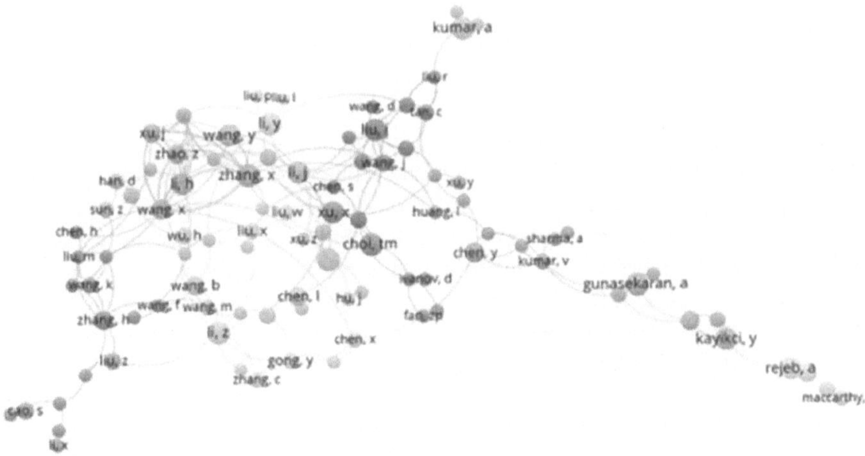


**FIGURE 6.2** (a) Author network clusters.

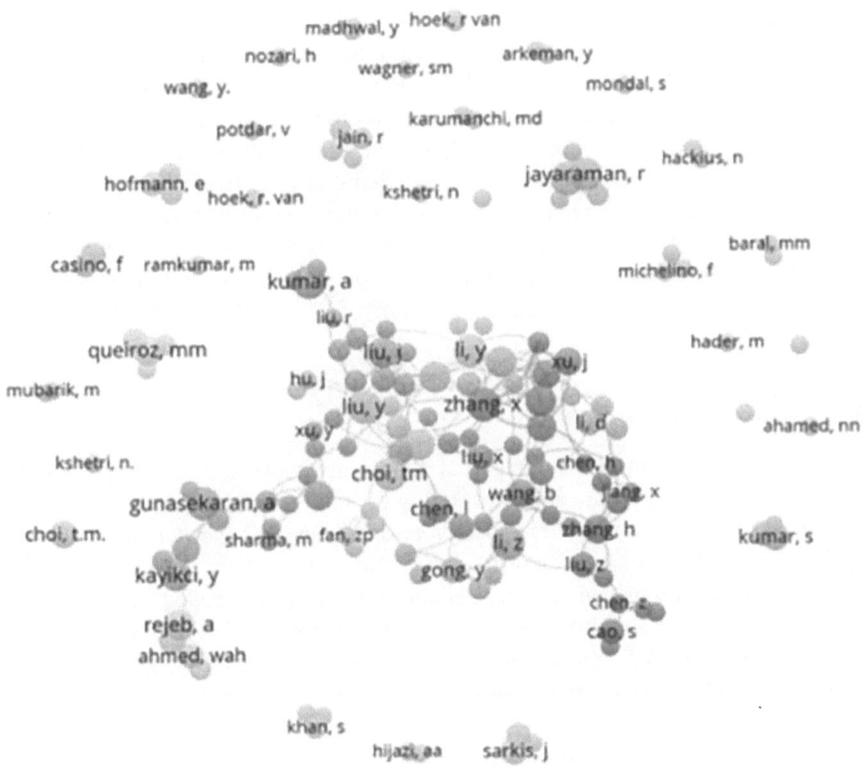


**FIGURE 6.2** (b) All the author networks.

## 6.5 BLOCKCHAIN, PERFORMANCE ANALYTICS, AND OPTIMIZATION IN THE SUPPLY CHAIN

### 6.5.1 Demand Forecasting and Planning

Demand forecasting and planning in supply chain management involve helping organizations anticipate customer demand and efficiently allocate resources (Perboli et al., 2018). Demand forecasting aims to guide ongoing production, inventory management, and distribution decisions by projecting future product and service demand based on relevant factors such as historical data, market trends, and consumer preferences (Dutta et al., 2020). However, knowing customer preferences and demand patterns may help businesses avoid understocking, which loses revenue and upsets customers, and overstocking, which wastes resources and boosts storage expenses (Julsvoll and Sickenberger, 2022). Modern technologies like machine learning, blockchain, and artificial intelligence (AI) are often used to improve demand forecasting and planning accuracy. According to Reddy et al. (2021) and Kashem et al. (2024), these technologies use sophisticated algorithms to evaluate enormous volumes of data in real time, spot patterns and relationships, and provide incredibly accurate projections.

In addition, enterprises are employing demand-sensing capabilities to collect and address fluctuations in demand more effectively. These functionalities incorporate live data from several sources, including online platforms, social media, and point-of-sale systems (Leng et al., 2018). With the advancement of blockchain technology, firms may utilize sophisticated tools and analytics to enhance the precision and agility of demand forecasting and planning. This enables them to gain a competitive edge in today's fast-paced business environment (Chang et al., 2019). Demand forecasting and planning in supply chain management entails using historical data, market trends, customer behavior, and advanced technologies such as blockchain to predict customer demand, allocate resources effectively, prevent shortages and excess inventory, and achieve higher precision and adaptability in a constantly changing business environment.

### 6.5.2 Route Optimization

With the increasing complexity of global supply networks and growing consumer expectations, corporations are actively looking for solutions to streamline their transportation routes to save costs, shorten delivery times, and enhance customer satisfaction (Pu et al., 2022). Businesses utilize blockchain technology to leverage modern technologies and data-driven research, enabling them to make well-informed decisions on route planning and resource allocation (Cole et al., 2019). This route optimization algorithm takes into account several variables, including distance, traffic conditions, vehicle capacity, and delivery priority, to develop the most efficient routes (Wang et al., 2019a). Organizations can cut transportation expenses by limiting fuel consumption, eliminating vehicle wear and tear, and improving fleet usage (Jiang et al., 2023). Companies use modern technologies and data-driven analysis, such as blockchain, to optimize transportation routes. This optimization considers aspects like distance, traffic, and delivery priority. Consequently, it reduces costs, improves delivery times, and increases customer satisfaction.

In addition, route optimization enables organizations to overcome logistical obstacles and adjust to changing market conditions (Chang et al., 2019). Business establishments utilize real-time traffic pattern data and other pertinent aspects to modify routes, as Bottoni et al. (2023) demonstrated. By employing this strategy, organizations circumvent any potential setbacks, optimize the timing of deliveries, and guarantee prompt order completion. The degree of adaptability is highly advantageous in sectors that prioritize prompt deliveries and rapid reaction, such as e-commerce, perishable commodities, and urgent medical supplies (Li et al., 2019; Nathani and Singh, 2020). Furthermore, route optimization enables enterprises to overcome logistical obstacles, adjust to ever-changing market conditions, and accomplish prompt order fulfillment.

### 6.5.3 Data-Driven Decision-Making

Organizations can make informed decisions by employing the data-driven decision-making strategy, which involves utilizing objective data, insights, and various data sources and analytical tools to collect, analyze, and interpret pertinent information regarding an organization's performance (Korpela et al., 2017). Decision-makers can enhance the comprehension of their operations and pinpoint areas for enhancement by utilizing data instead of relying on intuition or guesswork. This data can be derived from various sources, including customer feedback, sales figures, operational metrics, and market trends (Esmaeilian et al., 2020). Through the ongoing surveillance and examination of performance data, firms are able to identify early signs of market fluctuations, evolving client preferences, or possible threats (Etemadi et al., 2021). By adopting a proactive strategy, decision-makers may promptly respond, implement strategic modifications, and maintain a competitive advantage (Treiblmaier, 2018). Organizations may make timely and impactful decisions by utilizing data-driven decision-making to optimize marketing campaigns, refine product offers, or enhance operational efficiency (Etemadi et al., 2021; Shamsuddoha et al., 2023). Therefore, in a market characterized by competition, using data to assist decision-making allows companies to make educated choices based on unbiased facts and insights, thereby improving performance, adaptability, and strategic benefits.

### 6.5.4 Supply Chain Optimization

It is crucial to employ efficient tactics and optimal methods to mitigate supply chain risks related to delivery schedules, order precision, stock quantities, and transportation expenses. The supply chain management approach focuses on several key strategies to optimize operations and reduce risk, including open lines of communication with suppliers, distributors, and logistics partners and regular exchange of information about order volumes, delivery schedules and inventory levels. This process includes accurate response to demand forecasting, optimized inventory levels, and better production and procurement planning. However, in keeping with modern technology, the traditional view of the supply chain, its drivers and metrics are revived

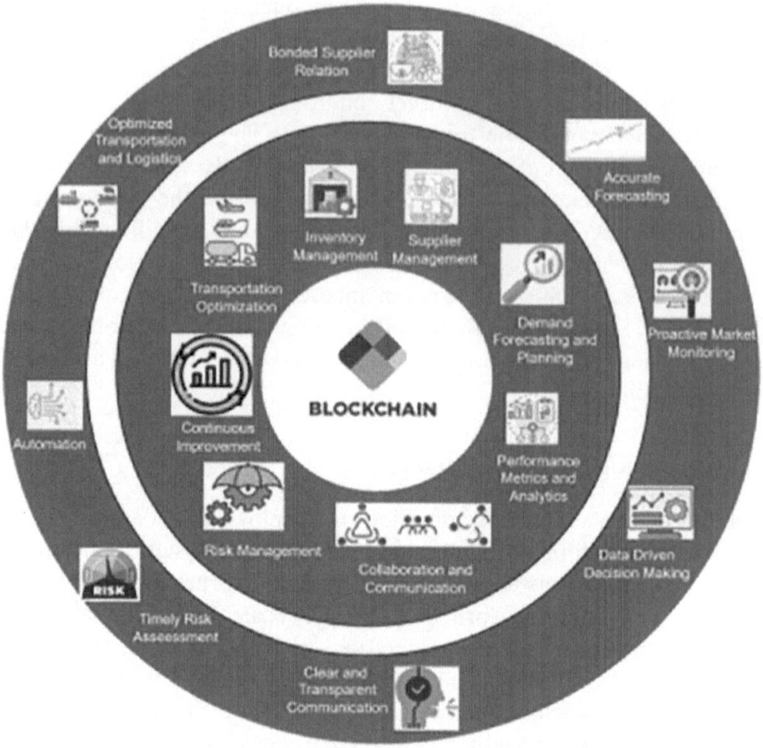


**FIGURE 6.3** Revolution of supply chain through blockchain technology.

to a greater extent with trust, immutability and decentralization with blockchain technology (see Table 6.6 and Figure 6.3). The literature suggests that Dutta et al., 2020, Esmaeilian et al., 2020, Perboli et al., 2018 and Saberi et al., 2019 idealize this process.

Once more, researchers contribute to the development of the previous frameworks and models (Figure 6.3) that enable efficient risk management, optimization, and performance analytics (Dutta et al., 2020; Schmidt and Wagner, 2019; Wong et al., 2020) by looking at the theoretical underpinnings of blockchain and its integration with supply chains. The research suggests that risk management strategies should be established in several phases, and the literature of Petratos and Faccia (2023), Jiang et al. (2023), Fu and Zhu (2019), Etemadi et al. (2021), and Alkhudary et al. (2020) idealizes this process. Overall, this unique approach ensures a continuous focus on improvement, adaptability to changing market conditions, and proactive risk management within the supply chain. By integrating blockchain into supply, organizations significantly reduce risks related to delivery time, order accuracy, inventory levels, and transportation costs in the supply chain instead of an ongoing process by regular monitoring, evaluation, and adaptation to ensure long-term success.

**TABLE 6.6**
**Traditional and Blockchain Integrated Supply Chain**

| Supply Chain Aspect | Traditional Approach | Blockchain-Integrated Approach |
|---|---|---|
| **Demand forecasting** | Uses historical data and market analysis for forecasting (Reddy et al., 2021). | Leverages real-time data and smart contracts for accurate demand forecasting (Kadadevaramth et al., 2020). |
| **Communication channels** | Relies on manual communication methods such as phone calls and emails (Chang et al., 2019). | Utilizes blockchain-based communication platforms for secure and transparent communication (Salah et al., 2019). |
| **Relationships with stakeholders** | Establishes relationships based on trust and contractual agreements (Treiblmaier, 2018). | Enables direct and transparent interactions with suppliers through smart contracts and decentralized platforms (Etemadi et al., 2021). |
| **Invest in technology and automation** | Adopts traditional software solutions and automation tools (Chang et al., 2019). | Incorporates blockchain technology for enhanced automation, data integrity, and process efficiency (Dutta et al., 2020). |
| **Supplier base** | Engages with a limited number of suppliers (Kshetri, 2018). | Explores a broader supplier network with decentralized marketplaces and smart contract-based transactions (Shamsuddoha et al., 2022). |
| **Monitor and analyze performance indicators** | Relies on manual data collection and analysis (Korpela et al., 2017). | Utilizes blockchain-enabled analytics tools for real-time performance monitoring and data-driven insights (Kashem et al., 2023b). |
| **Optimization of transportation and logistics** | Uses traditional logistics methods with limited visibility (Perboli et al., 2018). | Implements blockchain-based tracking systems for enhanced visibility, traceability, and efficiency in logistics operations (Dutta et al., 2020). |
| **Continuous improvement process** | Depends on manual process improvement initiatives (Cole et al., 2019). | Utilizes blockchain-enabled smart contracts and consensus mechanisms for collaborative process improvement (Shamsuddoha and Kashem, 2022). |
| **Monitor market and industry trends** | Relies on market research and industry reports (Tian, 2016). | Leverages blockchain-based platforms for real-time market data and industry trend analysis (Dutta et al., 2020). |
| **Regular risk assessments** | Performs risk assessments based on historical data (Caro et al., 2018). | Utilizes blockchain for secure data storage, traceability, and risk assessment in real time (Tian, 2016). |

### 6.5.5 Overview of Traditional versus Blockchain Integrated Supply Chain

The integrated supply chains using conventional and blockchain technologies are summarized in the following (Table 6.7) about risk management, performance analytics, and optimization:

Traditional supply chain optimization, performance analytics, and risk management methods rely on centralized systems and intermediaries, making them prone to inefficiencies, delays, and vulnerabilities. In contrast, blockchain-integrated approaches revolutionize these aspects by leveraging decentralized networks, smart contracts, and immutable ledgers. With blockchain, supply chain optimization becomes more transparent, efficient, and cost-effective, enabling real-time tracking of goods, streamlined transactions, and automated workflows. Performance analytics is enhanced by obtaining precise and unalterable data across the supply chain, enabling thorough insights, predictive analytics, and proactive decision-making. Additionally, blockchain's inherent security features enhance risk management by reducing fraud, counterfeiting, and data breaches while ensuring compliance and traceability throughout the SCM operations. Ultimately, SCM operations revolutionize through BT technology, which improves analytics, boosts risk management procedures, and maximizes efficiency.

**TABLE 6.7**
**Traditional versus Blockchain-Integrated Supply Chain**

| Aspects | Traditional Supply Chain | Blockchain-Integrated Supply Chain |
|---|---|---|
| **Supply Chain Optimization** | Primarily relies on centralized systems and manual processes for data collection, analysis, and decision-making (Korpela et al., 2017). | Utilizes a decentralized network of nodes to securely record and validate transactions, enabling real-time visibility and automated execution of supply chain processes (Abeyratne and Monfared, 2016). |
| **Performance Analytics** | Limited visibility into real-time data leads to delayed insights and reactive decision-making (Julsvoll and Sickenberger, 2022). | Provides transparent and immutable records of transactions and events, enabling real-time data sharing and analysis across supply chain participants for proactive decision-making (Reddy et al., 2021). |
| **Risk Management** | Relies on traditional risk assessment methods, often based on historical data and subjective judgments (Nathani and Singh, 2020). | Uses smart contracts and consensus mechanisms to enforce predefined rules and automate risk assessment and mitigation processes and provides enhanced traceability and auditability, reducing fraud and enhancing supply chain security (Julsvoll and Sickenberger, 2022). |

## 6.6 THEORETICAL AND MANAGERIAL IMPLICATIONS

For practitioners and scholars in the future, integrating BT technology into SCM risk management, optimization, and performance analytics bears significant theoretical and managerial implications. This revived blockchain technology might change how supply chains are maintained since it would provide more efficiency, security, and transparency. Organizations may reduce risks, streamline processes, and obtain insightful data about the performance of their supply chains by utilizing blockchain. Researchers investigate the intricacies of BT and its uses in various aspects of SCM from a theoretical standpoint, particularly in developing and executing smart contracts, decentralized networks, and consensus mechanisms to tackle specific challenges like supply chain visibility, traceability, and trust.

Integrating blockchain technology has several management consequences for practitioners, including automating compliance verification, assuring adherence to quality standards, and expediting dispute resolution processes. The decentralized blockchain structure enhances accessibility to real-time data on inventory levels, demand trends, and supplier performance through automated procurement procedures, decreasing paperwork and eliminating intermediaries. ChainLink facilitates supply chain automation by seamlessly connecting smart contracts with external data sources and APIs (Schmidt and Wagner, 2019). VeChain is a blockchain network created to oversee supply chains, monitor product movements, and verify their genuineness (Kadadevaramth et al., 2020). Furthermore, blockchain facilitates sophisticated performance analytics in the supply chain and offers a substantial data repository for research. Through the utilization of this review, professionals may identify obstacles in the supply chain, enhance the efficiency of tasks, and get extensive knowledge about its functioning. Organizations may use data analytics and machine learning methodologies to use blockchain data to discern patterns, detect irregularities, and make informed decisions that enhance customer satisfaction, decrease expenses, and improve performance. Continued advancements in this technology will drive greater investigation and practical applications, ultimately influencing the SCM future. This will result in supply chains that are more robust, streamlined, and characterized by increased transparency.

## 6.7 CONCLUSION

This chapter investigates the creation of resilient risk management frameworks utilizing blockchain technology. Generally, since blockchain technology offers a trustworthy and impenetrable record of transactions and activities, its inherent immutability and transparency enhance supply chain risk assessment and mitigation. According to this report, companies may use blockchain technology to detect risks and weaknesses in their supply chain operations more efficiently by adopting a holistic approach. Nevertheless, supply chain risk management is even stronger when performance indicators are integrated into blockchain systems. Companies can find places where their supply chain performance needs to be improved and bottlenecks and inefficiencies using real-time data and advanced analytics approaches, resulting in improved decision-making and operational procedures. This chapter also reduces the likelihood of disputes and delays, resulting in improved efficiency and faster transaction settlement among supply chain participants, creating valuable assets in the dynamic and competitive business world. While integrating blockchain technology into supply chain risk management shows promise, further research and development are needed in several areas, such as scalability, interoperability, and the full potential of blockchain to revolutionize supply chain management. Future blockchain research should focus on creative ways to assist sustainable supply chains and track carbon emissions, among other ways to create a more sustainable future. Thus, continued research and collaboration between academia, industry, and technology providers will be critical to advancing blockchain integration into supply chain management practices.

## REFERENCES

Abeyratne, S. A., & Monfared, R. P. (2016). Blockchain ready manufacturing supply chain using distributed ledger. *International Journal of Research in Engineering and Technology*, *5*(9), 1–10.

Abidi, M. H., Alkhalefah, H., Umer, U., & Mohammed, M. K. (2021). Blockchain-based secure information sharing for supply chain management: Optimization assisted data sanitization process. *International Journal of Intelligent Systems*, *36*(1), 260–290.

Alkhudary, Rami, Brusset, Xavier, & Fenies, Pierre (2020). Blockchain and risk in supply chain management. In *Dynamics in Logistics: Proceedings of the 7th International Conference LDIC 2020*, Bremen, 159–165. Springer.

Azzi, R., Chamoun, R. K., & Sokhn, M. (2019). The power of a blockchain-based supply chain. *Computers & Industrial Engineering*, *135*, 582–592.

Bottoni, P., Ciccio, C. D., Pareschi, R., Tortola, D., Gessa, N., & Massa, G. (2023). Blockchain-as-a-service and blockchain-as-a-partner: Implementation options for supply chain optimization. *Blockchain: Research and Applications*, *4*(2), 100119.

Caro, Miguel Pincheira, Ali, Muhammad Salek, Vecchio, Massimo, & Giaffreda, Raffaele (2018). Blockchain-based traceability in agri-food supply chain management: A practical implementation. In *2018 IoT Vertical and Topical Summit on Agriculture-Tuscany (IOT Tuscany)*, 1–4, IEEE.

Chang, S. E., Chen, Y.-C., & Lu, M.-F. (2019). Supply chain re-engineering using blockchain technology: A case of smart contract based tracking process. *Technological Forecasting and Social Change*, *144*, 1–11.

Cole, R., Stevenson, M., & Aitken, J. (2019). Blockchain technology: Implications for operations and supply chain management. *Supply Chain Management: An International Journal*, *24*(4), 469–483.

Dutta, P., Choi, T.-M., Somani, S., & Butala, R. (2020). Blockchain technology in supply chain operations: Applications, challenges and research opportunities. *Transportation Research Part e: Logistics and Transportation Review*, *142*, 102067.

Esmaeilian, B., Sarkis, J., Lewis, K., & Behdad, S. (2020). Blockchain for the future of sustainable supply chain management in Industry 4.0. *Resources, Conservation and Recycling*, *163*, 105064.

Etemadi, N., Borbon-Galvez, Y., Strozzi, F., & Etemadi, T. (2021). Supply chain disruption risk management with blockchain: A dynamic literature review. *Information*, *12*, 70. Note: MDPI stays neutral with regard to jurisdictional claims in published?

Etemadi, Niloofar, Borbon-Galvez, Yari, Strozzi, Fernanda, & Etemadi, Tahereh (2021). Supply chain disruption risk management with blockchain: A dynamic literature review. *Information*, *12*(2), 70.

Francisco, K., & Swanson, D. (2018). The supply chain has no clothes: Technology adoption of blockchain for supply chain transparency. *Logistics*, *2*(1), 2.

Fu, Y., & Zhu, J. (2019). Big production enterprise supply chain endogenous risk management based on blockchain. *IEEE Access*, *7*, 15310–15319.

Guan, P. (2021). Supply chain optimization of agricultural products in the internet environment with blockchain. *Informatica*, *45*(6).

Hackius, Niels, & Petersen, Moritz (2017). Blockchain in logistics and supply chain: Trick or treat? In *Digitalization in Supply Chain Management and Logistics: Smart and Digital Solutions for an Industry 4.0 Environment: Proceedings of the Hamburg International Conference of Logistics (HICL)*, Vol. 23, 3–18. Berlin: Epubli GmbH.

Hastig, G. M., & Sodhi, M. S. (2020). Blockchain for supply chain traceability: Business requirements and critical success factors. *Production and Operations Management*, *29*(4), 935–954.

Hong, L. (2022). *Blockchain Technology in Supply Chain Management: Performance and Risk Management Perspective*. University of Rhode Island.

Hossain, N. U. I., Fazio, S. A., Lawrence, J. M., Gonzalez, E. D. S., Jaradat, R., & Alvarado, M. S. (2022). Role of systems engineering attributes in enhancing supply chain resilience: Healthcare in context of COVID-19 pandemic. *Heliyon*, *8*(6).

Hossain, N. U. I., Dayarathna, V. L., Nagahi, M., & Jaradat, R. (2020). Systems thinking: A review and bibliometric analysis. *Systems*, *8*(3), 23.

Jiang, F., Jiang, C.-X., & Li, J. X. (2023). Deep learning for blockchain in medical supply chain risk management. *International Journal of Grid and Utility Computing*, *14*(2), 250–263.

Hu, Y., & Ghadimi, P. (2022). A review of blockchain technology application on supply chain risk management. *IFAC-PapersOnLine*, *55*(10), 958–963.

Julsvoll, A. A., & Sickenberger, M. (2022). How can blockchain technology be used to support supply chain risk management in drug supply chains? (Master's thesis, Handelshøyskolen BI).

Kadadevaramth, Rajeshwar S., Sharath, D., Ravishankar, B., & Kumar, P. Mohan (2020). A review and development of research framework on technological adoption of blockchain and IoT in supply chain network optimization. In *2020 International Conference on Mainstreaming Block Chain Implementation (ICOMBI)*, 1–8, IEEE.

Kalbouneh, N., Bataineh, K., Al-Hamad, A., Dwakat, M., Abualoush, S., Almasarweh, M., & Al-Smadi, R. (2023). The effects of the blockchain technology and big data analytics on supply chain performance: The mediating effect supply chain risk management. *Uncertain Supply Chain Management*, *11*(3), 903–914

Kamble, S., Gunasekaran, A., & Arha, H. (2019). Understanding the blockchain technology adoption in supply chains-Indian context. *International Journal of Production Research*, *57*(7), 2009–2033.

Kamilaris, A., Fonts, A., & Prenafeta-Boldύ, F. X. (2019). The rise of blockchain technology in agriculture and food supply chains. *Trends in Food Science & Technology*, *91*, 640–652.

Kashem, M. A., Shamsuddoha, M., & Nasir, T. (2024). Digital-era resilience: Navigating logistics and supply chain operations after COVID-19. *Businesses*, *4*(1), 1–17.

Kashem, M. A., Shamsuddoha, M., Nasir, T., & Chowdhury, A. A. (2023a). Supply chain disruption versus optimization: A review on artificial intelligence and blockchain. *Knowledge*, *3*(1), 80–96.

Kashem, M. A., Shamsuddoha, M., Nasir, T., & Chowdhury, A. A. (2023b). The role of artificial intelligence and blockchain technologies in sustainable tourism in the Middle East. *Worldwide Hospitality and Tourism Themes*, *15*(2), 178–191. https://doi.org/10.1108/WHATT-10-2022-0116

Khan, M. M., Bashar, I., Minhaj, G. M., Wasi, A. I., & Hossain, N. U. I. (2023). Resilient and sustainable supplier selection: An integration of SCOR 4.0 and machine learning approach. *Sustainable and Resilient Infrastructure*, 1–17.

Korpela, Kari, Hallikas, Jukka, & Dahlberg, Tomi (2017). Digital supply chain transformation toward blockchain integration. In *Proceedings of the 50th Hawaii International Conference on System Sciences*, 4182–4191. University of Hawai'i at Mānoa.

Kshetri, N. (2018). 1 Blockchain's roles in meeting key supply chain management objectives. *International Journal of Information Management*, *39*, 80–89.

Leng, K., Bi, Y., Jing, L., Fu, H.-C., & Nieuwenhuyse, I. V. (2018). Research on agricultural supply chain system with double chain architecture based on blockchain technology. *Future Generation Computer Systems*, *86*, 641–649.

Li, J., Wang, Y., Li, Y., & Li, Q.-L. (2019). A simple survey for supply chain finance risk management with applications of blockchain. *Stochastic Models in Reliability, Network Security and System Safety: Essays Dedicated to Professor Jinhua Cao on the Occasion of His 80th Birthday 1*, 116–133.

Mylrea, Michael, & Gourisetti, Sri Nikhil Gupta (2018). Blockchain for supply chain cybersecurity, optimization and compliance. In *2018 Resilience Week (RWS)*, 70–76, IEEE.

Munim, Z. H., Vladi, O., & Hossain, N. U. I. (2023). Data analytics applications in supply chain resilience and sustainability management: The state of the art and a way forward. *Data Analytics for Supply Chain Networks*, 1–13.

Nathani, M. U., & Singh, J. S. K. (2020). Using blockchain for effective risk management in supply chain: A qualitative study. *Global Business & Management Research*, *12*(3).

Nethravathi, K., Tiwari, Ankit, Uike, Dipesh, Jaiswal, Rachana, & Pant, Kumud (2022). Applications of artificial intelligence and blockchain technology in improved supply chain financial risk management. In *2022 5th International Conference on Contemporary Computing and Informatics (IC3I)*, 242–246, IEEE.

Perboli, G., Musso, S., & Rosano, M. (2018). Blockchain in logistics and supply chain: A lean approach for designing real-world use cases. *IEEE Access*, *6*, 62018–62028.

Petratos, P. N., & Faccia, A. (2023). Fake news, misinformation, disinformation and supply chain risks and disruptions: Risk management and resilience using blockchain. *Annals of Operations Research*, 1–28.

Pu, Wei, Ma, Shuang, & Yan, Xiangbin (2022). Blockchain-Enabled E-Commerce Supply Chain Optimization Considering Consumer Service Level Under Disruption Risks. Available at SSRN 4227663

Queiroz, M. M., & Wamba, S. F. (2019). Blockchain adoption challenges in supply chain: An empirical investigation of the main drivers in India and the USA. *International Journal of Information Management*, *46*, 70–82.

Queiroz, M. M., Telles, R., & Bonilla, S. H. (2020). Blockchain and supply chain management integration: A systematic review of the literature. *Supply Chain Management: An International Journal*, *25*(2), 241–254.

Rasi, R. Z., Rakiman, U. S. B., Radzi, R. Z. R. M., Masrom, N. R., & Sundram, V. P. K. (2021). A literature review on blockchain technology: Risk in supply chain management. *IEEE Engineering Management Review*, *50*(1), 186–200, IEEE

Rahman, S., Ibne Hossain, N. U., Roy, A., & Saiful Islam, M. (2023). Enhancing the viability of green supply chain management initiatives leveraging data fusion technique. In *Data Analytics for Supply Chain Networks* (pp. 15–47). Cham: Springer International Publishing.

Reddy, K. R. K., Gunasekaran, A., Kalpana, P., Sreedharan, V. R., & Kumar, S. A. (2021). Developing a blockchain framework for the automotive supply chain: A systematic review. *Computers & Industrial Engineering*, 157, 107334.

Saberi, S., Kouhizadeh, M., Sarkis, J., & Shen, L. (2019). Blockchain technology and its relationships to sustainable supply chain management. *International Journal of Production Research*, *57*(7), 2117–2135.

Salah, K., Nizamuddin, N., Jayaraman, R., & Omar, M. (2019). Blockchain-based soybean traceability in agricultural supply chain. *IEEE Access*, *7*, 73295–73305.

Schmidt, C. G., & Wagner, S. M. (2019). Blockchain and supply chain relations: A transaction cost theory perspective. *Journal of Purchasing and Supply Management*, *25*(4), 100552.

Shamsuddoha, M., & Kashem, M. A. (2022). A revolutionary paradigm shift in supply chain management: The blockchain mechanism. *Exploring the Latest Trends in Management Literature*, *1*, 15–33.

Shamsuddoha, M., Kashem, M. A., & Koul, S. (2022, December). Systemic review of AI reshaped blockchain applications. In *International Conference on Advanced Network Technologies and Intelligent Computing* (pp. 470–494). Cham: Springer Nature Switzerland.

Shamsuddoha, M., Kashem, M. A., & Nasir, T. (2023). Smart transportation logistics: Achieving supply chain efficiency with green initiatives. In *Data Analytics for Supply Chain Networks* (pp. 243–258). Cham: Springer International Publishing.

Singh, A., Dwivedi, A., & Agrawal, D. (2023). A qualitative study on supply chain risk management adopting blockchain technology. In *Supply Chain Risk and Disruption Management: Latest Tools, Techniques and Management Approaches* (pp. 141–153). Springer.

Tian, Feng (2016). An agri-food supply chain traceability system for China based on RFID & blockchain technology. In *2016 13th International Conference on Service Systems and Service Management (ICSSSM)*, 1–6, IEEE.

Tian, Feng (2017). A supply chain traceability system for food safety based on HACCP, blockchain & internet of things. In *2017 International Conference on Service Systems and Service Management*, 1–6, IEEE.

Treiblmaier, H. (2018). The impact of the blockchain on the supply chain: A theory-based research framework and a call for action. *Supply Chain Management: An International Journal*, *23*(6), 545–559.

Wamba, S. F., Mishra, D. (2017). Big data integration with business processes: A literature review. *Business Process Management Journal*, *23*, 477–492.

Wang, K., Yan, X., & Fu, K. (2020). Research on risk management of agricultural products supply chain based on blockchain technology. *Open Journal of Business and Management*, *8*(6), 2493–2503.

Wang, Y., Han, J. H., & Beynon-Davies, P. (2019a). Understanding blockchain technology for future supply chains: A systematic literature review and research agenda. *Supply Chain Management: An International Journal*, *24*(1), 62–84.

Wang, Y., Singgih, M., Wang, J., & Rit, M. (2019b). Making sense of blockchain technology: How will it transform supply chains? *International Journal of Production Economics*, *211*, 221–236.

Wong, L.-W., Leong, L.-Y., Hew, J.-J., Tan, G. W.-H., & Ooi, K.-B. (2020). Time to seize the digital evolution: Adoption of blockchain in operations and supply chain management among Malaysian SMEs. *International Journal of Information Management*, *52*, 101997.

# Section 2

## Environmental Sustainability and Digitalization

# 7 Environmental Sustainability, Digitization, and Big Data

*Gediminas Lipnickas, Bora Qesja, and Svetlana De Vos*

## 7.1 INTRODUCTION

Environmental sustainability has been increasingly viewed as a crucial issue on a global scale and has been highlighted as a paramount consideration in the 21st century. At the core, environmental sustainability is about protecting human life (Goodland, 1995) by sustaining the global life supporting systems, which provide us with raw materials including food and energy. Rapid depletion of resources, coupled with the degradation of land and atmospheric quality, inflict harm on the global supporting ecosystems (Daily & Ehrlich, 1992). In order to maintain current consumption patterns and lifestyles, production in most parts of the global economy is causing an exhaustion of crucial resources (Goodland, 1995), subsequently making life unsustainable in the long run. Environmental concerns are therefore at the forefront of solving challenges of the future (Secundo et al., 2020). This highlights the need of finding a sustainable solution to meeting the consumption needs of the present without compromising future generations (Charles et al., 2017).

Despite the environmental concerns that have been raised, the global economy continues to grow. Figure 7.1 reflects a continuous upward trend of the global gross domestic product (GDP) reflecting the growth of production and consumption.

The backbone of this growth is, however, energy. As output continues to increase, so does consumption of energy that is required to produce that output or, in many cases, to also use the products and/or services that are produced. Figure 7.2 thus reflects the production and consumption of various forms of energy across the world.

While alternative energy sources are on the rise, oil, coal, and gas make up the major share of energy sources, continuing to deplete these resources as well as polluting the environment in the process. This is reflected in Figure 7.3, showcasing the annual global $CO_2$ emissions. Figures 7.1 and 7.3 display a similar pattern over time. As global production continues to increase, so does $CO_2$ emission, hence contributing towards global warming and adversely altering the environment.

While it is unlikely production and consumption trends will reverse, there is a need for environmental sustainability to become a strategic priority for businesses (Engert et al., 2016; Tomšič et al., 2015). As highlighted by UNESCO, businesses

DOI: 10.1201/9781003438748-9

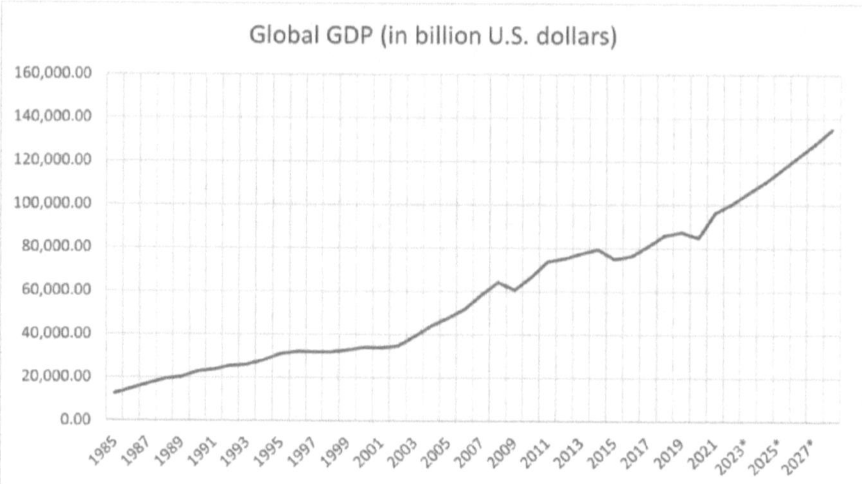


**FIGURE 7.1** Growth over time in global GDP in billions of U.S. dollars.
**(Source: IMF, 2023).**

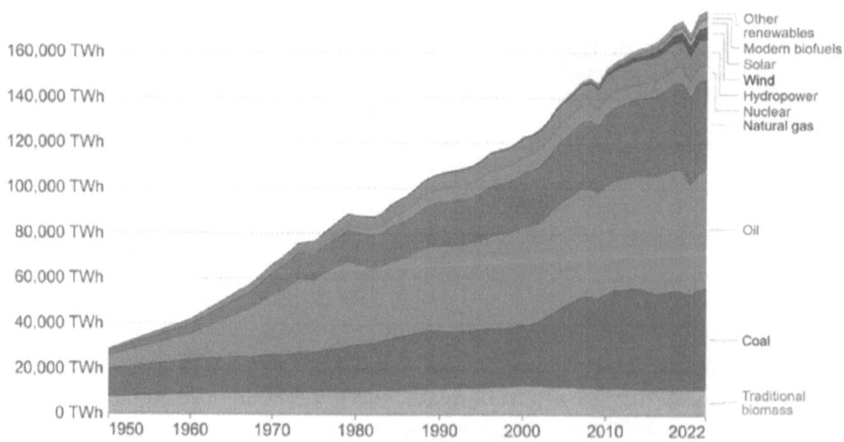


**FIGURE 7.2** Global primary energy consumption by source.
**(Source: Ritchie et al., 2022).**

have a pivotal role to play in ensuring environmentally sustainable practices. UNESCO outlined a range of sustainable development goals that are key in attaining environmental sustainability, among which were innovation, sustainable manufacturing, and sustainable consumption (UNESCO, 2020). This places a significant onus on businesses worldwide to work and improve environmental sustainability. In accordance with the sustainable development goals, future business considerations

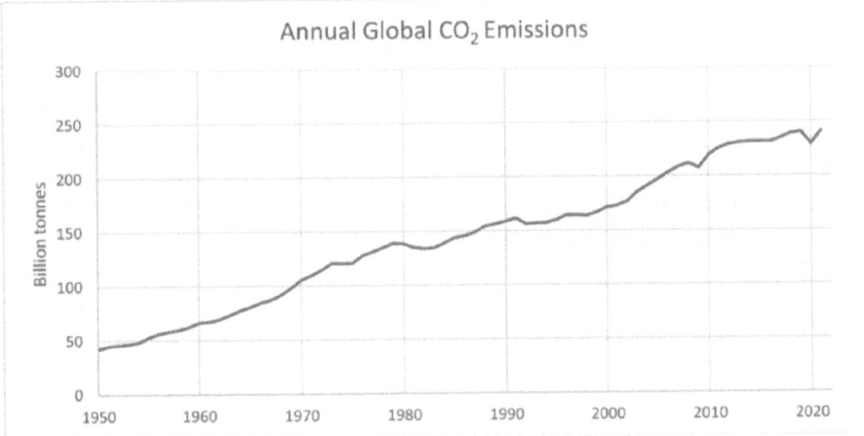


**FIGURE 7.3** Annual global $CO_2$ emissions.

**(Source: Ritchie et al., 2020).**

must thus include environmentally sustainable production processes (Hart & Dowell, 2011) while balancing profitability and consumer demand (Steurer et al., 2005).

According to the OECD, use of digital technologies is key in attaining sustainable goals outlined by UNESCO (Camodeca & Almici, 2021). Businesses should seek to transition towards digital business models which have been shown to improve performance and efficiency (Ash & Burn, 2003; Kauffman & Walden, 2001; Li et al., 2018). Beyond improvements to business performance, digital business models enable scalability and provide a new avenue for businesses to attain a sustainable competitive advantage (Melo et al., 2019). Moreover, digital business models can be further enhanced through the use of big data, enabling opportunities to improve environmental sustainability (Feroz et al., 2021). In addressing the UNESCO sustainability goals, businesses should thus move towards digitization and employ big data to optimize their operations.

## 7.2 LITERATURE REVIEW

### 7.2.1 Sustainability and Competitive Advantage

While environmental sustainability is crucial, businesses transitioning towards a more environmentally sustainable model is not without difficulties. Often mandated by governments, the desire for businesses to become more sustainable ignores the core economic principle driving many businesses. Research has shown that pursuits of environmental sustainability can negatively impact the financial success of companies (Baumgartner & Rauter, 2017; Kurucz et al., 2008). As firms are guided by economic performance, willingness to incorporate sustainable practices can be affected by a perceived decrease in performance (Baumgartner & Rauter, 2017) that would result from implementation of such practices. Transitioning towards a more sustainable model may also put a business's competitive advantage at risk. Having a sustainable competitive advantage is a crucial factor in business survival and success in a competitive market. Competitive advantage is achieved when an organization acquires or develops attributes that allow it to outperform its competitors. These may include access to natural resources, inexpensive energy required for production, or highly skilled human resources. At the core, competitive advantage stems from one of two areas, being able to deliver the same benefits as competitors but at a lower cost or delivering benefits exceeding those of competitors.

The possibility of losing competitive advantage is therefore the main barrier preventing a more sustainable approach for many businesses (Nidumolu et al., 2009). Attempts to make businesses more environmentally sustainable would incur costs associated with transitioning all business operations to a greener, sustainable model. Even if costs were not a factor, businesses also build up expertise which would be impacted as processes change, rendering some skills and expertise obsolete. Improving environmental sustainability can thus inflate costs when compared to companies that do not face such pressures (Nidumolu et al., 2009), as well as rendering much of the built-up knowledge obsolete. This is why global initiatives such as UNESCO, encouraging global transition towards a more sustainable future, are necessary to put firms on equal footing. Transformation towards a digital business model can help overcome challenges associated with becoming more environmentally sustainable. A transition towards a digital business model has been associated with improvements in performance, such as improvements in efficiency, reduction of costs, and optimization of business processes (Ash & Burn, 2003; Kauffman & Walden, 2001; Li et al., 2018). Beyond performance improvements, digital business models enable scalability and a sustainable competitive advantage (Melo et al., 2019), which is a primary roadblock preventing firms from transitioning.

### 7.2.2 Digital Business, Competitive Advantage, and Sustainability

With the advancement of technology in the last few decades and the rapid price decline of that technology (Boland et al., 2007), the global economy has rapidly transitioned to the digital space with businesses increasingly embracing a digital business model, where resource use is optimized through digital technology (Bharadwaj et al., 2013). These digital business models are, however, not mere counterparts to traditional business operations or to be seen as separate. Digital business models encompass all the strategic functions and processes of a business and enhance them through the use of digital resources (Bharadwaj et al., 2013). The integration of digital business models, complemented by a digital business strategy, embodies a 'transfunctional' approach to business, encompassing diverse functional areas such as marketing, logistics, operations, and others (Bharadwaj et al., 2013).

Transforming businesses to digital involves fundamental changes in business processes, operations, and capabilities (Markus, 2004; Chen et al., 2014; Tan et al., 2015; Li et al., 2018). A transition towards a digital business model will involve a technological shift, either transitioning wholly into a service-oriented business, or making physical products programmable, communicable, and traceable (Yoo, 2010; Yoo et al., 2010). This transition thus fundamentally reshapes traditional business strategy, expanding it across numerous boundaries like time, distance, and function (Straub & Watson, 2001; Subramaniam & Venkatraman, 2001; Sambamurthy et al., 2003; Ettlie & Pavlou, 2006; Kohli & Grover, 2008; Rai et al., 2012). Numerous opportunities are created in the process to develop new, future-proof competitive advantages (Burgelman & Grove, 2007). For one, competitive advantage can be achieved or enhanced through optimized use of resources using digital technology (Bharadwaj et al., 2013). By transitioning to a digital business model, businesses can innovate, creating novel products and services through new combinations of digital and physical components (Feroz et al., 2021; Yoo et al., 2010). Take for example, Amazon redefining the book-selling ecosystem via moving away from purely physical locations to selling online, as well as creating a digital book ecosystem. By doing this, Amazon redefined the competitive advantage in the book category while continuing to build upon Amazon Web Services and developing other business streams. Similarly, companies like Microsoft or Adobe transitioned towards software as a service (SaaS) model, moving away from distributing software through physical medium towards it being accessible online, thus eliminating traditional forms of distribution and cutting down on pollution. These transformations impact not only the relative competitive advantages of companies as costs can be passed down to consumers or benefits can be enhanced but also have implications for sustainability (Schallmo et al., 2017; Matt et al., 2015).

Transforming businesses into digital counterparts will significantly impact short- and long-term resource allocation and use, creating more sustainable market activities (Li et al., 2020; Minatogawa et al., 2019). Transitioning to digital can help to significantly reduce waste, as in, for example, SaaS business models or other alternatives like Netflix. Elimination of physical products that must be distributed globally eliminates the potential waste that is created by manufacturing, shipping, and disposing of the physical goods, in addition to eliminating the energy consumption required for this process. These practices may even contribute towards the ideas of a circular economy, where products are reused instead of disposed of (Centobelli et al., 2020; Kirchherr et al., 2017), significantly reducing wastage. Through a more efficient allocation and use of resources (Li et al., 2020; Minatogawa et al., 2019), businesses can not only become more scalable and competitive but also become more environmentally friendly (Centobelli et al., 2020). In addition to enabling businesses to create new forms of sustainable competitive advantage while helping the environment, this transition can also foster more efficient consumption practices (Bican & Brem, 2020).

However, digital businesses rely on rich information sources inside and outside of the organization (Rai et al., 2012). Due to this reliance on data, digital business strategies are broader, more embedded, and more encompassing (Bharadwaj et al., 2013). In the view of Bharadwaj et al. (2013), in time '*digital business*' will be ubiquitous. Because of the transition already occurring, products and services are increasingly embedded with digital technologies (El Sawy, 2003), where digital business models can be even seen as a hybrid by incorporating traditional IT practices (Bharadwaj et al., 2013). As all businesses transition towards digital, data will become a crucial factor in success, enabling optimization of business practices and creating more sustainable avenues for value creation (Centobelli et al., 2020). Efficient use of data will thus be key in deriving a competitive advantage in the future.

### 7.2.3 Emergence of Big Data

In 2006, Clive Humby had an extraordinary vision when he coined the phrase "Data is the new oil". Since the dot-com era, the digital revolution has resulted in technology becoming inseparable from the daily lives of many, from the creation of ubiquitous social platforms connecting people across the globe and cloud computing revolutionizing the way information is stored and accessed to the widespread adoption of smartphones. With the advent of the Internet of Things (OECD, 2012), the interconnected network of devices only further expands the continuous data generation that has been occurring since the invention of the internet. This abundance of data has the potential to be a treasure trove of insights and opportunities and should be used as a source of value creation in the modern digital economy. The incredible wealth of information that is created in the digital economy is often available for everyone to access. Businesses can thus gain access to valuable information that was previously out of reach and employ data-based decision making in their operations. The continuous generation of data in the digital economy is often referred to as 'big data' (Ohlhorst, 2012). It is referred to as such due to the size and complexity, requiring a specialized approach to view, store, manage, and analyze (Chen et al., 2012; Erevelles et al., 2016).

Due to its size and complexity, big data is often viewed in accordance with the 5V model, categorizing the composition of big data into volume, velocity, variety, veracity, and value (Demchenko et al., 2014). Within this categorization, *volume* is viewed as the amount of data that exists and can be accessed. Due to the composition and size, specialized knowledge is needed to process and analyze the data. The volume of data will differ across different businesses and/or industries, where, for example, financial service providers will have access to data generated by their platforms/systems and government data, as well as consumer/user data. Social media companies, on the other hand, will have an abundance of data generated by consumers themselves (such as social media posts of users). *Velocity* represents real-time processing of data, including collection and analysis. Businesses of various sizes may differ in their ability to generate and use big data. Social media companies such as X (formerly Twitter) have tools to draw insights from data collected in real time. *Variety* represents the type of data that is generated and collected, including internal and external to the organization. This data may be structured or unstructured, where structured data is quantitative, often encompassing financial or performance information such as sales figures. Unstructured data is often qualitative, for example, data that is generated by consumers, such as contents of an email or social media posts. *Veracity* represents the authenticity of data that is collected. For the data to be useful and to enable data-driven decision making, it should be of high quality and accuracy. Gaps in data resulting from failure to capture it, or due to errors in the system, render it less accurate and useful. Finally, *value* represents the usefulness that can be generated from big data supporting the decision making of businesses (Watson, 2014).

With digitization, data permeates every facet of life, enabling businesses to utilize large volumes of data from a range of sources (Rowley, 2002). Digital transformation brings together firms from previously unrelated industries (Yoo et al., 2010) and enables data exchange, generation, and analysis to develop insights that were previously not possible. Where previously businesses relied on research or scarce data, there is now an overwhelming abundance of information that is generated by businesses, governments, and consumers and which can be used to optimize business practices given the know-how. Most importantly, the ability to draw insights from data can become one of the core competencies of a company (Schallmo et al., 2017). Developing such competencies can lead to improved business results and an overall competitive advantage derived through efficiency or enhanced benefits for consumers. While difficulty in utilizing big data has been noted, it can partially be attributed to lack of vision and understanding of the capabilities that can be extracted from its use. Just as competitive advantage requires skills and expertise, so does the use of big data, and companies must develop this expertise if they are to take advantage of this incredible resource.

### 7.2.4 Role of Big Data in Competitive Advantage

Transitioning towards a digital business model not only fundamentally alters the strategic approach of the business but has implications for all business aspects and stakeholders (Schallmo et al., 2017; Matt et al., 2015). The abundance of data can be utilized to reach solutions that were not possible before (Chen et al., 2012; Xu et al., 2016; Wang et al., 2016); that is, with access to various sources of data both internally and externally to the organization, as well as real-time data from various platforms, businesses can facilitate continuous improvement and achieve a sustainable competitive advantage. In this context, marketplace and customer behavior information can be derived from big data (Erevelles et al., 2016; Wang et al., 2016). It can also assist in informing a course of action (Picciano, 2012; Provost & Fawcett, 2013) and be the basis for decision making in the digital sphere (Davenport, 2014). This can lead to customer journey mapping as well as improved relationships between companies and customers (Chen et al., 2012; Xu et al., 2016).

Big data has been increasingly used to improve relationships between business and customers. While customer relationship management (CRM) is not a new process, strategic use of big data can significantly enhance its effectiveness. CRM was born in the IT sector and is underpinned by use of technology (Payne & Frow, 2005). It is only natural that big data would further the purpose of CRM. At the core, CRM is about identifying different types of customers and designing strategies for interacting with them, with the aim of maximizing customer value (Kumar & Reinartz, 2018). That is, businesses should aim to develop better relationships with profitable customers, locate and attract new customers with the aim of making them profitable, and find strategies to deal with unprofitable ones. For example, data may reveal unprofitable customers occupying valuable staff time with simple queries that could be addressed by a frequently asked questions (FAQ) page, or a chatbot. These insights may come from internal company data or external, for example, interactions via social media profiles/pages.

With digitization and the growth of Web 2.0, traditional CRM techniques are becoming outdated, with companies increasingly relying on big data channels such as social media (Anshari et al., 2019). Using social media data, customer conversations can be analyzed, enabling more accurate profiling and tailoring products/services for specific interests/preferences. Big data can thus be used to not only profile and service customers better, but it can help achieve a sustainable competitive advantage over rivals due to creating better customer benefits (Chen et al., 2012). Alternatively, competitive advantage can be gained from reducing costs. Utilizing big data, firms can identify inefficiencies in their operations and employ technology to overcome them (Payne & Frow, 2005). For example, technology could be used to establish automated telephone services, web services, self-service facilities, chat bots, or other resources enabling more efficient customer service/interaction. Even traditionally brick-and-mortar businesses can enhance their operations utilizing big data, from optimizing in-store customer interactions via self-service to online services via apps. McDonald's is an example of a business utilizing various channels to optimize their operations. From in-store self-service to their web presence, as well as using apps to enhance customer experience and offer additional benefits. All of these sources generate data and can be used to continuously improve and refine the services offered, creating a sustainable competitive advantage.

With many industries moving towards the digital space, they are becoming more service-based, and consumer oriented (Weill & Woerner, 2013). With the transition towards digital, traditional value creation via products shifts towards the platform, content, and experience in the digital space (Weill & Woerner, 2013). While digital transformation can help in attaining efficiency, it often comes with increased complexity (Parida et al., 2019; Nambisan et al., 2017), requiring new skills and capabilities. As highlighted by the CRM discussion, customers are at the core of ensuring business profitability and sustainable competitive advantage.

Hence, customer journey needs to be considered as a key factor in delivering value (Kuehnl et al., 2019). Defined as the process the customer goes through, it encompasses all stages of company–customer interactions (pre-purchase, purchase, and post purchase) and touchpoints (either company or non-company controlled, digital or non-digital) (Lemon & Verhoef, 2016). The customer journey therefore requires understanding in order to cater to customer preferences and habits, and create desired customer value (Minatogawa et al., 2019).

Due to big data encompassing various sources and types of data, it can be extremely useful in examining the totality of customer journey. Utilization of data from various sources can be traced through the customer journey stages inside and outside of the organization. First, organizations own large sets of data internally, typically segregated by departments. As a result, different departments have partial views of the whole customer journey and a fragmented view of the total business operations. Instead of aimlessly combining datasets with no real intent or purpose to their use, the first step should be understanding what behaviors/processes should be optimized. Effectively, what is driving the customer experience? And what can be optimized to reduce costs and provide a smoother journey? In one form or another, all businesses will arrive at a similar core question, how can we improve customer experience and achieve higher financial return?

Relevant data must then be collated to enable the required answers to be drawn from it. While most digital business models will have platform related data, it must be combined with other forms such as financial as well as social media data to create a more encompassing picture. While individual sources of data may convey usage statistics of the platform or number of likes on social media posts, adequate combination and use of this data enables understanding of customer habits that lead to revenue generation. For example, inclusion of sales data as well as social media sentiment data may yield unexpected insights. Social media sentiment may drive traffic to the site, while specific sentiment may prompt higher sales or specific page use behaviors. Over time, many businesses can over incumber their digital presence. With big data, online services can be streamlined accounting for not only what customers want but how they want it. That is, businesses often develop numerous services, accompanied by web pages and files containing information or instructions and FAQ sections. But are the resources used? Take, for example, an over-incumbered website. Whilst data might show high usage/engagement statistics, consumers with these statistics might be spending less money on the platform or not completing purchases. On the other hand, consumers with higher expenditure might take a streamlined path to purchasing, ignoring most pages/resources. Sales being the ultimate goal, big data use can indicate that website bloat is preventing sales. Effort and resources, however, go into developing these pages and maintaining them, with likely social media posts mirroring the content. Resources that are not useful in making sales or maximizing customer profitability are therefore inflating costs and could be optimized.

Use of data in a such a way can deliver customer-driven services (Davenport, 2014) and enhance performance, creating more sustainable businesses (Minatogawa et al., 2019). Furthermore, big data could be used to identify trends or gaps in the market so that businesses are able to better take advantage of them. For example, Nidumolu et al. (2009) outlines how Procter & Gamble identified high energy use by U.S. households when washing clothes in hot water. The result was a development of cold-water products, helping customers save money and providing P&G with a competitive advantage in the respective category. Big data, whether from within the organization or outside of it, can thus be used in a variety of ways, offering numerous avenues for companies to create a sustainable competitive advantage (Bican & Brem, 2020). The scope and applicability of big data is only limited by the scope of the goal and ingenuity of the user.

### 7.2.5 Role of Big Data in Environmental Sustainability

The transition towards a digital business model enables big data generation while creating unique opportunities for fostering and enabling environmental sustainability (Feroz et al., 2021). A digital business model and big data usage allow the optimization of a range of processes (Rosário & Dias, 2022) and resources (Bharadwaj et al., 2013) in business operations, as well as in understanding and attaining customers, managing their journey, and offering desirable benefits. Understanding customers better and serving their needs throughout the customer journey, or CRM may shorten their journey and enable creation of resources that would significantly cut down the time consumers require to achieve a certain outcome. Optimizing the process and reducing the time consumers spend to attain the desired benefits reduces energy consumption related to activities. Resource optimization within the business confines translates to fewer resources wasted, thus improving environmental sustainability. For example, Amazon utilizes big data across all their operations. Data is obtained from suppliers, warehouses, and customers and is used to predict customer interests and suggest products that they may be looking for, streamlining the purchasing process (Galea-Pace, 2020). Inventory and location are also tracked to ensure efficient order fulfillment, resulting in dispatch from the closest warehouse, cutting down on shipping times and associated costs (Galea-Pace, 2020). Through streamlining the process, environmental sustainability is enhanced across all operations. Consumers being able to complete purchases faster results in reduced power usage for the consumer, as well as via Amazon data centers that power the platform. Furthermore, optimization of the supply chain and order fulfillment similarly reduces pollution resulting from distribution activities.

Another retailing giant, Walmart, places significant strategic emphasis on pushing sustainability goals and utilizes big data to do so (Walmart, 2023a). However, instead of optimizing only their operations, their strategic priorities extend to their suppliers, as well as other retailers and organizations. Walmart's approach attempts to create a circular economy concept, where their efforts extend to changing supplier practices via certification requirements, offering useful tools, resources, and expertise in driving sustainable practices to engaging customers in recycling and sustainable consumption. Products compliant with certification requirements are tagged with RFID chips that enable monitoring of their movement across the supply chain, identifying inefficiencies and allowing for improvements in performance. Consumer data is combined with a range of other data such as economic, social media, or even weather to estimate and predict consumer behavior and appropriate response across Walmart's operations. Walmart seeks to design products to enable recycling and incentivizes consumers to do so with incentives before and after purchase (Walmart, 2023b).

As another company that optimizes both the company and consumer side of operations, Tesla utilizes data from consumer vehicles in optimizing their battery management and energy consumption, allowing vehicles to travel further (Tesla, 2023). Not only does this provide valuable insights for Tesla but improves the product, as well as enabling the customer to make more informed decisions in relation to their product usage and travel plans.

Energy smart grids also utilize a range of technologies and data to examine electricity demand and supply on the grid. This enables efficient electricity supply, minimizing costs and the environmental impact (IEA, 2023a). For example, Italian energy company Enel uses big data to optimize their power grids, balancing supply and demand across them and enabling more efficient integration of renewable sources of power into the grid, cutting down on the reliance on coal and other polluting sources (Enel, 2020). Smart grids powered by big data also enable consumer-side optimization. Consumers can track their power consumption and the associated costs, which allows optimization of usage. Similarly, consumers who have solar panels can track their solar usage, generation, and how much gets fed back into the grid, together with any associated financial incentives. Feeding back solar into the grid also allows the energy provider to better optimize power usage.

Farming can also be enhanced using big data. Through development of digital agriculture platforms, big data can be gathered from a range of sources (i.e., sensors) to optimize farming practices. This enables real-time tracking and optimization of resource use and practices based on weather conditions, soil health, crop performance, and other indicators (CSIRO, n.d.). This extends to the aquaculture industry, where companies such as Aquabyte offer big data–powered solutions to enhance fish farming. Utilizing a range of data points via cameras and sensors, Aquabyte enables optimization of the fish farming practices, including feeding, and improves fish welfare while reducing environmental impact (Aquabyte, 2023).

Going beyond farming, big data solutions are also used to optimize and eliminate food wastage. Wasteless is a company that uses real-time data within supermarkets to dynamically adjust pricing of perishable items based on demand, inventory, and expiration dates, with the aim of eliminating food wastage (Wasteless, 2023). Further examples can be drawn from the software as a service category. Whether it is Amazon revolutionizing book sales or Netflix creating an entertainment platform over internet, much of the wastage that was previously created within the supply chain is eliminated. While all these services can be optimized using big data, for some businesses data serves as the backbone of their operations. For example, car parking apps are increasingly utilizing big data to track parking space availability, directing consumers to empty spaces that they can often pre-book, thus reducing the time spent in the vehicle and by proxy reducing pollution and energy consumption. Another example relates to sharing economy and companies such as Uber. Car sharing has contributed to urban sustainability, among which is reduction in air pollution, energy use, improvement in biodiversity, and other benefits (Roblek et al., 2021). Similarly, car sharing can have a drastic impact on reducing energy consumption and $CO_2$ emissions (Baptista et al., 2014). As ownership of vehicles decreases, so does resource consumption required to manufacture and maintain them. Use of big data in optimizing such business models can help in moving away from business models sustained by planned obsolescence (Minatogawa et al., 2019) towards the use of digital platforms where products are shared, reducing wastage in manufacturing as well as disposal. Further examples of benefits extend to reduction in traffic congestion, creating more efficient traffic flow and reducing pollution (Schaller, 2021). In Uber's case (and other rideshare companies), big data is also used in real time to ensure efficiency of service. An example would include so-called 'surge pricing', where prices are adjusted based on supply and demand. Depending on the availability of drivers and the demand for the service, dynamic adjustments to pricing ensure the service effectively optimizes itself based on data alone. With fewer drivers in the area and high demand, prices would increase, drawing more drivers to the area to serve the surge in demand.

Digital business models and big data also enable creation of businesses or organizations with the sole purpose of environmental sustainability (Feroz et al., 2021). For example, the Ocean Cleanup is an organization that utilizes a range of data sources across the world to prevent and reduce ocean pollution, remove plastic from the oceans, and improve environmental sustainability (The Ocean Cleanup, 2023a). The organization tracks plastic pollution across the globe, intercepting it in rivers before it can enter the ocean, while also removing plastic from the ocean. This is enabled by the use of big data sources from a range of sensors, government agencies, and even individuals across the world (The Ocean Cleanup, 2023b). Similarly, the Freshwater Trust utilizes a range of data sources to monitor and manage water usage and river conservation and overall ecological outcomes (The Freshwater Trust, 2023).

Big data therefore has a significant role to play in making businesses more sustainable in the 21st century and solving the challenges of the future, from optimizing business practices and operations to optimizing delivery of customer benefits via an improved understanding of customers and their behaviors (Wang et al., 2016). Through effective use of data, a feedback loop can be created between sustainable business and sustainable environmental impact.

## 7.3 PRACTICAL IMPLICATIONS

### 7.3.1 Small Companies and Big Data

With the advent of big data, "timely and cost-effective analytics over Big Data is now a key ingredient for success in many businesses" (Rodríguez-Mazahua et al., 2016, p. 3078).

While this data is typically utilized by larger organizations who have the necessary resources and expertise, small businesses can also benefit from big data. According to the estimates of the World Bank, small to medium enterprises account for up to 90% of businesses worldwide (World Bank, n.d.). It is thus imperative that the transition towards digital and subsequent big data use be employed by small companies to attain the desired sustainable outcomes.

First, disparity in resources must be acknowledged between large and small businesses. Lacking an abundance of resources, smaller organizations will need to be more strategic and focused in their approach to big data analysis. Defining clear objectives and goals will enable businesses to obtain the relevant data and draw data-driven conclusions. While all businesses generate data, smaller organizations might be limited in the type of data generated (or captured) or its quantity. While internal data should be captured, companies should consider other relevant sources to develop robust insights. For example, social media offers data that can be collected, ranging from insights stemming from advertising campaigns to consumer interactions, likes, comments, and discourse. Website traffic can be captured via Google Analytics, providing insights into web resource usage. Further data can be obtained from government organizations that consistently release data pertaining to the economy, for example, via the Census Bureau or Bureau of Labor Statistics (for US companies). Other digital sources that constantly publish large sets of data or indicate data sources include Data Reportal, Data Is Plural, and Google Dataset Portal. Finally, data is increasingly being published by academic researchers (for example, open-source journal *Data In Brief*).

As by definition, big data is a large quantity of data, often continuously generated in real time, businesses require the capacity to store and retrieve this data. Small businesses are constrained by their limited resources. However, with digitization, numerous companies offer services that could be utilized in order to take advantage of big data. Services such as Amazon Web Services or Google Cloud Platform to store and use data affordably eliminate the need for expensive hardware acquisition. An abundance of software solutions exist that make big data analytics relatively simple, for example, Q Research Software, Tableau, Power Bi, and many others.

In addition, programing languages such as R or Python can also be used. With the numerous tools available to businesses, the advent of cloud computing has made big data analytics cost-effective, reliable, and readily accessible. Small businesses can thus also take advantage of the wealth of information available online and similarly optimize their business models like their larger counterparts.

## 7.4 CHALLENGES AND FUTURE DIRECTIONS IN ENVIRONMENTAL SUSTAINABILITY AND BIG DATA USE

Digitization and use of digital technologies enable unique opportunities to create new business models that can focus on environmental sustainability (Feroz et al., 2021). For example, Goralski and Tan (2020) discuss big data–powered solutions such as smart water management systems or big data–trained artificial intelligence solutions to identify waterborne diseases. While these are an option, businesses seeking to transition to digital business models often do not place environmental considerations at the core of their strategic intent (Feroz et al., 2021). The challenge will thus be incentivizing digital transformation that also contributes to environmental sustainability. Moreover, among the biggest challenges of big data use is building an environmentally sustainable infrastructure that can facilitate storage and analysis of data (Villars et al., 2011). While big data use by businesses can enable more sustainable environmental practices, the very use of big data has a significant environmental impact that requires mitigation. As data generation continues to grow, this challenge will extend to many companies seeking to utilize or facilitate big data use. Data centers required to facilitate big data storage and use (in addition to other services) require significant resources and are estimated to account for up to 1.5% of global electricity use (IEA, 2023b). As big data use leads to efficiency gains, it will also likely result in increased economic activity (Beier et al., 2020; Camodeca & Almici, 2021). The use of big data in the digital economy can thus inadvertently contribute to higher consumption and pollution stemming from the use of digital technologies (Piscicelli, 2023).

As the digitization trend continues, accompanied by exponential growth in data generation, sustainable solutions must be found to accommodate the digital economy and reduce environmental impact. Microsoft, for example, plans on reducing the impact and reach carbon negativity in their data centers by 2030 via achieving operational efficiency via optimization of energy and water use (Walsh, 2022).

### 7.4.1 Big Data Use and AI

Through the use of big data, artificial intelligence (AI) tools can be trained to simulate human intelligence, as for example in the case of OpenAI and ChatGPT (OpenAI, 2023a). With the advent of large language models such as ChatGPT, many uses have already been outlined for artificial intelligence. These uses range from applications in healthcare (Haleem et al., 2022), such as the use of chatbots, diagnosis, and treatment recommendations, to similar applications across education, business, financial, legal services, and data analytics (Ray, 2023; Taecharungroj, 2023). Services such as ChatGPT, and others built upon it, are increasingly completing tasks more efficiently than some professions might be able to. While currently there are challenges with the reliability and accuracy of such models, they are continuously improving as they are used, generating data in the process (Ray, 2023). As AI application becomes increasingly multimodal, being able to utilize a range of data formats, such as text, video, and others (Ray, 2023), applications of AI will begin replacing standalone software and will be used to optimize whole systems based on data, as in the example of Amazon and their platform and distribution.

### 7.4.2 Big Data and the Circular Economy

The use of data and digital technologies reduces the use of resources and enhances circular models (García-Muiña et al., 2021; Rosário & Dias, 2022). Optimization attained from the use of big data and digitization can facilitate a circular economy, where products are reused instead of disposed of (Centobelli et al., 2020; Kirchherr et al., 2017), significantly reducing wastage. Such optimization can also lead to a reduction of resource and energy consumption and higher productivity (Piscicelli, 2023). Examples include Walmart optimizing their supply chain, incentivizing better manufacturing practices as well as recycling; Amazon reducing pollution via optimized order fulfilling; and others. Big data is thus a powerful tool that can bring significant benefits to its users.

## REFERENCES

Anshari, M., Almunawar, M. N., Lim, S. A., & Al-Mudimigh, A. (2019). Customer relationship management and big data enabled: Personalization & customization of services. *Applied Computing and Informatics*, *15*(2), 94–101. https://doi.org/10.1016/j.aci.2018.05.004

Aquabyte. (2023). *Radically Better Fish Farming*. https://aquabyte.ai/

Ash, C. G., & Burn, J. M. (2003). Assessing the benefits from e-business transformation through effective enterprise management. *European Journal of Information Systems*, *12*(4), 297–308. https://doi.org/10.1057/palgrave.ejis.3000476

Baptista, P., Melo, S., & Rolim, C. (2014). Energy, environmental and mobility impacts of car-sharing systems. Empirical results from Lisbon, Portugal. *Procedia-Social and Behavioral Sciences*, *111*, 28–37. https://doi.org/10.1016/j.sbspro.2014.01.035

Baumgartner, R. J., & Rauter, R. (2017). Strategic perspectives of corporate sustainability management to develop a sustainable organization. *Journal of Cleaner Production*, *140*, 81–92. https://doi.org/10.1016/j.jclepro.2016.04.146

Beier, G., Fritzsche, K., Kunkel, S., Matthess, M., Niehoff, S., Reißig, M., & van Zyl-Bulitta, V. (2020). A green digitalized economy? Challenges and opportunities for sustainability. *Institute for Advanced Sustainability Studies (IASS) Fact Sheet*, *1*, 1–11. https://doi.org/10.2312/iass.2020.028

Bharadwaj, A., El Sawy, O. A., Pavlou, P. A., & Venkatraman, N. V. (2013). Digital business strategy: Toward a next generation of insights. *MIS Quarterly*, *37*, 471–482.

Bican, P. M., & Brem, A. (2020). Digital business model, digital transformation, digital entrepreneurship: Is there a sustainable "digital"? *Sustainability*, *12*(13), 5239–5254. https://doi.org/10.3390/su12135239

Boland Jr, R. J., Lyytinen, K., & Yoo, Y. (2007). Wakes of innovation in project networks: The case of digital 3-D representations in architecture, engineering, and construction. *Organization Science*, *18*(4), 631–647. https://doi.org/10.1287/orsc.1070.0304

Burgelman, R. A., & Grove, A. S. (2007). Cross-boundary disruptors: Powerful interindustry entrepreneurial change agents. *Strategic Entrepreneurship Journal*, *1*(3-4), 315–327. https://doi.org/10.1002/sej.27

Camodeca, R., & Almici, A. (2021). Digital transformation and convergence toward the 2030 agenda's sustainability development goals: Evidence from Italian listed firms. *Sustainability*, *13*(21), 11831. https://doi.org/10.3390/su132111831

Centobelli, P., Cerchione, R., Chiaroni, D., Del Vecchio, P., & Urbinati, A. (2020). Designing business models in circular economy: A systematic literature review and research agenda. *Business Strategy and the Environment*, *29*(4), 1734–1749. https://doi.org/10.1002/bse.2466

Charles Jr, O. H., Schmidheiny, S., & Watts, P. (2017). *Walking the talk: The business case for sustainable development*. Routledge.

Chen, H., Chiang, R. H., & Storey, V. C. (2012). Business intelligence and analytics: From big data to big impact. *MIS Quarterly*, *36*(4), 1165–1188.

Chen, J. E., Pan, S. L., & Ouyang, T. H. (2014). Routine reconfiguration in traditional companies' e-commerce strategy implementation: A trajectory perspective. *Information & Management*, *51*(2), 270–282. https://doi.org/10.1016/j.im.2013.11.008

CSIRO. (n.d.). *Rural Intelligence Platform*. https://research.csiro.au/cceo/rural-intelligence-platform/

Daily, G. C., & Ehrlich, P. R. (1992). Population, sustainability, and Earth's carrying capacity. *BioScience*, *42*(10), 761–771. https://doi.org/10.2307/1311995

Davenport, T. H. (2014). How strategists use "big data" to support internal business decisions, discovery and production. *Strategy & Leadership*, *42*(4), 45–50. https://doi.org/10.1108/SL-05-2014-0034

Demchenko, Y., De Laat, C., & Membrey, P. (2014, May). *Defining architecture components of the big data ecosystem*. 2014 International conference on collaboration technologies and systems (CTS). https://doi.org/10.1109/CTS.2014.6867550

El Sawy, O. A. (2003). The IS Core IX: The 3 faces of IS identity: Connection, immersion, and fusion. *Communications of the Association for Information Systems*, *12*, Article 39. https://doi.org/10.17705/1CAIS.01239

Enel. (2020). *i-EM: Empowering a More Sustainable Future with Data Analytics*. https://openinnovability.enel.com/stories/articles/2020/07/smart-grid-analysis-with-i-em-big-data

Engert, S., Rauter, R., & Baumgartner, R. J. (2016). Exploring the integration of corporate sustainability into strategic management: A literature review. *Journal of Cleaner Production*, *112*, 2833–2850. https://doi.org/10.1016/j.jclepro.2015.08.031

Erevelles, S., Fukawa, N., & Swayne, L. (2016). Big Data consumer analytics and the transformation of marketing. *Journal of Business Research*, *69*(2), 897–904. https://doi.org/10.1016/j.jbusres.2015.07.001

Ettlie, J. E., & Pavlou, P. A. (2006). Technology-based new product development partnerships. *Decision Sciences*, *37*(2), 117–147. https://doi.org/10.1111/j.1540-5915.2006.00119.x

Feroz, A. K., Zo, H., & Chiravuri, A. (2021). Digital transformation and environmental sustainability: A review and research agenda. *Sustainability*, *13*(3), 1530–1550. https://doi.org/10.3390/su13031530

Galea-Pace, S. (2020). *How Amazon uses Big Data to Transform Operations*. SupplyChain. https://supplychaindigital.com/technology/how-amazon-uses-big-data-transform-operations

García-Muiña, F., Medina-Salgado, M. S., González-Sánchez, R., Huertas-Valdivia, I., Ferrari, A. M., & Settembre-Blundo, D. (2021). Industry 4.0-based dynamic social organizational life cycle assessment to target the social circular economy in manufacturing. *Journal of Cleaner Production*, *327*, 129439. https://doi.org/10.1016/j.jclepro.2021.129439

Goodland, R. (1995). The concept of environmental sustainability. *Annual Review of Ecology and Systematics*, *26*(1), 1–24. https://doi.org/10.1146/annurev.es.26.110195.000245

Goralski, M. A., & Tan, T. K. (2020). Artificial intelligence and sustainable development. *The International Journal of Management Education*, *18*(1), 1–9. https://doi.org/10.1016/j.ijme.2019.100330

Haleem, A., Javaid, M., & Singh, R. P. (2022). An era of ChatGPT as a significant futuristic support tool: A study on features, abilities, and challenges. *BenchCouncil Transactions on Benchmarks, Standards and Evaluations*, *2*(4), 100089. https://doi.org/10.1016/j.tbench.2023.100089

Hart, S. L., & Dowell, G. (2011). Invited editorial: A natural-resource-based view of the firm: Fifteen years after. *Journal of Management*, *37*(5), 1464–1479. https://doi.org/10.1177/0149206310390219

Humby, C. (2006). *Data is the new oil*. ANA Senior Marketer's Summit. Evanston, IL, USA.

IEA. (2023a). *Smart Grids*. www.iea.org/energy-system/electricity/smart-grids

IEA. (2023b). *Data Centres and Data Transmission Networks*. www.iea.org/energy-system/buildings/data-centres-and-data-transmission-networks

International Monetary Fund (IMF). (2023). *World Economic Outlook Database April 2023 Edition*. www.imf.org/en/Publications/WEO/weo-database/2023/April

Kauffman, R. J., & Walden, E. A. (2001). Economics and electronic commerce: Survey and directions for research. *International Journal of Electronic Commerce*, *5*(4), 5–116. https://doi.org/10.1080/10864415.2001.11044222

Kirchherr, J., Reike, D., & Hekkert, M., 2017. Conceptualizing the circular economy: An analysis of 114 definitions. *Resources, Conservation and Recycling*, *127*, 221–232. https://doi.org/10.1016/j.resconrec.2017.09.005

Kohli, R., & Grover, V. (2008). Business value of IT: An essay on expanding research directions to keep up with the times. *Journal of the Association for Information Systems*, *9*(1), 23–39. https://doi.org/10.17705/1jais.00147

Kuehnl, C., Jozic, D., & Homburg, C. (2019). Effective customer journey design: Consumers' conception, measurement, and consequences. *Journal of the Academy of Marketing Science*, *47*, 551–568. https://doi.org/10.1007/s11747-018-00625-7

Kumar, V., & Reinartz, W. (2018). *Customer relationship management*. Springer-Verlag GmbH Germany.

Kurucz, E. C., Colbert, B. A., & Wheeler, D. (2008). The business case for corporate social responsibility. In A. Crane, D. McWilliams, J. Matten, Moon, & D. Siegel (Eds.), *The Oxford Handbook on Corporate Social Responsibility* (pp. 83–112). Oxford University Press, Oxford. https://doi.org/10.1093/oxfordhb/9780199211593.003.0004

Lemon, K. N., & Verhoef, P. C. (2016). Understanding customer experience throughout the customer journey. *Journal of Marketing*, *80*(6), 69–96. https://doi.org/10.1509/jm.15.0420

Li, L., Su, F., Zhang, W., & Mao, J. Y. (2018). Digital transformation by SME entrepreneurs: A capability perspective. *Information Systems Journal*, *28*(6), 1129–1157. https://doi.org/10.1111/isj.12153

Li, X., Cao, J., Liu, Z., & Luo, X. (2020). Sustainable business model based on digital twin platform network: The inspiration from haier's case study in China. *Sustainability*, *12*(3), 936–962. https://doi.org/10.3390/su12030936

Markus, M. L. (2004). Technochange management: Using IT to drive organizational change. *Journal of Information Technology*, *19*, 4–20. https://doi.org/10.1057/palgrave.jit.2000002

Matt, C., Hess, T., & Benlian, A. (2015). Digital transformation strategies. *Business & Information Systems Engineering*, *57*, 339–343. https://doi.org/10.1007/s12599-015-0401-5

Melo, S., Macedo, J., & Baptista, P. (2019). Capacity-sharing in logistics solutions: A new pathway towards sustainability. *Transport Policy*, *73*, 143–151. https://doi.org/10.1016/j.tranpol.2018.07.003

Minatogawa, V. L. F., Franco, M. M. V., Rampasso, I. S., Anholon, R., Quadros, R., Durán, O., & Batocchio, A. (2019). Operationalizing business model innovation through big data analytics for sustainable organizations. *Sustainability*, *12*(1), 277. https://doi.org/10.3390/su12010277

Nambisan, S., Lyytinen, K., Majchrzak, A., & Song, M. (2017). Digital innovation management. *MIS Quarterly*, *41*(1), 223–238.

Nidumolu, R., Prahalad, C. K., & Rangaswami, M. R. (2009). Why sustainability is now the key driver of innovation. *Harvard Business Review*, *87*(9), 56–64.

OECD. (2012). *Machine-to-machine communications: Connecting billions of devices*. OECD Digital Economy Papers, No. 192. www.oecd-ilibrary.org/science-andtechnology/machine-to-machine-communications_5k9gsh2gp043-en

Ohlhorst, F. J. (2012). *Big data analytics: Turning big data into big money*. John Wiley & Sons.

OpenAI. (2023a). *Creating Safe AGI that Benefits All of Humanity*. https://openai.com/

Parida, V., Sjödin, D., & Reim, W. (2019). Reviewing literature on digitalization, business model innovation, and sustainable industry: Past achievements and future promises. *Sustainability*, *11*(2), 391–399. https://doi.org/10.3390/su11020391

Payne, A., & Frow, P. (2005). A strategic framework for customer relationship management. *Journal of Marketing*, *69*(4), 167–176. https://doi.org/10.1509/jmkg.2005.69.4.167

Picciano, A. G. (2012). The evolution of big data and learning analytics in American higher education. *Journal of Asynchronous Learning Networks*, *16*(3), 9–20.

Piscicelli, L. (2023). The sustainability impact of a digital circular economy. *Current Opinion in Environmental Sustainability*, *61*, 101251. https://doi.org/10.1016/j.cosust.2022.101251

Provost, F., & Fawcett, T. (2013). Data science and its relationship to big data and data-driven decision making. *Big Data*, *1*(1), 51–59. https://doi.org/10.1089/big.2013.1508

Rai, A., Pavlou, P. A., Im, G., & Du, S. (2012). Interfirm IT capability profiles and communications for cocreating relational value: Evidence from the logistics industry. *MIS Quarterly*, *36*(1), 233–262.

Ray, P. P. (2023). ChatGPT: A comprehensive review on background, applications, key challenges, bias, ethics, limitations and future scope. *Internet of Things and Cyber-Physical Systems*, *3*, 121–154. https://doi.org/10.1016/j.iotcps.2023.04.003

Ritchie, H., Roser, M., & Rosado, P. (2020). *$CO_2$ and Greenhouse Gas Emissions*. OurWorldInData. https://ourworldindata.org/co2-and-greenhouse-gas-emissions

Ritchie, H., Roser, M., & Rosado, P. (2022). *Energy*. OurWorldInData. https://ourworldindata.org/energy

Roblek, V., Meško, M., & Podbregar, I. (2021). Impact of car sharing on urban sustainability. *Sustainability*, *13*(2), 905–924. https://doi.org/10.3390/su13020905

Rodríguez-Mazahua, L., Rodríguez-Enríquez, C. A., Sánchez-Cervantes, J. L., Cervantes, J., García-Alcaraz, J. L., & Alor-Hernández, G. (2016). A general perspective of Big Data: Applications, tools, challenges and trends. *The Journal of Supercomputing*, *72*, 3073–3113. https://doi.org/10.1007/s11227-015-1501-1

Rosário, A. T., & Dias, J. C. (2022). Sustainability and the digital transition: A literature review. *Sustainability*, *14*(7), 4072. https://doi.org/10.3390/su14074072

Rowley, J. E. (2002). Reflections on customer knowledge management in e-business. *Qualitative Market Research: An International Journal*, *5*(4), 268–280. https://doi.org/10.1108/13522750210443227

Sambamurthy, V., Bharadwaj, A., & Grover, V. (2003). Shaping agility through digital options: Reconceptualizing the role of information technology in contemporary firms. *MIS Quarterly*, *27*(2), 237–263.

Schaller, B. (2021). Can sharing a ride make for less traffic? Evidence from Uber and Lyft and implications for cities. *Transport Policy*, *102*, 1–10. https://doi.org/10.1016/j.tranpol.2020.12.015

Schallmo, D., Williams, C. A., & Boardman, L. (2017). Digital transformation of business models—best practice, enablers, and roadmap. *International Journal of Innovation Management*, *21*(08), 1–17. https://doi.org/10.1142/S136391961740014X

Secundo, G., Ndou, V., Del Vecchio, P., & De Pascale, G. (2020). Sustainable development, intellectual capital and technology policies: A structured literature review and future research agenda. *Technological Forecasting and Social Change*, *153*, 1–21. https://doi.org/10.1016/j.techfore.2020.119917

Steurer, R., Langer, M. E., Konrad, A., & Martinuzzi, A. (2005). Corporations, stakeholders and sustainable development I: A theoretical exploration of business–society relations. *Journal of Business Ethics*, *61*, 263–281. https://doi.org/10.1007/s10551-005-7054-0

Straub, D. W., & Watson, R. T. (2001). Research commentary: Transformational issues in researching IS and net-enabled organizations. *Information Systems Research*, *12*(4), 337–345. https://doi.org/10.1287/isre.12.4.337.9706

Subramaniam, M., & Venkatraman, N. (2001). Determinants of transnational new product development capability: Testing the influence of transferring and deploying tacit overseas knowledge. *Strategic Management Journal*, *22*(4), 359–378. https://doi.org/10.1002/smj.163

Taecharungroj, V. (2023). "What can ChatGPT do?" Analyzing early reactions to the innovative AI Chatbot on Twitter. *Big Data and Cognitive Computing*, *7*(1), 35. https://doi.org/10.3390/bdcc7010035

Tan, B., Pan, S. L., Lu, X., & Huang, L. (2015). The role of IS capabilities in the development of multi-sided platforms: The digital ecosystem strategy of Alibaba.com. *Journal of the Association for Information Systems*, *16*(4), 248–280. https://doi.org/10.17705/1jais.00393

Tesla. (2023). *Range Tips*. www.tesla.com/en_au/support/range

The Freshwater Trust. (2023). *Tools & Technology*. www.thefreshwatertrust.org/services/research-technology/

The Ocean Cleanup. (2023a). *The Largest Cleanup in History*. https://theoceancleanup.com/

The Ocean Cleanup. (2023b). *Research*. https://theoceancleanup.com/research/

Tomšič, N., Bojnec, Š., & Simčič, B. (2015). Corporate sustainability and economic performance in small and medium sized enterprises. *Journal of Cleaner Production*, *108*, 603–612. https://doi.org/10.1016/j.jclepro.2015.08.106

UNESCO. (2020, January 14). *UNESCO and Sustainable Development Goals*. https://en.unesco.org/sustainabledevelopmentgoals

Villars, R. L., Olofson, C. W., & Eastwood, M. (2011). *Big data: What it is and why you should care* [White paper]. IDC. www.admin-magazine.com/HPC/content/download/5604/49345/file/IDC_BigData_whitepaper_final.pdf

Walmart. (2023a). *Product Supply Chain Sustainability*. https://corporate.walmart.com/purpose/esgreport/environmental/product-supply-chain-sustainability

Walmart. (2023b). *Waste: Circular Economy*. https://corporate.walmart.com/purpose/esgreport/environmental/waste-circular-economy

Walsh, N. (2022). Sharing the latest improvements to efficiency in Microsoft's datacenters. *Microsoft*. https://azure.microsoft.com/en-us/blog/sharing-the-latest-improvements-to-efficiency-in-microsoft-s-datacenters/

Wang, G., Gunasekaran, A., Ngai, E. W., & Papadopoulos, T. (2016). Big data analytics in logistics and supply chain management: Certain investigations for research and applications. *International Journal of Production Economics*, *176*, 98–110. https://doi.org/10.1016/j.ijpe.2016.03.014

Wasteless. (2023). *Sell More, Waste Less with Dynamic Pricing and Smart Markdowns*. www.wasteless.com/

Watson, H. J. (2014). Tutorial: Big data analytics: Concepts, technologies, and applications. *Communications of the Association for Information Systems*, *34*(1), 1247–1268. https://doi.org/10.17705/1CAIS.03462

Weill, P., & Woerner, S. L. (2013). Optimizing your digital business model. *MIT Sloan Management Review*, *54*, 71–78. https://doi.org/10.1109/EMR.2015.7059380

World Bank. (n.d.). *Small and Medium Enterprises (SMEs) Finance*. www.worldbank.org/en/topic/smefinance#:~:text=SMEs%20account%20for%20the%20majority,than%2050%25%20of%20employment%20worldwide.

Xu, Z., Frankwick, G. L., & Ramirez, E. (2016). Effects of big data analytics and traditional marketing analytics on new product success: A knowledge fusion perspective. *Journal of Business Research*, *69*(5), 1562–1566. https://doi.org/10.1016/j.jbusres.2015.10.017

Yoo, Y. (2010). Computing in everyday life: A call for research on experiential computing. *MIS Quarterly*, *34*(2), 212–231.

Yoo, Y., Henfridsson, O., & Lyytinen, K. (2010). Research commentary—the new organizing logic of digital innovation: An agenda for information systems research. *Information Systems Research*, *21*(4), 724–735. https://doi.org/10.1287/isre.1100.0322

# 8 Green Logistics
## *A New Frontier for Economic Growth and Social Development in Türkiye and Poland*

*F. Talya Temizceri, and A. Zafer Acar*

## 8.1 INTRODUCTION

Sustainability emerges as a priority concept for business as a result of a combination of frequent environmental crises, rising environmental awareness, regulatory pressure, reputation and risk management, and consumer demand. Due to these circumstances, to preserve their existing place in the market, businesses have included environmental considerations as one of the primary factors in their business strategy (Danış et al., 2022). Since sustainability is a very broad concept, containing many subtopics, and is relatively new, it is essential to first define it accurately in order not to cause any confusion but also to determine the boundaries of this study. According to the 1987 Commission Report from the United Nations titled "Our Common Future," the term "sustainability" is frequently used and is defined as the capacity to satisfy both present and future demands (Daly, 1996: 1). Accordingly, UNESCO describes sustainability as a future-oriented and integrated approach that balances economic development, human well-being, and ecosystem integrity (UN, 2020), which points out the triple bottom line of sustainability. Therefore, the three dimensions of sustainability are economic, social, and environmental sustainability. Indeed, those three dimensions are directly dependent on each other. Promoting the well-being of the people, supporting environmental feasibility for social engagement, and providing equal opportunities to access resources are the crucial echelons of social sustainability that encompass ecological and economic targets. Briefly, sustainability is understanding this interdependency and keeping these dimensions in balance.

The World Bank measures the logistics performance index (LPI) bi-yearly to evaluate and compare countries' logistics performance with international indices and identify the countries that are genuinely ready for the advancements necessitated by commercial globalization. The index, which is an essential tool that reveals the efficiency of the logistics supply chain, is accepted as a macro indicator that has significant effects on the environment, such as greenhouse gas and carbon emissions

 DOI: 10.1201/9781003438748-10

and energy consumption; reflects the environmental performance of the logistics industry; and, in this context, measures the green logistics performance at the national level. Green logistics refers to sustainable policies to reduce environmental impact. As in the sustainability concept, green logistics practices have economic, social, and environmental components. Following the aim of competitive advantage, social, environmental, and economic sustainability studies have revealed the 'green logistics' approach. Calculating carbon emissions by obtaining certificates in international standards; consumption of low-emission fuels, including environmentally friendly vehicles in fleets; and investments in automation systems are the solutions produced in green logistics. According to recent literature, green logistics activities include green procurement, distribution, warehousing, and packaging and all other activities in forward and reverse logistics (e.g., Akbal, 2022; Mücevher, 2021; Zengin & Akunal, 2017).

Besides the sustainability targets, the European Commission presented "The EU Green Deal", a set of strategies that reinforces Europe's goal of reducing carbon emissions by 55% by 2030 and becoming the world's first carbon-neutral continent by 2050. For economic and environmental sustainability, transportation corridors and core networks are designed as Trans-European Transport Networks (TEN-T) within the scope of the Green Deal. The objective and policy of TEN-T are currently being revised for greener, more efficient, and more resilient transportation networks in Europe comprising multimodal transportation (European Commission, 2013, 2019).

In line with the sustainable transportation goal of the EU Green Deal, Türkiye has prepared an action plan for 2021 (Republic of Türkiye Ministry of Trade, 2021). The core points of the action plan can be summarized as the following:

- "Combined Transportation" Regulation and "Logistics Center" Regulation will be implemented to support the balanced development of transportation types and methods.
- Efforts to improve and develop the railway infrastructure between the EU and Türkiye will continue.
- It is planned to carry out studies to reduce harmful emissions from the maritime sector and to support green shipping.

Minimizing the environmental impacts of logistics activities is considered greening and sustainability in the supply chain. According to Climate Action Plans prepared yearly by UNCC (2022), shifting to more environment-friendly and low-carbon public and freight transportation modes is the key to sustainability. Dependency on roads in freight transportation plays an especially vital role in the zero-carbon target (European Commission, 2020; European Environment Agency, 2023).

The logistics industry is among the essential industries determining the EU's economic cooperation and global competitiveness. According to Özgenç (2021), transportation is a strategic industry vital for all economic activities. The inclusion of environmental concerns in the Common Transportation Policy has led to the convergence of the term sustainable transportation and the adoption of transportation standards. The transportation policy of the European Union not only supports economic development and trade but also plays a vital role in the political and economic cohesion of the member states.

Examining the logistics performance of Türkiye and Poland reveals their shared status as rapidly developing countries facing similar challenges yet operating within distinct environments. The World Bank's ongoing assessment of logistics performance across 160 countries since 2007 serves as a crucial benchmark for global logistics stakeholders and businesses. This ranking, essential for superior logistic services delivery, particularly in remote areas, stimulates foreign direct investment (FDI) attraction in developing economies. The impact of enhanced logistics performance extends to strategic management, especially for multinational corporations, while inadequate performance escalates operational risks and doubts about investment decisions (Acar et al., 2020; Bentyn, 2021). Notably, the 2022 LPI index global rankings highlight Türkiye at 40th and Poland at 34th place with 3.5 and 3.6, respectively (The World Bank, 2023). Türkiye went up its position from the 34th to the 33rd rank, and Poland at the same time stayed at the same level for this year. However, both countries' LPI scores are 3.4 and 3.6. Geographic positioning accentuates their potential as international logistics hubs, with Türkiye's access to North Africa, the Middle East, Russia, Central Asia, and Europe and Poland's advantageous location for Western and Eastern European connections. Both nations' involvement in the Belt and Road Initiative (BRI, formerly OBOR), along distinct corridors, holds the promise of influencing further logistics infrastructure growth, underlining the pivotal role of geography in their divergent logistic trajectories (Acar et al., 2020). In short, both countries have potential roles and are considered the middle corridor and northern Europe gateway of this project.

Logistics is a cornerstone of Poland's national economy, serving as a vital conduit for goods and financial transactions that underpin the market economy. Polish road freight transportation swiftly and substantially impacted market dynamics upon EU integration. Polish carriers align closely with client expectations regarding prices and quality, leading to a significant and expanding share in the EU's international road transportation sector's volume and turnover. This success stems from broad access, thanks to transportation liberalization spanning international and domestic routes. Despite limitations in cabotage services, Polish and other new member state carriers could have achieved even more substantial market shares. The ascendant presence of carriers from Central and Eastern Europe in the EU's road freight transportation market is further propelled by the mounting role of international transportation, where cost-effective companies thrive naturally (Burnewicz, 2017). The nation's geographic positioning bestows paramount significance upon its freight transportation sector, positioning it as a pivotal transit country bridging Western and Eastern European trade routes. The transportation and logistics industry's escalating importance within the Polish economy is underscored, predominantly governed by micro and small enterprises, particularly in road and inland waterway transport. Medium and large entities dominate the air, sea, and rail transport sectors. In Poland, road transportation holds a substantial share of nearly 85%, followed by rail transportation at over 15%, while maritime transportation and air transportation remain of marginal significance, both constituting less than 0.1%. The expansion of transportation infrastructure, crucial for enhancing accessibility, plays a pivotal role in fortifying

Poland's regional and national competitiveness across diverse economic sectors. Foremost priorities encompass finalizing the road network through motorways and expressways linking major centers, bolstering cross-border connections with Baltic and Southern countries (Via Baltica and Via Carpatia), and revitalizing water routes like the Odra and lower Vistula (Nedeliaková & Stasiak-Betlejewska, 2019).

Poland's logistics performance is highly influential on the European economy. Since it is a member of the European Union, a product nationalized in Poland can be shipped quickly to the other EU member. Since Poland is adjacent to Germany, the leading country in the EU, German companies minimize their supply lead times by moving their production processes to Poland. The geographical location of the country, located in the region where Eastern and Western Europe meet, causes it to be a bridge between Northern and Southern European countries. Thus, Poland is an ideal distribution center on West–East and North–South Europe axes. It is possible to deliver goods from a logistics center in Poland to developed regions of Europe very fast. Furthermore, Poland has reasonable labor and land rental costs compared to other Western and Northern European countries.

Among the developed countries, European countries stand out, especially in green logistics. In line with the European Green Deal, companies in Türkiye are also developing and implementing essential strategies in the field of green logistics, in line with not restricting their access to the European Union market and preserving environmental-friendly economic integrity.

It has been observed that Turkish logistics companies are members of national and international platforms and have obtained certificates, mostly ISO 14001 Environmental Management System, for the registered sustainable service quality. To this end, the aim of the study is to reveal whether Turkish logistics firms act in line with contemporary developments in the field that have been recorded in the literature. The following chapter gives the methodology of the study in order to reach this aim. Then a comprehensive literature review is given to reveal and collect the data related to the developments in logistics regarding sustainability. After elaborating on the green applications in the Turkish logistics industry, finally, the study's results and implications have been given.

In Poland's context, efforts toward clean energy and green logistics are primarily shaped by the European Union's (EU) regulatory mandates. Notably, the EU's "A New EU Circular Economy Action Plan for a Cleaner and More Competitive Europe" adopted in 2020 is a pivotal framework (European Commission, 2020). This comprehensive plan encompasses green logistics activities such as waste management, utilizing renewable energy sources, and analyzing and properly managing customer behavior in the supply chain. The convergence of these EU mandates propels Poland's trajectory towards environmentally sustainable practices, thereby underscoring the crucial role of regulatory frameworks in steering green initiatives (Lis & Pawęska, 2023).

## 8.2 RESEARCH METHOD

Poland and Türkiye have distinct geographic locations, influencing their logistical challenges, transportation routes, and trade patterns. Analyzing their green logistics practices allows us to understand how each country adapts to its unique circumstances. Both Poland and Türkiye are rapidly developing economies with growing international trade. Comparing their green logistics efforts provides insights into how these emerging markets integrate sustainability measures into their economic growth strategies. Poland is an EU member state, subject to EU regulations and initiatives like the Circular Economy Action Plan. Comparing its green logistics applications with Türkiye, a candidate for EU membership, highlights the influence of EU policies on sustainability practices. Türkiye's strategic location as a bridge between Europe and Asia and Poland's role in Eastern and Central Europe make them key players in regional and international trade networks. A comparison can show how their logistics systems support global trade flows. Both countries have been investing in transportation and logistics infrastructure. Comparing these two OECD member countries' green logistics initiatives showcases their commitment to sustainable infrastructure development (Acar et al., 2020).

Environmental challenges such as air quality, carbon emissions, and waste management differ between the two countries. A comparison helps identify how green logistics addresses these concerns. Analyzing the adoption of eco-friendly technologies, alternative fuels, and digital solutions in logistics provides insights into each country's innovative capacity. Studying the approaches of logistics companies, manufacturers, and supply chain partners in each country can reveal how the private sector contributes to green logistics efforts. Comparing the economic and social outcomes of green logistics applications regarding job creation, cost savings, and environmental improvements provides a holistic view of their benefits.

Thus, comparing green logistics applications between Poland and Türkiye offers a nuanced understanding of how different economic, geographic, and regulatory contexts shape their sustainability efforts. It facilitates knowledge sharing, best practice identification, and the potential for mutual learning to drive more effective green logistics strategies in both countries.

As mentioned, this research aims to expose how Polish and Turkish logistics companies define their sustainability objectives based on the European Green Deal criteria and how they act for those objectives. To reach this aim, the following research questions were designed.

1. Which developments does international literature highlight within the scope of sustainability?
2. What kind of practices do logistics companies in two OECD countries adopt to adapt to these developments that can provide them with a strategic competitive advantage?

The research methodology structure has three phases. The first phase presents the data collection protocol regarding the current studies in the literature. The second phase establishes the analysis of the studies, and the last one examines the green logistics applications in Poland and Türkiye. Figure 8.1 presents a scheme of the data collection protocol. First, a comprehensive search of papers was carried out with specific keywords in databases, ScienceDirect, Web of Science (WoS), and Google Scholar, and official websites of the Commissions, Ministries, and well-known logistics companies. More than 10,000 articles were found on those topics.

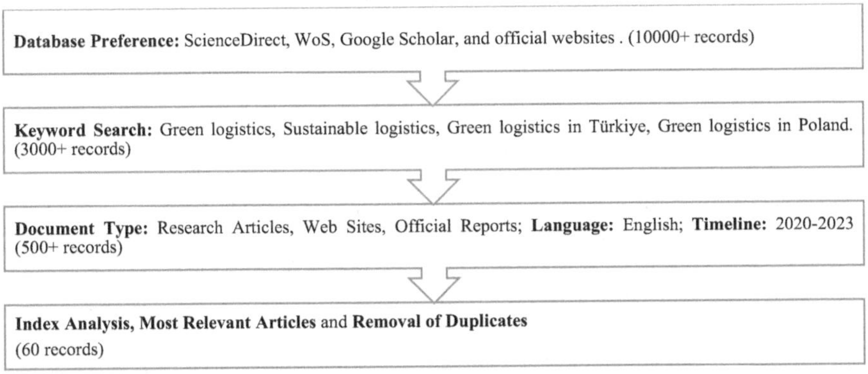


**FIGURE 8.1** Data collection protocol.

This review focuses on the economic and social impacts of green logistics. Therefore, the filtering criteria consisted of the following keywords within the 'title' and 'body': ("economic impacts of sustainability in logistics" OR "social impacts of sustainability in logistics") AND ("green logistics" OR "sustainable logistics") AND ("green logistics in Türkiye" OR "sustainable logistics in Türkiye") AND ("green logistics in Poland" OR "sustainable logistics in Poland").

This filtering process aims to identify research limitations and gaps in the literature in the context of green multi- and intermodal transportation. For this purpose, the most relevant studies are briefly explained with a theoretical framework based on the most common characteristics, such as objectives, data collection, and solution methodologies of problems in the last three years.

The last three years were refined to present the current studies and developments about green and sustainable logistics with multiple modes to examine the latest discussions in the current literature. Thus, papers published before 2019 were excluded, and unrelated and limited works were discarded. Another inclusion criterion was language. The search protocol was limited to the English language. For the academic journals, papers were categorized as (Research Area = supply chain and logistics), including freight transportation. Recent developments and actions taken by well-known companies in Türkiye and Poland were emphasized to investigate practical examples.

## 8.3 LITERATURE REVIEW AND DATA COLLECTION

Logistics traditionally has two main objectives of cost minimization and profit maximization. However, international trade has constantly been growing, and demand for international logistics has increased. Therefore, logistics service providers have been trying to decrease the transit time and increase service quality. However, these kinds of objectives still serve the economic targets. On the other hand, taking actions for sustainable logistics has become a mandatory choice for logistics service providers and supply chain managers. Because today's intense competition has directed businesses to organize green logistics strategies, the three dimensions of sustainability

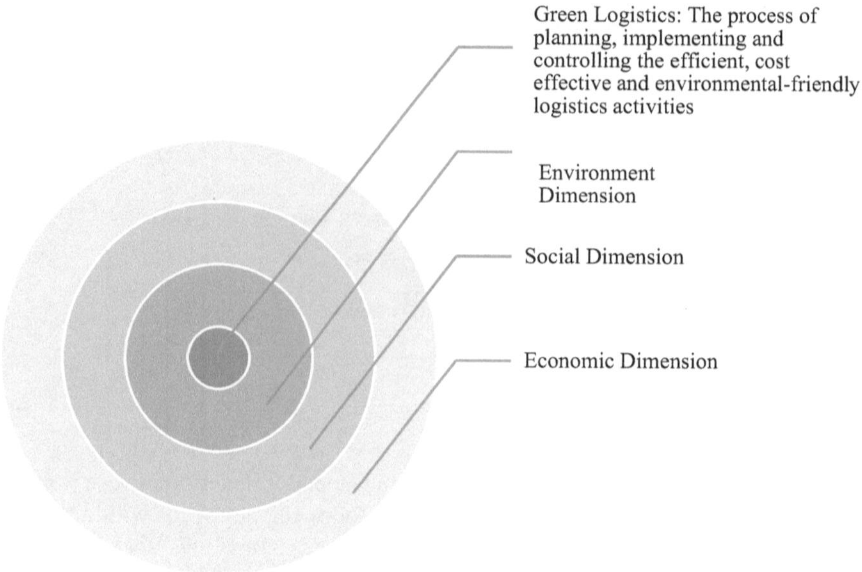


**FIGURE 8.2** Interrelation among sustainable developments and green logistics.

**(Source: Authors).**

discussed in the previous section can be applied to green logistics. Since the companies coordinate their logistics activities, including freight transportation, transferring, warehousing, material handling, and meeting customers' demands at minimum cost, the environmental dimension also contributes to handling the external costs associated with environmental issues. Thus, environmental awakening and environmental impact are the current concerns.

Figure 8.2 displays the interrelations among all pillars of sustainable development with green logistics. Green logistics is a sophisticated discipline incorporating economic, social, and environmental elements to balance sustainability. Developing efficient logistics services is one of the fundamental requirements for economic sustainability to survive the current competitiveness. Furthermore, it is only possible to focus on social and environmental issues with a stable economy. It is significant for all members of a society to ensure equal access and opportunities because social sustainability is defined as the development of the ability to meet the requirements of present and future generations without compromising anyone or anything (World Commission on Environment and Development, 2017). On the one hand, utilizing natural resources without harm is key for maintaining environmental integrity. We can conclude that green logistics is an evolutionary version of sustainability in all terms and encourages the logistics service providers to optimize environmental impacts with logistics-related costs and social considerations (El-Berishy et al., 2013).

It is known that logistics activities in the supply chain cause a carbon footprint, and transportation is the main emitter among the others (EPA, 2020). Although environmental awareness reached the top in 2020 among the published studies, we noticed that publications on green and sustainable freight transportation systems with cost-oriented studies take a significant part in the literature.

Once the European Commission approved a policy called "Climate-Neutral" to zero out the GHG emissions step-by-step targets until 2050, logistics companies started acting to reduce the emission for freight transportation (EPA, 2020). Green logistics literature dates back to the 1970s with the idea of sustainability. Existing literature investigates green logistics in different phases, such as tactical, strategic, and operational, through subdivided industries considering closed loop, forward, and reverse logistics (Mohsin et al., 2022).

Logistics' environmental effects can be assessed from a strategic and operational perspective (Mckinnon et al., 2015). The strategic perspective comprises restricting the logistics systems and supply chains; transferring greener modes of transportation; developing green vehicles, ships, and aircraft; and reducing the environmental impacts at the warehousing stage in logistics activities. From the operational perspective, managers should optimize vehicle routing; make the correct decisions on vehicle purchases, fleet management, and required green investments; and correctly operate reverse logistics activities. Additionally, it is crucial to understand that governments should promote the sector's energy efficiency through policies and regulations.

Some existing literature studies have generally discussed sustainable logistics strategies such as green, reverse, and lean logistics (Mücevher, 2021). On the other hand, green logistics incorporates lean and reverse logistics concepts in both practice and literature. The main aim is always to decrease environmental impacts due to logistics activities through the supply chain while maintaining economic and social sustainability. Table 8.1 presents an overview of current academic studies.

**TABLE 8.1**
**An Overview of the Current Literature**

| Reference | Scope | Application Area |
|---|---|---|
| **Khan (2019)** | The effect of green logistics on sustainability in all terms | Developing countries in Asia |
| **Peng & Wu (2019)** | Relationships between economy and transportation emissions | China |
| **Seroka-Stolka, & Ociepa-Kubicka. (2019)** | Green logistics and circular economy | Generic |
| **Agyabeng-Mensah et al. (2020)** | Exploring the influence of green logistics management practices | Ghanaian business perspective |
| **Amin et al. (2020)** | Role of renewable energy in green transportation | European Countries |
| **Mete (2020)** | Green logistics and sustainability | EU and Türkiye |
| **Habib et al. (2021)** | Emissions related to road transportation | G20 countries |
| **Mücevher (2021)** | Strategies for sustainable logistics | Generic |

*(Continued)*

**TABLE 8.1**
**(Continued)**

| Reference | Scope | Application Area |
|---|---|---|
| **Akbal (2022)** | Green logistics and sustainability | Generic |
| **Bardakçı (2022)** | Trends in sustainable logistics | Türkiye |
| **Çitil & Görgün (2022)** | Effects of green logistics on competitiveness | Developing countries |
| **De Souza et al. (2022)** | Performance evaluation technique in green logistics | Brazil |
| **Erdogan et al. (2022)** | Transportation demand and environment | G7 countries |
| **Hussain (2022)** | Economic- and environment-transportation efficiency | Generic |
| **Hussain et al. (2022)** | Effects of various indicators on transportation emissions | OECD countries |
| **Jayarathna et al. (2022)** | Sustainable logistics in perspective of value creation | Singapore, Mexico, UK, Australia, UAE, Sri Lanka |
| **Kaya Samut (2022)** | Relationship among green logistics, energy, environment, and health economy | OECD countries |
| **Liu et al. (2022)** | Green innovations and influence of logistics performance carbon emissions | Asian economies |
| **Mohsin et al. (2022)** | Relationship between the green logistics, environment, and economic growth | Countries along Belt and Road Initiative |
| **Taş and Aylak (2022)** | Effects of Green Deal on logistics | TEN-T Network |
| **Balcilar et al. (2023)** | Investments for sustainability | Türkiye |
| **Dai et al. (2023)** | Effect of green energy consumption in transportation | USA |
| **Layaoen et al. (2023)** | Economic and social outcomes of environmental performance | Philippines |
| **Sharma et al. (2023)** | Green logistics practices and Industry 4.0 era | North India |
| **Noorliza (2023)** | Human–environment connection within the LSP context | Generic |
| **Romagnoli et al. (2023)** | Circular supply chain | Europe, Asia, USA |
| **Zeeshan et al. (2023)** | Corporate social responsibility | Generic |

### 8.3.1 Green Logistics for Sustainability

Environmental sustainability improves the monetary aspect pf companies due to increasing market share and customer loyalty. Besides improving economic performance, green logistics operations using renewable assets and fuel-efficient and green vehicles and providing mode shifts decrease the carbon footprint. However, it requires policy implications and significant investment. Policies and regulations influence transportation efficiency. Policymakers should focus on reducing the use of diesel and petrol vehicles, decreasing dependency on road transportation, and agreements among countries for sustainability-based trade (Hussain, 2022; Hussain et al., 2022). Specifically, road transportation dependency should be decreased. Habib et al. (2021) show that the intensity of road transportation positively correlated with carbon emissions. Besides, although the investment in renewable energy or logistics infrastructure is considered a considerable cost, the long-term benefits are obvious. Through continuous economic policies and incentives regulated by governments, obtaining competitive returns will be easier for the supply chain and logistics managers and service providers. It has become essential for countries to develop their energy efficiency and logistics infrastructure to get a larger share of world trade and thus increase their competitiveness (Amin et al., 2020; Erdogan et al., 2022; Balcilar et al., 2023; Sharma et al., 2023). Besides contributing to the national economy, protecting the environment and natural resources is essential to ensure sustainability (Mete, 2020). The effect of green logistics on economic growth can be measured by the efficiency of customs clearance processes, quality of logistics services, and trade and transportation-related infrastructure because they are all related and positively correlated with income, trade openness, and value-added activities in the supply chain. Khan (2019) indicated that poor trade and logistics infrastructure are the primary causes of economic instability.

Green logistics is an inherent fact of a circular economy. Sustainable supply chains and green logistics management are the fundamental concepts of the circular economy because these terms directly relate to the degree of environmental degradation. Seroka-Stolka and Ociepa-Kubicka (2019), De Souza et al. (2022), and Jayarathna et al. (2022) investigate economically feasible and environmentally safe methods for green logistics applications. Sharma et al. (2023) and Romagnoli et al. (2023) discuss the importance of technological developments for sustainable and green transportation. The Internet of Things (IoT) and machine learning techniques in the Logistics 4.0 era can effectively implement circular supply chain activities under the circular economy principle.

Green logistics activities aim to minimize the undesired effects of logistics activities on the environment. In this sense, logistics service providers have started to install noise and sound-preventing devices in vehicles, prefer environmentally friendly recycling methods, and enhance their fleet vehicles with EURO 5 standards or give importance to utilizing low-impact transportation modes. Basically, green logistics provides a strong bottom line without sacrificing customer satisfaction or the welfare of the planet. Procurement, warehousing, packaging, and transportation operations are composed of green logistics as well as common logistics practices but from a sustainable point of view, as shown in Figure 8.3.

Logistics activities have a positive performance on economic growth. However, increasing logistics activities are the major contributor to environmental degradation. Since logistics activities play a significant role in $CO_2$ emission, efficient environmental control in the supply chain and logistics will keep logistics-driven pollution under control (Peng & Wu, 2019; Liu et al., 2022; Dai et al., 2023). Engaging

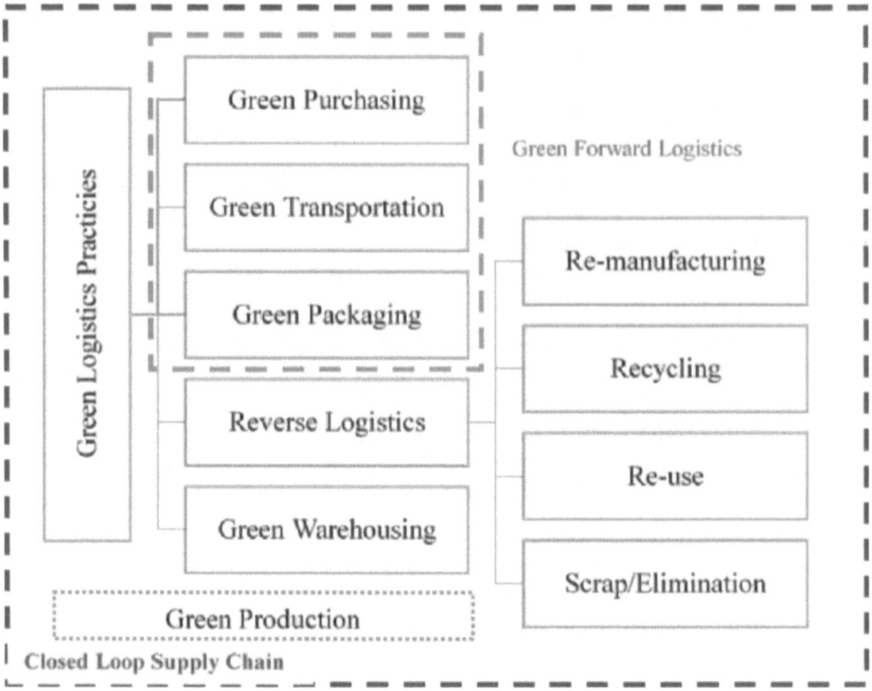


**FIGURE 8.3** Green logistics practices.

**(Source: Authors).**

green logistics activities with sustainable transportation and supply chain activities positively influences business performance (De Souza et al., 2022). Economic growth through green logistics practices requires more prominence for sustainable development (Mohsin et al., 2022). The Belt Road Initiative is one of the most significant developments for grated sustainability in economics and the environment. Investments in ports and infrastructure will support green logistics activities and increase economic growth with higher export opportunities. Agyabeng-Mensah et al. (2020) suggest that eco-centric logistics engages green logistics practices as a significant driver of a sustainable supply chain. Additionally, the transportation sector should be planned depending on the intervention and mechanism of the EU Green Deal, especially in the European continent. Correct design of networks and facilitation of logistics centers are directly about foreign trade for both economic and environmental sustainability (Taş & Aylak, 2022).

Providing ecological solutions with logistics actions improves logistics companies' social and economic performance (Çitil & Görgün, 2022). Jayarathna et al. (2022) indicate that environmental and social benefits follow the creation of economic value chains because social sustainability practices such as green administration, internal and external human resource practices, environment training, and information are directly linked with the economic and consequently the environmental performance of a company. Layaoen et al. (2023), Noorliza (2023), and Zeeshan

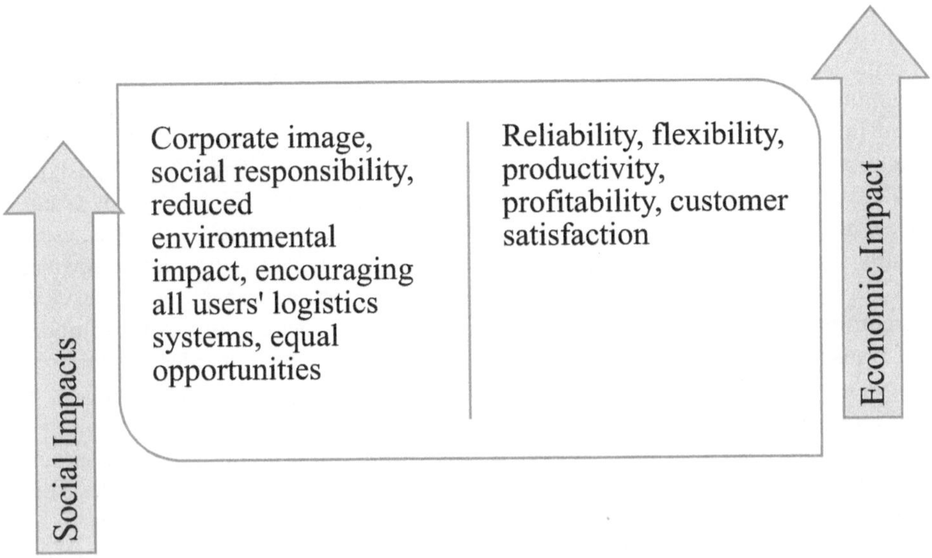


**FIGURE 8.4** Social and economic impacts of green logistics.

**(Source: Authors).**

et al. (2023) indicate that socially responsible and eco-friendly applications are the key for transportation and logistics companies to remain competitive.

In particular, the usage of renewable energy reduces carbon emissions during transportation and transfer operations of logistics companies, and this directly affects social sustainability due to the decreased environmental impact (Kaya Samut, 2022).

In short, the social impact of green logistics is that green logistics requires ethical sensitivity and corporate responsibility. Companies that successfully carry out green logistics activities can implement environmental and social sustainability concepts, including environmental, social, and managerial factors; working conditions; workplaces; employee relations; and social justice. According to the literature, we can conclude green logistics' economic and social impacts, as shown in Figure 8.4.

## 8.4 EXPLORING GREEN LOGISTICS APPLICATIONS IN CASE COUNTRIES

### 8.4.1 First Case: Türkiye

The competitive environment due to the increasing importance of green logistics worldwide has directed companies in Türkiye to change their operations and policies towards "greening" action. Turkish logistics companies focusing on green logistics practices have quality management systems that meet ISO standards, especially ISO 14001 environmental management systems. The Association of International Forwarding and Logistics Service Providers (UTIKAD, 2018), one of the most important institutions of the Turkish logistics industry, supports logistics service providers to contribute to sustainable growth and has given sustainable logistics certificates to the companies that are deemed suitable because of several pieces of training given within the program. On the other hand, the Ministry of Transportation and Infrastructure of the Republic of Türkiye prepared Combined Transportation Regulation to support integrated, balanced, environmentally friendly, and sustainable freight transportation in Türkiye (Official Gazette, 2022). Targets set by Green Deal and government incentives accelerate applying green logistics practices, especially freight transportation, in Türkiye. There are many active logistics companies in Türkiye. Logistics companies provide road, air, and sea transportation services according to their fields of activity. In addition to companies that use all three modes of transportation and perform intermodal logistics, some companies provide services only in the field they specialize in. Companies that provide only road, sea, or air logistics services are the majority.

Companies such as Ekol Logistics, Mars, and Borusan Logistics are among Türkiye's largest and leading logistics companies that provide integrated logistics services in international areas. Those logistics companies provide services in all transportation areas such as land, sea, rail, and air; warehouse management; and customs clearance customer services such as foreign trade and supply chain services that contribute to the sector's high-quality services (LojiPort, 2018).

Thus, ten leading companies operating in the Turkish logistics industry based on their market share and employment capacity, given in Table 8.2, are investigated according to their green practices, given in Table 8.3. The last and most updated total turnovers of the companies in TL in 2019 are calculated in Euro (1 Euro = 6.6621 TL) according to the exchange rate of the Central Bank of the Republic of Türkiye for the last day of 2019 (Central Bank of the Republic of Türkiye, 2019).

**TABLE 8.2**
**Numerical Details about Ten Leading Turkish Logistics Companies**

| Company | # of Employees | Annual Return (bn €) |
|---|---|---|
| **Arkas Holding** | 403 | >2 |
| **Borusan Logistics** | 1,568 | >3 |
| **Çobantur Boltaş Logistics** | 500 | >2 |
| **Ekol Logistics** | 6,834 | >3.5 |
| **Horoz Logistics** | 868 | >3.5 |
| **Mars Logistics** | 1,076 | >1 |
| **Netlog Logistics** | 15,624 | >10 |
| **Omsan Logistics** | 1,181 | >1 |
| **Reysaş Logistics** | 870 | >4 |
| **Sarp Intermodal** | 175 | >0.4 |

(*Source:* Capital, 2019)

**TABLE 8.3**
**Green Logistics Applications of Top Ten Turkish Logistics Companies**

| Company | Goal | Action | Membership/ Documentation |
|---|---|---|---|
| **Arkas Holding (2023)** | 25% reduction in carbon emission | • Low-sulfur-fuel-oil usage in vessels<br>• Speed and route optimization<br>• Waste management in warehouses<br>• Intermodal transportation operations and investment in railway connection projects | Clean Cargo Working Group, EcoVadis |
| **Borusan Logistics (2018)** | Energy efficiency and carbon emission minimization | • EURO 6 trucks in the fleet<br>• Renewable energy systems in warehouses<br>• Route optimizations | ISO 50001 Energy Management System and ISO 14064 Carbon Emission Verification Systems |
| **Çobantur Boltaş Logistics (2021)** | Minimizing carbon footprint | • Waste management and recycling<br>• EURO 6 trucks in the fleet and just-in-time (JIT) operations in logistics services | Environmental management system in line with ISO 14001:2015 standards |
| **Ekol Logistics** | Maintaining environmental protection, reducing the ecological footprint | • EURO 5 and 6 trucks<br>• Intermodal transportation operations<br>• Waste recycling and energy-efficient facilities<br>• Collaboration with TEMA for sapling planting | World Wildlife Fund (WWF) Green Office Program, Environmental management system in line with ISO 140001 standards |
| **Horoz Logistics (2022)** | Reduction in carbon emission | • EURO 6 and hybrid trucks in the fleet<br>• Smart and renewable energy systems in warehouses | Environmental management system in line with ISO 140001 standards |
| **Mars Logistics (2020)** | Environmental, economic, and social sustainability | • Energy savings and waste management in warehouse services<br>• Intermodal transportation operations<br>• Collaboration with TEMA for sapling planting | |
| **Netlog Logistics (2022)** | Reduction in carbon emission | • Waste management<br>• Multimodal transportation<br>• Energy-saving strategies | |

*(Continued)*

**TABLE 8.3**
**(Continued)**

| Company | Goal | Action | Membership/ Documentation |
|---|---|---|---|
| **Omsan Logistics (2021)** | Minimizing carbon footprint | • Waste management<br>• Water and waste recycling<br>• Intermodal transportation operations and railway transportation collaborations | Environmental management system in line with ISO 140001 standards; Zero-Waste Certificate from Minister of Environment, Urbanization and Climate Change, Türkiye |
| **Reysaş Logistics (2023)** | Maintaining energy efficiency, reduction in greenhouse gas (GHG) emission | • Renewable energy investments in warehouses<br>• Integrating railway transportation | ISO 140001:2015 certification |
| **Sarp Intermodal (2020)** | Maintaining Green Deal 2030 target | • Investment in high-cube containers and combined transportation systems<br>• Maritime and railway integration | |

As can be seen in Table 8.3, the importance of green logistics in the Turkish logistics industry is gradually increasing. Turkish logistics companies focused on environmentally friendly operations following the global "green" trend and the European Green Deal. Turkish logistics companies focus on waste management and energy efficiency to reduce costs.

Turkish logistics companies focusing on green practices are considered the industry's pioneers. These companies have quality management systems that meet ISO Standards, especially ISO 14001 Environmental Management Systems. In addition, companies take an active role in national and international projects with environmental protection, waste management, and energy efficiency. Turkish logistics companies declare themselves green and have a strategic management perspective and focus on long-term, sustainable development. Minimizing carbon footprint and emission is the common goal among companies' greening targets. On the other hand, recycling and utilizing renewable energy are mostly preferred applications under the waste management strategies of companies. Utilizing multimodal and intermodal transportation is another approach and investment area for environmental sustainability targets.

### 8.4.2 Second Case: Poland

As indicated before, the EU, advocating for ecological advancement, urges the integration of Green Deal regulations within members' legislation, spanning various economic sectors. With Polish road cargo transportation firms dominating the EU

**TABLE 8.4**
**Numerical Details about Ten Polish Logistics Companies**

| Company | # of Employees | Annual Turnover (m €) |
|---|---|---|
| Ewals Cargo Care | 2150 | >660 |
| Transmec Group | 1,500 | >500 |
| OMEGA Pilzno Group | 1000 | >125 |
| Spedimex Sp. z o. o. | 30 | >54 |
| Epo-Trans Logistics S.A. | 526 | >50 |
| Greencarrier Freight Services Poland Sp. z o.o. | ~50 | >45 |
| Don-Trucking | 223 | >35 |
| No Limit Sp. z o.o. | 91 | >21 |
| Tsl Transport Sp. z o.o. sp. k. | ~25 | >9 |
| Miratrans Transport Sp. z o.o. | ~25 | >5 |

(*Source*: ZoomInfo, 2023).

market, curbing exhaust emissions holds the potential to enhance air quality and the EU's carbon neutrality. Effective incentives are pivotal to aligning Polish road carriers with Green Deal objectives. This study spotlights the environmental policies of sizable Polish road transportation enterprises, aiming to identify pivotal factors to regulate the sector efficiently. Successful implementation of supportive measures to enforce standards and address companies' expectations yields systemic advantages for the state, environment, and society while enhancing economic benefits (Letkiewicz et al., 2023). Polish logistics service providers have also started green logistics activities, apart from the reduction of $CO_2$ emissions achieved through the augmentation of renewable and nuclear energy proportions within the energy sector. Following this achievement, the deputy minister emphasized the advancement of emission-free transportation methods on Polish roadways. Hence, an examination is conducted on ten prominent companies operating within the Polish logistics industry, chosen based on their market share and employment capacity, as detailed in Table 8.4. The investigation is centered on their eco-friendly initiatives, as outlined in Table 8.5.

As depicted in Table 8.5, the significance of eco-friendly logistics within the Polish logistics industry is progressively growing. Polish logistics enterprises align with eco-conscious practices in line with the global "green" movement and the European Green Deal. These companies prioritize enhanced energy efficiency as strategies to decrease expenses. Renowned for their pioneering role, Polish logistics firms also boasting green initiatives adhere to ISO 14001 Environmental Management Systems as part of their quality management systems. Additionally, these entities actively participate in national undertakings related to environmental preservation, waste management, and energy efficiency. Affirming their commitment to environmentally sound practices, these companies adopt a strategic management outlook,

**TABLE 8.5**
**Green Logistics Applications of Top Ten Polish Logistics Companies**

| Company | Goal | Action | Membership/ Documentation |
|---|---|---|---|
| **Ewals Cargo Care (2023)** | Smartest and most sustainable logistics platform by being part of 2030 Agenda | • Liquid natural gas (LNG) trucks<br>• Testing Duo Trailer, which can lead to a 27% $CO_2$ reduction<br>• Fleet featuring one-axle trailers with lower fuel consumption<br>• Investment in railway connection projects | Best Improver in Sustainability (2021) and Employee Experience award (2021), EcoVadis 2023 |
| **Transmec Group (2023)** | Reducing carbon emissions and providing a quality service which protects the environment and meets customers' needs | • EURO 6 trucks in the fleet<br>• Intermodal transportation operations<br>• Behavior-based safety principles | ISO 9001:2008 and ISO 14001 certification |
| **OMEGA Pilzno Group (2023)** | Reducing the emission of harmful substances | • Fleet of LNG-powered and EURO 6 trucks<br>• Photovoltaic installations in logistics facilities<br>• Calculating the carbon footprint and analyzing the impact of the implemented pro-ecological solutions | ISO 14001:2015, ISO 45001:2018, and ISO 9001:20015 certification |
| **Spedimex Sp. z o. o. (2023)** | Multilateral approach to taking actions supporting environmental protection | • Segregating waste (foil, paper, e-waste) and energy-efficient facilities<br>• Route optimization<br>• Integrating electric trucks into the fleet | |
| **Epo-Trans Logistics S.A. (2023)** | Reduction in carbon emissions | • Fleet of LNG-powered trucks | ISO 9001:2009 certification |
| **Greencarrier Freight Services Poland Sp. z o.o. (2023)** | Promoting environmental sustainability | • Utilizing biofuels<br>• Integrating multimodal transportation solutions | ISO 140001:2015 and ISO 45001:2018 certification, Ecovadis Silver Rating |

(*Continued*)

**TABLE 8.5**
**(Continued)**

| Company | Goal | Action | Membership/ Documentation |
|---|---|---|---|
| **Don-Trucking (2022)** | First transportation company seeking full carbon neutralization in Europe | • Fleet of LNG-powered and Bio-LNG trucks<br>• "Ecobal" European forests conservation project | |
| **No Limit Sp. z o.o. (2023)** | Reduction in carbon emission | • Fleet of LNG-powered and EURO 6 trucks<br>• Fleet of electric minivans<br>• Telematics systems to monitor fuel consumption<br>• Energy-saving strategies | Member of Polish Alternative Fuels Association, ISO 9001:2015 and ISO 14001:2015 certification |
| **Tsl Transport Sp. z o.o. sp. k. (2023)** | Minimizing carbon footprint | • Investment to RO-RO and railway transportation<br>• Fleet of EURO 6 trucks | ISO 9001:2015 and ISO 14001:2015 certification |
| **Miratrans Transport Sp. z o.o. (2023)** | Minimizing pollutants emitted to the atmosphere | • Development in intermodal transportation services<br>• Fleet of Euro 5 or Euro 6 emission trucks<br>• Monitoring monthly fuel consumption of every vehicle<br>• Route optimization | Ecovadis Silver Rating, ISO 9001:2015 and ISO 14001:2015 certification |

concentrating on sustainable, long-term growth congruent with Green Deal objectives. Among preferred strategies underpinning sustainability goals, companies prioritize route optimization, fuel consumption monitoring, and emission calculations besides minimizing carbon emissions. Furthermore, adopting multimodal and intermodal transportation emerges as an avenue and investment domain aligned with their ecological sustainability aims.

## 8.5 RESULTS AND CONCLUSION

Green practices are an undeniable phenomenon that has the potential to gain a competitive advantage for today's logistics industry, both by the economic and social impact it creates and by responding to the related legal regulations that are increasing its pressure. In parallel with the increase in the level of awareness about the importance of the subject, the number of creative green practices in the relevant sector is increasing, and the related theory that has been developed by these practices takes its place in the academic literature. Acar et al. (2016) argued that green issues need to be considered strategically if long-term improvements and changes in the business environment are to be sustained in any industry. In this respect, the current chapter has been designed to reveal whether Turkish logistics firms act in line with contemporary developments in the green logistics field that have been recorded in the literature or not.

The results of this study highlight the attention on sustainability targets in the logistics industry in both theoretical and practical ways in the context of the Turkish logistics industry. To reveal the interrelations among economic, social, and environmental sustainability, international and national works on green logistics over the last three years were investigated (see Table 8.1). Our analysis showed that there is increasing attention and investment in using sustainable and alternative fuels in the field of logistics, reducing pollution to zero, and maintaining economic and social sustainability besides the environmental one due to the European Green Deal. These findings are also compatible with other studies in the literature and show that the leading companies of the Turkish logistics industry have started to adopt green practices, especially reducing carbon emissions and energy efficiency, in order to maintain their international competitiveness.

In this context, with the rising importance of environmental issues today, decision-makers need to consider a trade-off between conflicting traditional cost-based and green-focused approaches. In the meanwhile, considering the increasing resistance of customers to emerging profit-oriented services and products that ignore environmental issues, companies that adapt their business environments accordingly by implementing green-based strategies will have the opportunity to gain a competitive advantage over their rivals. Overcoming the trade-off between cost-based and green-focused approaches in strategic management can be a challenging task, but it is possible to find a balance, and there are several ways to overcome this trade-off. One is to focus on efficiency. By streamlining their operations and reducing waste, companies can save money and reduce their environmental impact. Another way to overcome the trade-off is to innovate. By developing new technologies and processes that are more efficient and environmentally friendly, companies can create new opportunities for growth and profitability.

It is possible to generalize the conclusions drawn in this chapter to the activities of companies in other countries. The principles of efficiency and innovation are universal and can be applied to businesses of all sizes and industries. However, it is essential to note that the specific challenges and opportunities facing companies in different countries will vary. Thus, it is essential to consider each country's specific context and circumstances. While balancing cost and environmental considerations can be universally applicable, the specific challenges, regulations, and cultural factors may differ across countries. For example, companies in developing countries may face different regulatory and market conditions than those in developed countries. Therefore, conducting country-specific analyses and adapting the strategies accordingly is crucial to ensure their effectiveness in different business environments.

Instead of viewing cost reduction and environmental sustainability as mutually exclusive, adopt an integrated approach. Companies should look for opportunities where cost reductions and green initiatives can align and complement each other. Several technologies can help companies to reduce their costs and environmental impact. For example, renewable energy sources, energy-efficient equipment, and waste-management systems can help companies save money while being more environmentally friendly.

Companies can also achieve more by engaging with various stakeholders, including employees, customers, suppliers, and the local community. By involving stakeholders in decision-making processes, companies can identify opportunities for win-win solutions that balance cost and environmental considerations. For example, companies can partner with government agencies, non-profit organizations, and other businesses to develop and implement green initiatives.

Moreover, employees often have the most direct impact on a company's environmental impact. By educating employees about the importance of sustainability, companies can encourage them to change their daily work habits, which can significantly impact them.

Although this chapter provides an up-to-date international aspect of green logistics application in literature and the Turkish logistics industry, it has some limitations that have potential opportunities for future research. Green logistics applications are generally considered reverse logistics in literature. Thus, future studies can evaluate reverse logistics applications or compare the sustainability impacts in different countries. Considering more companies with comprehensive field research and extending this concept within the scope of green logistics applications in European countries can be another example for future research direction.

The European Union's push for ecological progress has prompted the integration of Green Deal regulations across its member states, spanning diverse economic sectors. Given Poland's dominance in the EU's logistics market, reducing exhaust emissions is poised to enhance air quality and the EU's carbon neutrality. Effective incentives are crucial for aligning, especially for Polish road carriers, with Green Deal objectives, while Polish logistics companies also engage in green logistics activities. These firms prioritize energy efficiency to cut costs and exhibit pioneering efforts, often adhering to ISO 14001 Environmental Management Systems. Participation in national initiatives focused on environmental preservation, waste management, and energy efficiency further underscores their commitment to sustainable practices. A strategic management approach aligns them with the Green Deal's long-term growth goals. Strategies like route optimization, fuel consumption monitoring, and emission calculations aid their sustainability targets and a drive to minimize carbon emissions.

Moreover, adopting multimodal and intermodal transportation emerges as a strategy aligned with ecological sustainability goals. In contrast, its geographic and economic context influences Türkiye's approach to green logistics. While both countries are integrating environmentally friendly practices, their unique circumstances shape the execution and emphasis of these initiatives.

While both countries emphasize green logistics, Poland's approach focuses on curbing emissions due to its dominance of road cargo. At the same time, Türkiye aims at broad sustainability across road, air, and sea transportation.

Several key distinctions emerge when comparing green logistics applications in Poland and Türkiye. Poland's emphasis on emission reduction, primarily within road cargo transportation as a logistics activity, aligns with its market dominance, thereby addressing the significant environmental impact. In contrast, Türkiye adopts a more holistic approach, encompassing road, air, and sea transportation modes, indicating a broader industry commitment to sustainability. While both countries acknowledge the importance of incentives in driving green practices, Poland's focus on large road transportation enterprises contrasts with Türkiye's emphasis on diverse logistics service providers and integrating multiple transportation methods. Overall, both countries' efforts reflect their respective positions and priorities in the global green logistics landscape, influenced by their market strengths and regional significance.

In brief, Poland and Türkiye are responding to the global shift towards green logistics practices, driven by the EU's advocacy for ecological progress. In Poland, Green Deal integration is urged to span various economic sectors, with logistics companies dominating the EU market and emission reduction seen as vital for decarbonization improvement. Similarly, a competitive environment spurred by global green logistics importance prompts operational and policy changes in Türkiye. Turkish logistics companies align with green practices, obtaining ISO 14001 environmental management system accreditation. UTIKAD, a crucial institution in the Türkiye logistics industry, supports sustainable growth by awarding sustainable logistics certificates. Government efforts, like the Combined Transportation Regulation, enhance green logistics.

## REFERENCES

Acar, A. Z., Bentyn, Z., & Kocaoğlu, B. (2020). Comparing the development of Turkey and Poland as potential logistics hubs in context of international logistics. *International Journal of Logistics Systems and Management*, 35(3), 410. DOI: 10.1504/ijlsm.2020.105921.

Acar, A. Z., Önden, İ., & Gürel, Ö. (2016). Evaluation of the parameters of the green supplier selection decision in textile industry. *Fibres & Textiles in Eastern Europe*, 24(5–119), 8–14. DOI: 10.5604/12303666.1215520.

Agyabeng-Mensah, Y., Afum, E., Acquah, I. S., Dacosta, E., Baah, C., & Ahenkorah, E. (2020). The role of green logistics management practices, supply chain traceability and logistics ecocentricity in sustainability performance. *The International Journal of Logistics Management*, 32(2), 538–566. DOI: 10.1108/ijlm-05-2020-0187.

Akbal, H. (2022). Sürdürülebilir lojistik kapsamında yeşil lojistik uygulamaları (Green logistics applications within sustainable logistics). *Kapadokya Akademik Bakış*, 6(1), 15–23.

Amin, A., Altinoz, B., & Dogan, E. (2020). Analyzing the determinants of carbon emissions from transportation in European countries: The role of renewable energy and urbanization. *Clean Technologies and Environmental Policy*, 22(8), 1725–1734. DOI: 10.1007/s10098-020-01910-2.

Arkas Holding. (2023). *Arkas holding achieves 99.58% carbon savings by migrating to Microsoft Azure*. Retrieved April 25, 2023, from https://arkasnews.com/arkas-holding-achieves-99-58-carbon-savings-by-migrating-to-microsoft-azure/

Balcilar, M., Uzuner, G., Nwani, C., & Bekun, F. V. (2023). Boosting energy efficiency in Türkiye: The role of public–private partnership investment. *Sustainability*, 15(3), 2273. DOI: 10.3390/su15032273.

Bardakçi, H. (2022). Sürdürülebilir lojistikte Dünya'nın iki yeni trendi tersine lojistik ve yeşil lojistik uygulamaları: Türkiye için stratejiler (Two new trends in sustainable logistics in the world reverse logistics and green logistics applications: Strategies for Türkiye). *Avrasya Dosyası*, 13(2), 9–37.

Bentyn, Z. (2021). Comparing of logistic performance in Turkey and Poland: Explaining divergence in development. *European Proceedings of Social and Behavioural Sciences*. DOI: 10.15405/epsbs.2021.12.04.23.

Borusan Logistics. (2018). *Corporate*. Retrieved April 25, 2023, from www.borusanlojistik.com/en/corporate/sustainability

Burnewicz, J. (2017). Competitiveness of Polish road transport; on the European market. *ResearchGate* [Electronic Source]. Retrieved November 15, 2019, from www.researchgate.net/publication/315656322

Capital. (2019). Türkiye'nin en büyük 500 şirketi (Türkiye's largest 500 companies). *Capital*. Retrieved April 25, 2023, from www.capital.com.tr/listeler/capital-500

Central Bank of the Republic of Türkiye. (2019). *Indicative exchange rates*. Retrieved April 25, 2023, from www.tcmb.gov.tr/wps/wcm/connect/EN/TCMB+EN/Main+Menu/Statistics/Exchange+Rates/Indicative+Exchange+Rates

Çitil, M., & Görgün, M. R. (2022). Yeşil lojistiğin rekabetçilik üzerindeki etkisi: Gelişmekte olan ülkelere yönelik bir panel veri analizinden kanıtlar (The impact of green logistics on competitiveness: Evidence from a panel data analysis for developing countries). *Gümüşhane Üniversitesi Sosyal Bilimler Dergisi*, 13(2), 740–756.

Çobantur Boltaş Logistics. (2021). *Corporate social responsibility*. Retrieved April 25, 2023, from https://cobanturboltas.com/about/corporate-social-responsibility/

Dai, J., Alvarado, R., Ali, S., Ahmed, Z., & Meo, M. S. (2023). Transport infrastructure, economic growth, and transport $CO_2$ emissions nexus: Does green energy consumption in the transport sector matter? *Environmental Science and Pollution Research*. DOI: 10.1007/s11356-022-25100-3.

Daly, H. E. (1996). *Beyond Growth: The Economics of Sustainable Development*. Boston: Beacon Press.

Danış, M., Acar, A. Z. & Arslan, A. (2022). Is the use of green port certification strategic or not? *TransNav: The International Journal on Marine Navigation and Safety of Sea Transportation*, 16(2), 203–209. DOI: 10.12716/1001.16.02.01.

De Souza, E., Kerber, J., Bouzon, M., & Rodriguez, C. (2022). Performance evaluation of green logistics: Paving the way towards circular economy. *Cleaner Logistics and Supply Chain*, 3, 100019. DOI: 10.1016/j.clscn.2021.100019.

Don-Trucking. (2022). Retrieved August 15, 2023, from https://don-trucking.eu/sustainability-ecobal/

Ekol Logistics. (2013). *Green logistics*. Retrieved August 15, 2023, from www.ekol.com/en/logistics/green-logistics/

El-Berishy, N., Rügge, I., & Scholz-Reiter, B. (2013). The interrelation between sustainability and green logistics. *IFAC Proceedings*, 46(24), 527–531. DOI: 10.3182/20130911–3-br-3021.00067.

EPA—European Commission. (2020). *European green deal*. Retrieved March 27, 2023, from https://commission.europa.eu/strategy-and-policy/priorities-2019-2024/european-green-deal_en

Epo-Trans. (2023). Retrieved August 15, 2023, from www.epo-trans.com.pl/en/lng/

Erdogan, S., Sarkodie, S. A., Adedoyin, F. F., Bekun, F. V., & Owusu, P. A. (2022). Analyzing transport demand and environmental degradation: The case of G-7 countries. *Environment, Development and Sustainability*. DOI: 10.1007/s10668-022-02729-1s.

European Commission. (2013). *Trans-European Transport Network (TEN-T)*. https://transport.ec.europa.eu/transport-themes/infrastructure-and-investment/trans-european-transport-network-ten-t_en

European Commission. (2019). *The European green deal.* https://ec.europa.eu/info/sites/default/files/european-green-deal-communication_en.pdf

European Environment Agency. (2023). *Transport and mobility.* Retrieved from https://www.eea.europa.eu/en/topics/in-depth/transport-and-mobility?activeTab=e3e6b879-fef4-4a88-9436-5f0064698270

Ewals Cargo Care. (2023). Retrieved August 15, 2023, from www.ewals.com/en/about-us/esg-environment-social-governance/?country=TR

Greencarrier Freight Services. (2023). www.jas.com/corporate-responsibility

Habib, Y., Xia, E., Hashmi, S. H., & Ahmed, Z. (2021). The nexus between road transport intensity and road-related $CO_2$ emissions in G20 countries: An advanced panel estimation. *Environmental Science and Pollution Research*, 28(41), 58405–58425. DOI: 10.1007/s11356-021-14731-7.

Horoz Logistics. (2022). *Green logistics: Horoz Lojistik.* Retrieved April 25, 2023, from www.horoz.com.tr/en/corporate/social-responsibility/green-logistics

Hussain, Z. (2022). Environmental and economic-oriented transport efficiency: The role of climate change mitigation technology. *Environmental Science and Pollution Research*, 29(19), 29165–29182. DOI: 10.1007/s11356-021-18392-4.

Hussain, Z., Khan, M. K., & Shaheen, W. A. (2022). Effect of economic development, income inequality, transportation, and environmental expenditures on transport emissions: Evidence from OECD countries. *Environmental Science and Pollution Research*, 29(37), 56642–56657. DOI: 10.1007/s11356-022-19580-6.

Jayarathna, C. P., Agdas, D., & Dawes, L. (2022). Exploring sustainable logistics practices toward a circular economy: A value creation perspective. *Business Strategy and the Environment*, 32(1), 704–720. DOI: 10.1002/bse.3170.

Kaya Samut, P. (2022). OECD Ülkelerinin yeşil lojistik performansı ile enerji, sağlık ekonomisi ve çevre ilişkisinin analizi (Analysis of the green logistics performance of OECD countries and the relationship of energy, health economics and environment). *Verimlilik Dergisi.* DOI: 10.51551/verimlilik.1112577.

Khan, S. A. (2019). The effect of green logistics on economic growth, social and environmental sustainability: An empirical study of developing countries in Asia. *Preprints.org.* DOI: 10.20944/preprints201901.0104.v1.

Layaoen, H. D., Abareshi, A., Abdulrahman, M. D., & Abbasi, B. (2023). Sustainability of Transport and logistics companies: An empirical evidence from a developing country. *International Journal of Operations and Production Management.* DOI: 10.1108/ijopm-12-2021-0799.

Letkiewicz, A., Ławreszuk, M., & Majecka, B. (2023). The green deal—Activity and expectations of Polish large and medium-sized road freight carriers in relation to potential regulatory solutions. *Research in Transportation Business and Management*, 48, 100844. DOI: 10.1016/j.rtbm.2022.100844.

Liu, L., Wu, H., Hafeez, M., Albaity, M. S., & Ullah, S. (2022). Carbon neutrality through supply chain performance: Does green innovation matter in Asia? *Economic Research-Ekonomska Istraživanja*, 36, 1–13. DOI: 10.1080/1331677x.2022.2149588.

LojiPort. (2018). *Hizmet ihracatında sıralamaya giren 50 lojistik firması.* Retrieved April 25, 2023, from www.lojiport.com/hizmet-ihracatinda-siralamaya-giren-50-lojistik-firmasi-103938h.htm

Mars Logistics. (2020). *Sustainability.* Retrieved April 25, 2023, from www.marslogistics.com/en/sustainability

Mckinnon, A., Browne, M., Piecyk, M., & Whiteing, A. (2015). *Green Logistics Improving the Environmental Sustainability of Logistics.* London, UK: Kogan Page.

Mete, E. (2020). Sürdürülebilir kalkınma kapsamında yeşil lojistik: Avrupa Birliği ve Türkiye örneği (Green logistics context of sustainable development: The case of the European Union and Türkiye). *Karadeniz Sosyal Bilimler Dergisi*, 101–114. DOI: 10.38155/ksbd.790740.

MiraTrans. (2023). Retrieved August 15, 2023, from https://miratrans.pl/en/about-us/

Mohsin, A., Tushar, H., Abid Hossain, S. F., Shams Chisty, K. K., Masum Iqbal, M., Kamruzzaman, M., & Rahman, S. (2022). Green logistics and environment, economic growth in the context of the belt and road initiative. *Heliyon*, 8(6). DOI: 10.1016/j.heliyon.2022.e09641.

Mücevher, M. H. (2021). Sürdürülebilir lojistik için üç öncelikli strateji: Yeşil lojistik, tersine lojistik ve yalın lojistik (Three priority strategies for sustainable logistics: Green logistics, reverse logistics and lean logistics). *Enderun*, 5(1), 39–54. www.dergipark.org.tr/tr/pub/enderun

Nedeliaková, E., & Stasiak-Betlejewska, R. (2019). Transport management in Polish cities in the context of the European sustainable transport concept. *Transportation Research Procedia*, 40, 1150–1157. DOI: 10.1016/j.trpro.2019.07.160.

Netlog Logistics. (2022). *Sustainability*. Retrieved April 25, 2023, from www.netloglogistics.com/en/surdurulebilirlik

No Limit. (2023). Retrieved August 15, 2023, from https://en.nolimit.pl/eco-logistics/our-green-technologies/eko-fleet

Noorliza, K. (2023). Determinants of an environmentally sustainable model for competitiveness. *Sustainability*, 15(2), 1444. DOI: 10.3390/su15021444.

Official Gazette. (2022, May 25). Kombine Taşımacılık Yönetmeliği (Combined transport regulation). Official Journal Dated 25.05.2022 and Numbered 31848.

OMEGA Pilzno Group. (2023). Retrieved August 15, 2023, from https://omega-pilzno.com.pl/en/ecologistics/

Omsan Logistics. (2021). *Environmental policy*. Retrieved April 25, 2023, from www.omsan.com/en/sustainability/policies/environmental-policy/

Özgenç, Ş. (2021). Avrupa Birliği'nin Ortak Ulaştirma Politikasinin Kuramsal Yapida İncelenmesi (A theoretical examination of the European Union's Common Transportation Policy). Retrieved from https://www.researchgate.net/publication/368246611_AVRUPA_BIRLIGI'NIN_ORTAK_ULASTIRMA_POLITIKASININ_KURAMSAL_YAPIDA_INCELENMESI/citations

Peng, Z., & Wu, Q. (2019). Evaluation of the relationship between energy consumption, economic growth, and $CO_2$ emissions in China' Transport Sector: The FMOLS and VECM approaches. *Environment, Development and Sustainability*, 22(7), 6537–6561. DOI: 10.1007/s10668-019-00498-y.

Republic of Türkiye Ministry of Trade. (2021). *Yeşil Mutabakat Eylem Planı (Green Deal Action Plan)*. T.C. Ticaret Bakanlığı (Republic of Türkiye Ministry of Trade). Retrieved April 4, 2023, from https://ticaret.gov.tr/dis-iliskiler/yesil-mutabakat/yesil-mutabakat-eylem-plani-ve-calisma-grubu/eylem-plani

Reysaş Logistics. (2023). *Green logistics project*. Retrieved April 25, 2023, from www.reysas.com/green-logistics-project_0

Romagnoli, S., Tarabu', C., Maleki Vishkaei, B., & De Giovanni, P. (2023). The impact of digital technologies and sustainable practices on circular supply chain management. *Logistics*, 7(1), 1. DOI: 10.3390/logistics7010001.

Sarp Intermodal. (2020, May 1). *Sustainability*. Retrieved April 25, 2023, from https://sarpintermodal.com/en/corporate/sustainability/

Seroka-Stolka, O., & Ociepa-Kubicka, A. (2019). Green logistics and circular economy. *Transportation Research Procedia*, 39, 471–479. DOI: 10.1016/j.trpro.2019.06.049.

Sharma, M., Luthra, S., Joshi, S., Kumar, A., & Jain, A. (2023). Green logistics driven circular practices adoption in industry 4.0 era: A moderating effect of institution pressure and supply chain flexibility. *Journal of Cleaner Production*, 383, 135284. DOI: 10.1016/j.jclepro.2022.135284.

Spedimex. (2023). Retrieved August 15, 2023, from https://spedimex.pl/en/think-eco-logistics

Taş, M. A., & Aylak, B. L. (2022). Analyzing the impacts of European Green deal on logistics through CIMO-logic. *European Journal of Science and Technology*. DOI: 10.31590/ejosat.1082778.

Transmec Group. (2023). Retrieved August 15, 2023, from www.transmecgroup.com/en/the-group/environmental-policy

Tsl Transport. (2023). Retrieved August 15, 2023, from https://tsltrans.pl/

UN. (2020). *Global sustainable development report (GSDR) 2023*. United Nations, Department of Economic and Social Affairs. Retrieved March 24, 2023, from https://sdgs.un.org/gsdr/gsdr2023

UNCC—United Nations Climate Change. (2022). *Climate action pathways*. Retrieved March 28, 2023, from https://unfccc.int/climate-action/marrakech-partnership/reporting-and-tracking/climate_action_pathways

UTİKAD. (2018). *Commercial documents*. Uluslararası Taşımacılık ve Lojistik Hizmet Üretenleri Derneği (Association of International Forwarding and Logistics Service Providers). Retrieved April 14, 2023, from www.utikad.org.tr/En/TicariBelge.aspx?DataID=1

The World Bank. (2023). Retrieved July 20, 2023, from https://lpi.worldbank.org/international/global

World Commission on Environment and Development. (2017). *UN World Commission on Environment and Development*. Report of the World Commission on Environment and Development: Our Common Future. Retrieved from https://sustainabledevelopment.un.org/content/documents/5987our-common-future.pdf

Zeeshan, M., Lilong, Z., Fakhrullah, A. U., & Khan, S. N. (2023). The green supply chain model implications for economic and environmental sustainability via corporate social responsibility. *Multicultural Education*, 9(1), 42–61.

Zengin, E., & Akunal, E. O. (2017). Green logistics practices in Türkiye. *Journal of Management, Marketing and Logistics*, 4(2), 116–124. DOI: 10.17261/pressacademia.2017.456.

ZoomInfo, O. (2023). Retrieved August 15, 2023, from www.zoominfo.com/features/contact-company-search Company Insights

# 9 Energy-Efficient Design Strategies for Buildings

## *Lessons from High-Performance Buildings*

*Ritu Bala Garg and Gurpreet Singh*

## 9.1 INTRODUCTION

Many countries are working to cut their emissions of carbon dioxide in an effort to keep the rate of global warming below 2°C (Liu& Raftery, 2021; Anika et al., 2022; Napp et al., 2019). If we do nothing to address climate change, the economy might lose between 15 and 20 percent of its gross domestic product (Rochlin, 2021; Xuan et al., 2023; Gu & Hale, 2023). As the fourth-largest producer of greenhouse gases, India has committed to bringing its carbon emissions down to levels not seen since 2005 by the year 2030 (Akash et al., 2017; Sarwar et al., 2022; Patidar et al., 2023). India is implementing three initiatives in order to reduce its greenhouse gas emissions: increasing the amount of clean renewable energy it generates, increasing the number of trees it plants to act as carbon sinks, and increasing its investment in energy efficiency programs (Dey et al., 2022). In order to combat the effects of climate change, India has launched a number of different missions, including the National Solar Mission, the National Mission for Enhanced Energy Efficiency, and the National Water Mission (Shaw& Mukherjee, 2022; Pradhan & Ghosh, 2021). These missions address topics like as the cultivation of sustainable agriculture, the production of new information, and the administration of energy resources (Kremsa, 2021). This chapter overviews the steps that have been implemented by the Indian government in order to increase the energy efficiency of buildings, decrease reliance on fossil fuels, and encourage the use of renewable energy sources (Ikram et al., 2021). The chapter also identifies obstacles and barriers that limit the implementation of energy efficiency programs in residential buildings and recommends solutions to overcome them to accelerate energy efficiency programs and decrease $CO_2$ emissions (Mondal & Palit, 2022). Additionally, the chapter highlights challenges and barriers that inhibit the implementation of energy efficiency programs in commercial buildings (McLaughlin et al., 2023; Ramirez-Contreras & Faaij, 2018).

DOI: 10.1201/9781003438748-11

## 9.2 BACKGROUND AND SIGNIFICANCE OF ENERGY-EFFICIENT BUILDING DESIGN

The building and construction sector is responsible for the highest overall use of energy and produces the greatest quantity of carbon dioxide emissions worldwide (Norouzi et al., 2023; Anika et al., 2022; Bherwani et al., 2022). This is largely due to the fact that heating and cooling systems in buildings rely on fossil fuels (Fournier et al., 2022). A crucial aspect to consider when designing buildings is energy efficiency, which not only promotes sustainability but also ensures the provision of optimal thermal conditions for human health and performance (Gambhir et al., 2014). For instance, the use of air conditioners in various types of enclosures provides comfort to humans and can improve their work efficiency (Akhai et al., 2016a, 2017, 2020, 2021; Tanwar & Akhai, 2017). When buildings are designed to be more energy efficient, the amount of energy they consume to function is reduced, so it is essential that this be taken into consideration at an early stage of construction, when the building is in service (Rakesh, 2023; Harputlugil, 2017). Meeting the high demand for energy services while keeping costs affordable is crucial to ensuring a high quality of life for people (Ozbalta et al., 2021; Huang et al., 2022). In order to achieve this, it is important to improve energy efficiency, which can help reduce one's carbon footprint and promote sustainable practices (Ighravwe & Oke, 2019; Soussi et al., 2023; Hauashdh et al., 2022).

## 9.3 ENERGY CONSUMPTION SCENARIO OF INDIA AND NEED FOR REGULATORY PROCEDURE

The Ministry of Power is responsible for generating and distributing electrical energy throughout India, and they've been supported by the Energy Conservation Act and the Bureau of Energy Efficiency (BEE). The BEE rates equipment to encourage energy-efficient usage (Anika et al., 2022). It's predicted that over 75% of India's buildings will be constructed by 2030, offering a promising opportunity to limit the rise in energy consumption by ensuring that new buildings are built energyefficiently (Haldar et al., 2023). With India's rapid social and economic growth, population expansion, and migration to cities, more building projects are expected, which also means an increase in energy usage (Jain et al. 2017). Therefore, it's crucial for India and other developing countries to focus on optimizing their energy savings and constructing energy-efficient buildings (Lai et al., 2023; Khosla & Janda, 2019).

To achieve this, India needs to implement energy efficiency measures and frameworks for newly constructed buildings. Additionally, residential areas accounted for 30% of total final energy consumption in 2017. Figure 9.1 shows that bioenergy is the leading source of energy (68%) and then followed by oil and electricity (Shankar & Singh, 2022). There are significant possibilities available to limit the future rise in consumption and demand for energy, as shown in Figure 9.2, by ensuring that new buildings are becoming more energy efficient (Chaturvedi et al., 2014).

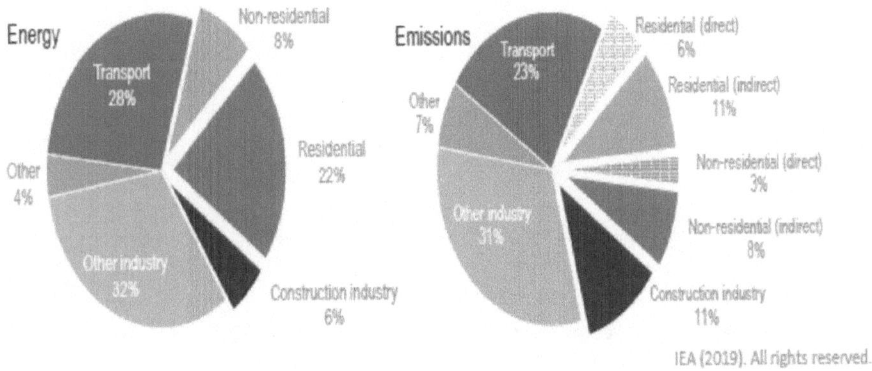


**FIGURE 9.1** Global share of buildings and construction final energy and emissions, 2018.

**Note: The construction industry is the portion (estimated) of overall industry devoted to manufacturing building construction materials such as steel, cement, and glass. Indirect emissions are emissions from power generation for electricity and commercial heat.**

**(Source: Adapted from IEA (2019a), World Energy Statistics and Balances (database), www.iea.org/statistics and IEA (2019b), Energy Technology Perspective, building model, www.iea.org/buildings).**

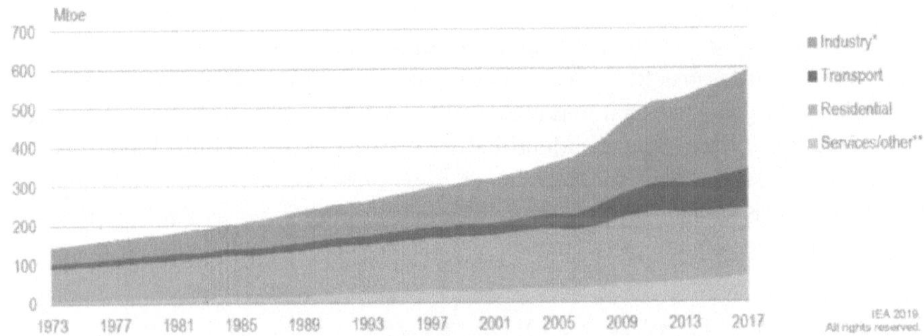


**FIGURE 9.2** Total fuel consumption by India in various sectors.

*** Industry includes non-energy consumption.**

**** Services/other includes commercial and public services, agriculture, and forestry.**

**Note: Years run from 1 April to 31 March.**

**(Source: IEA (2019a). World Energy Balances 2019. www.iea.org/statistics).**

According to Figure 9.3, the residential sector was the second-largest user of energy in 2017, accounting for 30% of total final consumption. According to estimates supplied by the IEA, residential energy consumption is substantially led by the conventional use of bioenergy, which accounts for 68% of the sector's total. However, the proportion provided by oil and notably electricity is expanding.

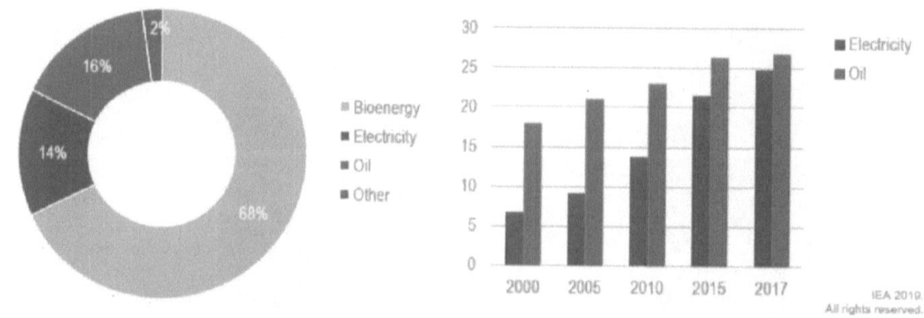


**FIGURE 9.3** Residential sector energy consumption by source, 2017, and growth in electricity and oil, 2000–17.

**Notes: Bioenergy data are estimated by the IEA.**

**(Source: IEA (2019b). World Energy Balances 2019. www.iea.org/statistics).**

## 9.4 LEGISLATIVE GUIDELINES FOR ENERGY EFFICIENCY IN INDIA

The Indian government is promoting the development of energy-efficient buildings and habitats through regulations, as stated in the national policy on housing in metropolitan regions (Khare et al., 2022). The national solar mission aims to construct structures that generate their own energy, while these regulations encourage the construction of energy-efficient buildings and habitats (Abbas et al., 2023). The smart city initiative prioritizes environmentallyfriendly and energy-efficient technologies across various infrastructure levels (Razmjoo et al., 2022; Fakhimi et al., 2021). To achieve this goal, national and state-level rules have been established to ensure clean and green cities. India has identified 100 "smart cities" to implement this plan, applying the latest methods for reducing energy consumption and maximizing available resources (Kandpal, 2018). Energy efficiency codes serve as guiding principles for implementing these regulations.

## 9.5 ENERGY EFFICIENCY CODES, RULES, AND GUIDELINES IN INDIA

The Indian government has supported renewable energy sources like solar and wind power. In addition, India is making progress toward achieving one of the United Nations' Sustainable Development Goals, which is to provide access to sustainable forms of energy.

- The prime minister's Council on Climate Change approved the National Mission on Sustainable Habitat in June 11, 2010; the mission's most important objective is the creation of sustainable habitat criteria that lead to robust development plans while addressing climate change–related issues. The formulation of municipal development plans that comprehensively address adaptation and mitigation concerns. There is enhancement of capabilities to execute mission-related actions on climate change.

- The Energy Conservation Act (EC Act) was enacted in 2001 with the intention of reducing India's energy consumption. It was revised in 2010 and now functions as the foundation for India's energy efficiency policies.
- The Building Energy Efficiency Act is responsible for establishing the Residential Energy Conservation Building Code (ECBC). This law states that modest enforcement measures, such as the use of passive systems, can improve the thermal comfort of building occupants. The Bureau of Energy Efficiency (BEE) of the Ministry of Power undertook a number of initiatives to enhance energy efficiency. These initiatives included establishing standards for energy consumption and residential illumination, among others. The government of India enacted the Energy Conservation Building Code for new commercial buildings on May 27, 2007. The ECBC establishes the minimum energy requirements that newly constructed commercial buildings with a connected load of 100kW or more must meet.
- Green classification for Integrated Habitat Assessment (GRIHA) is a classification system devised in India to assess and determine the performance of structures.
- The Energy and Resources Institute (TERI), a research organization, and the Ministry of New and Renewable Energy (MNRE), a government agency in India, both play a significant role in the management and promotion of green buildings in India. It has adopted the Leadership in Energy and Environmental Design (LEED) green building classification system, which assigns "Platinum", "Gold", "Silver", and "Certified" ratings to different categories of structures.

The objective of the planning, design, construction, and operation of a green building is to maximize occupant comfort and productivity while minimizing the building's overall energy consumption (Kandpal, 2018). In India, the construction of environmentally sustainable buildings is expected to acquire momentum. In addition to these landmarks, however, there are obstacles to energy conservation that will be discussed in greater detail in the following and must be surmounted. As per India Energy Agency, the Paradigm of Indian Green Building is due to a number of impediments and voids in the current system, it is notoriously difficult to realize the construction industry's enormous potential for cost and energy savings (Barbar et al., 2023).

Coal for power generation has been the primary source of energy demand in India (Mathivathanan et al., 2022), driven by a growing demand for air conditioning and electricity access in buildings. More than half of new structures to be built by 2060 will be developed in the next 20 years, making it urgent to reduce energy consumption and emissions from buildings to limit global warming to 2°C or lower. Residential and commercial buildings account for 24% and 9% of India's overall power consumption, respectively (Panicker et al., 2023), with commercial buildings using 71 billion units of electricity annually (Gandhi et al., 2022). To achieve the goal of reducing global energy usage by at least 10% per person by 2025, it is necessary to deploy energy-efficient technologies in building design and construction that balance environmental responsibility, resource efficiency, occupant comfort, and community sensitivity (Awadh, 2017). Definitions of sustainable buildings vary, but they all aim to minimize environmental impact while enhancing user comfort and productivity (Cureau et al., 2022).

**Key attributes of sustainable buildings**—Sustainable buildings are characterized by several factors. Sustainability considerations are integrated into all stages of building design and planning, as well as throughout the construction process and the manufacture of building materials (Berges-Alvarez et al., 2022). Sustainable building materials that are ecologically conscious and non-hazardous are used (Sanderson, 2022). Time-saving systems and simple-to-service structures and systems are implemented to protect high levels of functionality, adaptability, and flexibility. User, occupant, and visitor health and comfort are prioritized (Dong et al., 2023). Sustainable buildings are widely accepted by the general public due to their excellent aesthetic and urban design quality and convenient location with easy access to public transportation services and networks (Meerow& Newell, 2017).

## 9.6 BENEFITS OF SUSTAINABLE BUILDING DESIGN FOR OCCUPANTS AND THE ENVIRONMENT

Environmentally responsible architecture can bring tangible and intangible benefits to both the inhabitants of the building and the environment (Hafez et al., 2023). Sustainable construction, design, and operations of a building can result in lower carbon emissions, reduced energy consumption and waste, better water conservation, and less exposure to harmful pollutants (Golušin et al., 2013). Besides, buildings constructed with eco-friendly materials and strategic landscaping can help maintain better indoor air quality and increase the productivity of the occupants (Gupta& Chakraborty, 2021). Passive design measures like strategic site planning, orientation of the building, daylighting, facade design, use of appropriate construction materials, strategic landscaping, and natural ventilation can improve the energy efficiency of commercial and residential buildings in a cost-effective manner (Santamouris& Vasilakopoulou, 2021). The following benefits demonstrate the importance and advantages of incorporating sustainable building design practices not just for the occupants' well-being but also for the long-term health and sustainability of our environment:

**Tangible benefits**—Building sustainably offers many benefits, some of which are tangible. Through sustainable design, construction, and operations, buildings can have a positive impact on the environment, even if tenants and tourists do not immediately see the visible advantages (Lassen et al., 2021). Sustainable buildings produce fewer carbon emissions, use less energy and water, generate less waste, prioritize the use of better materials, and reduce exposure to harmful pollutants (Açikkalp et al., 2023). As a result of these sustainable design concepts, energy consumption may be reduced by up to 50% and water use can be reduced by up to 35%, resulting in significant cost savings for building owners and occupants (Sethi et al., 2020).

**Intangible benefits**—Sustainable buildings not only give real advantages but also offer intangible benefits to its occupants. As occupants spend the majority of their time inside a building, it is essential that the indoor air quality be kept in good condition (Abokersh et al., 2021). Sustainable buildings have the potential to enhance indoor air quality, which may lead to a working environment that is both healthier and more pleasant. Increases in productivity of up to 10% are possible as a result of this (Omer, 2008). Other intangible benefits provided by sustainable buildings include daylighting, the monitoring of $CO_2$ levels for air quality, and views of the exterior environment (Whittemet al., 2022). These qualities not only provide for a more enjoyable and productive working environment, but they also contribute to the overall welfare of the occupants.

## 9.7 PASSIVE DESIGN STRATEGIES FOR ENERGY EFFICIENCY

Passive design strategies are an effective and cost-efficient way to optimize energy efficiency in both commercial and residential buildings (Parvin et al., 2021). These strategies focus on minimizing heat gain in buildings through passive cooling measures and utilizing natural environmental cooling methods like vegetation, landscaping, and shading (Balali et al., 2023). Since buildings are designed primarily to provide comfortable indoor environments for occupants, the architectural design should take into account the surrounding site environment. The following arc some key passive design measures that can be implemented:

**Site Planning and Orientation**—When designing a new building on a green field site, it is crucial to consider site planning and orientation. Proper building orientation in equatorial regions aims to prevent excessive solar radiation exposure through openings in the structure as the sun moves from east to west. As a general guideline, the building should be oriented so that the main long axis with more windows faces north to south, and the narrow ends face east to west (Renuka et al., 2022). This minimizes exposure to the east–west solar travel direction as much as possible (Schweiker, 2017). Building orientation can also have an impact on the immediate microclimate of open spaces by casting shadows and providing shade to nearby areas, which can benefit interior rooms located near the open space. See Figure 9.4.

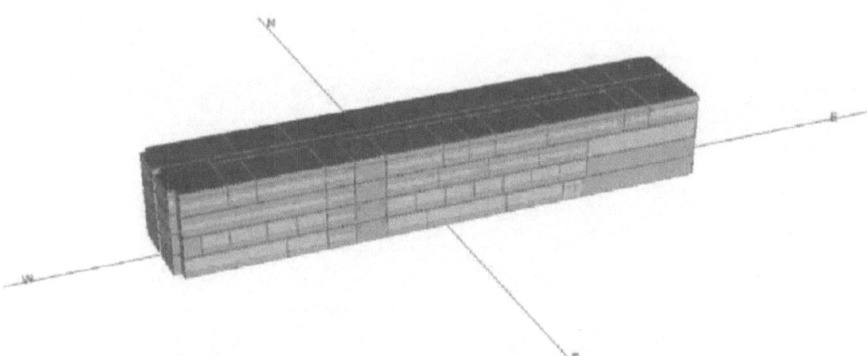

**FIGURE 9.4** Orientation of the building along the longer axis.

**Daylighting**—Daylighting is basically the practice of using natural light to meet illumination requirements during daylight hours. It should be the first consideration when designing a building, even before discussing energy-efficient electrical lighting (Shahid& Karimi, 2021). With an appropriately designed daylight harvesting system, building occupants can enjoy an improved working environment and increased energy efficiency (Reffat & Ahmad, 2020). However, if sunlight is not properly harvested, it can result in uncomfortable glare, excessive heat gain, thermal discomfort, and high energy usage. Therefore, proper daylighting design is critical for achieving optimal building performance and occupant comfort (Zhou et al., 2023).

**Facade Design/Building Envelope**—The building envelope and facade design are both essential components of passive design measures (Wuet al., 2023). Architects can use their creative ideas to reduce the amount of heat gained by buildings from the sun (Ihara et al., 2015). The exterior building envelope, which is the facade of a structure, determines the shape and beauty of the building (Mirrahimi et al., 2016). A well-designed facade that uses appropriate architectural treatments and materials can minimize solar heat gain and optimize daylighting and thermal comfort (Luo et al., 2019). It is crucial for the building envelope to be constructed in a way that prevents heat from entering the building via conduction and solar radiation (Shareef et al., 2023). With a thoughtfully planned envelope, a building's energy consumption and cooling load can be significantly reduced. See Figure 9.5.

**FIGURE 9.5** Façade gaining maximum solar heat.

**Building Materials**—Using appropriate construction materials that have insulating capabilities can significantly reduce a building's overall energy usage (Hasan et al., 2021). During the day, the outside air temperature is typically higher than the inside temperature of air-conditioned spaces, which can result in heat transfer into the building through convection. However, during the night and early morning, the outside air temperature is often cooler than the inside temperature, causing heat to move in the opposite direction (Yang & Chen, 2022). By selecting construction materials with good insulation properties, the amount of heat transfer through the building envelope can be reduced, resulting in lower energy consumption for cooling and heating (Fadye et al., 2021). This can include materials such as insulation boards, double-glazed windows, and reflective roof coatings, among others (Cao et al., 2015). Choosing sustainable and eco-friendly materials can also minimize the environmental impact of the building over its lifecycle (Sedaghat & Sharif, 2022).

**Strategic Landscaping**—This type of landscaping is particularly relevant for highly urbanized areas that are heavily built up and lack a significant amount of vegetation in the surrounding area of a structure (Tafakkori & Fattahi, 2021). By strategically planting the area, it is possible to limit the amount of heat gained in a building development. The urban heat island effect can be mitigated to some extent by strategically landscaping the area surrounding the building to provide a cooler microclimate (Balali et al., 2023). Typically, the air temperature in highly urbanized and built-up areas is significantly higher than in rural and less built-up areas. Strategic landscaping can also improve the aesthetic appeal of a building development and provide a more pleasant environment for occupants and visitors Additionally, landscaping can play a crucial role in stormwater management by reducing the amount of runoff and improving water quality.

**Natural Ventilation**—Natural ventilation, or the flow of air, is crucial for providing clean air, maintaining comfortable indoor temperatures, and cooling the interior when outside air is cooler. In recent years, many solutions have been proposed to combat rising global temperatures and increasing energy consumption in the building sector (Kuzman et al., 2013). Regulations have been developed alongside innovative technology to address this issue. The primary objective of passive construction is to provide a comfortable living environment without requiring a separate active heating system (Hee et al., 2015). Passive structures can meet their minimum energy requirements by striking a balance between heat losses and gains in response to climatic conditions. When designing a passive building, various aspects of the structure's design should be taken into consideration, including thermal insulation, thermal mass, glazing, window size, shape, and location, color of exterior surfaces, external shading devices, building orientation, and structure form (Sánchez-Quiles & Tovar-Sánchez, 2015). Thus, natural ventilation is an effective way to achieve sustainable and energy-efficient buildings. By incorporating passive design measures that prioritize natural ventilation, designers can create comfortable and healthy living environments while reducing energy consumption and minimizing the carbon footprint of buildings.

## 9.8 OBJECTIVES OF THE CHAPTER

This chapter explores how energy-efficient building design in India reduces carbon emissions and promotes sustainability. It asks how India might apply and improve energy-efficient building design solutions to minimize carbon emissions and promote sustainability. Specifically, the chapter examines India's energy-efficient building design legislation, regulations, guidelines, and codes, with a focus on the Ministry of Power, the Bureau of Energy Efficiency, and GRIHA and TERI.

The chapter also examines energy use in India, including residential and commercial trends. It stresses the importance of energy-efficient construction methods and examines their possible influence on energy usage. We also evaluate passive design tactics, including site planning, daylighting, facade design, building materials, strategic planning, and natural ventilation.

The research points out the reduction of carbon emissions, lower energy consumption, improved indoor air quality, and improved occupant well-being as compelling reasons to prioritize energy efficiency in building design. This chapter examines legislative frameworks, energy consumption patterns, passive design strategies, and associated benefits to contribute to the discourse on energy-efficient building design in India and provide actionable insights for promoting sustainability and reducing carbon emissions in the building sector.

## 9.9 CASE STUDIES DEMONSTRATING THE EFFECTIVENESS OF PASSIVE DESIGN STRATEGIES

Passive design solutions are effective in reducing energy usage and increasing comfort in buildings, as demonstrated by various case studies.

Albayyaa et al. (2019) used a computer simulation to analyze the impact of different design strategies on energy consumption in a two-story, four-bedroom residential home in Sydney. The study tested various exterior wall and floor systems, as well as passive solar and energy efficiency design strategies (PSEEDS), to determine how they affected the amount of energy needed to maintain a comfortable temperature inside the home. The study found that using PSEEDS and passive solar design strategies during construction could reduce winter heating energy usage by 37% and 36%, respectively, in typical fibro and brick veneer homes. Additionally, energy consumption could be reduced by as much as 58% by increasing the thermal mass of the building through different walls and flooring systems, switching from a fibro to a brick veneer house, and implementing PSEEDS. These strategies can lead to significant savings in energy costs over the 50-year lifespan of a building. The simulation results were compared to actual data and found to be within an allowable margin of error of 16%. See Figure 9.6.

Randjelovic et al. (2021), in their research on the retrofitting of educational institutions to make them more energy efficient, investigated the impact of different passive design solutions on building energy efficiency. As part of the research, the author tries to improve various passive design systems that are connected to the building envelope. These systems include green tools, double-skin facades, and Trombe walls. The researchers analyzed the effect of these modifications using modeling software

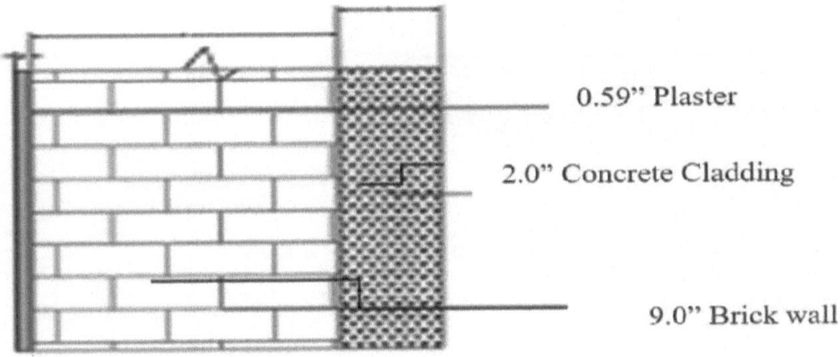


**FIGURE 9.6** Brickwall with concrete cladding.

to determine how much more energy would be needed to heat and cool the building as a result of these changes. According to the results, a combination of these initiatives has the potential to result in considerable energy savings of up to 77% for heating and 79% for cooling. Buildings are responsible for more than one third of the world's energy consumption, with the residential sector being the largest contributor in the Gaza Strip. This study focuses on reducing energy consumption by implementing passive design techniques in a standard house. The researchers conducted a comprehensive review of relevant literature and used simulation software to analyze the impact of shading devices, natural ventilation, and thermal insulation on the building's internal temperature and cooling load. The results show that incorporating these passive design features can lead to a 59% reduction in energy consumption, which is equivalent to 9.89 kW. This study highlights the potential of passive measures in building design and their significance in cities with limited natural resources.

In their study, Gamero-Salinas et al. (2021) discuss how high temperatures and humidity in tropical areas can have negative effects on people's health and increase energy demand and carbon emissions, especially with the exacerbation of climate change. The study examines how passive cooling design strategies can reduce the risk of indoor overheating in warm tropical cities like Tegucigalpa and San Pedro Sula, with a focus on apartment-style housing. The study found that natural ventilation, wall absorptance, solar heat gain coefficient, and semi-outdoor areas are effective in minimizing risks. The research emphasizes the need to incorporate passive cooling and overheating prevention design solutions into tropical building regulations and building design while considering future risks.

Pajek and Košir (2021), in their article, explored the potential effects of passive design techniques on the energy required to heat and cool single-family homes in Europe. The study examined various values of thermal transmittance, window to floor ratio, window distribution, shape factor, diurnal heat storage capacity, external opaque surface solar absorptivity, and natural ventilation cooling rates, which were combined in 496,800 building energy models and simulated at eight different locations. The study found that a lower window-to-floor ratio is the most universally applicable design measure for achieving low energy usage in buildings. However,

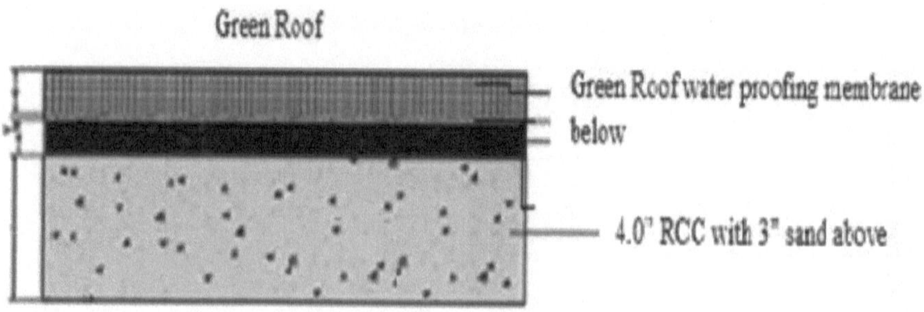


**FIGURE 9.7** Cross section of green roof.

some indicators vary depending on the climate and time period analyzed. The study also found that even with the most effective combination of passive design measures, it is difficult to counteract the anticipated effects of climate change on building energy consumption. Nevertheless, passive building design can still ensure reasonably low energy consumption, particularly in oceanic, warm, and some temperate climate locations. These findings can inform strategies and guidelines for energy-efficient buildings.

Díaz-López et al. (2022) conducted research on passive intervention tactics in schools, which is a complex and diverse field due to the many academic disciplines, cultural contexts, and research methods involved. The study analyzed 537 publications from 1982 to 2020 using the Sci-MAT program and identified the 24 most effective passive intervention options for schools based on their use in 42 different countries. The study found that natural ventilation was a popular option in the early years of the research, while green roofs have become increasingly popular since 2010 due to concerns about urban heat islands and energy efficiency. There has also been a focus on occupant comfort, indoor air quality, and cost-effectiveness of passive vs. active design. The study emphasizes the importance of using climate-resilient passive design principles and provides valuable insights for scholars, practitioners, and policymakers in the field. See Figure 9.7.

Alhuwayil et al. (2023) investigated the use of passive design strategies, such as shade and thermal insulation, in multistory hotel buildings in the northern hemisphere to save energy. The study used Design Builder version 4.5.0.148 to model and simulate the structures in cities with different climates and geographic locations. The results showed that outdoor climate and geographic location significantly affected the energy-saving potential of both shade and insulation options. Factors such as cooling degree days, share of cooling and heating energy demands, global horizontal irradiance, and solar shading effectiveness were identified as important. The study found that the potential for conserving energy by passive solar shading was greatest in cities with warm exterior climates like Khamis Mushait, while high-performance insulation was more effective in cities with high heating needs like Athens. The research highlights the importance of considering external factors such as climate and geography in the design phase to maximize the energy efficiency of buildings.

Zoure et al. (2023) evaluated natural ventilation and daylighting improvements in office buildings in Burkina Faso with the aim of increasing thermal comfort and energy efficiency. The study utilized meteorological data from 2004 to 2018 for climatic analysis and used the Integrated Environmental Solutions Virtual Environment (IESVE) for natural ventilation and daylight simulations, as well as DesignBuilder (DB) for simulations of thermal comfort and energy consumption.

Four rounds of simulation were conducted to identify the impacts of building shape, glazing type, window shading type, and window-to-wall ratio on the performance of the office building model. The results showed that the optimal design for a building with a south–north axis orientation is a structure in the form of a "H" with blue-tinted glass, horizontal louvers for shading, and a window-to-wall ratio of 30%. This design reduced yearly energy consumption by 14.9%, cooling energy demand by 25.3%, and solar gains by 68.3% while increasing the hourly average airflow by 64.8% during warm seasons and 2% during cold seasons. The study also found that natural ventilation and daylighting solutions can significantly improve thermal comfort and energy efficiency in buildings, especially in sub-Saharan nations that are rapidly urbanizing. By including climate-based design modelling (CBDM) and natural ventilation factors in the design process, the study provided a framework for a sustainable bioclimatic office building model in a landlocked poor country like Burkina Faso. The study highlights the need for a paradigm shift in the construction industry towards eco-friendly buildings that reduce energy usage and increase thermal comfort and the scientific contributions that natural ventilation and daylighting solutions can make towards achieving this goal.

## 9.10 ACTIVE DESIGN STRATEGIES FOR ENERGY EFFICIENCY

Active design solutions take advantage of the building's existing electrical infrastructure in order to maintain occupant comfort. The use of artificial lighting as well as heating, ventilation, and air conditioning systems are included in these techniques (Zoure& Genovese, 2023). Architects play an important part in establishing whether or not a project will be sustainable, and they are required to make use of technology to improve their designs rather than allowing technologies to dictate what those designs will look like (Akhai et al., 2016b).

- Installing renewable energy systems in a building, such as solar panels or wind turbines, may drastically cut a building's reliance on fossil fuels and contribute to the structure's energy self-sufficiency (Zoure & Genovese, 2023). These systems have the potential to create energy that is clean and renewable, which can then be put to a variety of uses, including lighting, heating, and air conditioning. It is possible to gain credits for energy performance under the GRIHA by including renewable energy systems into the design of the building.
- Energy-efficient heating, ventilation, and air conditioning (HVAC) systems. HVAC systems are essential components of the energy consumption of a structure (Akhai et al., 2016a). Energy recovery ventilation (ERV) systems, variable refrigerant flow (VRF) systems, and geothermal heat pumps are

examples of high-efficiency HVAC systems that have the potential to dramatically cut energy usage and receive points under GRIHA for energy efficiency.

- Energy-efficient lighting systems: Lighting is yet another critical component of building design that has the potential to influence overall energy usage. It is possible to dramatically cut energy use by implementing energy-efficient lighting systems such as LED lights, occupancy sensors, and controllers for daylight harvesting. If you do this, you will get points under GRIHA for energy performance. Occupancy sensors have the capacity to switch lights off automatically in unoccupied areas, while daylight harvesting controls have the ability to modify the amounts of artificial lighting depending on the amount of natural sunshine that is available.
- Building automation systems: Building automation systems, often known as BAS, are sophisticated control systems that improve the operation of various building systems, such as lighting, heating, ventilation, and air conditioning, and security (Zoure & Genovese, 2023). The building automation system has the ability to monitor, regulate, and optimize a variety of building characteristics in real time, which ultimately leads to improvements in energy efficiency, comfort, and security. The use of a BAS in building design may result in credits being awarded under GRIHA for the building's energy efficiency and indoor environmental quality.

The use of active design techniques is an essential step toward attaining and promoting sustainability in the building design and construction industries. The incorporation of renewable energy systems; high-efficiency heating, ventilation, and air conditioning systems; efficient lighting systems; and building automation systems may considerably minimize the negative effect that buildings have on the surrounding environment and help to the creation of a more sustainable built environment. Buildings' commitment to environmentally responsible architecture and building practices may be realized if these ideas are put into reality.

## 9.11 CASE STUDIES DEMONSTRATING THE EFFECTIVENESS OF ACTIVE DESIGN STRATEGIES

Xin and Rao (2013) provided examples of how active design methods can be effective in reducing a building's energy usage. The Diamond Building was designed to minimize direct sunlight entering the structure, which significantly reduced its cooling load, the main contributor to energy usage in tropical buildings. The building's shape was designed to shade lower levels with higher ones, using passive design methods to reduce cooling loads. However, this resulted in trade-offs, as potential areas for active design strategies (such as installing photovoltaic panels) had to be sacrificed to implement the passive design strategy. As a result, the building's form contributed to a higher building energy index (BEI), a metric used to assess energy performance in buildings. While the slanted facade blocked direct sunlight, it also took up space where more PV panels could have been installed to produce more energy and achieve a lower BEI. Using a straight facade and more PV panels may have been a better choice for energy performance, but it would have resulted in the building losing its distinctiveness and visual appeal.

Luo et al. (2019) aimed to develop control strategies for smart windows and investigate how these strategies affect the energy consumption and comfort levels in office buildings. Smart windows can adjust the amount of heat and sunlight that enters a building by controlling shutters, blinds, and openings. The goal is to optimize indoor comfort while minimizing the energy used for heating, cooling, lighting, and ventilation. The researchers used a simplified method to calculate the yearly energy and comfort performance of an office room with a smart window. This method was verified using a dynamic building simulation tool and experimental tests. The results showed that using a smart window can reduce the annual energy consumption of an office building by about 60% compared to using a static window. Additionally, using smart windows can help buildings comply with Danish building regulations and improve indoor comfort levels.

Munoz-Liesa et al. (2022) investigated into how combining greenhouses and buildings could help improve energy efficiency. They studied the potential of using the waste airflows from buildings to heat and ventilate rooftop greenhouses, specifically in a Mediterranean climate. By analyzing monitored data, they found that the integrated system could act as a solar collector and recycle thermal energy back into the building. The study showed that the integrated rooftop greenhouse (IRTG) could harvest 205.2 $kWh/m^2\ y^{-1}$ of thermal energy, limited by the greenhouse's low transmissivity. Additionally, the building's exhaust air at temperatures sufficient to heat and cool the greenhouse was measured to be 198 $kWh/m^2\ y^{-1}$. By incorporating active ventilation strategies, energy benefits doubled compared to a passive ventilated configuration. The results indicate that coupling greenhouse and building functionalities can contribute to urban circularity and sustainability by reducing energy consumption and resource waste.

Rana and Rahman (2020) conducted a study on a two-story residential building in the Okanagan Valley, Canada, to evaluate the impact of COVID-19 lockdowns on HVAC and non-HVAC loads. They used steady-state building energy simulations with the HOT2000 tool to assess the effectiveness of energy efficiency measures during a full lockdown. They analyzed three years of monitored energy data and observed a significant change in energy use patterns, particularly in May 2020, compared to the previous two years. The researchers then used calibrated energy models to investigate the effects of pre-COVID-19 (old normal occupancy) and post-COVID-19 (new normal occupancy) on energy upgrade performance. They found that higher occupancy times led to a 16.4% increase in annual electricity use, while natural gas use decreased by 7.6%. The study highlights that buildings following pre-COVID-19 occupancy schedules have greater potential for energy savings than those following new normal occupancy schedules. The results also show that different occupancy levels and stakeholder preferences affect the ranking of energy efficiency measures. The study identifies low-cost energy efficiency measures that provide flexibility for decision-makers under various occupancy schedules.

Perwez et al. (2023) proposed a new method to evaluate the feasibility of making commercial building stock carbon neutral. They combined a geographic information system (GIS) with a building energy model (UBEM) and a physical-based approach for building integrated photovoltaics (BIPV) potential estimation. They used this model to study the commercial building stock in Tokyo and evaluated the impact of various active and passive design measures and BIPV on carbon neutrality. The results showed that a combination of all measures can reduce annual $CO_2$ emissions by 84%, and passive design measures alone are not enough to achieve carbon neutrality in commercial buildings. The study also found that BIPV contributes 16–40% to the decarbonization potential, although less than efficiency measures. Additionally, the analysis showed that total BIPV potential changes with the scale of the study, and more self-sufficiency is observed at the building level compared to district and city scales. This study can help energy modelers and policymakers to make informed decisions on achieving carbon neutrality in commercial building stock at different levels.

## 9.12 INTEGRATION OF PASSIVE AND ACTIVE DESIGN STRATEGIES

Sung et al. (2019) conducted a simulation study to reduce heating and cooling demand in a school building located in the Seoul metropolitan area, Korea. They compared the effectiveness of passive measures, such as better insulation and improved building shape and orientation, versus active measures, such as lighting control, heat exchangers, and geothermal heat pumps. They found that the most significant energy savings were achieved through lighting control. Additionally, the energy savings increased by 32% when they improved the building design. They also discovered that the potential energy savings of each room were influenced by factors like window-to-wall ratio, internal heat generation, ventilation requirements, and orientation. Therefore, energy-saving analyses should be conducted at the individual room level, not at the entire building level. They also recommended that simulation studies be used to inform design decisions. Overall, they concluded that passive building design strategies, such as building orientation and solar radiation control, are more effective for improving energy efficiency than active measures, and should be prioritized in urban, architectural, and engineering design.

Prafitasiw et al. (2022) conducted research on building occupants' awareness of energy efficiency in green buildings, specifically at the Institute Teknologi Sepuluh Nopember (ITS) campus in Indonesia. The study aimed to identify the criteria for energy efficiency awareness among lecturers, staff, and students in three campus buildings. A questionnaire survey was conducted, and 181 data were analyzed. The study found that the overall performance of the awareness variables was good, but the students' awareness was moderate and needed improvement to achieve energy efficiency. The research highlighted three critical variables that require improvement: recognizing the issue of electricity resource depletion, resource sharing to reduce the use of electrical energy consumption, and understanding how to make efforts and prepare for electricity resource depletion. This study emphasizes the importance of managing building occupants' awareness of energy efficiency in the context of campus buildings.

Knuutila et al. (2022) studied that buildings and construction contribute to around 40% of the energy and $CO_2$ emissions worldwide. However, the rate of renovating existing buildings is quite low. This study looks at 12 city-owned buildings in two Finnish cities to see how energy efficiency improvements can help reduce energy costs and emissions. By analyzing the buildings' heat and electricity usage, the researchers developed a model to see the impact of different energy efficiency improvements, such as heat pumps, solar panels, and improving the building's envelope. The study found that such investments can reduce energy-related costs by an average of 35%. This highlights the importance of considering energy efficiency improvements in existing buildings to reach the goal of a carbon-neutral world by 2050.

The benefits of integrating passive and active design strategies—Buildings contribute significantly to greenhouse gas emissions and are a major focus of carbon reduction programs (Mekhilef et al., 2012). Organizations such as GRIHA, TERI, and IGBC are promoting sustainable design practices to address this issue (Elavarasan et al., 2022). Architects and engineers must re-evaluate their standard practices for building design, construction, and operation to mitigate the negative impacts of climate change. To achieve this, research and development programs must be accelerated, and policies modified to reflect the realities of climate change (Lazarevic et al., 2022). Professional architecture and engineering organizations can change sample contract documents, rating systems, and design standards to drive the industry forward. With leadership from these professions, recent research on climate change and the building sector can be turned into action.

## 9.13 IMPORTANCE OF CONSIDERING LONG-TERM ENVIRONMENTAL AND SOCIAL IMPACTS IN BUILDING DESIGN

When designing a building, it is essential to take into account the long-term environmental and social impacts for a number of reasons.

- First, structures have a substantial influence on the surrounding ecosystem, since they are responsible for a sizeable proportion of the world's greenhouse gas emissions, energy consumption, and trash production. It may be possible to alleviate some of these problems and advance sustainability by designing buildings that have a less detrimental influence on the surrounding environment.
- Second, the people who utilize buildings are also influenced by the structures themselves. The health and well-being of tenants may be negatively impacted by factors such as a lack of proper illumination, unpleasant temperatures, and poor indoor air quality (Elavarasan et al., 2022). These challenges may be overcome with the help of environmentally responsible architectural design that incorporates elements such as adequate ventilation, natural lighting, and thermal comfort.
- Third, buildings are considered to be investments for the long term because they are designed to last for a number of years. When a structure is planned without taking into account the long-term environmental and social implications it will have, the building may wind up having high maintenance and operation costs, which may have a detrimental effect on the owner's bottom line (Zandvoort & van der Vlist, 2020). It is possible to minimize running costs, boost property values, and improve the overall performance of a building by incorporating sustainable design principles into the design process.

Thus, when it comes to advancing environmentally friendly architectural practices, architects, engineers, and politicians all have an important part to play. They are able to advocate for environmentally responsible design techniques, encourage the use of environmentally friendly building materials, and integrate sustainability into building laws and regulations. They may also collaborate to explore new possibilities for sustainable design and search for creative solutions to problems related to sustainability. They may assist in guaranteeing that the design of buildings satisfies the demands of both the environment and the people who use them, both now and in the future, if they cooperate with one another and work together.

## 9.14 THE ROLE OF ARCHITECTS, ENGINEERS, AND POLICYMAKERS IN PROMOTING SUSTAINABLE BUILDING DESIGN

When it comes to advancing environmentally friendly architectural practices, architects, engineers, and politicians all play an important part.

- They may collaborate to reduce the detrimental impacts that the construction sector has on the surrounding environment by implementing environmentally friendly policies and procedures into the design, construction, and operation of buildings (Zutshi& Creed, 2015). This may be accomplished by making use of materials that are kind to the environment; maximizing the orientation and form of the structure; and putting in place energy-efficient systems for things like lighting, heating, ventilation, and air conditioning (Beyene et al., 2018).
- To further the cause of sustainable building design, architects and engineers should make it a priority to go above and beyond the minimum requirements established by laws and regulations (Condori Fernandez & Lago, 2018). They may seek diverse viewpoints and perspectives in order to tackle difficulties related to sustainability, such as consulting with specialists in fields such as green construction technology, water conservation, and renewable sources of energy (Sarkis et al. 2011).
- In addition, architects and engineers may make efficient and effective use of resources by evaluating the life cycle of building materials and choosing those that are durable, simple to maintain, and friendly to the environment (Hernandez et al., 2019). They may also include passive design strategies like natural ventilation, daylighting, and landscaping in order to reduce the amount of energy and resources they require (Gagnon et al., 2019).
- The implementation of legislation, incentives, and standards that support sustainable practices in the construction sector is another key role that policymakers may play in encouraging sustainable building design (Aggarwal et al., 2022). They are able to provide financial incentives in the form of tax credits, subsidies, and grants to constructors and developers that include environmentally friendly methods into their projects (Ohene et al., 2022).

Thus architects, engineers, and policymakers need to collaborate in order to promote sustainable building design (Kumar & Akhai, 2022). This can be done by incorporating environmentally friendly practices, seeking multiple perspectives, efficiently using resources, and implementing regulations and incentives that encourage sustainable practices in the construction industry (Hafez et al., 2023; Giorgi et al., 2022).

## 9.15 DISCUSSION

The chapter delves into the important results of energy-efficient building design in India, with an emphasis on legal frameworks, energy usage trends, and passive design solutions. Through the Energy Conservation Act, the Building Energy Efficiency Act, and projects like as GRIHA, the Indian government has adopted a proactive approach to encourage energy-efficient building design. The efficacy of these policies, however, is dependent on their implementation and modification at different levels of government.

The Indian energy consumption landscape urgently requires interventions in both the home and business sectors. Residential energy use accounted for 30% of total final energy consumption in 2017, providing a potential for energy reductions. Bioenergy continues to be the dominating source, emphasizing the need for cleaner, more sustainable alternatives. A more sophisticated knowledge of geographical variances and socioeconomic variables impacting energy usage, on the other hand, is critical for developing successful measures.

The rising demand for power, particularly in metropolitan areas, mandates an emphasis on energy efficiency in the business sector. Passive design solutions offer enormous potential for improving building energy efficiency, but their performance is dependent on variables such as local temperature, building type, and tenant behavior. These solutions' economic feasibility and scalability must be carefully considered.

The chapter recommends a more thorough, evidence-based approach to reaching conclusions. While the assumptions about the possible influence of energy-efficient design on carbon emissions and economic well-being are consistent with the larger sustainability discourse, the speculative character of these results must be noted. Improving results via improved data collecting, comparative research, stakeholder participation, and longitudinal analysis is critical for turning speculative ideas into evidence-based recommendations.

Finally, the chapter provides a foundational exploration of energy-efficient building design in India, with identified challenges and opportunities paving the way for future studies that delve deeper into policy implementation, regional variations in energy consumption, and the efficacy of passive design strategies.

## 9.16 CONCLUSIONS AND FUTURE SCOPE

The chapter investigates the significance of carbon emission reduction and energy-efficient building design in promoting sustainable development. The chapter examines a range of passive and active design approaches that can be utilized to enhance the energy efficiency of buildings. These include the utilization of shading devices, window treatments, natural lighting and ventilation, energy-efficient HVAC systems, intelligent lighting systems, and renewable energy technologies.

The research highlights the advantages of incorporating both passive and active design methodologies and provides case studies that illustrate their efficacy in diminishing energy usage. In addition, it emphasizes the significance of long-term social and environmental impacts in building design and proposes directions for future research.

Researchers, engineers, architects, and policymakers can all benefit from this exhaustive knowledge of energy-efficient building design. The prioritization of sustainable design parameters should include materials with low embodied energy, optimal utilization of natural illumination, and decreased reliance on artificial lighting in order to mitigate air conditioning loads and carbon emissions. Passive solar energy efficiency design involves modifying the roof, walls, and facade to decrease heating and cooling burdens.

The implementation of shading mechanisms, natural ventilation systems, and thermal insulation can substantially enhance energy efficiency and reduce energy usage. By optimizing the window-to-wall ratio and installing double glazing with energy-efficient glassware, it is possible to achieve a 15% reduction in energy consumption. Energy-efficient glazing, HVAC systems, artificial lighting, and solar heating and cooling devices are examples of active energy-saving strategies that can reduce consumption by 50 to 60%.

Energy efficiency improvements such as envelope design, daylighting optimization, and HVAC system optimization have the potential to decrease annual energy consumption by 35 to 40%, contingent upon climatic conditions. Utilizing simulation tools to evaluate energy efficiency and investigate alternative design alternatives prior to construction presents numerous benefits.

Over time, investments in energy-efficient building design have the potential to yield substantial financial benefits, such as decreased energy usage, simplified maintenance obligations, and elevated property valuations.

Further areas of investigation encompass broadening the scope of data collection and analysis, employing sophisticated analytical techniques, undertaking comparative studies, incorporating the viewpoints of stakeholders, and carrying out longitudinal investigations. The integration of these components into subsequent investigations seeks to convert conjectures into solid, empirically supported understandings, thereby furnishing policymakers, practitioners, and scholars with an interest in promoting sustainable and energy-efficient building design methodologies in India with actionable suggestions.

## REFERENCES

Abbas, A., Sajid, M. B., Sajid, J., & Ahmed, N. (2023). Forecasting environmental and social benefits of adopting cleaner technologies in Indian brick manufacturing industry. *Energy for Sustainable Development*, *72*, 202–211.

Abokersh, M. H., Gangwar, S., Spiekman, M., Vallès, M., Jiménez, L., & Boer, D. (2021). Sustainability insights on emerging solar district heating technologies to boost the nearly zero energy building concept. *Renewable Energy*, *180*, 893–913.

Açikkalp, E., Hepbasli, A., Palmero-Marrero, A. I., & Borge-Diez, D. (2023). Application of net zero extended energy buildings concept for sustainable buildings analysis. *Journal of Building Engineering*, *68*, 106095.

Aggarwal, P., Rana, M., & Akhai, S. (2022). Briefings on e-waste hazard until COVID era in India. *Materials Today: Proceedings*, *71*, 389–393.

Akash, A. R., Rao, A. B., & Chandel, M. K. (2017). Relevance of carbon capture & sequestration in India's energy mix to achieve the reduction in emission intensity by 2030 as per INDCs. *Energy Procedia*, *114*, 7492–7503.

Akhai, S., Mala, S., & Jerin, A. A. (2020). Apprehending air conditioning systems in context to COVID-19 and human health: A brief communication. *International Journal of Healthcare Education & Medical Informatics (ISSN: 2455–9199)*, *7*(1&2), 28–30.

Akhai, S., Mala, S., & Jerin, A. A. (2021). Understanding whether air filtration from air conditioners reduces the probability of virus transmission in the environment. *Journal of Advanced Research in Medical Science & Technology (ISSN: 2394–6539)*, *8*(1), 36–41.

Akhai, S., Singh, V. P., & John, S. (2016a). Human performance in industrial design centers with small unit air conditioning systems. *Journal of Advanced Research in Production Industrial Engineering*, *3*(2), 5–11.

Akhai, S., Singh, V. P., & John, S. (2016b). Investigating indoor air quality for the split-type air conditioners in an office environment and its effect on human performance. *Journal of Mechanical Civil Engineering*, *13*(6), 113–118.

Akhai, S., Thareja, P., & Singh, V. P. (2017). Assessment of indoor environment health sustenance in air-conditioned class rooms. *Advanced Research in Civil and Environmental Engineering*, *4*(1&2), 1–9.

Albayyaa, H., Hagare, D., & Saha, S. (2019). Energy conservation in residential buildings by incorporating passive solar and energy efficiency design strategies and higher thermal mass. *Energy and Buildings*, *182*, 205–213.

Alhuwayil, W. K., Almaziad, F. A., & Mujeebu, M. A. (2023). Energy performance of passive shading and thermal insulation in multistory hotel building under different outdoor climates and geographic locations. *Case Studies in Thermal Engineering*, *45*, 102940.

Anika, O. C., Nnabuife, S. G., Bello, A., Okoroafor, R. E., Kuang, B., & Villa, R. (2022). Prospects of low and zero-carbon renewable fuels in 1.5-degree net zero emission actualization by 2050: A critical review. *Carbon Capture Science & Technology*, 100072.

Awadh, O. (2017). Sustainability and green building rating systems: LEED, BREEAM, GSAS and Estidama critical analysis. *Journal of Building Engineering*, *11*, 25–29.

Balali, A., Yunusa-Kaltungo, A., & Edwards, R. (2023). A systematic review of passive energy consumption optimisation strategy selection for buildings through multiple criteria decision-making techniques. *Renewable and Sustainable Energy Reviews*, *171*, 113013.

Barbar, M., Mallapragada, D. S., & Stoner, R. J. (2023). Impact of demand growth on decarbonizing India's electricity sector and the role for energy storage. *Energy and Climate Change*, *4*, 100098.

Berges-Alvarez, I., Sanguinetti, C. M., Giraldi, S., & Marín-Restrepo, L. (2022). Environmental and economic criteria in early phases of building design through Building Information Modeling: A workflow exploration in developing countries. *Building and Environment*, *226*, 109718.

Beyene, H. D., Werkneh, A. A., & Ambaye, T. G. (2018). Current updates on waste to energy (WtE) technologies: A review. *Renewable Energy Focus*, *24*, 1–11.

Bherwani, H., Nair, M., Niwalkar, A., Balachandran, D., & Kumar, R. (2022). Application of circular economy framework for reducing the impacts of climate change: A case study from India on the evaluation of carbon and materials footprint nexus. *Energy Nexus*, *5*, 100047.

Cao, X., Liu, J., Cao, X., Li, Q., Hu, E., & Fan, F. (2015). Study of the thermal insulation properties of the glass fiber board used for interior building envelope. *Energy and Buildings*, *107*, 49–58.

Chaturvedi, V., Eom, J., Clarke, L. E., & Shukla, P. R. (2014). Long term building energy demand for India: Disaggregating end use energy services in an integrated assessment modeling framework. *Energy Policy*, *64*, 226–242.

Condori Fernandez, N., & Lago, P. (2018). The influence of green strategies design onto quality requirements prioritization. In *Requirements Engineering: Foundation for Software Quality: 24th International Working Conference, REFSQ 2018, Utrecht, The Netherlands, March 19–22, 2018, Proceedings 24* (pp. 189–205). Springer International Publishing.

Cureau, R. J., Pigliautile, I., Pisello, A. L., Bavaresco, M., Berger, C., Chinazzo, G., Belafi, Z. D., Ghahramani, A., Heydarian, A., Kastner, D., Kong, M., Licina, D., Luna-Navarro, A., Mahdavi, A., Nocente, A., Schweiker, M., Vellei, M., &Wang, A. (2022). Bridging the gap from test rooms to field-tests for human indoor comfort studies: A critical review of the sustainability potential of living laboratories. *Energy Research & Social Science*, *92*, 102778.

Dey, S., Sreenivasulu, A., Veerendra, G. T. N., Rao, K. V., & Babu, P. A. (2022). Renewable energy present status and future potentials in India: An overview. *Innovation and Green Development*, 100006.

Díaz-López, C., Serrano-Jiménez, A., Lizana, J., Lopez-Garcia, E., Molina-Huelva, M., & Barrios-Padura, A. (2022). Passive action strategies in schools: A scientific mapping towards eco-efficiency in educational buildings. *Journal of Building Engineering*, *45*, 103598.

Dong, Y., Ng, S. T., & Liu, P. (2023). Towards the principles of life cycle sustainability assessment: An integrative review for the construction and building industry. *Sustainable Cities and Society*, 104604.

Elavarasan, R. M., Pugazhendhi, R., Irfan, M., Mihet-Popa, L., Khan, I. A., & Campana, P. E. (2022). State-of-the-art sustainable approaches for deeper decarbonization in Europe–An endowment to climate neutral vision. *Renewable and Sustainable Energy Reviews*, *159*, 112204.

Fadye, F. A., Maref, W., & Awad, M. M. (2021). Review of white roofing materials and emerging economies with focus on energy performance cost-benefit, maintenance, and consumer indifference. *Sustainability*, 13, 9967.

Fakhimi, A. H., Khani, A. H., & Sardroud, J. M. (2021). Smart-city infrastructure components. In *Solving Urban Infrastructure Problems Using Smart City Technologies* (pp. 17–54). Elsevier.

Fournier, E. D., Federico, F., Cudd, R., Pincetl, S., Ricklefs, A., Costa, M., Jerrett, M., & Garcia-Gonzales, D. (2022). Net GHG emissions and air quality outcomes from different residential building electrification pathways within a California disadvantaged community. *Sustainable Cities and Society*, *86*, 104128.

Gagnon, R., Gosselin, L., & Decker, S. A. (2019). Performance of a sequential versus holistic building design approach using multi-objective optimization. *Journal of Building Engineering*, *26*, 100883.

Gambhir, A., Napp, T. A., Emmott, C. J., & Anandarajah, G. (2014). India's $CO_2$ emissions pathways to 2050: Energy system, economic and fossil fuel impacts with and without carbon permit trading. *Energy*, *77*, 791–801.

Gamero-Salinas, J., Monge-Barrio, A., Kishnani, N., López-Fidalgo, J., & Sánchez-Ostiz, A. (2021). Passive cooling design strategies as adaptation measures for lowering the indoor overheating risk in tropical climates. *Energy and Buildings*, *252*, 111417.

Gandhi, H. H., Hoex, B., & Hallam, B. J. (2022). Strategic investment risks threatening India's renewable energy ambition. *Energy Strategy Reviews*, *43*, 100921.

Giorgi, S., Lavagna, M., Wang, K., Osmani, M., Liu, G., & Campioli, A. (2022). Drivers and barriers towards circular economy in the building sector: Stakeholder interviews and analysis of five European countries policies and practices. *Journal of Cleaner Production*, *336*, 130395.

Golušin, M., Dodić, S., & Popov, S. (2013, January). Strategic priorities of sustainable energy development. *Sustainable Energy Management*, 243–333. DOI: 10.1016/B978-0-12-415978-5.00007-2.

Gu, G. W., & Hale, G. (2023). Climate risks and FDI. *Journal of International Economics*, 103731.

Gupta, J., & Chakraborty, M. (2021). Energy efficiency in buildings. In *Sustainable Fuel Technologies Handbook* (pp. 457–480). Academic Press.

Hafez, F. S., Sa'di, B., Safa-Gamal, M., Taufiq-Yap, Y. H., Alrifaey, M., Seyedmahmoudian, M., Stojcevski, A., Horan, B., & Mekhilef, S. (2023). Energy efficiency in sustainable buildings: A systematic review with taxonomy, challenges, motivations, methodological aspects, recommendations, and pathways for future research. *Energy Strategy Reviews*, *45*, 101013.

Haldar, S., Peddibhotla, A., & Bazaz, A. (2023). Analyzing intersections of justice with energy transitions in India: A systematic literature review. *Energy Research & Social Science*, *98*, 103010.

Harputlugil, T. (2017, February). Energy efficient building design development: A retrospective approach. In *International Symposium on Energy Efficiency in Buildings, Ankara, Turkey* (pp. 296–303).

Hasan, A. S., Ali, O. M., & Hussein, A. A. (2021). Comparative study of the different materials combinations used for roof insulation in Iraq. *Materials Today: Proceedings*, *42*, 2285–2289.

Hauashdh, A., Jailani, J., & Rahman, I. A. (2022). Strategic approaches towards achieving sustainable and effective building maintenance practices in maintenance-managed buildings: A combination of expert interviews and a literature review. *Journal of Building Engineering*, *45*, 103490.

Hee, W. J., Alghoul, M. A., Bakhtyar, B., Elayeb, O., Shameri, M. A., Alrubaih, M. S., & Sopian, K. (2015). The role of window glazing on daylighting and energy saving in buildings. *Renewable and Sustainable Energy Reviews*, *42*, 323–343.

Hernandez, P., Oregi, X., Longo, S., & Cellura, M. (2019). Life-cycle assessment of buildings. In *Handbook of Energy Efficiency in Buildings* (pp. 207–261). Butterworth-Heinemann.

Huang, H., Wang, H., Hu, Y. J., Li, C., & Wang, X. (2022). The development trends of existing building energy conservation and emission reduction—A comprehensive review. *Energy Reports*, *8*, 13170–13188.

Ighravwe, D. E., & Oke, S. A. (2019). A multi-criteria decision-making framework for selecting a suitable maintenance strategy for public buildings using sustainability criteria. *Journal of Building Engineering*, *24*, 100753.

Ihara, T., Gustavsen, A., & Jelle, B. P. (2015). Effect of facade components on energy efficiency in office buildings. *Applied Energy*, *158*, 422–432.

Ikram, M., Ferasso, M., Sroufe, R., & Zhang, Q. (2021). Assessing green technology indicators for cleaner production and sustainable investments in a developing country context. *Journal of Cleaner Production*, *322*, 129090.

Jain, M., Hoppe, T., & Bressers, H. (2017). Analyzing sectoral niche formation: The case of net-zero energy buildings in India. *Environmental Innovation and Societal Transitions*, *25*, 47–63.

Kandpal, V. (2018, April). A case study on smart city projects in India: An analysis of Nagpur, Allahabad and Dehradun. In *Companion Proceedings of the the Web Conference 2018* (pp. 935–941).

Khare, V., Khare, C. J., Nema, S., & Baredar, P. (2022). Path towards sustainable energy development: Status of renewable energy in Indian subcontinent. *Cleaner Energy Systems*, *3*, 100020.

Khosla, R., & Janda, K. B. (2019). India's building stock: Towards energy and climate change solutions. *Building Research & Information*, *47*(1), 1–7.

Knuutila, M., Kosonen, A., Jaatinen-Värri, A., & Laaksonen, P. (2022). Profitability comparison of active and passive energy efficiency improvements in public buildings. *Energy Efficiency*, *15*(6), 38.

Kremsa, V. Š. (2021). Sustainable management of agricultural resources (agricultural crops and animals). In *Sustainable Resource Management* (pp. 99–145). Elsevier.

Kumar, P., & Akhai, S. (2022). Effective energy management in smart buildings using VRV/VRF systems. In *Additive Manufacturing in Industry 4.0* (pp. 27–35). CRC Press.

Kuzman, M. K., Grošelj, P., Ayrilmis, N., & Zbašnik-Senegačnik, M. (2013). Comparison of passive house construction types using analytic hierarchy process. *Energy and Buildings*, *64*, 258–263.

Lai, F., Zhou, J., Lu, L., Hasanuzzaman, M., & Yuan, Y. (2023). Green building technologies in Southeast Asia: A review. *Sustainable Energy Technologies and Assessments*, *55*, 102946.

Lassen, N., Hegli, T., Dokka, T. H., Løvold, T., Edwards, K., Goia, F., & Andresen, I. (2021). Enabling holistic design for high energy efficient office buildings through the use of subjective occupant feedback. *Sustainable Cities and Society*, *69*, 102867.

Lazarevic, D., Salo, H., & Kautto, P. (2022). Circular economy policies and their transformative outcomes: The transformative intent of Finland's strategic policy programme. *Journal of Cleaner Production*, *379*, 134892.

Liu, P. R., & Raftery, A. E. (2021). Country-based rate of emissions reductions should increase by 80% beyond nationally determined contributions to meet the 2 C target. *Communications Earth & Environment*, *2*(1), 29.

Luo, Y., Zhang, L., Bozlar, M., Liu, Z., Guo, H., & Meggers, F. (2019). Active building envelope systems toward renewable and sustainable energy. *Renewable and Sustainable Energy Reviews*, *104*, 470–491.

Mathivathanan, D., Mathiyazhagan, K., Khorana, S., Rana, N. P., & Arora, B. (2022). Drivers of circular economy for small and medium enterprises: Case study on the Indian state of Tamil Nadu. *Journal of Business Research*, *149*, 997–1015.

McLaughlin, H., Littlefield, A. A., Menefee, M., Kinzer, A., Hull, T., Sovacool, B. K., Bazilian, M. D., Kim, J., & Griffiths, S. (2023). Carbon capture utilization and storage in review: Sociotechnical implications for a carbon reliant world. *Renewable and Sustainable Energy Reviews*, *177*, 113215.

Meerow, S., & Newell, J. P. (2017). Spatial planning for multifunctional green infrastructure: Growing resilience in Detroit. *Landscape and Urban Planning*, *159*, 62–75.

Mekhilef, S., Saidur, R., & Kamalisarvestani, M. (2012). Effect of dust, humidity and air velocity on efficiency of photovoltaic cells. *Renewable and Sustainable Energy Reviews*, *16*(5), 2920–2925.

Mirrahimi, S., Mohamed, M. F., Haw, L. C., Ibrahim, N. L. N., Yusoff, W. F. M., & Aflaki, A. (2016). The effect of building envelope on the thermal comfort and energy saving for high-rise buildings in hot–humid climate. *Renewable and Sustainable Energy Reviews*, *53*, 1508–1519.

Mondal, S., & Palit, D. (2022). Challenges in natural resource management for ecological sustainability. In *Natural Resources Conservation and Advances for Sustainability* (pp. 29–59). Elsevier.

Munoz-Liesa, J., Royapoor, M., Cuerva, E., Gassó-Domingo, S., Gabarrell, X., & Josa, A. (2022). Building-integrated greenhouses raise energy co-benefits through active ventilation systems. *Building and Environment*, *208*, 108585.

Napp, T. A., Few, S., Sood, A., Bernie, D., Hawkes, A., & Gambhir, A. (2019). The role of advanced demand-sector technologies and energy demand reduction in achieving ambitious carbon budgets. *Applied Energy*, *238*, 351–367.

Norouzi, M., Haddad, A. N., Jiménez, L., Hoseinzadeh, S., & Boer, D. (2023). Carbon footprint of low-energy buildings in the United Kingdom: Effects of mitigating technological pathways and decarbonization strategies. *Science of the Total Environment*, 163490.

Ohene, E., Chan, A. P., & Darko, A. (2022). Prioritizing barriers and developing mitigation strategies toward net-zero carbon building sector. *Building and Environment*, 109437.

Omer, A. M. (2008). Renewable building energy systems and passive human comfort solutions. *Renewable and Sustainable Energy Reviews*, *12*(6), 1562–1587.

Ozbalta, T. G., Yildiz, Y., Bayram, I., & Yilmaz, O. C. (2021). Energy performance analysis of a historical building using cost-optimal assessment. *Energy and Buildings*, *250*, 111301.

Pajek, L., & Košir, M. (2021). Strategy for achieving long-term energy efficiency of European single-family buildings through passive climate adaptation. *Applied Energy*, *297*, 117116.

Panicker, K., Anand, P., & George, A. (2023). Assessment of building energy performance integrated with solar PV: Towards a net zero energy residential campus in India. *Energy and Buildings*, *281*, 112736.

Parvin, K., Lipu, M. H., Hannan, M. A., Abdullah, M. A., Jern, K. P., Begum, R. A., Mansur, M., Muttaqi, K. M., Mahlia, T. I., & Dong, Z. Y. (2021). Intelligent controllers and optimization algorithms for building energy management towards achieving sustainable development: Challenges and prospects. *IEEE Access*, *9*, 41577–41602.

Patidar, A. K., Singh, R. K., & Choudhury, T. (2023). The prominence of carbon capture, utilization and storage technique, a special consideration on India. *Gas Science and Engineering*, 204999.

Perwez, U., Shono, K., Yamaguchi, Y., & Shimoda, Y. (2023). Multi-scale UBEM-BIPV coupled approach for the assessment of carbon neutrality of commercial building stock. *Energy and Buildings*, 113086.

Pradhan, B. K., & Ghosh, J. (2021). COVID-19 and the Paris Agreement target: A CGE analysis of alternative economic recovery scenarios for India. *Energy Economics*, *103*, 105539.

Prafitasiwi, A. G., Rohman, M. A., & Ongkowijoyo, C. S. (2022). The occupant's awareness to achieve energy efficiency in campus building. *Results in Engineering*, *14*, 100397.

Rakesh, P. (2023). Influence of various parameters on lifecycle cost of buildings with active energy efficiency measures. In *IOP Conference Series: Earth and Environmental Science* (Vol. 1130, No. 1, p. 012048). IOP Publishing.

Ramirez-Contreras, N. E., & Faaij, A. P. (2018). A review of key international biomass and bioenergy sustainability frameworks and certification systems and their application and implications in Colombia. *Renewable and Sustainable Energy Reviews*, *96*, 460–478.

Rana, J., & Rahman, M. (2020). Building energy performance optimization towards net zero energy building using passive solution technologies and building information modeling at design phase. *Journal of Construction Engineering, Management & Innovation*, *3*(1), 29–51.

Randjelovic, D., Vasov, M., Ignjatovic, M., Stojiljkovic, M., & Bogdanovic, V. (2021). Investigation of a passive design approach for a building facility: A case study. *Energy Sources, Part A: Recovery, Utilization, and Environmental Effects*, 1–19.

Razmjoo, A., Gandomi, A. H., Pazhoohesh, M., Mirjalili, S., & Rezaei, M. (2022). The key role of clean energy and technology in smart cities development. *Energy Strategy Reviews*, *44*, 100943.

Reffat, R. M., & Ahmad, R. M. (2020). Determination of optimal energy-efficient integrated daylighting systems into building windows. *Solar Energy*, *209*, 258–277.

Renuka, S. M., Maharani, C. M., Nagasudha, S., & Priya, R. R. (2022). Optimization of energy consumption based on orientation and location of the building. *Materials Today: Proceedings*, *65*, 527–536.

Rochlin, C. (2021). A case study: California utilities and the progressive business model. *The Electricity Journal*, *34*(5), 106959.

Sánchez-Quiles, D., & Tovar-Sánchez, A. (2015). Are sunscreens a new environmental risk associated with coastal tourism? *Environment International*, *83*, 158–170.

Sanderson, K. (2022). The path towards more-sustainable building construction. *Nature, 611*(7936), S18–S19.

Santamouris, M., & Vasilakopoulou, K. (2021). Present and future energy consumption of buildings: Challenges and opportunities towards decarbonization. *e-Prime-Advances in Electrical Engineering, Electronics and Energy*, *1*, 100002.

Sarkis, J., Meade, L., & Presley, A. (2011). Sustainability in the built environment: Factors and a decision framework. In *Handbook of Corporate Sustainability*. Edward Elgar Publishing.

Sarwar, S., Waheed, R., Aziz, G., & Apostu, S. A. (2022). The nexus of energy, green economy, blue economy, and carbon neutrality targets. *Energies*, *15*(18), 6767.

Schweiker, M. (2017). Understanding occupants' behavior for energy efficiency in buildings. *Current Sustainable/Renewable Energy Reports*, *4*, 8–14.

Sedaghat, A., & Sharif, M. (2022). Mitigation of the impacts of heat islands on energy consumption in buildings: A case study of the city of Tehran, Iran. *Sustainable Cities and Society*, *76*, 103435.

Sethi, P., Chakrabarti, D., & Bhattacharjee, S. (2020). Globalization, financial development and economic growth: Perils on the environmental sustainability of an emerging economy. *Journal of Policy Modeling*, *42*(3), 520–535.

Shahid, M., & Karimi, M. N. (2021). Optimization of energy transmittance through building envelope for hot dry climate. *Journal of Thermal Engineering*, *8*(5), 595–605.

Shankar, R., & Singh, S. (2022). Development of smart grid for the power sector in India. *Cleaner Energy Systems*, *2*, 100011.

Shareef, S., Mushtaha, E. S., Abu Dabous, S., & Alsyouf, I. (2023). Holistic analysis for the efficiency of the thermal mass performance of precast concrete panels in hot climate zones. *Engineering, Construction and Architectural Management*. Ahead-of-print.

Shaw, R., & Mukherjee, S. (2022). The development of carbon capture and storage (CCS) in India: A critical review. *Carbon Capture Science & Technology*, 100036.

Soussi, Y., Bahi, H., Mastouri, H., & El Bouazouli, A. (2023). An embedded concept for sustainable building. *Materials Today: Proceedings*, *72*, 3556–3563.

Sung, U. J., & Kim, S. H. (2019). Development of a passive and active technology package standard and database for application to zero energy buildings in South Korea. *Energies*, *12*(9), 1700.

Tafakkori, R., & Fattahi, A. (2021). Introducing novel configurations for double-glazed windows with lower energy loss. *Sustainable Energy Technologies and Assessments*, *43*, 100919.

Tanwar, N., & Akhai, S. (2017). Survey analysis for quality control comfort management in air-conditioned classroom. *Journal of Advanced Research in Civil and Environmental Engineering*, *4*(1&2), 20–23.

Whittem, V., Roetzel, A., Sadick, A. M., & Kidd, A. N. (2022). How comprehensive is post-occupancy feedback on school buildings for architects? A conceptual review based upon Integral Sustainable Design principles. *Building and Environment*, 109109.

Wu, X., Li, X., Qin, Y., Xu, W., & Liu, Y. (2023). Intelligent multi objective optimization design for NZEBs in China: Four climatic regions. *Applied Energy*, *339*, 120934.

Xin, H. Z., & Rao, S. P. (2013). Active energy conserving strategies of the Malaysia energy commission diamond building. *Procedia Environmental Sciences*, *17*, 775–784.

Xuan, D., Jiang, X., & Fang, Y. (2023). Can globalization and the green economy hedge natural resources? Functions of population growth and financial development in BRICS countries. *Resources Policy*, *82*, 103414.

Yang, Y., & Chen, S. (2022). Thermal insulation solutions for opaque envelope of low-energy buildings: A systematic review of methods and applications. *Renewable and Sustainable Energy Reviews*, *167*, 112738.

Zandvoort, M., & van der Vlist, M. J. (2020). Planning infrastructure replacements: Restructuring and exerting partial control over the environment. *Environmental Science & Policy*, *103*, 67–76.

Zhou, Y., Ma, M., Tam, V. W., & Le, K. N. (2023). Design variables affecting the environmental impacts of buildings: A critical review. *Journal of Cleaner Production*, 135921.

Zoure, A. N., & Genovese, P. V. (2023). Implementing natural ventilation and daylighting strategies for thermal comfort and energy efficiency in office buildings in Burkina Faso. *Energy Reports*, *9*, 3319–3342.

Zutshi, A., & Creed, A. (2015). An international review of environmental initiatives in the construction sector. *Journal of Cleaner Production*, *98*, 92–106.

# *Section 3*

## *Sustainability Strategies and Management*

# 10 From Industry 4.0 to Industry 5.0

## *Transitioning to Circular Business Paradigms—A Review*

*Mohammad Shamsuddoha, Mohammad Abul Kashem, Tasnuba Nasir, and Niamat Ullah Ibne Hossain*

## 10.1 INTRODUCTION

The circular economy, a sustainable economic mode, promotes the continual use and regeneration of resources to minimize waste, maximize resource efficiency, and create a closed-loop system where products, materials, and resources circulate in a perpetual cycle (Giovanni, 2023; Shamsuddoha et al., 2023; Velenturf and Purnell, 2021). In the circular economy, the resources flow seamlessly through a system designed to eliminate waste and retain their value (Atif, 2023). Rather than being discarded after use, resources are prudently orchestrated to circulate through different stages, such as product design, production, consumption, and recovery (MacArthur, 2013). This continuous circulation ensures that resources maintain their usefulness and reduce the need for extracting new resources from the Earth (Corvellec et al., 2022). Once more, materials are revitalized through cutting-edge recycling, upcycling, and purposive procedures and are prepared to take center stage in new products and systems (Leng et al., 2022). This regeneration process lessens the need for new resources, conserves energy, and lessens the adverse effects of resource extraction and manufacture on the environment.

In order to maximize productivity, creativity, and overall efficiency, Industry 5.0, the cutting-edge revolution of manufacturing and production processes, emphasizes the seamless integration of human intelligence with modern automation technology (Leng et al., 2022). The capacity to recognize the unrivaled importance of human intuition, problem-solving abilities, and flexibility is one of Industry's 5.0's distinctive characteristics (George and George, 2023). However, Industry 5.0 eventually defines a future where technology complements rather than replaces humanity by unleashing new levels of innovation, productivity, and sustainability (Alojaiman,

DOI: 10.1201/9781003438748-13

2023; Huang et al., 2022; Turner et al., 2022; Xu et al., 2021). Even so, converging sophisticated automation with human cognition results in improved outcomes and ongoing development (Fraga-Lamas et al., 2021). Nevertheless, people may obtain real-time data insights, make knowledgeable decisions, and increase their productivity by incorporating AI and machine learning into work processes (Akundi et al., 2022). Industry 5.0, as opposed to Industry 4.0, which concentrated on digitizing manufacturing, aspires to combine technology with human labor to create intelligent and networked production systems (Adel, 2022; Maddikunta et al., 2022). To foster a circular economy, these practices encourage reuse and recycling while minimizing the consumption of new resources.

The harmonic interplay of reducing, reusing, recycling, and rethinking generates a virtuous cycle that transcends the constraints of the conventional linear economy, conserves valuable resources, reduces environmental impact, and promotes sustainable economic growth (Ciliberto et al., 2021). Additionally, the fusion of Industry 4.0 technology with circular economy (CE) principles allows the creation of circular goods in the Industry 5.0 age, fostering sustainability and resilience in all social, economic, and environmental spheres (Atif, 2023). For instance, human intellect and knowledge are included in both the smart and intelligent manufacturing paradigms, where they work and cooperate with robots rather than being displaced by automation (George and George, 2023). In order to help businesses overcome uncertainties and risks, it is necessary to better investigate the complex and diverse relationship between CE and Industry 5.0 within the context of the circular paradigm (Gagnidze, 2023; Piscicelli, 2023; Rajput and Singh, 2019).

A people-centered society will continue to evolve because of the significance of the "physical to digital to physical loop" made possible by technology in Industry 5.0 (Turner et al., 2022). In contrast, Industry 4.0 (largely employed for automation) would be further modified to improve the productivity of human employees by using their expertise and creativity and filling in any skill shortages with the advent of Industry 5.0 (Akundi et al., 2022; Alves et al., 2023; Ivanov, 2023; Mukherjee et al., 2023). Surprisingly, human aspects are now considered in the design of production line machinery and robots, aiming to achieve human–robot collaboration in job fulfillment (Maddikunta et al., 2022). Although employees will play more significant roles in high-tech, automated workplaces in the future, digitally connected systems may also align with human-centricity and sustainability objectives, particularly regarding the customer base and product manufacture (MacArthur, 2013). So, verification and real-time data monitoring are essential for controlling resilient and sustainable production and consumption flows.

Likewise, mass personalization addresses environmental issues by reducing waste and material use in manufacturing, while robotic disassembly increases recycling efficiency (Adel, 2022). So, in the spirit of innovation, collaboration and sustainability, this research leads us to the dawn of a new era where Industry 5.0 and the circular economy harmonize to shape a world of limitless possibilities. Consequently, the authors highlight the great potential of Industry 5.0 to accelerate the adoption and implementation of the circular economy. However, this study shows strong alliances between the information economy and Industry 5.0 create a spillover effect on industries by analyzing real-world case studies, reviewing data and findings, and increasing awareness and sustainable innovation.

## 10.2 INDUSTRY 5.0 AND ITS KEY FEATURES

Industry 5.0, the human-centered revolution, aspires to combine human employees' special skills and creativity with the efficiency and productivity advantages of automation and artificial intelligence. However, Industry 5.0 encourages the development of sustainable materials and circular economy techniques, reducing waste and promoting environmental stewardship (Gagnidze, 2023; Piscicelli, 2023). Factories reduce their negative environmental effects by using resource-saving technologies and environmentally friendly procedures (Giovanni, 2023; Korhonen et al., 2018; Sarkar et al., 2022; Murray et al., 2017; Morseletto, 2020). Real-time monitoring of waste production, emissions, and energy consumption through using Internet of Things (IoT) devices and smart sensors facilitates improved resource allocation, preventive maintenance to lower the frequency of interruptions, and optimized energy use. Figure 10.1 and Table 10.1 list the characteristics of Industry 5.0 and its wide-ranging effects.

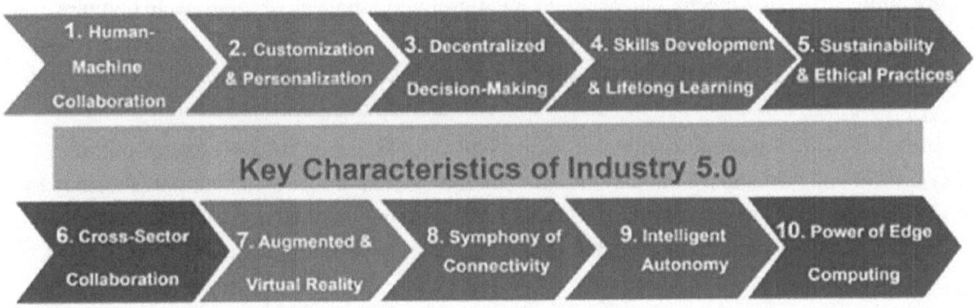


**FIGURE 10.1** Key characteristics of Industry 5.0.

**TABLE 10.1**
**Industry 5.0 Key Features and Impacts**

| Features | Impacts |
|---|---|
| **Human–Machine Collaboration** | Create a symbiotic relationship between humans and machines to enhance productivity and innovation (Coelho et al., 2023; Ciliberto et al., 2021; Stahel, 2016).<br>Design machines to support and augment human abilities rather than replacing humans (Hazen et al., 2021). |
| **Customization and Personalization** | Focus on providing personalized and customized products and services to meet individual customer needs (Alves et al., 2023).<br>Permit the efficient and cost-effective production of tailored goods, allowing businesses to cater to diverse consumer preferences with advanced technologies such as 3D printing, robotics, and big data analytics (Maddikunta et al., 2022). |
| **Decentralized Decision-Making** | Promote decentralized decision-making processes, empowering workers at all levels to contribute their ideas and expertise (Giovanni, 2023; Voulgaridis et al., 2022).<br>Foster a more agile and responsive organization by leveraging technologies like artificial intelligence, machine learning, and the Internet of Things; employees access real-time data and make informed decisions (Akundi et al., 2022). |

*(Continued)*

**TABLE 10.1**
**(Continued)**

| Features | Impacts |
| --- | --- |
| **Skill Development and Lifelong Learning** | Create a scope for companies to invest in upskilling and reskilling programs to equip their workforce with the necessary competencies to work alongside advanced technologies effectively (Mukherjee et al., 2023).<br>Maintain employability and drive innovation to adapt to evolving technologies and job roles (Alojaiman, 2023). |
| **Sustainability and Ethical Practices** | Emphasize environmentally sustainable practices and ethical considerations by adopting green manufacturing techniques, energy-efficient processes, and waste reduction strategies (Mukherjee et al., 2023).<br>Integrate ethical considerations into developing and deploying technologies to ensure fairness, privacy, and accountability (Giovanni, 2023). |
| **Cross-Sector Collaboration** | Encourage collaboration and integration across different sectors and industries by leveraging synergies and developing innovative solutions to complex challenges (Rajput and Singh, 2019).<br>Foster interdisciplinary approaches and drive economic growth and societal benefits with collaborative ecosystems (Leng et al., 2022). |
| **Augmented Reality and Virtual Reality** | Leverage augmented reality (AR) and virtual reality (VR) technologies to enhance human–machine interaction and improve operational efficiency by enabling workers to visualize and interact with digital information in real time, facilitating training, maintenance, and remote collaboration (Adel, 2022). |
| **The Symphony of Connectivity** | Form a symphony of data and information where machines, devices, and sensors are intricately interconnected to make intelligent decisions, optimize processes, and unlock new levels of efficiency and productivity (Alojaiman, 2023). |
| **Intelligent Autonomy** | Empowerment of intelligent autonomy to create self-aware, self-optimizing, and self-adapting systems in manufacturing systems with cyber-physical systems (CPS), IoT, and AI (Xu et al., 2021). |
| **Power of Edge Computing** | Enable the emergence of powerful edge computing capabilities with CPS, IoT, and AI by eliminating the need for extensive data transmission to centralized cloud servers (Fraga-Lamas et al., 2021). |

In brief, Industry 5.0 advocates a shift towards a human-centric approach to industrialization, emphasizing collaboration, customization, decentralized decision-making, skill development, sustainability, cross-sector collaboration, and the use of augmented and virtual reality. In addition, the combination of CPS, IoT, and AI provides a foundation for human–machine interaction that transcends conventional boundaries. This combination has improved manufacturing ecosystem productivity, quality and safety. Thus, this singular confluence opens up a universe where connectivity, autonomy, edge computing and human–machine collaboration work together to transform industrial processes.

## 10.3 METHODOLOGY

This chapter uses a systematic and rigorous approach to examine these two interrelated concepts' complex relationships and implications and focuses on how Industry 5.0 advances the circular economy. In this chapter, the researchers chose to delve

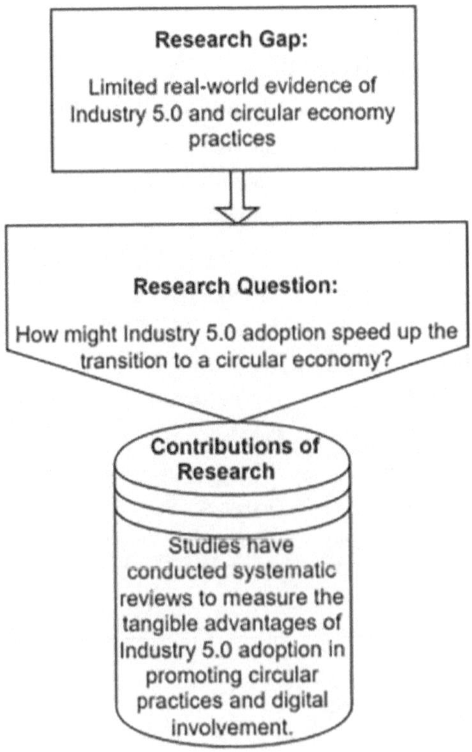


**FIGURE 10.2** Research gap, research question, and contribution of this research.

deeper into various aspects of Industry 5.0 and the circular economy, offering a complete understanding of Industry 5.0 and the circular economy and how they interact and influence each other. Ultimately, this data-driven study provides a reliable perspective on the practical implications of Industry 5.0 in the circular economy, guaranteeing that research findings are accurate and supported. Likewise, this chapter provides a basis for future projections and scenario planning, enabling decision-makers to anticipate long-term risks and opportunities under the assumption that research findings will increase credibility, reliability, and impact on policymakers and business executives. The following research questions (Figure 10.2) are clearly stated in the study that the systematic review seeks to answer.

Due to the ease of citation and the large readership it has for various types of research, the authors primarily concentrated on the academic publications and research register and databases from the Google Scholar, Scopus, and PubMed databases, which produced about 1184 publications from the deduction on the initial search and considered the citation metrics data from there. Extensive research has discovered that works on circular economy and Industry 5.0 have been added to the sources between 1996 and 2022. Once again, the search terms "industry 5.0" AND "circular economy" yielded a total of four entries (Table 10.2). After focusing on keywords, abstracts and manuscripts, 47 papers were considered. Again, publications are evaluated through the rigorous review on inclusion and exclusion criteria set for this review (Table 10.3).

**TABLE 10.2**
**Systematic Literature Search and Bases**

| Search Keywords | Sources | Number of Publications |
|---|---|---|
| Industry 5.0 | Google Scholar | 655 |
| Circular Economy | Google Scholar | 15,100 |
| Industry 5.0 | PubMed | 22 |
| Circular Economy | PubMed | 767 |
| Circular Economy | Scopus | 2619 |
| Industry 5.0 | Scopus | 4 |
| Industry 5.0 OR Circular Economy | Scopus | 2623 |
| Circular Economy OR Industry 5.0 | Google Scholar | 14,600 (after deleting citations) |
| Circular Economy OR Industry 5.0 | PubMed | 789 |
| Industry 5.0 AND Circular Economy | PubMed | 2 |
| Industry 5.0 AND Circular Economy | Google Scholar | 4 |
| Industry 5.0 AND Circular Economy | Scopus | Not yet |

**TABLE 10.3**
**Inclusion and Exclusion Criteria**

| Inclusion Criteria | Exclusion Criteria |
|---|---|
| i. The research must explicitly investigate the relationship between Industry 5.0 and the circular economy.<br>ii. Inclusion is limited to studies involving case studies, surveys, experiments, or real-world implementations of Industry 5.0 principles to enhance circular economy practices.<br>iii. Selected research should directly address circular economy concepts and key aspects.<br>iv. Preference will be given to studies published ensuring up-to-date and relevant findings.<br>v. Research that has undergone rigorous peer review to ensure the quality and credibility of the findings will be considered.<br>vi. Studies have been conducted to capture diverse perspectives on how Industry 5.0 impacts the circular economy.<br>vii. The research include various industries (e.g., manufacturing, agriculture, energy, transportation) to explore the broad applicability of Industry 5.0 in promoting circular practices.<br>viii. Research written in English will be included for better accessibility and understanding. | i. Research published in languages other than English without sufficient translation will be excluded due to language constraints.<br>ii. Studies that do not explicitly relate to the circular economy or Industry 5.0 will be excluded.<br>iii. Research published more than five years ago will be excluded as it may not reflect recent advancements in Industry 5.0 and the circular economy.<br>iv. Research with inadequate or ambiguous methodologies, insufficient data, or incomplete results will be excluded to maintain the credibility and reliability of the analysis.<br>v. Studies lacking proper peer review will be excluded to ensure the research validity and accuracy.<br>vi. In cases where multiple studies cover the same datasets, only the most comprehensive or recent publication will be included to avoid duplication. |

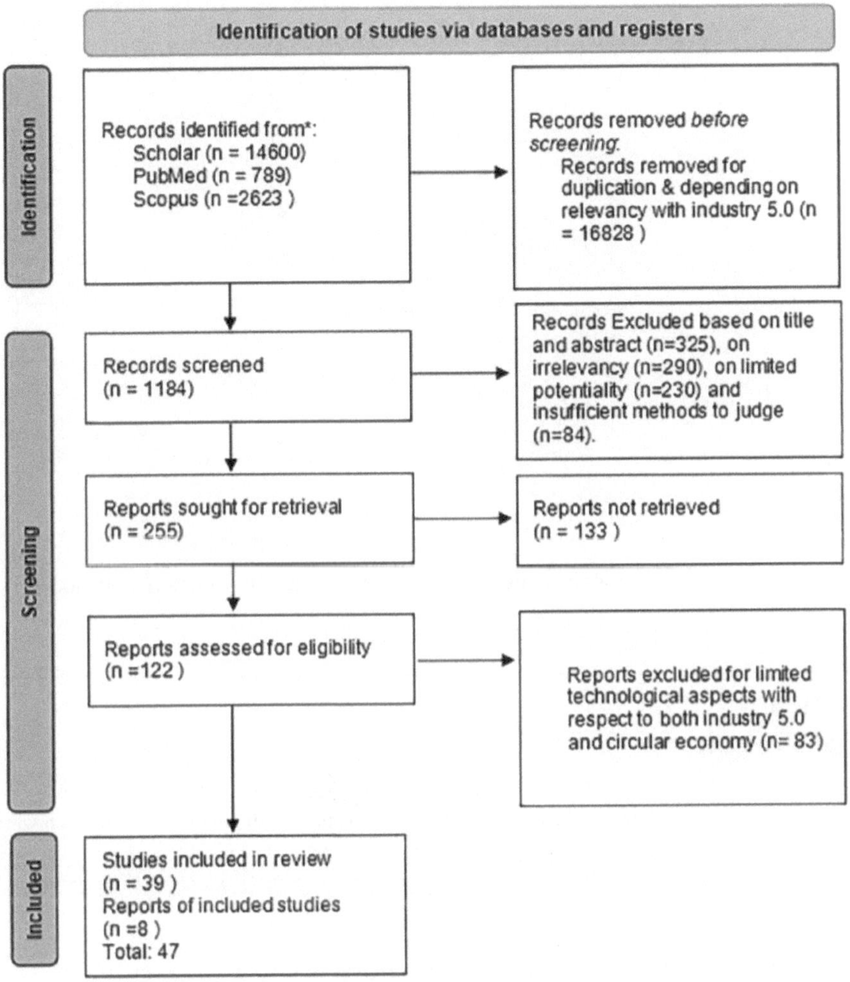


**FIGURE 10.3** Flowchart for systematic review.

Table 10.4 from the 1996–2023 citation study reveals that the average number of citations per year was 9128.52, with 208.17 citations per paper, an h-index of 255, a g-index of 447, and an hA-index of 103. This qualitative research focuses on technology-driven platforms, specifically changes based on Industry 5.0 and the circular economy. More specifically, this work adheres to the updated systematic review concept for earlier research from Ross et al. (2016), and Kashem et al. (2022). The initial stages of this research were mostly concerned with finding and selecting potential and contributing papers for this review. When abstract screening is considered, context, purpose, data source, and other general factors are assumed. Last but not least, the most-cited research was prioritized for this review (see Figure 10.3).

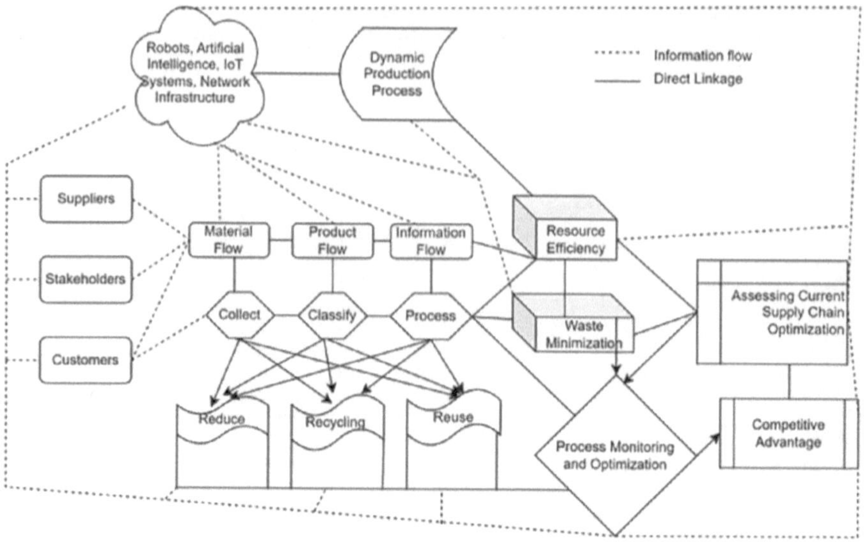


**FIGURE 10.4** Circular supply chain production process and information flow model.

**TABLE 10.4**
**Citation Metrics**

| Papers | Citations | Years | Cites per Year | Cites per Paper | Authors_ Paper | h_index | g_index | hA-index |
|---|---|---|---|---|---|---|---|---|
| 1184 | 246470 | 27 | 9128.52 | 208.17 | 2.76 | 255 | 447 | 103 |

## 10.4 COLLABORATION BETWEEN INDUSTRY 5.0 AND THE CIRCULAR ECONOMY—THE CASE OF THE CIRCULAR SUPPLY CHAIN LINKAGE MODEL

By reusing and recycling current resources, circular supply chain practices aim to reduce the need for new materials and promote a circular economy. This strategy has become more popular due to worries about resource depletion and the sustainability of the environment, which has decreased greenhouse gas emissions (Velenturf and Purnell, 2021). These procedures also help businesses cut costs and enhance their image for sustainability, bringing in more clients and fostering loyalty (Ciliberto et al., 2021). Furthermore, circular supply chain practices provide economic benefits and creative potential by extending product lifespans, lowering material and production costs, and creating income through resale or recycling operations (Figure 10.4).

Industry 5.0 facilitates real-time data collecting, agile decision-making, and dynamic modifications in production processes in the industrial sector by increasing productivity, lowering waste, and fostering sustainability (Giovanni, 2023).

However, putting Industry 5.0 into practice requires investments in sophisticated robots, artificial intelligence, IoT sensors, and networking infrastructure (Ivanov, 2023). Once again, deploying closed-loop supply chains includes salvaging, restoring, and recycling damaged goods or materials, resulting in a sustainable production cycle that includes sustainable procurement to lessen their carbon footprint and support sustainable manufacturing operations.

On the contrary, cross-chain collaboration and partnerships promote information exchange, technical advancement, and the development of new business concepts (Alves et al., 2023). Thus, implementing circular supply chain practices necessitates a comprehensive strategy that includes product redesign, process optimization, and network reconfiguration (Dwivedi et al., 2023; Hazen et al., 2021). This procedure comprises assessing current supply chain operations, identifying potential problem areas, and setting waste minimization and resource efficiency objectives.

Collaboration with suppliers, customers, and other stakeholders is crucial to connecting material, information, and product flows with circular principles (Lindgreen et al., 2020). But fostering an environment of cooperation, ongoing education, and innovation is also essential for a successful adoption (Angelis, 2021; Gagnidze, 2023; Geissdoerfer et al., 2017; Kirchherr et al., 2017; Shevchenko et al., 2023). Although adopting circular supply chain practices has drawbacks like high upfront costs, demanding organizational change, and opposition from stakeholders, the long-term gains in terms of enhanced resource efficiency, improved environmental performance, and competitive advantage make the transition highly advantageous for businesses committed to sustainability.

Process monitoring and optimization need investments in data analytics, tracking systems, and recycling systems. Reverse logistics companies and recycling facilities may be partnered with to collect, classify, and process materials and products efficiently at the end of their life cycles (Saidani et al., 2019). Using circular supply chain practices also creates new revenue streams for businesses by allowing them to sell used or recycled goods and provide trash management and recycling services (Ciliberto et al., 2021). Businesses could obtain a competitive edge in the market by catering to customers' environmental values and preferences by providing sustainable goods and services, leading to customer happiness and loyalty. However, businesses have been able to add intelligence to their processes and products while giving consumers and stakeholders access and knowledge over pertinent data by incorporating digital technology into CE principles (Ciliberto et al., 2021; King and Locock, 2022). This integration has increased the reuse and recycling of components, made it easier to foresee maintenance needs, built consumer trust, and encouraged healthy competition among enterprises (MacArthur, 2013). Smart technologies may automate tedious activities by monitoring human activity while workers oversee manufacturing and create value via inventiveness (Kashem et al., 2022; Maddikunta et al., 2022). A cloud-based infrastructure is used in cloud manufacturing to store and manage data, manage supplies and equipment, and improve scheduling and service coordination (Leng et al., 2022). Digital replicas of an industry's infrastructure can be useful for supply chain management. They allow for real-time simulation, process prediction, and optimization through smart re-manufacturing, smart recycling, and reuse to provide new and sustainable solutions.

## 10.5 CIRCULAR ECONOMY AND INDUSTRY 5.0—A SYNERGY

### 10.5.1 Supply Chain

Organizations become less vulnerable to supply chain disruptions and pricing changes by supporting local procurement and closing material loops (Atif, 2023). By lowering risks and promoting a more sustainable and secure future, this greater resilience ensures steady access to resources. Additionally, the contributions range from real-time, collaborative decision-making to smart production. In contrast to linear systems that create enormous volumes of trash, causing overflowing landfills and environmental deterioration, the circular economy promotes creating goods with greater longevity, repairability, and modularity (Hazen et al., 2021). However, it is feasible to dramatically reduce trash output and its related environmental and social effects by encouraging prolonged product life cycles, supporting product sharing and leasing models, and allowing efficient recycling and waste management infrastructure (Ciliberto et al., 2021; Stahel, 2016). On the other hand, Industry 5.0 technologies, including the Internet of Things, blockchain, and artificial intelligence, allow for real-time material traceability and tracking where every component and product can be uniquely identified and tracked, giving manufacturers total visibility and control over their supply chains (Atif, 2023). In addition to improving the effectiveness of material recovery and recycling, this openness makes optimizing product life extension and re-manufacturing plans possible.

### 10.5.2 Employment

Contrary to popular belief, sustainability does not impede economic growth. In fact, the circular economy presents a rare chance for both environmental protection and economic success (Velenturf and Purnell, 2021). The shift to a circular economy is intended to open up new economic opportunities, spur innovation, and promote employment growth in various industries (Kirchherr et al., 2018). There are many job possibilities available, for example, due to the demand for resource recovery, recycling facilities, re-manufacturing factories, and reverse logistics networks (Cioffi et al., 2020; ElMaraghy et al., 2021; Rathi et al., 2022). Businesses may also lower manufacturing costs, improve resource resilience, and access new markets by adopting circular business practices promoting sustainable economic growth.

### 10.5.3 Sustainability

The circular economy has considerable advantages for environmental protection, economic growth, and a large decrease in waste production. Organizations reduce pollution, habitat loss, and environmental damage by encouraging ethical production and consumption practices and minimizing resource exploitation (Rathi et al., 2022). Circular business methods lessen the strain on ecosystems, cut back on greenhouse gas emissions, and support biodiversity preservation. In addition, by lowering the carbon footprint associated with resource extraction, production, and waste treatment, the circular economy helps to mitigate climate change. Moreover, the circular economy helps to diversify resource streams by minimizing reliance on scarce or geopolitically unstable commodities, such as rare metals or fossil fuels (Dang et al., 2021).

### 10.5.4 Industrial Movement

Advanced robotics, additive manufacturing, and machine learning technologies encourage closed-loop production cycles and waste conversion into reusable materials or new products to minimize environmental impact and reduce reliance on virgin resources (Atif, 2023). By sharing knowledge, ideas, and best practices, Industry 5.0 encourages collaboration and co-creation with stakeholders such as manufacturers, researchers, designers, and consumers. This promotes continuous improvement and hastens the adoption of circular principles across industries.

### 10.5.5 Energy Saving

By gathering information on how much energy, water, and other valuable resources are used, IoT-enabled sensors and devices are being deployed across industries and households to monitor resource consumption in real-time, side-by-side analyze patterns, and identify areas of inefficiency (Turner et al., 2022). For instance, predictive algorithms may forecast demand patterns and modify the energy supply appropriately, resulting in considerable energy savings.

### 10.5.6 Waste Collection

Waste collection services can now optimize routes and schedules based on real-time data by integrating IoT devices, such as smart bins and sensors, with AI and data analytics (Maddikunta et al., 2022). Artificial intelligence systems analyze and examine previous data to forecast trends of trash creation, enabling proactive waste management methods. Additionally, AI-powered image recognition technology can automate sorting procedures at recycling plants, enhancing precision and effectiveness.

### 10.5.7 Recycling

The convergence of IoT, AI, and data analytics can benefit recycling. Using IoT devices, recyclable materials are tracked and traced throughout their lifecycles to ensure accountability and transparency (ElMaraghy et al., 2021). To make recycling operations more efficient, AI systems can analyze enormous volumes of data, including material composition, recycling rates, and market demand (King and Locock, 2022). This may make it possible to find new opportunities for recycling, develop better sorting methods, and increase material recovery rates, lessening the need for raw materials.

### 10.5.8 Culture of Sustainability

Collaborative platforms establish thorough policies and promote a culture of sustainability to encourage data sharing, the sharing of best practices, and the co-creation of novel solutions to maximize the potential of IoT, AI, and data analytics in resource optimization, waste management, and recycling (Leng et al., 2022). Embracing this unique and holistic approach will contribute to a more sustainable future contribute to a more sustainable future and unlock numerous economic and environmental benefits for society as a whole. Stakeholders must collaborate and invest in these technologies to unlock their full potential and pave the way for a greener and more efficient world.

### 10.5.9 Consumption Model and Revenue Source

By involving consumers in the product development process, the resulting products can be tailored to meet their specific needs and desires, increasing customer satisfaction. Additionally, this co-creation approach facilitates the shift towards product-as-a-service models, where consumers can access and utilize products on demand, reducing the need for ownership (Leng et al., 2022). By implementing blockchain technology, secure and transparent transactions can be facilitated, ensuring trust and accountability in collaborative consumption models.

Peer-to-peer transactions are made possible by the sharing economy marketplaces, which operate as platforms for people and companies to share their underutilized assets (Hazen et al., 2021). Anyone may rent or access things through such markets when needed, and the product owners get new revenue sources. These platforms might also include user reviews, ratings, and verification methods to increase trust and dependability among the sharing community.

Finally, Industry 5.0 strongly emphasizes incorporating supplementary services with physical products to promote the adoption of product-as-a-service models. Businesses provide consumers with a comprehensive solution without the hassle of product ownership by combining value-added services like maintenance, updates, and customer support. Thus, companies can also work with educational institutions and non-profit organizations to create training programs that provide future professionals with the abilities they need to succeed in Industry 5.0.

## 10.6 OVERCOMING CHALLENGES AND POTENTIAL RISKS

While the benefits of the circular economy are widely acknowledged, numerous challenges need to be addressed before its full adoption can be realized. A fundamental shift in perspective, encompassing ideas like waste reduction, product lifetime, and resource regeneration, is necessary to transition from the conventional linear economy to a circular one. This transformation calls for significant investments in education, public awareness campaigns, and promoting sustainable culture in all spheres of society (Leng et al., 2022). These investments must include rethinking supply chains, creating regulatory frameworks, removing financial obstacles, educating consumers, and fostering international cooperation. The major obstacles were:

1. When integrating Industry 5.0 technology into current circular economy systems, compatibility issues concerning interoperability and data protection are warned against (Fraga-Lamas et al., 2021).

2. In order to shift from a traditional circular economy to an Industry 5.0–driven circular economy, it is necessary to balance the integration of automation with a competent and adaptive workforce (Voulgaridis et al., 2022).
3. Circular economy networks may have weaknesses as they grow increasingly linked, leaving them open to cyber threats to protect private data and uphold public confidence (Fraga-Lamas et al., 2021).
4. Closed-loop supply chains can need a lot of raw materials, which might burden the availability of resources and exacerbate environmental problems (Moktadir et al., 2020; Morseletto, 2020; Sarkar et al., 2022).
5. Widespread adoption of cutting-edge technologies creates new types of inequality concerning differences in vulnerable people's connectivity, access to technology, and digital literacy despite the promise of a more just and inclusive economy (Giovanni, 2023).
6. Technological breakthroughs necessitate flexible, progressive policy and regulatory frameworks that stimulate innovation and manage any concerns related to newly developed technology (Alojaiman, 2023; Cioffi et al., 2020).

As a result, issues linked to significant investments, data security, and privacy concerns may be solved by the ethical implications of automation and providing a seamless workforce transition through training and assistance.

## 10.7 THEORETICAL IMPLICATIONS

This chapter aims to foster multidisciplinary research, where experts from various fields collaborate, and the exponential growth of data-driven research, as interconnected systems generate vast amounts of information that can be harnessed for analysis and innovation. This will result in more comprehensive and creative solutions to complex problems. Concurrently, adopting circular economy concepts, prioritizing minimizing waste and fostering resource reuse and recycling will influence research goals, elevating sustainability to the top. This change will spur research into environmentally friendly products, green technology, and sustainable production techniques. The research community may anticipate revolutionary developments that solve global concerns while fostering economic success and environmental stewardship when Industry 5.0 and circular economy concepts come together.

## 10.8 MANAGERIAL IMPLICATIONS

Likewise, the emergence of Industry 5.0, characterized by the incorporation of cutting-edge technology like artificial intelligence, the Internet of Things, and robots, has significantly altered the commercial environment. Parallel to this, the circular economy idea could gain popularity, supporting a regenerative resource use and waste minimization strategy. Managers are crucial in influencing the research environment for Industry 5.0 and the circular economy in this disruptive period. They must coordinate organizational aims with environmental objectives and use cutting-edge technology to improve industrial procedures. A culture of innovation and cooperation may be promoted as a result of this review, encouraging cross-disciplinary research to investigate novel solutions, speed up the adoption of circular practices, maximize the long-term advantages of sustainable practices, and encourage research-driven decision-making to move industries towards a more sustainable and economically viable future. The conclusions ultimately have significance for academics, managers, practitioners, and politicians, offering insights into the condition of the field right now and directing future research and decision-making.

## 10.9 CONCLUSION

A culture of innovation and cooperation may be promoted as a result of this review, encouraging cross-disciplinary research to investigate novel solutions, speed up the adoption of circular practices, maximize the long-term advantages of sustainable practices, and enable research-driven decision-making to move industries towards a more sustainable and economically viable future. The conclusions ultimately have significance for academics, managers, practitioners, and politicians, offering insights into the condition of the field right now and directing future research and decision-making. The presentation focused on the advantages of the circular economy, arguments for its future integration, and the transition in the Industry 5.0 era from an economic and environmental focus to a societal focus, focusing on human-centric values in resilient and sustainable movements. To close the gap between research and practice, the authors advised conducting further empirical research on market orientation, compliance, and risk management. Based on the findings presented in this paper, future research could examine various application fields and propose a potential framework for integrating the circular economy into business models, along with architectural frameworks and necessary components and technologies for a fully adaptive and sustainable solution. We expect this chapter will invite new perspectives, push boundaries, and conjure up a future where waste is nothing but an illusion.

## REFERENCES

Adel, Amr (2022). Future of Industry 5.0 in society: Human-centric solutions, challenges and prospective research areas. *Journal of Cloud Computing*, *11*(1), 1–15, SpringerOpen.

Akundi, Aditya, Euresti, Daniel, Luna, Sergio, Ankobiah, Wilma, Lopes, Amit, & Edinbarough, Immanuel (2022). State of Industry 5.0—Analysis and identification of current research trends. *Applied System Innovation*, *5*(1), 27, MDPI.

Alojaiman, Bader (2023). Technological modernizations in the Industry 5.0 era: A descriptive analysis and future research directions. *Processes*, *11*(5), 1318, MDPI.

Alves, Joel, Lima, Tânia M., & Gaspar, Pedro D. (2023). Is Industry 5.0 a human-centred approach? A systematic review. *Processes*, *11*(1), 193, MDPI.

Angelis, Roberta De (2021). Circular economy and paradox theory: A business model perspective. *Journal of Cleaner Production*, *285*, 124823, Elsevier.

Atif, Sehrish (2023). Analysing the alignment between circular economy and industry 4.0 nexus with Industry 5.0 era: An integrative systematic literature review. *Sustainable Development*, *31*(4), 2155–2175, Wiley Online Library.

Ciliberto, Cristina, Szopik-Depczyńska, Katarzyna, Tarczyńska-Łuniewska, Małgorzata, Ruggieri, Alessandro, & Ioppolo, Giuseppe (2021). Enabling the circular economy transition: A sustainable lean manufacturing recipe for Industry 4.0. *Business Strategy and the Environment*, *30*(7), 3255–3272, Wiley Online Library.

Cioffi, Raffaele, Travaglioni, Marta, Piscitelli, Giuseppina, Petrillo, Antonella, & Parmentola, Adele (2020). Smart manufacturing systems and applied industrial technologies for a sustainable industry: A systematic literature review. *Applied Sciences*, *10*(8), 2897, MDPI.

Coelho, Pedro, Bessa, Catarina, Landeck, Jorge, & Silva, Cristovão (2023). Industry 5.0: The arising of a concept. *Procedia Computer Science*, *217*, 1137–1144, Elsevier.

Corvellec, Herv, Stowell, Alison F., & Johansson, Nils (2022). Critiques of the circular economy. *Journal of industrial ecology*, *26*(2), 421–432, Wiley Online Library.

Dang, Duc Huy, Thompson, Karen A., Ma, Lan, Nguyen, Hong Quan, Luu, Son Tung, Duong, Minh Thao Nguyen, & Kernaghan, Ashlyn (2021). Toward the circular economy of rare earth elements: A review of abundance, extraction, applications, and environmental impacts. *Archives of Environmental Contamination and Toxicology*, *81*(4), 521–530, Springer.

Dwivedi, Ashish, Agrawal, Dindayal, Jha, Ajay, & Mathiyazhagan, K. (2023). Studying the interactions among Industry 5.0 and circular supply chain: Towards attaining sustainable development. *Computers & Industrial Engineering*, *176*, 108927, Elsevier.

ElMaraghy, Hoda, Monostori, Laszlo, Schuh, Guenther, & ElMaraghy, Waguih (2021). Evolution and future of manufacturing systems. *CIRP Annals*, *70*(2), 635–658, Elsevier.

Fraga-Lamas, Paula, Lopes, Sérgio Ivan, & Fernández-Caramés, Tiago M. (2021). Green IoT and edge AI as key technological enablers for a sustainable digital transition towards a smart circular economy: An Industry 5.0 use case. *Sensors*, *21*(17), 5745, MDPI.

Gagnidze, Ineza (2023). Industry 4.0 and Industry 5.0: Can clusters deal with the challenges? (A systemic approach). *Kybernetes*, *52*(7), 2270–2287, Emerald Publishing Limited.

Geissdoerfer, Martin, Savaget, Paulo, Bocken, Nancy M. P., & Hultink, Erik Jan (2017). The circular economy—A new sustainability paradigm? *Journal of Cleaner Production*, *143*, 757–768, Elsevier.

George, A. Shaji, & George, A. S. Hovan (2023). Revolutionizing manufacturing: Exploring the promises and challenges of Industry 5.0. *Partners Universal International Innovation Journal*, *1*(2), 22–38.

Giovanni, Pietro De (2023). Sustainability of the metaverse: A transition to Industry 5.0. *Sustainability*, *15*(7), 6079, MDPI.

Hazen, Benjamin T., Russo, Ivan, Confente, Ilenia, & Pellathy, Daniel (2021). Supply chain management for circular economy: Conceptual framework and research agenda. *The International Journal of Logistics Management*, *32*(2), 510–537, Emerald Publishing Limited.

Huang, Sihan, Wang, Baicun, Li, Xingyu, Zheng, Pai, Mourtzis, Dimitris, & Wang, Lihui (2022). Industry 5.0 and Society 5.0—Comparison, complementation and co-evolution. *Journal of Manufacturing Systems*, *64*, 424–428, Elsevier.

Ivanov, Dmitry (2023). The Industry 5.0 framework: Viability-based integration of the resilience, sustainability, and human-centricity perspectives. *International Journal of Production Research*, *61*(5), 1683–1695, Taylor & Francis.

Kashem, Mohammad Abul, Shamsuddoha, Mohammad, Nasir, Tasnuba, & Chowdhury, Asma Akter (2022). The role of artificial intelligence and blockchain technologies in sustainable tourism in the Middle East. *Worldwide Hospitality and Tourism Themes*, *15*(2), 178–191, Emerald Publishing Limited.

King, Sarah, & Locock, Katherine E. S. (2022). A circular economy framework for plastics: A semi-systematic review. *Journal of Cleaner Production*, *364*, 132503, Elsevier.

Kirchherr, Julian, Piscicelli, Laura, Bour, Ruben, Kostense-Smit, Erica, Muller, Jennifer, Huibrechtse-Truijens, Anne, & Hekkert, Marko (2018). Barriers to the circular economy: Evidence from the European Union (EU). *Ecological Economics*, *150*, 264–272, Elsevier.

Kirchherr, Julian, Reike, Denise, & Hekkert, Marko (2017). Conceptualizing the circular economy: An analysis of 114 definitions. *Resources, Conservation and Recycling*, *127*, 221–232, Elsevier.

Korhonen, Jouni, Honkasalo, Antero, & Seppälä, Jyri (2018). Circular economy: The concept and its limitations. *Ecological Economics*, *143*, 37–46, Elsevier.

Leng, Jiewu, Sha, Weinan, Wang, Baicun, Zheng, Pai, Zhuang, Cunbo, Liu, Qiang, Wuest, Thorsten, Mourtzis, Dimitris, & Wang, Lihui (2022). Industry 5.0: Prospect and retrospect. *Journal of Manufacturing Systems*, *65*, 279–295, Elsevier.

Lindgreen, Erik Roos, Salomone, Roberta, & Reyes, Tatiana (2020). A critical review of academic approaches, methods and tools to assess circular economy at the micro level. *Sustainability*, *12*(12), 4973, MDPI.

MacArthur, Ellen (2013). Towards the circular economy. *Journal of Industrial Ecology*, *2*(1), 23–44.

Maddikunta, Praveen Kumar Reddy, Pham, Quoc-Viet, Prabadevi, B., Deepa, Natarajan, Dev, Kapal, Gadekallu, Thippa Reddy, Ruby, Rukhsana, & Liyanage, Madhusanka (2022). Industry 5.0: A survey on enabling technologies and potential applications. *Journal of Industrial Information Integration*, *26*, 100257, Elsevier.

Moktadir, Md Abdul, Kumar, Anil, Ali, Syed Mithun, Paul, Sanjoy Kumar, Sultana, Razia, & Rezaei, Jafar (2020). Critical success factors for a circular economy: Implications for business strategy and the environment. *Business Strategy and the Environment*, *29*(8), 3611–3635, Wiley Online Library.

Morseletto, Piero (2020). Targets for a circular economy. *Resources, Conservation and Recycling*, *153*, 104553, Elsevier.

Mukherjee, Abheek Anjan, Raj, Alok, & Aggarwal, Shikha (2023). Identification of barriers and their mitigation strategies for Industry 5.0 implementation in emerging economies. *International Journal of Production Economics*, *257*, 108770, Elsevier.

Murray, Alan, Skene, Keith, & Haynes, Kathryn (2017). The circular economy: An interdisciplinary exploration of the concept and application in a global context. *Journal of Business Ethics*, *140*, 369–380, Springer.

Piscicelli, Laura (2023). The sustainability impact of a digital circular economy. *Current Opinion in Environmental Sustainability*, *61*, 101251, Elsevier.

Rajput, Shubhangini, & Singh, Surya Prakash (2019). Connecting circular economy and industry 4.0. *International Journal of Information Management*, *49*, 98–113, Elsevier.

Rathi, Rajeev, Sabale, Dattatraya Balasaheb, Antony, Jiju, Kaswan, Mahender Singh, & Jayaraman, Raja (2022). An analysis of circular economy deployment in developing 'nations' manufacturing sector: A systematic state-of-the-art review. *Sustainability*, *14*(18), 11354, MDPI.

Ross, J., Stevenson, F., Lau, R., & Murray, E. (2016). Factors that influence the implementation of e-health: A systematic review of systematic reviews (an update). *Implementation Science*, *11*(1), 1–12.

Saidani, Michael, Yannou, Bernard, Leroy, Yann, Cluzel, François, & Kendall, Alissa (2019). A taxonomy of circular economy indicators. *Journal of Cleaner Production*, *207*, 542–559, Elsevier.

Sarkar, Biswajit, Debnath, Abhijit, Chiu, Anthony S. F., & Ahmed, Waqas (2022). Circular economy-driven two-stage supply chain management for nullifying waste. *Journal of Cleaner Production*, *339*, 130513, Elsevier.

Shamsuddoha, M., Nasir, T., & Ibne Hossain, N. U. (2023). Integrating circular economy and reverse logistics for achieving sustainable dairy operations. In *Data Analytics for Supply Chain Networks* (pp. 211–226). Cham: Springer International Publishing.

Shevchenko, Tetiana, Saidani, Michael, Ranjbari, Meisam, Kronenberg, Jakub, Danko, Yuriy, & Laitala, Kirsi (2023). Consumer behavior in the circular economy: Developing a product-centric framework. *Journal of Cleaner Production*, *384*, 135568, Elsevier.

Stahel, Walter R. (2016). The circular economy. *Nature*, *531*(7595), 435–438, Nature Publishing Group UK London.

Turner, Chris, Oyekan, John, Garn, Wolfgang, Duggan, Cian, & Abdou, Khaled (2022). Industry 5.0 and the circular economy: Utilizing LCA with intelligent products. *Sustainability*, *14*(22), 14847, MDPI.

Velenturf, Anne P. M., & Purnell, Phil (2021). Principles for a sustainable circular economy. *Sustainable Production and Consumption*, *27*, 1437–1457, Elsevier.

Voulgaridis, Konstantinos, Lagkas, Thomas, & Sarigiannidis, Panagiotis (2022). Towards Industry 5.0 and digital circular economy: Current research and application trends. In *2022 18th International Conference on Distributed Computing in Sensor Systems (DCOSS)* (pp. 153–158), IEEE.

Xu, Xun, Lu, Yuqian, Vogel-Heuser, Birgit, & Wang, Lihui (2021). Industry 4.0 and Industry 5.0—Inception, conception and perception. *Journal of Manufacturing Systems*, *61*, 530–535, Elsevier.

# 11 Industry 4.0 and Sustainability of Human Resources

*Farman A. Moayed*

## 11.1 INTRODUCTION

The concept of sustainability initially started around environmental challenges and has been predominantly applied to scarce and/or expensive natural resources, waste management, environmental conservation, and energy generation and consumption. A definition that is commonly used for sustainability is from the UN Brundtland Commission in 1987, which states, "sustainable development is development that meets the needs of the present without compromising the ability of future generations to meet their own needs" (Haslam and Waterson, 2013; Hanson, 2013).

As businesses started to adopt different models to become sustainable, Moray (1995) described the major ecological and social problems in the 21st century (i.e. population pressure, pollution, water shortage, urbanization, etc.) and explained how researchers in ergonomics and human factors (E/HF) could contribute in addressing "global problems" by changing human behavior (Moray, 1995; Haslam and Waterson, 2013). The expectation was that ergonomists might be able to use their expertise involving human characteristics, behavior, performance, and human interaction with technology to contribute to sustainability efforts at a global scale (Moayed et al., 2022). In order to achieve such a goal, the solutions must be accepted by general populations around the world and not just industries. To successfully promote sustainability at such scale, a broad-notion re-examination of E/HF philosophy and ethics is needed by experts (Dekker et al., 2013). However, instead of focusing on the general population, the E/HF experts stayed focused on industries to help create "sustainable businesses" (Hanson, 2013; Radjiyev et al., 2015).

Over the years, environmentally and socially responsible investments and decision making have become important factors for business sustainability. This trend means nonfinancial elements of operation have become important measurements for a company's values. Such measurements include employees' welfare (human resources) (Funk, 2003). Healthy, experienced, and knowledgeable human resources (HR) can be considered expensive and a scarce resource in the context of continuously changing technology, and thus it could be beneficial to businesses and society to develop sustainable HR in order for businesses to remain sustainable (Ehnert, 2008; Moayed et al., 2022).

 DOI: 10.1201/9781003438748-14

As described by Ehnert (2008), it is not easy to develop a comprehensive definition for sustainable HR. It is a complex concept, and it is intertwined with social responsibility and influenced by work systems. Work systems deal with the application of different dimensions of ergonomics and human factors (i.e. micro-, meso-, and macro-ergonomics) to study the fitness between a person and his/her job and work environment to increase wellness and performance (Moayed, 2016).

The wellness and performance of employees have been affected significantly throughout history every time there has been a major change or disruption in industry. However, before explaining how Industry 4.0 affects the sustainability of HR, it is important to develop a definition for HR sustainability.

## 11.2 DEFINITION AND ASSESSMENT OF HR SUSTAINABILITY

The following definition for human resources sustainability is suggested in order to explain the effects of Industry 4.0: "Sustainability of human resources is development and implementation of policies and practices in any organization to help its employees perform their jobs at or close to their highest capacity at any given time and maintain such performance throughout the term of their employment" (Moayed et al., 2022).

Considering a hypothetical employee who has a fluctuating performance throughout his/her employment, Figure 11.1 can be an example of his/her performance if measured and tracked over time. Every employee experiences a learning period once he/she is employed (A), which can be different depending on the job characteristics and environmental factors for each individual. Ideally, in an organization with sustainable human resources, employees achieve and maintain their performance close to 100% capacity without interruption. In reality, the performance fluctuates for different reasons. For example, once an employee starts his/her career at $t_0$ (Figure 11.1) and goes through the learning process (A), at some point ($t_1$), he/she reaches the practical highest performance and remains at that level ($B_1$). It is possible that the employee misses a few days of work for different reasons ($C_1$) such as injury, sickness, or even vacation. If the employee gets injured at some point during his/her career ($t_4$) but not so severely as to miss work days, his/her performance is

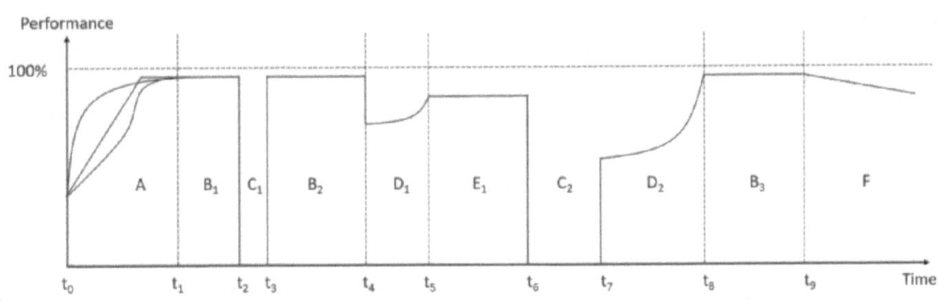


**FIGURE 11.1** Example of employee performance with low sustainability (Moayed et al., 2022).

negatively affected, and it takes a while to recover from the injury ($D_1$). In the case of chronic illnesses or injuries, the employee's performance remains lower than pre-injury levels even after the recovery ($E_1$). In the case of serious injury, he/she might get hospitalized and miss a few days of work ($C_2$), and when he/she comes back to work, ($t_7$) it takes a long time to recover and get back to the highest performance level ($D_2$). Even if the employee does not get injured, eventually performance starts to decline for different reasons such as aging, difficulty in learning/using new technology, and obsolete and outdated knowledge and skills sets (F). These types of events and changes in performance can happen frequently in someone's career until he/she leaves the company or retires (Moayed et al., 2022).

Based on the definition and as described in the example, the sustainability of HR is directly correlated to employees' performance, which could be affected and reduced by factors such as injury, absenteeism, and turnover rate. As the result, such changes in performance of human resources (in general terms) would increase a company's costs and expenses. Injuries or illnesses could be physical or psychological. Even if an injured or sick employee leaves a company to work at another one, he/she is still going to have performance lower than pre-injury or pre-sickness.

It might be reasonable to assess the sustainability of HR by measuring performances. Different variables such as task completion time, work time, motor performance, results of decision making, errors, and cognitive and physical workload can be used for modeling to quantify and predict human performance (Wu and Wada, 2018). In abstract, the sustainability of HR in a company could be equivalent to the average of employees' performance over a certain period of time. This can be a function of observable factors during the term of employment (Krishnaveni and Monica, 2018).

Moayed et al. (2022) developed a conceptual model to use work systems analysis to estimate the sustainability of HR in a company. This heuristic method uses E/HF variables in three different levels (micro-, meso-, and macro-ergonomics) to study the factors in a manufacturing company that can affect the performance of its employees, which can be an indicator of HR sustainability (Moayed, 2016; Bolis et al., 2014; Genaidy et al., 2009; Diamantidis and Chatzoglou, 2019; Muhs et al., 2018). This heuristic method allows experts to calculate three different indices for three dimensions of E/HF variables and one index for performance with values ranging from −1 to 1. The distribution of the performance index values across the micro-, meso-, and macro-ergonomic indices can reveal the level of sustainability of HR in the participating company. According to this method, a 3D presentation of data can help with the interpretation of results. The three indices of E/HF variables create a 3D space with eight octants (Table 11.1). Depending on where the performance index of a job/employee falls in this 3D space, it determines how each dimension of E/HF is contributing to the sustainability of HR (Moayed et al., 2022). For example, the performance indices in octant I are expected to have larger values compared to other octants because all three E/HF indices are contributing positively to the performance and as a result to HR sustainability, whereas the performance indices in octant VII are expected to have smaller values because all three E/HF indices are negatively affecting the performance and HR sustainability.

Figure 11.2 is one of the examples from Moayed et al. (2022) that shows the employees with larger values in performance indices are mostly distributed in octants

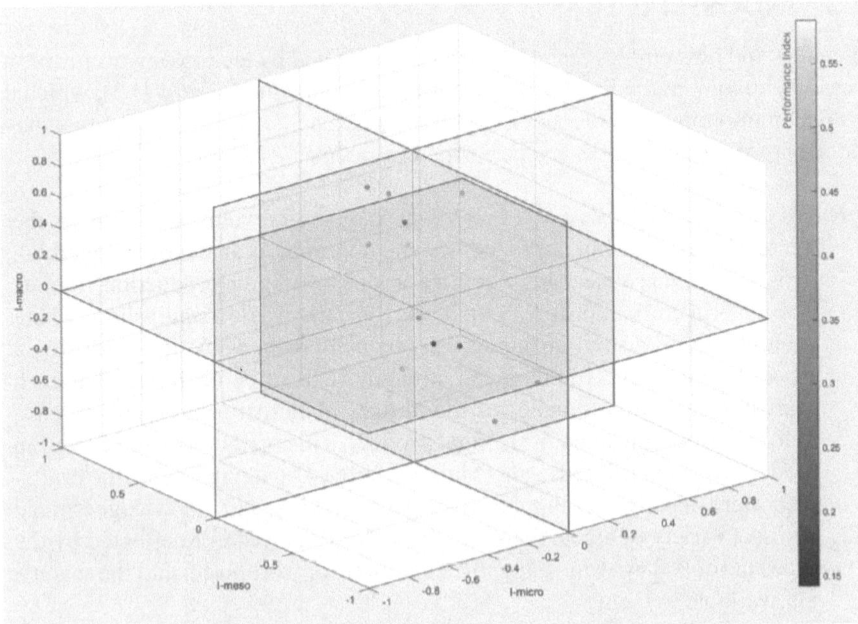


**FIGURE 11.2** The 3D presentation of data from a manufacturing company (Moayed et al., 2022).

**TABLE 11.1**
**Interpretation of E/HF Indices (Moayed et al., 2022)**

| Octant | $Index_{Micro}$ | $Index_{Meso}$ | $Index_{Macro}$ | Description |
|---|---|---|---|---|
| I | + | + | + | Most sustainable |
| V | + | + | - | |
| IV | + | - | + | Fairly sustainable |
| II | - | + | + | |
| VIII | + | - | - | |
| VI | - | + | - | Mostly unsustainable |
| III | - | - | + | |
| VII | - | - | - | Unsustainable |

I and V in a 3D space. Despite a few outliers, mainly due to small sample size, with some adjustments, this heuristic approach to assess the sustainability of HR might be possible in manufacturing companies. The authors admitted that the variables used to calculate the E/HF indices and performance index most likely should be customized for different industries in order to have a more valid and reliable assessment of HR sustainability.

## 11.3 INDUSTRIAL ERAS

To understand how employees' performance is affected by technology over time, it is necessary to look at the history of industrial developments (Figure 11.3), which can be divided into multiple eras. Each era represents specific characteristics that played important roles in how tasks are performed by people.

*Era of General Craftsmanship*: Prior to the industrial revolution in Europe, the general craftsmanship approach was the dominant manufacturing method. This approach produces high-quality products with low production volume. The general craftsmanship method was typically used in small shops where each worker or a small group of workers were responsible for producing a whole product. The fundamental strength of this approach was that each craftsperson is accountable for the design, manufacturing, quality, costs, delivery time, and so on of the final product. This method is still utilized in modern times in limited scale (Moayed, 2016). During this time, the craftsmen were mostly self-employed, and their skills and knowledge spread across a variety of areas; therefore their performances were affected by the types of tools they were using, the conditions of workplace, and the materials available to them.

*Era of Steam Power (Industry 1.0)*: With new developments and innovations in technology and growth in world population, the need to produce large volumes and varied types of goods increased. With the invention of steam power (which marked the beginning of the industrial revolution), the general craftsmanship approach transitioned to mass production mainly between the 1760s and 1870s. In order to meet demands, mechanized systems were developed to increase efficiency and reduce costs (Liu et al., 2020; Moayed, 2022). Employees' skills and knowledge were narrow and specialized, and the self-employment model changed to a corporation model. The organization and structure of work became more complicated and large. The type of tools used by employees changed significantly, and new hazards were introduced to workplaces, and as a result, employees' performance changed.

*Era of Electric Power (Industry 2.0)*: This era started in the 19th century (1870s–1940s) with the mass production of electricity. It led to an inexpensive

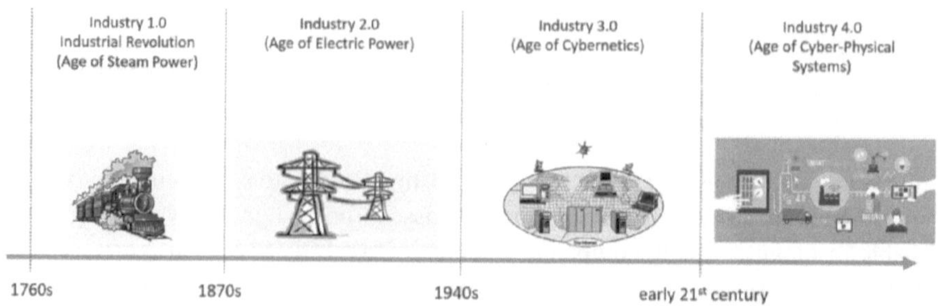


**FIGURE 11.3** Timeline of industrial eras (Moayed, 2022).

new source of energy which could be transferred across distances. Almost all manual tools and mechanized equipment changed to electrical ones, while new processes and (synthetic) materials were invented. The combination of such factors changed the characteristics of work environment and the type of hazards people were exposed to (Liu et al., 2020; Moayed, 2022). It required a new set of skills which, again, changed the way people performed their jobs. Certain types of jobs gradually disappeared, and some new ones were created. New and complex models were developed to design jobs and evaluate performance, which were supposed to comply with new standards and regulations. The production and consumption of fossil fuels accelerated, which contributed to air pollution and greenhouse emissions (environmental health hazards).

*Era of Cybernetics (Industry 3.0)*: As the use of electric power grows around the world, companies gradually expand their operation around the globe to other countries or continents primarily to have better access to resources such as labor, materials, or even a market for their products. This era (1940s—early 21st century) coincides with the space age, which enabled high-volume and high-speed data storage and transmission technology around the world. This allowed companies to enhance and expedite their decision-making process, which was a great advantage in highly competitive industries. Yet again, such major developments changed the way employers organize their work and how they evaluate the performance of their employees (Liu et al., 2020; Moayed, 2022). By this time, automation was part of the organization of work, and robots became one of the major inventions that significantly influenced employees' performance.

*Era of Cyber-Physical Systems (Industry 4.0)*: Starting the early 21st century and with major breakthroughs in microchip technology, the computational capacity of computers and digital equipment increased significantly, and their industrial and civilian application rapidly spread in almost every area. By implementing the new technology, companies are able to ideally collect large volumes of data from everything, everywhere, at any time and store/transfer them to large data processing centers. Traditional techniques were not efficient to analyze such large databases, and therefore new models were developed to analyze them. This allows organizations to understand highly complex systems and study the correlations between a large number of variables, which in return helps companies make better social and environmental decisions (Liu et al., 2020; Moayed, 2022). This industrial era involves the development and application of remote data sensing/collection, the Internet of Things, cloud computing, artificial intelligence, artificial neural networks, machine learning, highly complex simulation and modeling, automation, autonomous systems, and augmented reality, just to name a few. Such factors not only can change the scope, depth, and/or intensity of intellectual, mental, emotional, and physical performance of employees but also the quality of life for the general population. Therefore, it can have a broader effect in terms of environmental and HR sustainability.

## 11.4 INDUSTRY 4.0 AND THE FUTURE OF WORK

In order to understand how each element of Industry 4.0 can affect the sustainability of HR, it is better to organize them in a conceptual model. Figure 11.4 can explain what the future of work can look like once (if) Industrial 4.0 elements are fully implemented (Moayed, 2022). The model starts with data collection. Recent innovations and advancements in technology allow real-time remote data sensing/collection by using wireless communication. Such sensors can be installed everywhere, and they can be used to monitor or measure practically everything from physical and environmental parameters such as temperature, vibration, noise, and light to visual and voice recognition such as the presence or absence of a person/object, or they could understand messages. The proliferation of mobile and/or wearable devices such as cellphones, GPS, or exoskeleton suits, along with their accessories, allows high-volume and real-time data collection from users from movement/traffic tracking to biological measurements such as blood pressure, blood sugar, and cardiovascular factors. Access to reliable and high-speed internet enables companies to have access to other sources of data from financial transactions to climate data, theoretically from any region around the world. It is not difficult to imagine the volume of the data that can be collected every minute in such systems.

The second major component of this model involves the storage and processing of such massive data. Current technologies in material science and electronics enable companies to create data centers where very large volumes of data are stored inexpensively. Traditional methods for data analysis are not capable of analyzing such data for two major reasons: (a) the complexity of the data structure and (b) the massive volume of data. Therefore, new analytical methods are developed by mathematicians, computer scientists, and IT experts. The general term for this component is cloud computing. The Internet of Things (IoT) helps IT professionals to connect every sensor or device to processing units in order to simultaneously collect information, process it, and generate actions based on the results with no, or the least amount of, interaction with humans. Traditional statistical models are not enough to study the correlation between the large number of variables, and regular computers are incapable of

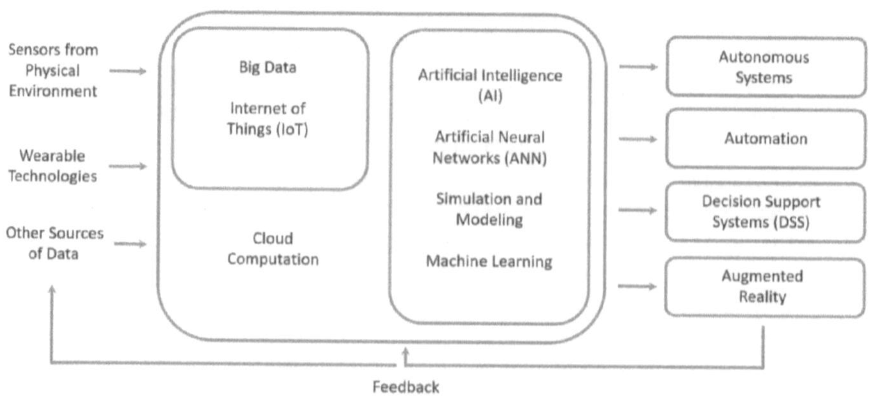


**FIGURE 11.4** Conceptual model for the future of work (Moayed, 2022).

handling the necessary calculations. New methods and models have been developed for data management such as big data to help humans make sense of such complex and massive data. These models (such as artificial intelligence (AI), artificial neural networks (ANNs), and simulation) are constantly being improved by scientists.

Another component of this model is about civil or specialized applications of this technology such as autonomous systems, automation, decision support systems (DSSs), and augmented reality, just to name a few. One example of an industrial application of autonomous systems are self-driving vehicles like drones or lift trucks that can be used for material handling or delivery of goods to customers. A civilian application can be public transportation to move people through pre-determined routes. Automation has already been used in industry for decades now, but with new technology, this can be done remotely. For example, an operator can be stationed in one physical location while operating or monitoring manufacturing operations in multiple different locations somewhere else. The DSS, with the use of AI, simulation, and modeling, is affecting mostly white-collar and top-management jobs where they need to understand the behavior of large and complex systems in order to make decisions in a timely manner. The most obvious application of augmented reality could be real-time and on-site instructions to operators where a hologram or a virtual character in the metaverse appears to provide instructions or feedback. This can be potentially very effective when the operator is in an emergency situation or performing a task that is very specialized or uncommon.

The final component of this model is the feedback data it generates for rapid adjustments and updates. This feature will eliminate or minimize the lag time between the execution of models and updates and increase the efficiency of operations and decision making. The feedback loop can increase the size and volume of data to be processed while it allows experts to optimize the models and algorithms used for data analysis and modeling.

## 11.5 HR SUSTAINABILITY DURING INDUSTRY 4.0

As described, HR sustainability is directly correlated with employees' performance based on the definition and example provided. The technologies that are, and will be, used during the Industry 4.0 era are affecting and will affect the way people work or even live for years to come. At one end of the spectrum, they can make a lot of complex and knowledge-based activities and tasks easier to do for people who have little training or skills, and at the other end of the spectrum, they will require new skill sets and training in order to use or implement them. Regardless of where an individual stands in such a spectrum, his/her performance can be affected as organizations and companies choose to move toward Industry 4.0. Certain aspects of work systems and employee performance which can be affected by the application of AI are explained in the following.

*Skill Sets and Knowledge about AI-Related Fields*: When automation and robotics started to proliferate in the 20th century, there was fear that humans would be completely replaced by robots or CNC machines and there would be no jobs for anyone. In reality, that did not happen. Instead, new jobs were created. Companies started to hire experts who can design robots, program CNC machines and robots, and perform maintenance on them. New interdisciplinary fields in engineering were created to design or modify production processes that can use and optimize robots or CNC machines. The implementation of such technology helped reduce or even eliminate certain occupational hazards and improve the safety and health of employees.

The employees who were willing and able to learn the new skills were able to stay in the job market and even increase their income, while those who resisted the implementation of automation and did not try to acquire new skills ended up unemployed or demoted to jobs that paid less. Because of better work conditions and more reliable equipment, productivity increased. In other words, elements of micro- and meso-ergonomics changed drastically (such as the type, severity, and frequency of occupational injuries and illnesses), which affected the sustainability of HR. This trend continued even when automation was extended to office work. The same fears were common among white-collar employees and eventually turned out to be baseless.

Employees need to learn about AI and its applications. If organizations want to improve the sustainability of their human resources, they should start training their existing employees now; otherwise they will experience larger turnover rates in the near future. Universities and colleges need to update their curricula and replace some of the outdated and traditional topics with new ones relevant to the applications of AI. All employees need to understand what algorithms are and how they are used in AI systems and other computer applications. Employees need to learn how machine learning works and how it can be implemented in their work. People should understand the limitations and potential pitfalls of AI systems in order to have realistic expectations and to avoid scams and misinformation (Brahma and Tripathi, 2021; Machado et al., 2020).

*HR Development and Management*: The future of work involves changes in professional relationships between employees and their supervisors and managers. There will be a substantial shift in the work culture (i.e. macro-ergonomic factors), and therefore the management style needs to adjust. New standards and regulations will be imposed on industries to comply with expectations, from HR to environmental and safety, as well as data security and privacy.

HR training and development will change in content and scope, which should include AI-related topics and prepare employees to accept and understand how such systems work and affect their jobs. Remote working will become more prevalent. Assessments of HR performance will be done more objectively using direct remote sensing and measurements instead of subjective methods such as self-reporting questionnaires or supervisory observations using Likert scale scoring methods.

As applications and implementation of AI-related systems spread, organizations might opt for integrated management systems (IMS) in order to avoid duplicities, share commonly needed resources, and achieve their strategic goals (Moayed and Yusuf, 2023). Industry 4.0 emphasizes IoT and big data analysis, which requires integrating data collection by different management system standards (MSS) such as quality control, environmental health and safety, and financial/accounting operation. In order to successfully implement all the elements of Industry 4.0, it is necessary to integrate all MSS that are used individually in an organization and develop models to optimize the operation. This cannot be achieved without changing organizational culture and engagement in continuous improvement by everyone, and such efforts should be led by management (Domingues et al., 2016).

*Occupational Safety*: In order to envision what the future of work will look like in terms of occupational safety, it is better to demonstrate potential changes in the form of a few examples, starting with personal protective equipment (PPE) and data collection (Figure 11.4). As the use of wearable technology spreads and remote or wireless data transmission techniques advance, smart PPE will be developed and used by employees in near future. Such smart PPE will collect data about environmental and occupational hazards in real time and, if necessary, provide on-time feedback, warning, and instructions to employees as they are performing their job. Smart safety goggles/glasses equipped with cameras can transmit images of the surrounding work environment. Gas detectors can measure the concentration of toxic gasses; noise and vibration sensors can detect and measure the intensity of exposures. Thermal sensors can track extreme temperatures, while other wearable equipment can track biological signals such as cardiac and respiratory functions of employees or even the level of activities in different parts of their brains. Exoskeleton suits not only can help employees avoid musculoskeletal disorders and injuries but also enable injured workers to perform at pre-injury levels. It can also allow objective measurements of biomechanical factors and variables. This massive volume of data can be compiled, integrated, and analyzed to protect employees from potential injuries and illnesses, which can improve the sustainability of HR (Leso et al., 2018; Mathis, 2018; Danko and Straka, 2022).

Stationary sensors and cameras placed in strategic locations can monitor the environment for potential hazards that can harm employees or cause property damage. Fire and smoke detectors integrated with cameras can differentiate false alarms from true hazards and initiate evacuation plans only in limited areas as necessary. Cameras with AI systems can identify moving or stationary objects to control and manage the movement of humans or autonomous systems. Drones can be integrated with AI systems and used to fly indoors and outdoors to conduct audits and inspections on hazardous materials or processes. The data collected by numerous systems and the information generated by analyzing them can help organizations to take preventive actions to eliminate accidents, injury, or illness and increase the reliability of processes (Leso et al., 2018; Mathis, 2018).

Implementation of such elaborate data collection systems in an organization that employs large numbers of people, and has complicated processes, can easily produce a high volume of data. This leads to implementation of cloud systems (Figure 11.4). First, all the wearable sensors, smart PPE, and stationary detectors should be integrated in one system (IoT) that has the capacity to store large volumes of data and also has the processing speed to analyze the data in real time. Second, advanced analytical methods and models should be customized based on different factors such as: (a) characteristics of data, (b) volume and frequency of data collection, (c) application of data, (d) desired level of accuracy of results, and (e) the rate/speed the results are needed. Machine learning and artificial neural network modeling can be used to understand the behavior of complex human–computer systems and utilize the knowledge for simulation under different scenarios for different purposes such as design, decision making, optimization, and prevention. This technological advantage can reduce workload or improve the performance for employees in respect to knowledge-based tasks, and, as a result, HR sustainability can improve.

There are other applications of AI in the occupational safety field such as autonomous systems and automation at any level where tasks are done in a procedural and algorithmic manner such as safety audits and inspections. An unmanned vehicle integrated with image and sound recognition and other sensors/detectors can be deployed to survey any given area to identify potential hazards and report them to management. Such vehicles can be airborne, ground-based, or submerged devices that can maneuver around obstacles or other vehicles. A variety of versions of such vehicles have been used in military and space explorations for years, and they can be used for commercial and civilian applications inexpensively. Augmented reality (or virtual reality) is already being used in educational applications, and it can be used in the occupational safety area as well. For example, smart goggles/glasses with cameras integrated with other remote sensors can be used to provide instructions and feedback to any operator during non-routine or complicated operations by projecting the information and/or images on a screen within their central vision. It can reduce the stress during high-risk operations and minimize human errors and, as a result, improve employee performance and sustainability of HR (Bower et al., 2014; Grabowski et al., 2018; Li et al., 2018).

*Environmental Health*: As mentioned at the beginning of this chapter, environmentally and socially responsible decision making has been one of the major factors motivating the implementation of the Industry 4.0 concept in industries. It would be wrong to envision the future of work without including environmental health (as a subset of the SHE domain) in the applications of AI technology. Organizations are increasingly mindful of how their operations and products affect the environment. Implementing AI-related technologies allows them to monitor and control their footprint on the environment. In the future, an IMS in any organization should not be limited to the systems within their boundaries. It is essential to collaborate with external organizations such as government, private, and educational institutions to network their centers and share information to track how the operation of each unit affects the environment. Since this aspect of operation involves different organizations that are mostly independent of each other, it is very important to standardize measurement methods and analysis for efficient and reliable exchange of data and information (Felknor et al., 2021; Moktadir et al., 2018).

Geospatial artificial intelligence (GeoAI) is the application of AI in spatial science, also known as geographic information science. It attempts to understand, analyze, and visualize real-world phenomena according to their locations. With the implementation of GeoAI, businesses can study and understand how harmful pollutants and emissions from their industries disperse in the environment (in atmosphere or oceans) and track the health effects on communities and the general population. Similar applications can be used during epidemics/pandemics to track how a disease spreads in a geographical region or around the world based on social and economic factors (VoPham et al., 2018). There are studies that show the application of AI can help the growth of green industries to reduce environmental pollutants, which can lead to improvement in the overall health of communities. It can contribute to more sustainable HR regardless of the type and location of occupation (Qian et al., 2023).

*Technology Alienation*: Implementation of advanced technology can have its own drawbacks. One known negative effect of advanced technology, such as automation, on society is "technology alienation". This refers to the perception that humans have no or very little control over their work or process and they have become part of it. This issue can lead to low morale and lack of motivation among employees. It can gradually diminish creativity and critical thinking among people (Goetsch, 2015). Another definition can be the decrease in one's adaptability to the social, cultural, and natural environment or one's loss of control over his/her surroundings and gradually being marginalized (Haga, 2022).

The factors contributing to technology alienation can be at micro-, meso-, and macro-levels of work systems and ergonomics. It is obvious that not only the content of work, tools, and the process itself can make humans feel insignificant, but also the organizational structure, team set-up, distribution and access to knowledge and skills, authority over decision making, and ability for problem solving are important for employees to feel that they are significant, in control of production processes, and contributing to the completion of the product or service. It is important to keep in mind that the work organization and work content are dependent upon other contextual factors such as the organization's status, market conditions and agility, competition, technology utilization, labor-management relationship, and organizational governance. These factors are constantly changing at different paces and are not static over time (Haga, 2022). Studies show that employees' alienation from technology can be in three different forms: (a) alienation from the product of their labor, (b) alienation from themselves (i.e. people find only extrinsic meaning in work which causes separation from their true selves), and (c) alienation from others (i.e. people are estranged from their colleagues) (Haga, 2022; O'Donohue and Nelson, 2014).

This problem can aggravate personnel turnover and absenteeism in organizations, which elevates the importance of HR development. If employees learn and understand how AI technology works and what their role is in the overall operation, it can help them find their position in the organization meaningful. The key factor will be for work systems designers to leave some room for humans to contribute to the overall operation of their organization. For example, employees should be allowed to control aspects of data collection or participate in developing algorithms for a variety of AI systems they are supposed to use or even get involved in decisions using DSS. Once people understand how their systems operate and what their role is in the process, it can alleviate the technology alienation problem.

*Breach of Privacy*: It is fair to say that violations of people's privacy is the one challenge that human rights advocates are most worried about as more advanced technologies are being used in daily life. Everyone should have the right to be left alone as long as they desire. The ability of an individual to selectively control access to themselves is an important aspect of privacy. Such control of access should be in two directions, that is, providing input from others to an individual and output from an individual to others. Even though people often claim privacy is important to them, some research shows that behavioral patterns contradict this notion (Kokolakis, 2017). Nowadays, private information is shared online in different ways, such as: an individual voluntarily shares personal information in exchange for a service or product, an individual voluntarily shares his/her (and probably others') personal information on social media, or government agencies share personal information of others as required by law, just to name a few. New technology allows companies to mine for data on publicly accessible domains (big data). The databases resulting from data mining are owned by the company that has conducted the mining even though the information is about individuals and most likely they are unaware that their personal information is being collected.

One of the definitions for privacy that is commonly used is the state in which a person is not observed or disturbed, or to be free from public attention (Curzon et al., 2021). This definition applies to individuals as well as groups. This definition must be adjusted for digital identities and according to geographical locations. For example, the North American perspective of legislation for digital privacy is less stringent, and it may be considered more of a commodity than a human right (Walsh et al., 2017, Curzon et al., 2021).

Cyber-physical systems encompass wearable devices, stationary and mobile remote sensing and measurements, and motion/image/sound detection and recognition. The fine line between work and private life gets blurred very quickly and easily. It is not difficult to imagine a world in which AI systems in the fields of medicine, sports, or even education get integrated with work in order to help people have healthy, safe, and productive lives. This likely scenario demonstrates how a breach of privacy can happen. As of now, there is no comprehensive and official regulation or even code of conduct for AI experts and companies to follow. Everything in this industry is self-regulated. One of the most recent developments to address ethical challenges and privacy protection has come from United Nations Educational, Scientific, and Cultural Organization (UNESCO). It drafted the *Recommendation on the Ethics of Artificial Intelligence* in September 2020. This recommendation sets up ten important principles of the ethics of AI (Zhang et al., 2021). Regardless of how the private data is shared or collected, the organization that owns it should be responsible to protect it. Any form of breach of privacy in an AI system will result in mistrust between employees and employers and can negatively affect the way employees interact with AI systems, with the possibility that users (intentionally) provide false information or disrupt the data collection and analysis processes. This type of flawed interaction between humans and AI systems will lead to lower HR sustainability because it cannot properly assess the performance or work systems variables contributing to it.

## 11.6 CONCLUSION

The future of jobs and workers in an environment with rapid social and technological changes is among the ten big ideas that National Science Foundation is interested to invest in for research. The future of work at the human–technology frontier aims at: (a) understanding and developing the human-technology partnership, (b) designing new technologies to augment human performance, (c) illuminating the emerging socio-technological landscape and understanding the risks and benefits of new technologies, (d) understanding and influencing the impact of artificial intelligence on workers and work, and (e) fostering lifelong and pervasive learning (NSF, 2023). The development and application of AI surely falls under this area of research. The complete scope of AI's effects on human resources and its sustainability is still unknown and the subject of ongoing research. Based on early evidence and outcomes produced by preliminary studies, it is clear that AI applications in HR sustainability can have both positive and negative effects. Implementation of AI systems will be vital to companies' competitiveness, and (a) it can cause the threshold for perfect performance to move higher, and (b) it helps less-experienced or disabled workers perform as well as (or close to) experienced and healthy workers. In both cases, it can help a company become more productive.

While companies rush to implement AI systems in their operation because of its benefits, they should not ignore its potential shortcomings. AI systems produce positive results only if the employees are involved in the process in meaningful ways; otherwise, the AI will not be able to deliver its potential benefits at the expense of alienating the employees and creating a work environment of mistrust and skepticism. It is better to start the HR development and training program about AI and its application before its implementation. It is very likely that AI-related legislation and codes of conduct will be developed and enforced in the near future, which will affect the research and implementation of AI systems. As more research in this area is needed, the full effects of AI systems in HR sustainability have yet to be seen, but it will happen sooner than what the general population may expect.

## REFERENCES

Bolis, I., Brunoro, C. M., & Sznelwar, L. I. (2014). Mapping the relationship between work and sustainability and the opportunity for ergonomic action. *Applied Ergonomics*, *45*(4), 1225–1239.

Bower, M., Howe, C., McCredie, N., Robinson, A., & Grover, D. (2014). Augmented reality in education—Cases, places and potentials. *Educational Media International*, *51*(1), 1–15.

Brahma, M., & Tripathi, S. S. (2021). Developing curriculum for Industry 4.0: Digital workplaces. *Higher Education, Skills and Work-Based Learning*, *11*(1), 144–163.
Curzon, J., Kosa, T. A., Akalu, R., & El-Khatib, K. (2021). Privacy and artificial intelligence. *IEEE Transactions on Artificial Intelligence*, *2*(2), 96–108.
Danko, A. D., & Straka, M. (2022). Exoskeletons—Robotic suits improving work in logistics. *Acta Logistica*, *9*(4), 405–410.
Dekker, S. W. A., Hancock, P., & Wilkin, P. (2013). Ergonomics and sustainability: Toward an embrace of complexity and emergence. *Ergonomics*, *56*(3), 357–364.
Diamantidis, A. D., & Chatzoglou, P. (2019). Factors affecting employee performance: An empirical approach. *International Journal of Productivity and Performance*, *68*(1), 171–193.
Domingues, J. P. T., Fonseca, L., Sampaio, P., & Arezes, P. M. (2016). Integrated versus non-integrated perspectives of auditors concerning the new ISO 9001 revision. In *IEEE International Conference on Industrial Engineering and Engineering Management*, Bali, Indonesia.
Ehnert, I. (2008). *Sustainable Human Resource Management*. Physical-Verlag.
Felknor, S., Streit, J. M. K., McDaniel, M., Schulte, P. A., Chosewood, L. C., Delclos, G., & Workshop Presenters and Participants. (2021). How will the future of work shape OSH research and practice? A workshop summary. *International Journal of Environmental Research and Public Health*, *18*(11), 5696.
Funk, K. (2003). Sustainability and performance. *MIT Sloan Management Review*, *44*(2).
Genaidy, A. M., Sequeira, R., Rinder, M. M., & A-Rehim, A. D. (2009). Determinants of business sustainability: An ergonomics perspective. *Ergonomics*, *52*(3), 273–301.
Goetsch, D. L. (2015). *Occupational Safety and Health for Technologists, Engineers, and Managers*. Pearson Education.
Grabowski, M., Rowen, A., & Rancy, J. P. (2018). Evaluation of wearable immersive augmented reality technology in safety-critical systems. *Safety Science*, *103*, 23–32.
Haga, T. (2022). Alienation in a digitalized world. *AI and Society*, *37*, 801–814.
Hanson, M. A. (2013). Green ergonomics: Challenges and opportunities. *Ergonomics*, *56*(3), 399–408.
Haslam, R., & Waterson, P. (2013). Ergonomics and sustainability. *Ergonomics*, *56*(3), 343–347.
Kokolakis, S. (2017). Privacy attitudes and privacy behavior: A review of current research on the privacy paradox phenomenon. *Computer and Security*, *64*(1), 122–134.
Krishnaveni, R., & Monica, R. (2018). Factors influencing employee performance: The role of human resource management practices and work engagement. *International Journal of Business Performance Management*, *19*(4), 450–475.
Leso, V., Fontana, L., & Iavicoli, I. (2018). The occupational health and safety dimension of Industry 4.0. *Medicina del Lavoro*, *109*(5), 327–338.
Li, X., Yi, W., Chi, H. L., Wang, X., & Chan, A. P. C. (2018). A critical review of virtual and augmented reality (VR/AR) applications in construction safety. *Automation in Construction*, *86*, 150–162.
Liu, Z., Xie, K., Li, L., & Chen, Y. (2020). A paradigm of safety management in Industry 4.0. *Systems Research and Behavioral Science*, *37*(4), 632–645.
Machado, C. G., Winroth, M. P., & Riberio da Silva, E. H. D. (2020). Sustainable manufacturing in Industry 4.0: An emerging research agenda. *International Journal of Production Research*, *58*(5), 1462–1484.
Mathis, T. L. (2018). Safety 4.0: Updating safety for Industry 4.0. *EHS Today*.
Moayed, F. A. (2016). Human factors measurement. In A. B. Badiru & L. Racz (Eds.), *Handbook of Measurements: Benchmarks for Systems Accuracy and Precision* (pp. 41–78). CRC Press.

Moayed, F. A. (2022, May 10–12). Preparing future generation of SHE experts. *BCSP Global Learning Summit*, Indianapolis, IN.

Moayed, F. A., Cribelar, K., & Smith, A. (2022, July 26–29). Sustainability of human resources: A heuristic approach. In *AHFE 2022 Conference*, New York, NY.

Moayed, F. A., & Yusuf, A. (2023). Benchmarking of three integrated management system models in respect to occupational health and safety management systems implementation. *SHIFT Global EHS Research to Practice*, *2*(1).

Moktadir, A., Ali, S. M., Kusi-Sarpon, S., & Shaikh, A. A. (2018). Assessing challenges for implementing Industry 4.0: Implications for process safety and environment protection. *Process Safety and Environment Protection*, *117*, 730–741.

Moray, N. (1995). Ergonomics and the global problems of the twenty-first century. *Ergonomics*, *38*(8), 1691–1707.

Muhs, K. S., Karwowski, W., & Kern, D. (2018). Temporal variability in human performance: A systematic literature review. *International Journal of Industrial Ergonomics*, *64*, 31–50.

NSF. (2023, May 17). *Future of Work at the Human-Technology Frontier*. National Science Foundation. www.nsf.gov/eng/futureofwork.jsp

O'Donohue, W., & Nelson, L. (2014). Alienation: An old concept with contemporary relevance for human resource management. *International Journal of Organizational Analysis*, *22*(3), 301–316.

Qian, Y., Liu, J., Shi, L., Forrest, J. Y., & Yang, Z. (2023). Can artificial intelligence improve green economic growth? Evidence from China. *Environmental Science and Pollution Research*, *30*(6), 16418–16437.

Radjiyev, A., Qiu, H., Xiong, S., & Nam, K. (2015). Ergonomics and sustainable development in the past two decades (1992–2011): Research trends and how ergonomics can contribute to sustainable development. *Applied Ergonomics*, *46*(Part A), 67–75.

VoPham, T., Hart, J. E., Laden, F., & Chiang, Y. (2018). Emerging trends in geospatial artificial intelligence (GeoAI): Potential applications for environmental epidemiology. *Environmental Health*, *17*(40).

Walsh, D., Parisi, J. M., & Passerini, K. (2017). Privacy as a right or as a commodity in the online world: The limits of regulatory reform and self-regulation. *Electronic Commerce Research*, *17*(2), 185–203.

Wu, C., & Wada, T. (2018). Introduction to the special issue of human performance modeling: Prediction of human performance to improve productivity and safety. *International Journal of Industrial Ergonomics*, *63*, 1–2.

Zhang, Y., Wu, M., Tian, G. Y., Zhang, G., & Lu, J. (2021). Ethics and privacy of artificial intelligence: Understandings from bibliometrics. *Knowledge-Based Systems*, *222*, 106994.

# 12 Model of Dual Management at Company Scale

*Vadim Lapidus*

## 12.1 INTRODUCTION

Nowadays, there has been extensive experience in applying various approaches to improvement, including Lean 6 Sigma, Kaizen, PDCA and DMAIC cycles, 8D and many others. All of them, with skillful application, showed very impressive results, but in the long term and throughout the company, positive results faded or disappeared. The author observed these phenomena in the framework of many projects on improvement of labor productivity (Grachev et al., 2021).

Despite significant improvements in labor productivity, many companies did not have an effect in performance indicators calculated throughout the company. The main reason—improvement approaches are usually local in nature, and changes in process capabilities and/or products and services characteristics are not reflected in planning methods—thus remains unknown to many employees of the organization. In particular, financial and economic departments often do not receive relevant information and, therefore, do not change planning standards.

Many of the provisions included in dual management (DM) systems have been put forward within other systems, as well as by various authors of publications. Note the influence of dual control work for technical systems proposed by A. Feldbaum in the middle of the 20th century. They are transferred into human-cybernetic-physical (HCP) systems (Feldbaum, 1963). The author also draws ideas from the approaches that have been developed in design of experiments (DOE) (Box and Draper, 1969); works by E. Deming (2018), J. Juran (1988) and Y. Kondo (1993); and institutional economics, as well as works on artificial intelligence (AI), digital twins and so on. DM is a collective concept with the goal of combining many rather well-known approaches and solutions into a single conceptual approach.

## 12.2 DEFINITION OF DUAL MANAGEMENT

*Brief: The concept of dual management, the definition of DM and an explanation of the need to build it on a company-wide basis. The relationship with improvement management is being considered.* (*Lapidus, 2021*)


DOI: 10.1201/9781003438748-15

The author calls dual management a unified approach and appropriate methodology for integrated management activities to achieve planned results and manage the capabilities of the business system through improvements in processes, product, quality, cost, time and management systems with synchronized cycles of transition to new standards and reference data, changed in accordance with the achieved improvements throughout the company (Lapidus and Tiuleneva, 2021; Lapidus and Ozerova, 2023). Important points of DM:

1. Improvements become the object of planning, indicating the expected results and the risks not to achieve them. They also have a PDCA cycle, but unlike outcome planning, which seeks guaranteed outcomes, the improvement cycle is risky and venture based, and the two cycles have different forms of responsibility and motivation.
2. DM requires an appropriate single-target scorecard combining effectiveness and efficiency targets.
3. DM opens the possibility of integrating the market goals of a business organization with the requirements of the ESG agenda, considering this agenda not as external restrictions and requirements but as embedded in effective business strategies and regular management (ESG agenda and the 17 goals. https://sdgs.un.org/goals).
4. DM management considers personnel at all levels, starting with managers involving all employees, as an asset and source of signals and suggestions for improvements.

This thesis requires the development of the social component as necessary to determine the rights, duties, powers and responsibilities of stakeholders in terms of making improvements and sharing the benefits from the previous combined dual activity. Note that the opportunities themselves do not bring additional benefits directly; they increase efficiency ratios.

## 12.3 CONCEPT OF DUAL MANAGEMENT

*Brief: The section discusses the main conceptual provisions of DM and its relationship with investments. Two types of investments: investments aimed at quantitative growth, scaling the business, and investments aimed at the effectiveness of the business system itself. The efficiency of this machine is determined by its processes, products, personnel and assets, including internal processes and value exchange processes. Such a "smart machine" is able to learn and improve. However, this requires the construction of a dedicated management that directs certain efforts and investments to study and change the characteristics of processes, products and systems and verifies these changes and introduces them into the corporate regulatory reference system in the form of new regulations that are used in further planning management cycles. The planning system is trained, and standards, reference data and planning methods in general are adapted to the changes that are made to the execution system. DM includes the use of performance indicators and the rate of change (growth) of efficiency as the results of investments in improvement projects.*

The concept of DM:

1. First, within the framework of the proposed concept, improvement activities are considered a certain type of smart investment, which is aimed at improving the capabilities of business management systems and is very effective, as shown by model calculations and the results of numerous practical examples. As a rule, the return on investment falls in the next calendar period. The effectiveness of joint dual investments is significantly increasing, that is, investments aimed both at growing the scale of the business (growth of resources and assets) and at increasing the efficiency of the business system itself (smart investment).
2. The transition from a permanent improvement model to a combined model, which can be determined based on the well-known slogan: "Do right things right way, just-in-time (JIT), first time through (FTT)". This slogan reflects the ideal of the performing part of the management system. They are based, on the one hand, on the requirements of technical documentation for products and processes and time plans and, on the other hand, on processes capabilities for their performance, availability factors and planning accuracy, as well as processes capabilities for quality, for example, $C_p$ indices.

All of these aspects can and shall be subject to improvements and system changes that allow all changes to be made synchronously and company-wide, communicating the changes to all system participants.

Thus, this slogan is supplied with improvement actions, getting the following slogan reflecting the essence of DM:

*Do right things, in the right time, think what's right and improve the understanding of the right.*

Figuratively speaking, this is a staircase with discrete steps, and it is continuous. Strictly comply with what is required, in accordance with the established plans and standards, exactly at the promised time, without defects and rework/repair. Further improve capabilities and, accordingly, increase requirements, regulations and standards for the next step. For example, improve processes, including reducing their flow time and variability, reducing the risks of inconsistencies. The results of the improvements are recorded and certified, and, based on these changed characteristics, alter the planning and remuneration standards. In this case, improvement processes become continuously discrete. That is, improvement strategy is continuous. It is a function of new experience, new knowledge, innovation and investment but at an established time within the calendar cycle to capture the levels and parameters of improvements.

3. Operational improvement cycles are synchronized with the execution cycles of plans, while the achieved results of types and cycles are recorded and entered into standard, passport, regulatory characteristics of processes, products, services and systems and are used as new standards for the next calendar execution period. This data is used to recalculate the planning and remuneration ratios. For example, improved production processes have increased productivity throughout the production system and reduced scrap levels, then changes to documents and information systems, labor standards and scrap and repair rates have been made, and wage and compensation rates are recalculated simultaneously.
4. Companies adopting the concept of dual management shall move to an appropriate culture and values, which combine the value of strict discipline and responsibility in employees for the implementation of all established procedures and regulations while maintaining critical thinking and the following principle: any work can be improved by constantly thinking about it and looking for new solutions, but no one can switch to new solutions without a synchronous solution throughout the management system. The combination of two attitudes in people at the same time: on the one hand, discipline; on the other hand, innovation and creativity. However, the experience of Japan, South Korea, and other countries shows that this combination is still possible.

*If you want to improve your effectiveness and efficiency, do it with everyone!*

5. DM clarifies the relationship between effectiveness and efficiency. Within the framework of execution planning cycles, the goal of management is to be effective with the opportunities achieved for this cycle; within the framework of improvement cycles, new opportunities are created allowing to obtain results with greater efficiency at the next cycle, that is, the following cycles: effectiveness—improvement—new effectiveness with increased efficiency. Efficiency is a consequence of the linkage of improvement cycles and effectiveness
6. It should be noted that in respect to efficiency, transformation of the money role reflects interests of owners and investors and corresponds to the model of the organization as a machine for turning cash flows into new more powerful cash flows.

However, at management levels, in particular in planning (Plan) and execution levels (Do), the role of money changes. For managers, money is a tool for achieving their goals. For performers, money ceases to be a tool; they become associated only with money as salary and remuneration. When performing, personnel uses physical characteristics of processes and products, for example, pieces, kg, seconds and m.

Thus, special actions are required to involve managers and performers in efficiency activities. This is allowed through the teamwork of functional managers and the involvement of them and performers in the implementation of improvement projects through special motivation, setting standards for remuneration and remuneration with characteristics of improving opportunities (Lapidus, 2021).

Performers do not understand the goals and objectives of improving performance characteristics well enough, but they understand the tasks of improving the physical capabilities of products and processes very well and see a connection between improvements and rewards.

### 12.3.1 Operational Model of DM

At the operational level, dual management considers the business an activity in two directions and two cycles:

- Results planning—execution (achievement of goals based on results).
- Capability planning—improvement—verification/validation of capability characteristics and introduction of new capability characteristics for the first cycle.

Management cycles are tied to calendar cycles: year, quarter, month, week. Let's designate them as a sequence of natural numbers: 1, 2, $n$, $n + 1$. Cycle 1 goes with a shift by one step. At the $n$-th step, the characteristics of process capabilities obtained at the $n - 1$ step are used.

Cycle 2 in the $n$-th step carries out work on improving processes and products through planned projects and initiatives on improvements in Lean, 6 Sigma and Kaizen. These improvements must be verified, and new characteristics must be measured, taking into account their variability and uncertainty. The values of the characteristics must be entered into the documentation and information system and used further in the next management cycle (see Figure 12.1).

Points (1–5) in Figure 12.1 are steps of standardization and verification of decisions and reference data changes and the introduction of new standards.

Changes made to the characteristics of physical processes and products are used to change planning standards. Planning standards are de facto characteristics of models (parametric mathematical models) used to calculate the necessary resources and process capabilities based on the target values of the expected results.

Planning is the conversion of raw information coming from stakeholders and information coming from content and context into necessary resources and implementation processes, with appropriate capabilities.

Currently, this is possible using various information systems, from Excel to AI, but the author suggests using mathematical parametric models and linear models for iterative planning systems that plan for relatively short intervals.

Master data management (MDM) systems in which process speeds are defined by the physical capabilities of machines and people are the determining ones, not the speeds of information technologies and AI (Allen and Cervo, 2015). The speed of the latter is needed to process huge amounts of data, but the speed of their use is limited to the first. In particular, the speed of improvement processes—the second important aspect of planning—is the factor of people's responsibility for decisions

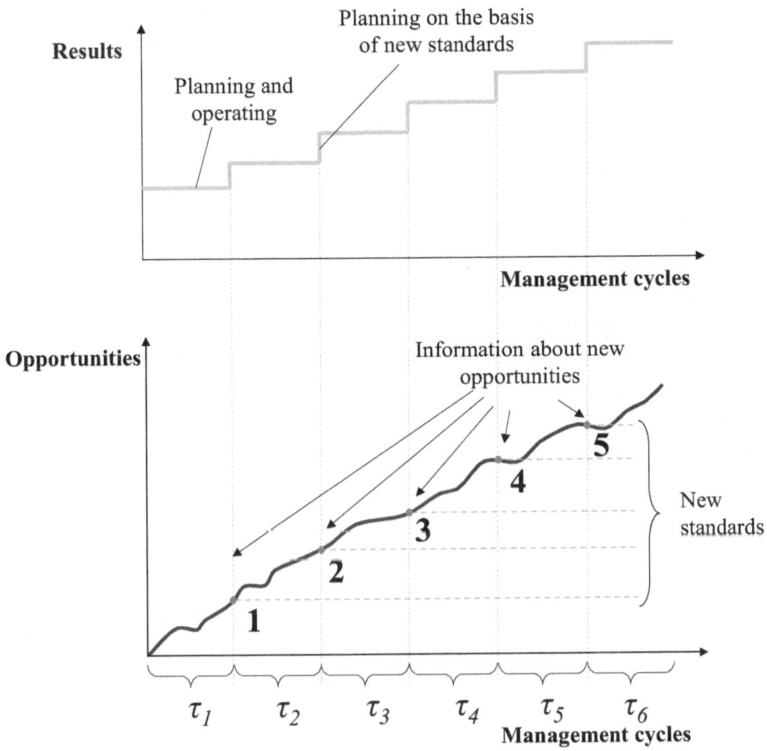


**FIGURE 12.1** Cyclic model of improvement of results and opportunities.

made. You can pass planning functions to AI, but it cannot be responsible for its decisions. Responsibility is the prerogative of people.

The use of parametric models, the parameters of which become characteristics of processes, products and services, allow them to be used as standards and build a self-learning system based on "mutations" (improvements, innovations) and signals that the system can "hear" from outside as well as from inside.

### 12.3.2 Signals, Feedback, Communication, Data

In a DM model, both the production system and the management system are considered the organization's trained network with signals coming from content and context.

These systems are considered signals from "teachers"; that is, all stakeholders are teachers who give the signals on the basis of which the system is trained; that is, it changes its algorithms in order to increase its efficiency and adaptability. Any sign, symbol or message that gives impetus for improvements is called a signal. It can be light, sound, body, taste signals or odors, that is, signals perceived by any sense. These can be messages and information. The essence of signals is to start internal processes in the organization and its systems but not to organize the processes and systems themselves. This role is performed by system members.

The term "signal", which originated in cybernetics and computer science, is poorly used in organizational systems. For example, feedback signals are often suppressed and/or ignored because of personal hazards.

A DM model includes special measures to protect signals, including feedback, but also others, for example, coming from competitors or society. This problem is due to the traditional organization of business, where order of the management dominates, that is, the system of one-man management. The order was accompanied by enforcement and quite often caused contradictions with the interests of consumers or the capabilities of personnel, but all signals from them were suppressed in order not to harm the principal one-man management. Very important signals come not only to bosses but to any employee at any level of the hierarchy, including people who do not have the authority to organize appropriate actions.

A DM system provides for two measures for detection, preservation and dispatching of signals. The first is to grant the right to act and initiate processes of improvement, correction and stop processes to ordinary employees who have received signals from any source. This is used in many systems, starting with TPS (Toyota), but in DM it is a fundamental part of the management system (Monden, 1993).

The second measure is the protection of signals from their suppression, distortion, untimely processing. This is particularly important for feedback signals. The main thing is to make these signals public, accessible to any participant in the system, as well as provide protection measures for signal messengers carrying "bad news", protect them from risks and penalties.

The "Diamond" concept is noteworthy. It proposes measures to protect workers' risks from damage caused by incorrect decisions (Tiuleneva and Kastorskaya, 2018).

DM proposes an insurance system for risks of damage to workers, the creation of insurance funds, which can be internal investment funds. Risk insurance in budgeting is an example of such a measure. In general, DM provides procedures for identifying signals aimed at possible improvements, including reducing the risks of loss, processing, preservation, publicity and monitoring of processes and projects launched by signals.

## 12.4 THREE FORMS OF DUALITY

*Brief: duality is considered, with three aspects and three horizons. The first aspect of DM is goal setting, which assumes that management sets two types of goals: to achieve certain financial results, for example, in terms of revenue and profit, as well as in terms of efficiency as a coefficient in the form of ratio of revenue to costs. The second aspect considers the issues of planning and execution as a mutually agreed-upon activity, and the third aspect considers the issues of investment in business growth, distributing investments between investments aimed at business growth and investments in business efficiency, in a certain sense, in business quality. Such a distribution of investments is very effective and gives at least two results: the overall efficiency of investments increases, and it becomes possible through investments in the efficiency ratio to control the growth rate, that is, growth dynamics.*

Looking at the business system as a money-making machine, there are three forms of duality:

## 1. Duality at the level of business goals

a. Profit maximization:

$$\Delta F = (F^{out} - F^{in}) \rightarrow max, \tag{1}$$

where $F^{in}$ —incoming money flow, $F^{out}$—outgoing flow, $\Delta F$—flow growth—profit.

b. Efficiency coefficient $R$ (as organization's efficiency measure) maximization:

$$R = \frac{F^{out}}{F^{in}} \rightarrow max, \tag{2}$$

where $R$ is considered a characteristic of the business system, reflecting its ability to create increased flows of money.

$R = \frac{F^{out}}{F^{in}}$ is a measure of efficiency. The organization's managers see it as an indicator of performance index, which they consider one of the management goals. In this form, $R$ affects the goal expressed in (1): it is given meaning. On the other hand, $R$ gives meaning to the characteristics of the internal structure of organization; it is used for the purpose of improving the business model of organization and is determined through the characteristics of business processes, products, resources and management systems.

To increase the $R$ value as a characteristic of a business organization, models are required that reveal its relationships with characteristics of business system elements.

## 2. Duality in operation planning and improvement level

a. Planning and operation.

*Purpose:* planning, execution of production, sales and profit plans.

b. Increased profitability ratio.

Two parallel cycles in time: planning—operation and improvement—information exchange. Information about all non-conformities and new opportunities is transmitted to the profitability ratio control unit, within which new standards for the parameters of the planning and execution system are determined based on this information and information obtained as a result of improvement projects (see Figure 12.2).

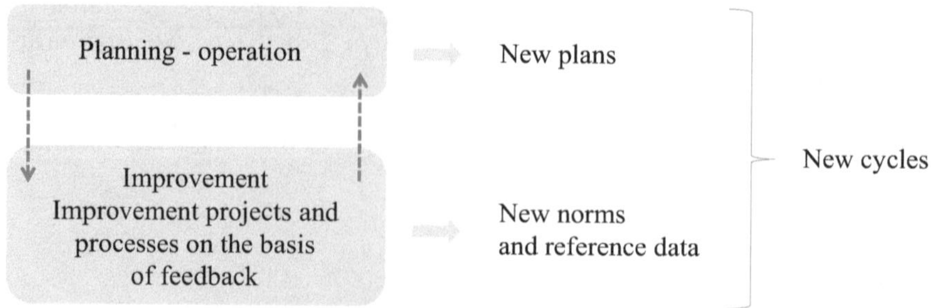


**FIGURE 12.2** Connection of the steps of two cycles: Plan–Do and Improvement.

### 3. Duality at investment level

Two types of investments:

a. **Money investment**—investments in the development of production and sales (these are ordinary investments in company growth).
b. **Smart investment**—smart investment to improve efficiency, including the profitability of money investment.

The first type of investment (money investment) has been worked out very well in a century (tracing the path of money "from the cash register to the cash register" through dozens of floors, thousands of people and millions of operations is a very difficult task).

The second type of investment (smart investment) is a new direction, in which there are also difficulties in bringing money to results, but it's somewhat easier, since it follows the trajectory of money investment.

Two business management cycles:

1. planning and execution of product manufacturing processes;
2. improvement of processes, products and systems; creating standards and new reference data for the next planning cycle.

1. **Plan and Do**

   *Purpose*: Planning, execution of production, sales and profit plans.
2. **Improvement**

   *Purpose:* Increasing profitability ratio due to reduction of unit costs:

   a. increase quality and technological productivity;
   b. increase product sales ("product at average price and for average customer flow") (see Figure 12.3).

The $\{X_n\}$, $\{Y_{n+1}\}$, $\{Z_{n+1}\}$ and $\{W_{n+1}\}$ sets are part of the regulatory reference systems (MDM systems) used in planning. The models that connect these sets,

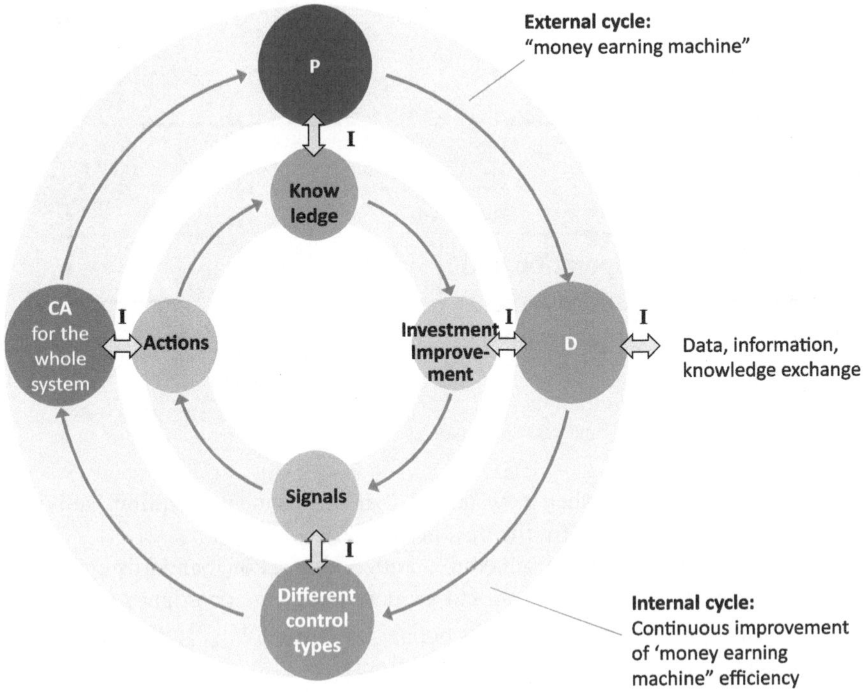


**FIGURE 12.3** Investment layer: two business management cycles.

including mathematical models, are part of the organization's knowledge system, which is self-learning and adaptable.

1. $\{X_n\}$—set of characteristics of processes, products and systems, including all modified ones, achieved at the $n$-th interval;
2. $\{Y_{n+1}\}$—set of characteristics of processes, products and systems installed in the $n + 1$ period as parameters and reference data;
3. $\{Z_{n+1}\}$—set of planning parameters established on the basis of $\{Y_{n+1}\}$ using planning methods as planning norms;
4. $\{W_{n+1}\}$—set of parameters of accruals for labor, for example, the standard of the time wage rate, bonus ratios and deductions for improvement programs for profit.

Figure 12.4 demonstrates model of dual cyclic business improvements, where, at each management cycle $n$:

1. Based on the results of improvement projects, actual $\{X_n\}$ characteristics of opportunities, quality and others are changed, checked and verified.
2. Based on the forecast of the variability of the internal environment (content), new standards are established $\{Y_{n+1}\}$ for the next period $n + 1$.

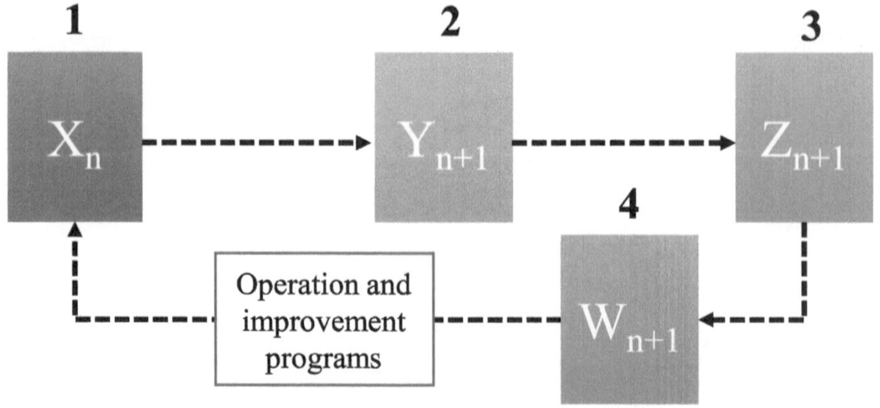


**FIGURE 12.4** Dual cyclic business improvement model.

3. Based on the established new norms that are used in planning, activity planning is carried out for the next period $\{Z_n +_1\}$.
4. New plans of labor and remuneration and remuneration standards creation; entails a change in the payroll and average salary in accordance with new business opportunities, as well as bonus standards $\left\{W_{n+1}\right\}$, deductions for projects, improvements and profit.
5. An extremely important point is to capture product/process improvements. After fixing the new/achieved parameters, it is necessary to verify and validate them.

The next example illustrates how the transformation system works:

$$\left\{X_n\right\} \rightarrow \left\{Y_{n+1}\right\} \rightarrow \left\{Z_{n+1}\right\} \rightarrow \left\{W_{n+1}\right\} \tag{3}$$

During the $n$-th calendar cycle, the organization carried out projects to improve processes and their maintenance and achieved a decrease in the level of inconsistencies from 5% to 2%; that is, the $X_n$ decreased from 5% to 2%. As a result of verification and prediction of process variability values for $n + 1$ period, it was decided to set the non-conformance level standard for the $n + 1$ period as $Y_{n+1}$ in the amount 2.5%. Based on the new standard $Y_{n+1} = 2.5\%$, the finance department reduced the loss rate in the form of "write-offs" for product scrap and repair by 2 times, from 8% to 4% of the cost price; as a result, 4% of the cost will be rescued and can be allocated for improvement investments, an increase in wages and in profits, including dividends for shareholders. The amount of money allocated for the above-mentioned items is denoted as $\left\{W_{n+1}\right\}$.

In fact, DM allows one to build a business system as a dynamic control system with two positive feedback loops.

The model in Figure 12.5 describes the actions of the business system without taking into account the distribution of profits between stakeholders, which reflects the actions on short cycles of capital turnover without profit recording.

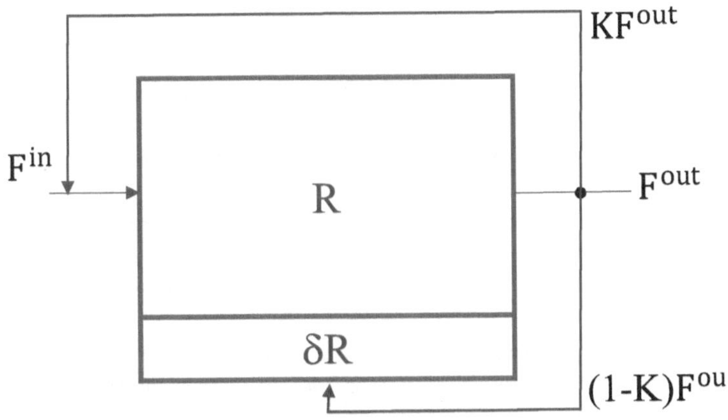


**FIGURE 12.5** Business system model with two positive feedback loops.

**Here, $\delta R = \Delta R = R_n - R_{n-1}$, and $K$ is the coefficient of distribution of reinvestment flows (cycle numbers ($n$) are omitted for simplification).**

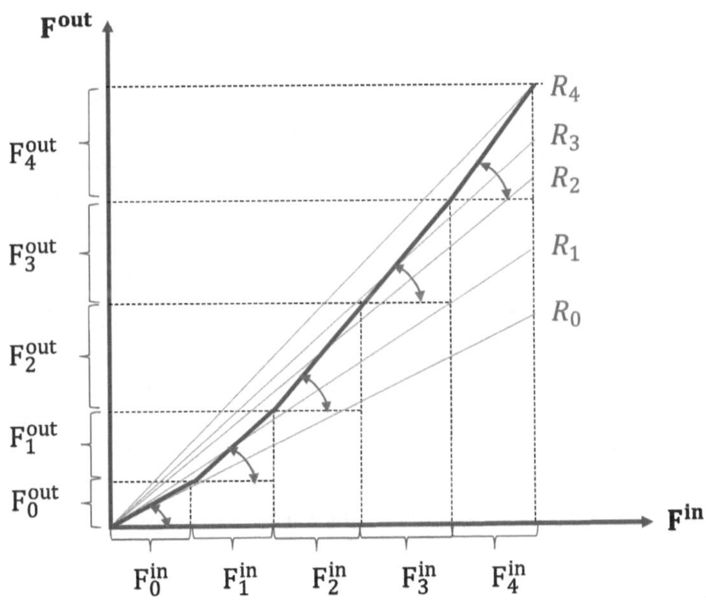


**FIGURE 12.6** Growth due to investment in improvement (growth of $F_{out}$ via dual investment into improvement and volume growth; increase of $F_{in}$).

As a result, there is significant growth of $F^{out}$ due to dual investment in improvements and $F^{in}$ volume growth, as shown in Figure 12.6.

Figure 12.6 shows how significantly $R$ coefficient growth can influence $F^{out}$ income growth.

## 12.5 STRATEGIES OF DUAL MANAGEMENT

*Brief: The organization's business is smart money-making machines. On the one hand, these are organizations that are able to ensure stable quality and strictly fulfill all obligations that have high reproducibility of processes, that is, the properties of machines with specified characteristics. On the other hand, these are organizations with the ability to learn and adapt to changes in the environment. They must combine characteristics sometimes perceived as contradictory, for example, strict discipline and a high level of creative activity. This is already being demonstrated by many companies.*

*In the medium term, organizations learn dynamism as a quick response to sudden changes in the environment and processes in order to maintain stability and continuity.*

*In the long-term horizon, organizations learn to predict and manage processes, systems, products and personnel, taking into account their life cycles, the physical and moral aging of all components and the processes of changing their generations.*

*These horizons require various kinds of investment programs. DM models allow you to evaluate these processes, taking into account performance indicators, and obtain recommendations for planning relevant investment programs.*

*From the point of view of training, improvement and adaptation, there are three horizons: short-term, medium-term and long-term, and they highlight aspects that distinguish strategies for these horizons. In the short-term horizon, knowledge is drawn from current processes and environmental characteristics. DM looks for hidden opportunities and implements changes in processes and systems.*

The assumption is that every business is created to make money. This is a goal that brings together all the stakeholders who have come together to increase their income or earn money for their current needs. Organizations operating as serial producers of certain values, using process approaches, standardization, methods and tools that ensure the reproducibility of processes with the same source data, resemble money-making machines. This metaphor was used by many authors; perhaps it was expressed better than others by E. Goldrat (Goldratt and Cox, 1992).

The close model of a smart money machine (SmMM) is used. It is important that managers use standardized, robust processes to ensure that inbound plans and requirements are met in the form of appropriate results with a low level of random variation.

Only organizations with such efficiency in the medium and long term can exist in a market economy.

Efficiency is seen as a concept for the short, medium and long term. Organizations can be effective in the short term, but the loss of efficiency in the medium and long term requires special consideration.

### 12.5.1 Organizations Can Be Effective Under Several Necessary Conditions

1. Organizations have knowledge, capabilities, technologies and people with relevant competencies that allow them to be effective at the moment.
2. Organizations have certain resources, abilities and knowledge to respond quickly to changes in the external and internal environment. This is dynamism, the ability to accelerate, slow down or change directions.
3. Organizations monitor the balance of interests of stakeholders and have the knowledge and resources to ensure the sustainability of these balances, for example, the sustainability of relationships between employees and owners or consumers and suppliers.
4. Organizations understand that their activities suffer many losses and have the will to identify them and the knowledge that can turn losses into new values and free resources.
5. Organizations are aware that they use the creative, innovative potential of employees and their intelligence only a little, which is a consequence of the choice in favor of a systematic approach.
6. Organizations are ready to create appropriate efficiency institutions based on their staff, seeing in it huge potential.
7. Organizations understand that losses are also untapped opportunities.

The three horizons for DM concept are improvement horizons and associated improvement planning cycles:

1. In the short term, with reliable forecasting of process context and content, planning relies on existing knowledge, including the characteristics of process capabilities. Opportunity improvement plans do not provide for changes in characteristics in the current cycle and are aimed at new changes in regulations and data describing opportunities in the next calendar periods. The changed target and reference data is used in the next cycle to plan new results.
2. In the medium term, it is necessary to foresight risks and plan appropriate management responses. This includes planning tasks, including planning resources to respond to risk, that is, creating risk funds and risk resources that allow you to quickly respond to emerging hazards, such as pandemics, natural disasters and changes in market conditions.
3. In the long-term strategic intervals of change, which are perceived as generational changes within the life cycles of products, technologies, knowledge, personnel, consumers, suppliers and investors, special attention is paid to maintaining the cyclical nature of the business, renewable resources and environmental conservation.

These are cycles of context, including global: climate change; ecology; and transition to the phase of irreproducibility and exhaustion of resources, such as oil, coal and metals. This is actually the ESG agenda, included in the general model of DM. To do this, the concepts of life cycles, including resources, are used as well as the associated concepts of physical and moral aging, wear and degradation.

The loss of moral aging is interpreted as the expected economic loss of future periods due to inaction or improper actions during current periods, which allow the use of a single approach to improvements: the identification, determination and calculation of *current* and *future* losses and their inclusion in the indices of economic efficiency introduced in DM. These indices create a single methodological basis for managing performance in the short-, medium- and long-term horizons.

Obviously, the actions of regulators and investors also have an impact on the long-term strategy, turning long-term goals, including those related to ESG, into current goals through restrictions and privileges, making the business more sensitive to future losses.

Regulators, through restrictions, norms and sanctions, create risks of current losses that prevent significantly greater risks of future losses.

In accordance with the identified horizons, the following cycles are considered (see Figure 12.7):

- Management and improvement cycles within short calendar periods that use investment to improve opportunities.
- Cycles of sustainability and continuity in the medium term, using investment in sustainability against the emergence of risky and unforeseen events and changes.
- Cycles of changing generations of products, processes and resources, as well as personnel and consumers, using investments in the development of new generations of all elements of business systems as a strategy for sustainable development.

Within each horizon, certain investments are made, differing in their orientation.

**Horizon I.** Cycle within the framework of regular calendar management. Investing in improvements to increase efficiency and business continuity at the moment (within the calendar cycle) to manage efficiency growth to increase agility, speed, acceleration, performance growth.

**Horizon II.** Cycles within the framework of maintaining sustainable efficiency under strong external influences and internal unforeseen problems. It is carried out through investments in proactive improvement projects (as preparation and response to possible and emerging impacts of external

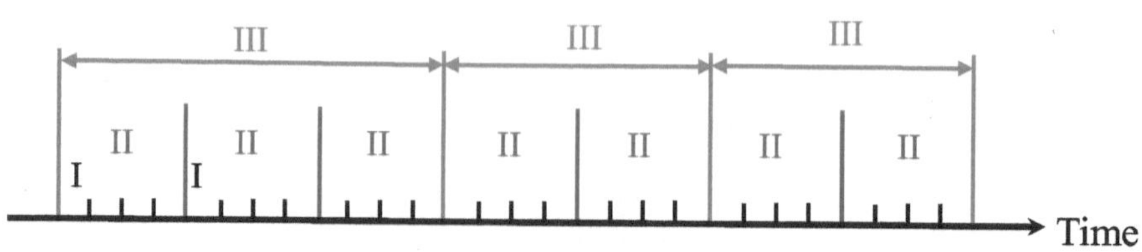


**FIGURE 12.7** Connection of management cycles with three horizons.

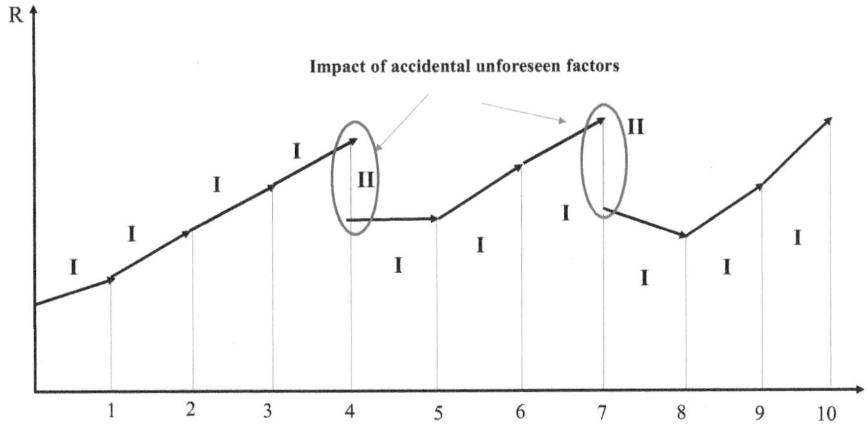


**FIGURE 12.8** Illustration of efficiency coefficients within Horizons I and II—Sustained dynamic growth in Horizon I, fall and recovery in Horizon II.

factors: the rise and fall of the situation, compensation for tax, regulatory impacts, pandemics, natural disasters, etc.) (see Figure 12.8).

Within the framework of cycles in Horizon 2, tasks and business continuity are solved.

**Horizon III.** Horizon III cycles are focused on generation processes related to life cycles, physical and moral aging of all elements of the business system: products, processes, technologies, resources and stakeholders, as well as shifts and life paradigms of values of new generations. Horizon III also considers ESG agenda issues that can be viewed from the perspective of generation change and business components, including its resource base, preparing a decision on the transition to renewable resources and environmentally friendly products and processes.

Within the framework of these cycles, management uses the idea of depreciation adopted in the life cycles of products, technology and education, including certain deductions from income received from existing generations of all types and directing them to the preparation of new generations. This practice is well known when upgrading and changing generations of products and technologies. However, it can be accepted as universal and is applied in the transition from non-renewable resources to renewable, from environmentally hazardous technologies to safe, to the preparation of new generations of employees, including special joint training programs in schools, colleges and universities. The DM concept extends organically to this type of investment, and considering future losses due to current inactivity, a unified model can be built.

The universal duality for all three horizons of the strategy means that by adopting the strategy of effective sustainable growth and development as the main, management strives to be effective in each individual management interval, as well as on medium- and long-term trajectories. Management also accepts that the company will

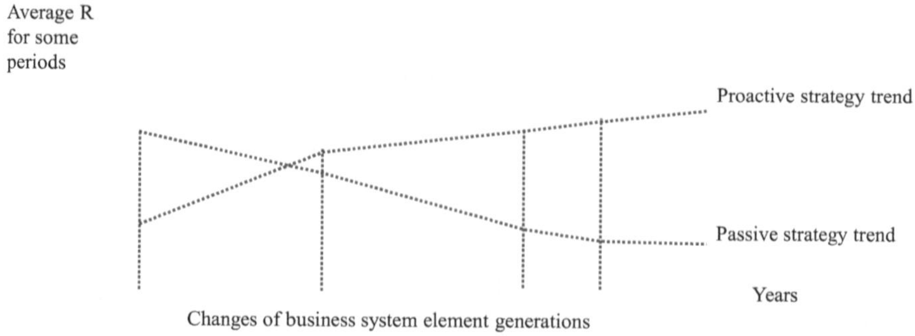


**FIGURE 12.9** Illustration of efficiency coefficient fall as a result of processes occurring in Horizon III.

be in constantly changing conditions, sometimes getting “blows” that can “knock it off the course”.

In the DM concept, management focuses on feedback signals and their precursors in order to constantly adjust and adapt strategy and tactics.

As a *universal feedback signal*, DM uses signals about past, current and future losses. It defines the latter as losses caused by inaction and/or omission of opportunities, including not paying attention to new technological trends and challenges of the ESG agenda. Taking into account the forecasts and estimates of the risks of losses in the future, management begins to act on a warning, while in addition to intuition, it has information and statistics related to the development of trends translated into future efficiency criteria.

The concept of efficiency includes aspects of continuity and sustainability in medium- and long-term dynamics. The latter will largely be determined by the ESG agenda (ESG agenda and the 17 goals).

Nevertheless, efficiency in any perspective requires dynamism, that is, changes in strategies, plans, procedures and technologies, which in turn require additional forces to accelerate, slow down or change the course of business. This means having and attracting some additional resources (see Figure 12.9).

## 12.6 PROCESS MODELS

*Brief: Modeling of processes and systems is the basic element of DM methodology. It is shown in the PDCA Deming-Shewhart cycle that planning is carried out on the basis of certain models of working with information, while planning uses models of physical execution processes but considers them in the opposite direction, from the desired output to the required input. In addition to execution process models, the planning function also uses stakeholder interest transformation models, including the needs of consumers, for the purpose of planning.*

Planning as a management tool converts information about the desired results at the output of processes into information about the necessary resources at the input of processes. From a mathematical point of view, planning solves the opposite problem—determining the optimal variables (at the input) to obtain a given value of the function (at the output).

When implementing the execution step (Do), the physical transformation of materials, energy, labor and the specified corresponding inputs of variable values into material information objects and the corresponding response values of the function at the output takes place.

Typically, planning departments use some of the mathematical models embedded in planning techniques to perform such calculations, using model parameters as reference data extracted from the MDM.

In DM, *step-by-step iterative procedures* are used, where the changes in the initial quantities at each step are small compared to the input and output quantities, which make the linear approximation of any function a universal approach, and information about past knowledge A is used to refine the definition of A for the current step.

Two initial assumptions are accepted in DM:

1. Model A parameters are not exactly known, and the error of measurement can be quite large (for example, there are cases when reference data from a decade ago was used as such values, although organizations have changed very significantly over a year);
2. Model A parameters can be changed under the influence of processes and improvement projects, that change physical process parameters.

In other words, DM proposes to use parametric models, the parameters of which are measured during control process and changed purposefully by improvement actions. Wherever possible, use linearization and make the parameters of these models the parameters of MDM systems.

When building process models, the black box (BB) approach is used, developed in cybernetics (see Figure 12.10) (Wiener, 1961). In fact, this is a basic approach for the process description and modeling of the activities of organizations.

The parameters of such models can be used in business planning or as the basis of MDM, as well as models of income distribution, including income obtained as a result of improvements.

Next, the "ZOOM" method is applied, in which the black box process is sequentially broken down into a sequence of other black boxes.

For example, as shown in Figure 12.11.

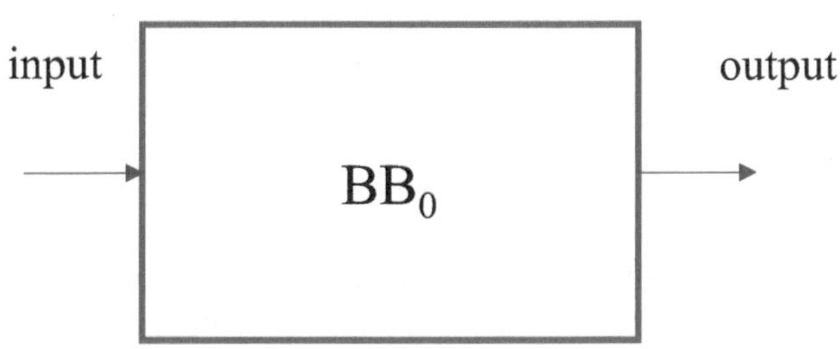


**FIGURE 12.10** Black box model.

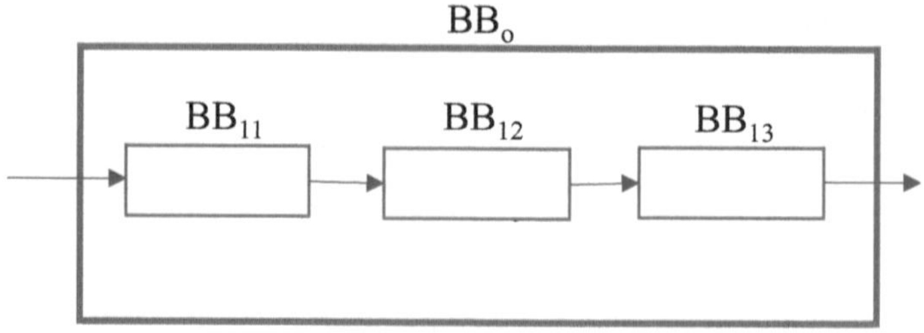


**FIGURE 12.11** Disclosed BB$_0$ model with BB levels.

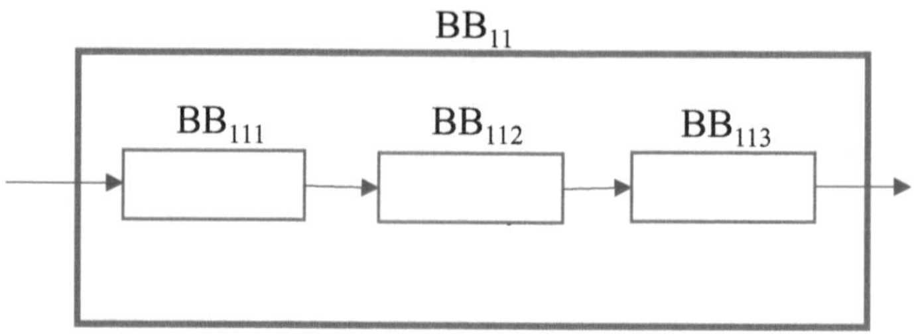


**FIGURE 12.12** Disclosed BB$_{11}$ model with lower levels.

And then process model is broken down into each BB$_{11}$, BB$_{12}$, BB$_{13}$ (see Figure 12.12). Such modeling allows you to reflect organizational structures based on reflecting the transformations that occur in the business processes described previously. The degree of detail is specified in certain cases.

Parameters of models reflect the real properties of physical systems (technological systems, equipment, machines) using information systems that allow organizing a cycle of self-study and targeted improvement while maintaining the responsibility and motivation of the personnel involved in improving. Responsibility for the reliability and accuracy of data reflecting the results of improvements at each step and the motivation to initiate and implement improvement projects should be established within the framework of the improvement management system.

The most important issue in management is how closely the process model reflects the real physical process and its changes. It is important that the elements of plans do not work with reality. They work with its reflection. You can, for example, improve real processes and products as you like, but if these actions are not reflected in the parameters of the models, and sometimes in their structure, none of the planners will know how good they have become and will plan business processes based on the established parameters in the form of standards and reference data. Accordingly, budgets will reflect old regulations and thereby generate losses of money, and the new actual results obtained will be perceived as happy accidents and not reflect people's efforts to improve. DM creates a special change management system, the task of which is to ensure the synchronicity, accuracy, reliability and relevance of changes in physical processes, products, systems and parameters of control models in the form of MDM.

## 12.7 KEY INDICATORS OF DUAL MANAGEMENT

*Brief: Indicators are built based on the modeling of transformation processes occurring in business systems. They allow managing programs and processes for improving the effectiveness of organizations.*

*The section presents performance models and performance improvement dynamics that allow expressing performance indicators through generally accepted performance and value indicators. Modeling uses a BB transformation process description approach. In this work, the example is limited by models of two levels, but the approach allows models at deeper and more detailed levels. It is important to emphasize the methodological aspect: models use aggregated indicators that reflect the financial and economic component of organizations' activities. Models reflecting technical transformations are not considered in this work. However, their consideration is possible using the same BB methodology.*

DM considers revenue, profit, efficiency ratios and indicators of their dynamics the main parameters. At the same time, efficiency factors are considered opportunities/business potential, as machines for increasing money flows. The internal structure of such SmMM is considered when moving to control tasks at deeper levels.

DM does not consider the goals and corresponding indicators related to the growth of companies' value, since this is more a matter of governance but not management. It's considered that management can influence those goals and indicators using those assets that are transferred to it for operation.

This model is considered a black box model (see Figure 12.13), with open and connected feedback. In this section, it is considered a system of three blocks that perform three processes in sequence:

1. the transformation of the input flow of money $F^{in}$ into the resources necessary for further transformation, where the main resources are the number of people endowed with the corresponding resources $N$, having a cost $C$ per person;
2. transformation of resources through work processes into products, where the total number of products produced during the management cycle equal to $\tau$ is $M$;
3. transformation of $M$ products through sales to customers at average price Z.

Thus, a chain is formed:

$$F^{in} = C \cdot N, -M = P_T N, -F_{out} = Z \cdot M \tag{5}$$

where $P_T$—technical productivity in pieces per person during cycle time, based on this chain, getting the expression for $R$

$$R = \frac{F^{out}}{F^{in}} = \frac{Z}{C} \cdot P_T \tag{6}$$

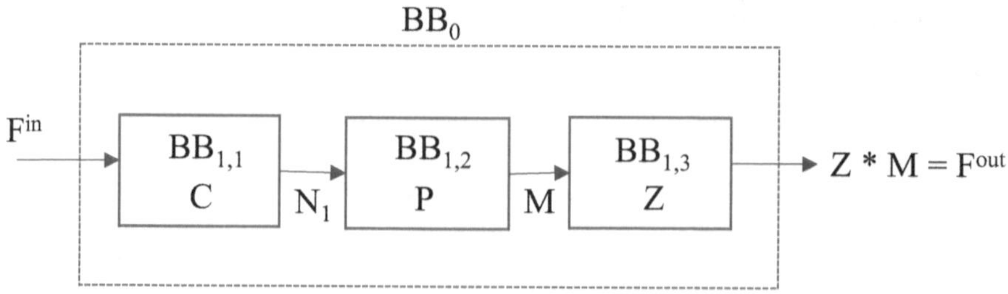


**FIGURE 12.13** Level B black box model that drops down through three next level B models, where BB1,1—planning process model; BB1,2—production system model; BB1,3—sales system model.

where $Z, C, P_T$ are characteristics of transformation blocks $BB_{11}$, $BB_{12}$ and $BB_{13}$, and $R$ is a characteristic of common $BB_0$. Formula (6) can also be represented as ratio of labor productivity by revenue to cost:

$$R = \frac{P_F}{C} \tag{7}$$

where $P_F = Z * P_T$,

$$P_F \text{ — output cash productivity (revenue).}$$

Next, the increment rate $R$ is defined as the increment of $\Delta R$ during $\tau$ as a result of the improvement processes and defined as the ratio of the increments of the output stream $\Delta F^{out}$ to the increment of the input stream $\Delta F^{in}$.

$$\Delta R = \frac{\Delta F^{out}}{\Delta F^{in}}, \text{ illustrated in Figure12.14.}$$

Omitting simple transformations, we give the next formula for

$$\Delta R = R \cdot \frac{\delta M}{\delta N} \cdot \frac{1 + \frac{\delta Z}{\delta M}}{1 + \frac{\delta C}{\delta N}}, \tag{8}$$

where $\delta M = \frac{\Delta M}{M}$—relative increment $M$—number of products;

$\delta N = \frac{\Delta N}{N}$—relative increment $N$—number of people;

$\delta C = \frac{\Delta C}{C}$—relative increment $C$—cost of all resources per person;

$\delta Z = \frac{\Delta Z}{Z}$—relative increment $Z$—average cost.

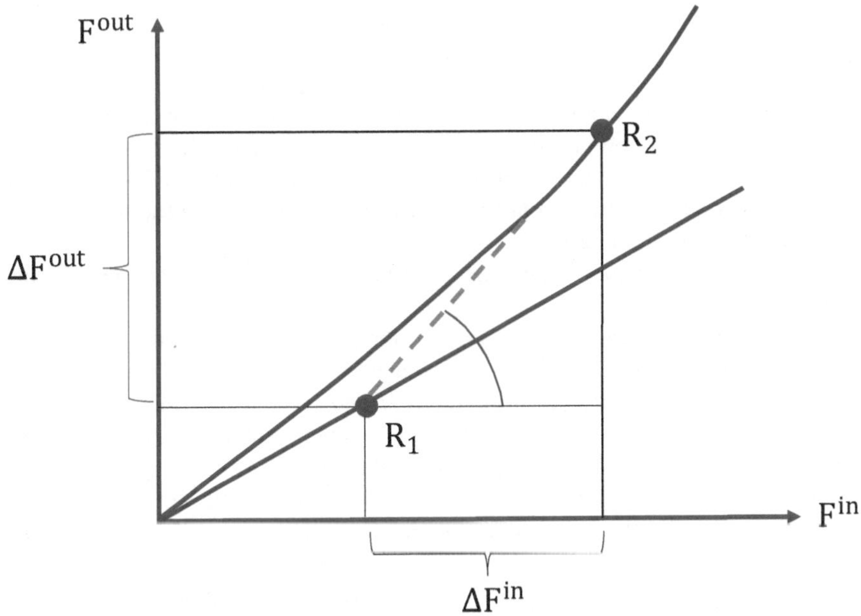


**FIGURE 12.14** Growth of $R$ coefficient due to improvements.

Let's introduce parameters $\beta = \dfrac{\delta M}{\delta N}$, $\gamma = \dfrac{\delta Z}{\delta M}$ and $\alpha = \dfrac{\delta C}{\delta N}$, where $\beta$, $\gamma$ and $\alpha$ are coefficients of the type elasticity coefficients and are determined by a joint response to execution and improvements.

Taking into account the accepted designations, we obtain the following expressions:

$$\Delta R = R \cdot \beta \cdot \frac{1+\gamma}{1+\alpha} \tag{9}$$

$$\delta R = \beta \cdot \frac{1+\gamma}{1+\alpha} \tag{10}$$

$\beta$, $\gamma$ and $\alpha$ have an economic sense.

$\beta$ shows the dynamics of the increase in technical labor productivity as a result of improvements.

$\gamma$ shows the sensitivity of the market price to increased product supply, taking into account the improvement in product quality.

$\Delta R$ is a parameter of efficiency growth in the interval of planning and execution and is a characteristic of efficiency management dynamism.

$\alpha$ shows how the cost of resources per person leads with an increase in numbers; note that $\alpha$ is usually a negative value.

In general, the formula is $\Delta R = R \cdot \psi$, where efficiency coefficient growth factor $\psi = \beta \cdot \frac{(1+\gamma)}{(1+\alpha)}$. It is an indicator of the implementation of the improvement program, which shows the potential for growth in connection with the improvements achieved.

Note: In the derivation of the formula, division by $\delta N$ and $\delta M$ is used, assuming that they are greater than zero, that is, $\delta N > 0$, $\delta M > 0$, as well as $\delta C + \delta N > 0$. An example of a model description at the level $BB_{1,2}$ is presented in the work with my colleagues (Lapidus et al., 2017).

Based on the indicators $R$ and $\Delta R$, a strategic dual performance map is built, shown in Figure 12.15.

The following diagrams are proposed for the purposes of analysis and regulation of efficiency investments. The blue zone is a zone that signals the need to increase investment in efficiency growth in horizons II and III. High $R$ values, achieved on the regular horizon, allow one to increase investments without the risk of their losses in the short-term horizon, including making $R$ estimates for subsequent periods in particular, assessing the reduction in output values, increasing the cost of negative outputs (for example, environmental), increasing costs for non-renewable resources and increasing losses.

DM implements a business system in the form of a dynamic system with two (dual) positive feedback loops.

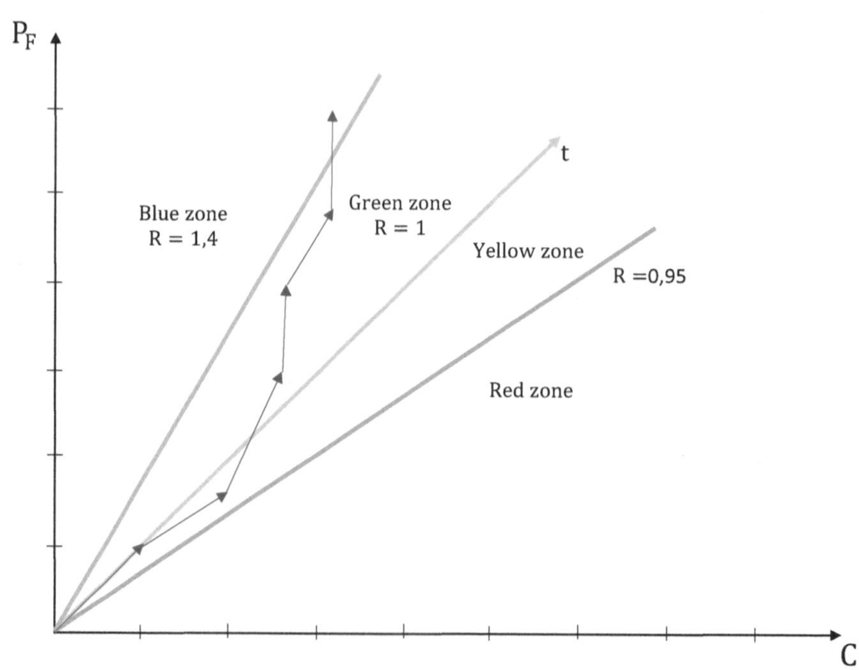


**FIGURE 12.15** Strategic dual control maps of DM.

## 12.8 RISKS, LOSSES AND ROLES OF DM

*Brief: The DM concept considers risks from the point of view of their insurance. Risks are always explicitly or non-explicitly insured, either individually through personal stocks and caution or collectively through insurance funds and personal risk protection through collective insurance. Collective system insurance is more efficient and fair.*

In DM, the essence of the risk insurance approach is that employees are exempted from liability for damage caused to the organization by non-conformities and failures.

As a rule, such damage takes place in organizations under the item write-off or write-off of inevitable losses. However, in fact, this is a recording of losses, which is implicitly distributed among all employees, making them non-responsible both personally and collectively.

The concept of improvement allows changing this approach and considers the debited funds insurance payments for the damage caused to the organization. It is reasonable to initially agree that the damage from non-compliance will be paid off collectively, but in fact it will be part of the salaries and remuneration that employees could receive but did not due to non-compliance or refusals. Damage amounts are generally well predicted and planned with a certain margin of variation for upcoming management periods.

The organization represented by managers offers the following motivational system. Employees are invited to actively participate in improvement processes, identifying the causes of non-conformities, including systemic ones, and offering new technical and organizational solutions to reduce the risks of non-conformities and failures.

Funds that were actually considered funds planned to write off damage become insurance funds and invite employees to direct efforts to general damage reduction. The reduction in write-off costs resulting from improvement processes and projects should be considered a bonus for all participants in this system and distributed between employees, management and owners, respectively, directing the "saved" funds to three areas:

1. remuneration fund and bonus fund;
2. internal investment fund, which can be managed by managers, providing budgets for improvement projects;
3. profit of the organization.

This approach allows you to include not only losses associated with nonconformities but also any other improvements that can be aimed at reducing all types of losses (Lapidus and Tiuleneva, 2018).

In general, determining losses based on comparison of applied solutions and possible alternatives, taking into account the costs of switching to new solutions, the creation of such non-conformance damage insurance funds and, de facto, missed opportunities, creates an approach to improving performance and opportunities across the spectrum of activities, including planning, whose errors are sometimes much more expensive than execution errors.

This approach also allows one to propose uniform approaches to administrative and cultural regulatory frameworks on the basis of various kinds of treaties, that is, to propose common institutional approaches.

As part of a Taylorism-based management approach, administrative regulation is built around the perception of the plan as an order and the obligation to execute it, using enforcement control and enforcement in accordance with the plan in the form of penalties for failure to comply with orders and their requirements. This approach does not work in organizing improvement processes; another approach is needed. It is important that not only signals, figuratively speaking, wrapped in a mandatory order, trigger certain actions, but also that feedback signals arising in a variety of processes within and outside the company trigger actions immediately from the level and place where they arose. The same should be said about signals coming from the context (consumers, competitors, regulators and a variety of sources, including the Internet and scientific publications).

The regulation function in the management systems of organizations is aimed at ensuring the balance sheets of the relationship of process owners, both among themselves and in the management system in terms of:

- responsibilities and powers;
- plans, requirements and interaction, including resources of different types;
- liability;
- the cost of tasks to rewards;
- rights and obligations;
- guilt and punishment;
- merits, achievements and rewards;
- individual, collective and systemic risks;
- complexity of tasks and competencies of employees.

The regulatory function of the organization also considers such issues of interaction between:

- internal suppliers and consumers;
- customers and performers;
- managers and subordinates;
- supervisors, auditors and owners of processes and products;
- experts and third parties and recipients of their services.

Note: In common sense, regulation establishes the rules of relations between employees and between them and the organization.

It is the efficiency, dynamism and stability of organizations that are especially sensitive to the quality of regulation. Regulation is a very important side of institutional approaches in economics and management. However, DM pays great attention to cultural regulation, based not only on requirements and norms but also on values, including trust and presumption, that is, a priori attitudes involving constructive behavior and initiatives of employees.

In particular, the presumption of integrity and responsibility is introduced, according to which all employees are considered honest, observing the laws, rules and orders of the organization, responsible for fulfilling the obligations and terms of any contracts.

This presumption significantly reduces the need to control people's behavior and transfers many issues to self-control and self-organization. As a result, this leads to a significant decrease in hierarchy layers in organizations, the number of which is largely determined by the tasks of controlling performers and their reporting.

As a result, the volume of administrative regulation is significantly reduced, and control tasks are shifted towards cultural regulation, while a minimum apparatus is required. It comes down to an analysis of people's actions in cases where the prerequisites of the presumption are violated: lies, crime or concealment of information.

A very important issue in DM is the problem of loss, suppression of feedback signals and desire of people to avoid responsibility in cases of failure. The control function, depending on what control principles are laid into the management system, can create multi-tiered hierarchical systems, suppress initiative and slow down processes or lead to the effective and rapid use of feedback chains, flat and simple organizational structures.

In general, the function of regulation in management includes principles, rules, norms and values of the behavior of subjects in the organization, which increase its efficiency and dynamism, as well as the stability of growth and development.

Controlling the symmetry of the parties in the Plan–Do–Check–Act relationship involves clarifying the role of the Check element. The regulation provides for checking the compliance of $\text{Check}_{in}$ plans (including requirements) with execution capabilities so that the plans comply with the current capabilities of processes, products and systems, while the capabilities are characterized by current regulatory and reference data. $\text{Check}_{in}$ also involves identifying risks in the execution of plans and taking them into account in resource planning. When the Do execution processes are completed, the Check role is to verify that the results meet the plans and requirements, which are designated as $\text{Check}_{out}$. The corrective actions of the $\text{Check}_{in}$ results, identified by $\text{Act}_{in}$, are manifested either in increasing capabilities by increasing resources and authority or adjusting the plans and requirements to align them with.

An example is action to align process requirements in the form of tolerances $\Delta$ with capabilities in $6\sigma$, where $\sigma$ is the standard deviation—a measure of variability. The industry uses the standard for the opportunity index $C_p$ as the value of the $\Delta/6\sigma$ ratio.

If the measured value of $\Delta/6\sigma$ is less than the standard set in the company or required by the customer, then either designers must increase $\Delta$ or technologists and manufacturers must reduce $\sigma$.

$\text{Check}_{in}$ requires $\Delta$ and $\sigma$ coordination between designers, technologists and production engineers. $\text{Check}_{out}$ output control requires other actions—this is the work to identify the causes of nonconformities and organize corrective actions, called feedback processes or short reverse processes in DM. If $\text{Act}_{in}$ is usually an activity with documentation, then $\text{Act}_{out}$ is an activity with physical, material processes and products. The $\text{Act}_{out}$ regulator's actions may include insurance for workers as discussed previously against damage caused by nonconformities.

Thus, the Plan–$Check_{in}$–$Act_{in}$–Do–$Check_{out}$–$Act_{out}$ cycle is a consideration of the requirements of the internal regulator in the organization, which not only identify inconsistencies in the Plan–Do–Check–Act chain but also implement actions that protect performers from two illegal risks: obtaining unfulfilled plans and requirements and the risks of assigning responsibility for nonconformities to the contractor.

The task of the regulator is to create procedures that protect the executor from risks not controlled by the executors.

## 12.9 PRACTICAL VERIFICATION OF DM CONCEPT

For several years the author and his colleagues (Centr "Prioritet") have been carrying out projects to increase labor productivity in about 50 companies in the region of Nizhny Novgorod (Russian Federation). These are relatively small companies with numbers ranging from hundreds to several thousand people and with a turnover of tens to hundreds of millions of dollars, and we have studied a number of other companies implementing lean approaches, including foreign companies (Shcherbinin et al., 2023).

Everywhere, the work center was shifted towards the use of lean tools and production systems. Several companies used the ideas of DM, but in a fragmentary way. DM methods are being tested on a group of three companies.

Based on the experience of this sample, the activities of these companies were analyzed in terms of DM indicators. The results of the analysis are shown in Figure 12.16.

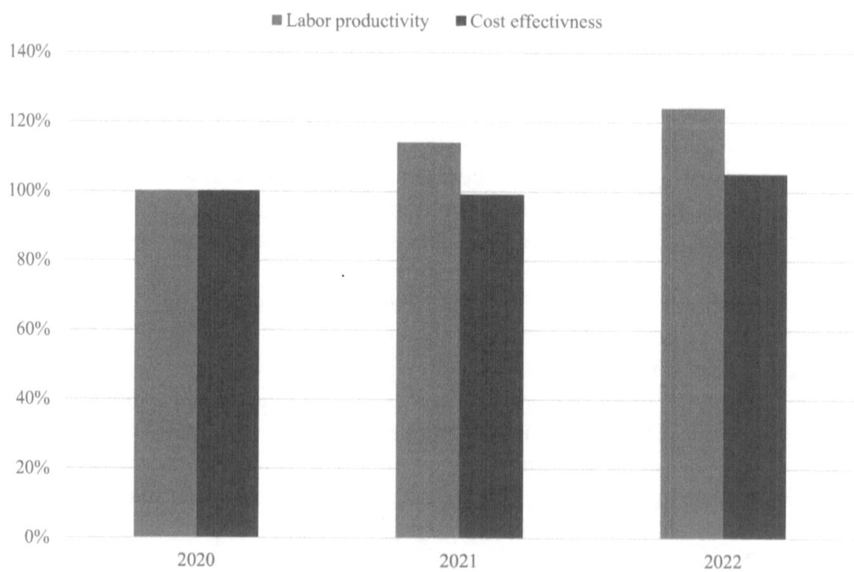


**FIGURE 12.16** Company dynamics in comparison to the base year 2020.

The conclusions are the following:

1. The application of approaches focused only on the growth of labor productivity using lean tools gives good results of increasing labor productivity but has little effect on efficiency in general. This is due to increasing costs and lower prices and demand. Companies focused on a more systematic approach, quality, performance and cost show a greater increase in efficiency.
2. DM tools, such as $R$ and $\delta R$ indicators and strategic dual maps make it possible to significantly improve efficiency growth strategies, as several companies changed their plans for staff growth and initially raised company's efficiency. This can be seen in Figure 12.16, which shows the trajectory of Sinergetic, which at the beginning of 2020 increased its headcount and expanded production, as a result of which the $R$ coefficient fell by 0.1. In the following year, the company implemented an efficiency program, as a result of which the number and costs were reduced, and at the same time the value was increased by $\delta R$ more than 0.2. The data in Figure 12.17 demonstrates the results of Russian companies implementing lean programs. Strategic maps of DM using benchmarking data, shown in Figure 12.17, significantly changed the views of company managers on efficiency growth strategies.

Figures 12.17 and 12.18 give an idea of the possibilities of this classical approach. All companies and their data are based on real information, but the results of each individual company are not the basis for their analysis. Information is taken from open sources for 2020, 2021 and 2022.

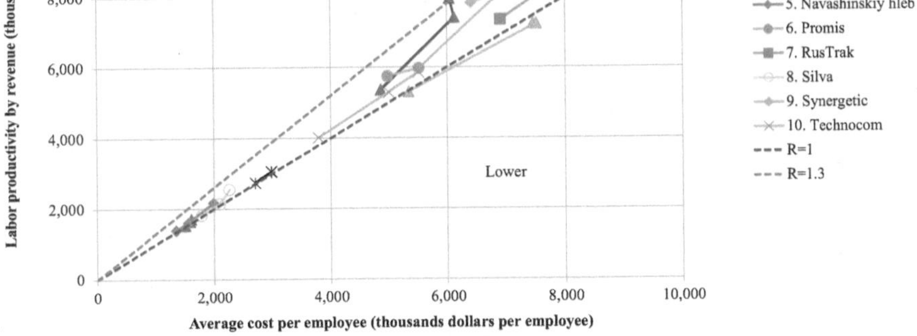


**FIGURE 12.17** Efficiency monitoring strategic chart (domestic companies of the Nizhegorodsky region, Russia, labor productivity growth project).

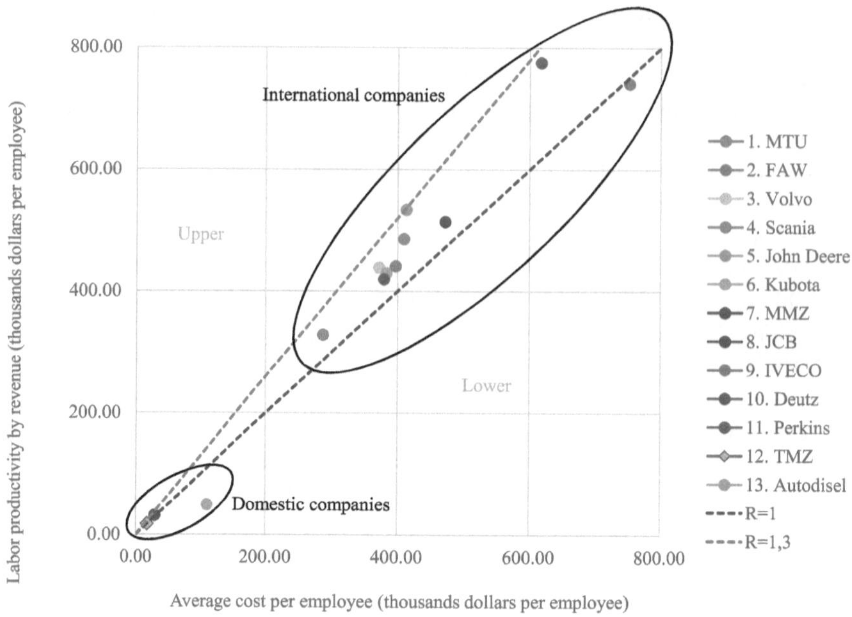


**FIGURE 12.18** Comparison of world leading car manufacturers and domestic companies.

## 12.10 CONCLUSIONS

*Brief: The chapter discusses the concept of dual management, which coordinates its activities both to achieve goals based on results and to increase its capabilities and, in general, to improve efficiency.*

This concept of the business system, as shown in the models and as a result of a number of practical studies, allows increasing the efficiency of companies both in the short and medium and long term to achieve dynamism and sustainability and pay attention to the fairness of relations and corporate culture values.

The concept proposes:

- dual management criteria and indicators;
- indicators for two types: to assess the efficiency of the business system and growth potential;
- a dual management approach and its principles;
- organizational structure of the dual management system;
- modification of the management cycle in the form of $Check_{in}$, $Act_{in}$, $Check_{out}$, $Act_{out}$;
- features of risk management and regulation;
- particularities of motivation and corporate culture.

In general, the DM concept allows building business system management as a dynamic system with two loops of positive connections, one of which is aimed at increasing cash flow and the second of which is aimed at increasing the efficiency of the business system itself as a machine that strengthens the input financial flow. Within the framework of the metaphor used, the latter is an increase in the efficiency of management.

The efficiency mechanism consists of four actions:

1. programs for improving the physical system: characteristics of resources, processes, products, systems;
2. verification and validation of these characteristics and establishment of new characteristics of the physical system on their basis;
3. establishment of new standards, norms and reference data for the planning system (as an information system);
4. establishment of new standards and norms for the system of remuneration.

DM concepts integrate dynamism, sustainability and stable development in efficiency management in accordance with ESG (ESG agenda and the 17 goals. https://sdgs.un.org/goals).

## REFERENCES

Allen, M., & Cervo, D. (2015). *Multi-Domain Master Data Management: Advanced MDM and Data Governance in Practice*. Morgan Kaufmann.

Box, G. E. P., & Draper, N. R. (1969). *Evolutionary Operation*. New York, London, Sydney, Toronto: John Wiley and Sons.

Deming, W. Edwards (2018). *Out of the Crisis, Reissue*. MIT Press. ISBN 978-0-262-35003-7.

ESG agenda and the 17 goals. https://sdgs.un.org/goals

Feldbaum, A. A. (1963, June 1). Dual control theory problems. *IFAC Proceedings Volumes*, 1(2): 541–550. 2nd International IFAC Congress on Automatic and Remote Control: Theory, Basle, Switzerland. DOI: 10.1016/S1474-6670(17)69687-3. ISSN 1474-6670.

Goldratt, E. M., & Cox, J. (1992). *The Goal: A Process of Ongoing Improvement (1984)*. North River Press; 2nd rev edition. ISBN 0-88427-061-0.

Grachev, A., Tiuleneva, G., & Tsvirkunov, D. (2021). National project of productivity improvement on principles of quality and lean management. In *Materials of 64th EOQ Congress*, Belgrade, Serbia.

Juran, J. M. (1988). *Juran's Quality Control Handbook*. McGraw-Hill. ISBN 0-07-033176-6.

Kondo, Y. (1995). *Companywide Quality Control: Its Background and Development*. 3A Corporation. ISBN 4-88319–048-X C3034.

Lapidus, V. (2021, June 29–30). The role of quality, lean and productivity in models of business growth. In *Materials of International Scientific and Business Quality Congress (EOQ, SMERK)*, Belgrade, Serbia.

Lapidus, V., & Ozerova, A. (2023, November 9–10). The new dual PDCA cycle in the modern context of organizational construction. In *Materials of 64th EOQ Congress*, Porto, Portugal.

Lapidus, V., Shcherbinin, V., Tsvirkunov, D., & Maltseva, M. (2017). RTM-diagrams and their application for business efficiency management. *Quality Management Methods*, 12: 48–54.

Lapidus, V., & Tiuleneva, G. (2018, June 21–22). Quality management in holdings. Objectives, tasks, management—Elements and main mechanisms. In *Materials of 62nd EOQ Congress*, Paris, France.

Lapidus, V., & Tiuleneva G. (2021). Model of dual management in company scale. In *Materials of 19th ANQ Congress*, Singapore.

Monden, Y. (1993). *Toyota Management System: Linking the Seven Key Functional Areas*. Portland, OR: Productivity Press.

Shcherbinin, V., Korolev, F., & Tsvirkunov, D. (2023). Diagnostics of business efficiency: From loss reduction to growth point. *Quality Management Methods*, 8: 14–20.

Tiuleneva, G., & Kastorskaya, L. (2018, September 19–20). Quality management in holdings. Increasing organization QMS maturity level on the basis of "DIAMOND" system. In *Materials of 16th ANQ Congress*, Almaty, Kazakhstan.

Wiener, N. (1961). *Cybernetics: Or Control and Communication in the Animal and the Machine*. Paris: Hermann & Cie & Cambridge, MA: MIT Press; 2nd rev edition (1961).

# 13 A Stochastic Programming Model for Managing Supply Chain Disruptions in the Post-COVID-19 Era

*Devika Kabe, Mucahit Cevik, and M. I. M. Wahab*

## 13.1 INTRODUCTION

Resilience is often defined as the ability of a supply chain to plan for unforeseen events, respond to disruptions, and recover from them. Resilience addresses a supply chain's ability to manage the damaging effects of unavoidable disturbances to return to its original state or move to a new, preferable state after being disturbed [1]. Such an ability is crucial for supply chains, as many catastrophic events have occurred in recent years, including earthquakes, political turmoil, and diseases. These events severely disrupt the ability of firms to produce and transport their goods [2, 3]. Hence, the idea of supply chain resilience was introduced in response to more frequent disasters occurring in the early 2000s.

The COVID-19 pandemic has changed the operational state of many firms and their supply chains at an unprecedented level. Many disruptions brought up by the pandemic forced these firms to learn to function in a highly volatile and unpredictable environment [4]. Specifically, the virus outbreak posed numerous problems for the supply chains. For instance, businesses have been put under pressure to meet customer demand while ensuring the protection and health of their workers [5]. In terms of manufacturing, many factories were forced to shut down due to lockdowns and faced production issues due to the lack of raw materials [6]. In addition, the travel restrictions imposed at airports, roads, and ports made it challenging to deliver finished products to customers [7].

Certain products had an increase in demand during the pandemic. The outbreak triggered spikes in stockpiling of emergency supplies such as sanitizers and disinfectants worldwide [8]. Healthcare supply chains for personal protection equipment, such as surgical masks and gloves, could not cope with the increased demand during the virus outbreak [9]. Similar issues have been observed in food supply chains as

DOI: 10.1201/9781003438748-16

well. For instance, meat products, which represent a core aspect of the U.S. food supply chain, have faced several disruptions since the beginning of the virus outbreak [10]. Thus, as a result of this pandemic, companies had to extensively deal with the concept of resilience, which ultimately became one of the primary supply chain management perspectives [9].

Research on supply chain resilience has become increasingly popular in the last decade [11]. These studies often focus on a particular resilience strategy or a metric, but they are rarely considered together in model development. A resilience strategy is typically defined as a strategy that strengthens the supply chain's ability to recover after a disruption [12]. It tends to be reactive, for example, switching to backup suppliers [11, 13] but may also be proactive. Different metrics have been proposed to quantify resilience. Some studies measure the level of demand satisfied after being disrupted [14, 15], which is known as the recovery level. Others quantify resilience by measuring the time to recovery, that is, the duration of time necessary for a supply chain network to return to routine operations after disruptions [16]. Another measure is lost profit during recovery, which corresponds to the lost profit from the beginning of the disruption until the end of the recovery period [17, 18]. The *p*-robust metric has also been used for supply chain resilience, which imposes an upper bound on the objective function value when considering different disruption scenarios [19].

Understanding these approaches regarding resilience strategies is metrics are essential to analyse supply chain resilience. Thus it would be beneficial to consider incorporating both simultaneously.

Aside from the lack of mathematical models that combine resilience metrics and strategies in the literature, many of the previous studies do not provide analysis with real-world datasets, for example, for a specific industry and/or product category. They instead model a general supply chain, which seldom considers uncertainty in various parameters. Moreover, while supply chains everywhere have already seen turmoil from epidemics, they have recently been struck by a large-scale, disruptive outbreak, COVID-19, and are still dealing with the aftermath of such destructive impacts. However, there is limited empirical research on the impacts of epidemic outbreaks on supply chains [20].

While minimizing the overall cost is typically the most important goal for companies, it is also crucial to guarantee a reasonable level of customer service. Accordingly, this chapter answers the following research question: how can resilience be achieved in supply chains by integrating both a metric that imposes requirements on the demand recovered and a strategy that helps the supply chain meet that recovery level?

We answer the aforementioned question and address the research gaps by investigating a supply chain that strategically integrates both a resilience strategy and a resilience metric. We analyse the results of a two-stage stochastic programming model applied to different product categories that were in high demand during lockdown periods for a particular supply chain. We consider uncertainty for multiple parameters and directly analyse pandemic impacts by observing how the supply chain operates when lockdowns are imposed and facilities are forced to close. For a resilience strategy, we focus on outsourcing, which is a way to obtain the finished products from an outside company when in-house operations are impeded. We use recovery level as the resilience metric and define it as the percentage of demand that must be satisfied to avoid facing a penalty cost. We impose a recovery level as a parameter and observe how the solution to the stochastic programming model changes at different recovery levels. The supply chain network decisions include: (1) which manufacturing plants and distribution centres to use for production and transportation, (2) the quantity of products to be produced at the active manufacturing plants, (3) the quantity of products to be shipped from manufacturing plants to distribution centres, (4) the quantity of products to be shipped from distribution centres to market zones, (5), the quantity of lost sales for each market zone, and (6) the quantity of products to be outsourced and shipped from the outsourcing facilities to the market zones. We define an outsourcing facility as a backup supplier that is able to meet a certain level of demand in the event of a lockdown when one or more of the company's manufacturing plants and distribution centres are forced to shut down. That is, the finished products can be outsourced and shipped directly to customers. The focus and objectives of this research include (1) measuring the value of stochasticity, (2) analysing the trade-off between the different objectives, (3) investigating the resilience achieved by incorporating outsourced facilities into the supply chain, and (4) analysing how the model behaves at different recovery levels. The main contributions of our chapter are summarized as follows.

- We propose a two-stage stochastic programming model for resilient supply chain management, which incorporates both a resilience strategy and a resilience metric. We analyse the use of recovery level as the metric and outsourcing as a specific strategy with a step-cost function to calculate the overall cost of outsourcing. Different from many studies that rely on an arbitrary number of scenarios in the sample average approximation, we systematically decide on a sufficient number of scenarios to include in our model to obtain high-quality solutions.
- We employ the epsilon-constraint method to analyse the trade-offs that exist between meeting demand in-house vs. outsourcing demand vs. not meeting demand and incurring penalties. While all three objectives are expressed in terms of monetary value, the multi-objective approach provides alternative solutions to the decision-makers, for example, to achieve a good service level by reducing lost sales costs and incurring higher outsourcing costs.
- We consider the circumstances brought on by the COVID-19 pandemic, where the uncertainties mainly stem from whether or not facilities are forced to go into a lockdown.
- We analyse how the supply chain reacts when these disruptions occur, impeding its ability to produce goods. We also consider relevant industries that involve products in high demand during the lockdown periods.
- We conduct a detailed numerical study based on realistic case studies to provide insights into the capabilities of stochastic programming-based approaches for resilient supply chain management. We quantify the value of our resilient strategy by comparing two models: one that has the option to outsource and one that does not. We then observe how the results from

these models differ in terms of cost and demand met at different recovery levels. In addition, we provide managerial insights through various visualizations of the model solutions that help better interpret the results.

These visuals clearly display how the decisions change under different scenarios.

The rest of this chapter is organized as follows. Section 13.2 discusses the relevant literature and highlights the research gap. Section 13.3 provides the problem definition and the mathematical model used throughout the chapter. The case study, dataset specifications, and parameter estimation are discussed in Section 13.4. We provide the results of our detailed numerical study in Section 13.5. Finally, we provide concluding remarks in Section 13.6, with a summary of contributions, key findings, study limitations, and directions for future research in this area.

## 13.2 LITERATURE REVIEW

In this section, we review the literature on supply chain resilience. A high-level summary of related papers in this field is provided in Table 13.1.

There has been significant research on resilient supply chain management in recent years. These studies mainly consist of a conceptual framework for resilience and explain different stages of a resilient risk management model, such as identifying, responding to, and recovering from risks and growing the supply chain [11]. Despite the numerous conceptual studies on resilient supply chains in the literature, quantitative studies in this context are limited. The few relevant quantitative studies mainly focus on either resilience strategies or resilience metrics [13].

A common resilience metric in the literature is the $p$-robust metric [19], which enforces an upper bound on the cost of each scenario. For instance, this metric can be used to enforce that the costs under each scenario may not be more than $100(1+p)\%$ of the optimal costs of the corresponding deterministic model for each scenario. Fahimnia et al. [26] develop a mixed integer programming (MIP) model to analyse the trade-offs between a resilient supply chain and an environmentally friendly supply chain in the clothing industry. They minimize only the cost of the nominal scenario, that is, the scenario with no disruptions, and enforce the $p$-robust measure as a constraint. They analyse the model outcomes for varying levels of $p$. The $p$-robust measure aims to keep the costs for each scenario relatively low by simply bounding the objective value of each scenario. However, it does not incorporate any other aspect of resilience.

Another resilience metric is the time to recovery (TTR), which is defined as the time for a supply chain to recover after it is disrupted. Losada et al. [16] incorporate this metric and present a bilevel mixed-integer linear programming (BMILP) model for protecting a facility network against worst-case losses. Their model aims to speed up the recovery time following a period of disruption. In order to lower the recovery time, the facilities are assigned to the $k$-nearest customers to minimize the weighted distance of delivering goods to all customers.

Lost profit during recovery (LPR) is another commonly used resilience metric. The LPR is typically defined as the lost profit from the start of the disruption period until the end of the recovery period. Hishamuddin et al. [17] use the LPR as a

**TABLE 13.1**
**Summary of the Most Relevant Papers Related to Supply Chain Resilience**

| Study | Technique | Uncertainty | Industry | Resilience | | Objective | |
|---|---|---|---|---|---|---|---|
| | | | | Strategy | Metric | Objective Function | Solution Approach |
| Peng et al. [21] | MIP solved with heuristics | | | | ✓ | Cost (opening, transp., storage) | Weighted |
| Losada et al. [16] | MILP solved with Benders | | | | ✓ | Distance | Weighted |
| Hishamuddin et al. [17] | MINLP | | | | ✓ | Cost during recovery period | Weighted |
| Garcia-Herreros et al. [14] | 2SP solved with multi-cut Benders | ✓ | | | ✓ | Cost (opening, transp.), penalties | Weighted |
| Medal et al. [22] | MIP solved with epsilon-constraint | | | ✓ | | Pre/post-disruption radius | Multiobjective |
| Pishvaee et al. [23] | PP solved with Benders | ✓ | Medical | | | Cost (opening, transp., processing), env./social impact | Multiobjective |
| Fahimnia and Jabbarzadeh [24] | Stochastic fuzzy goal programming | ✓ | Clothing | | ✓ | Cost (opening, prod., transp.), env./social impact | Multiobjective |
| Behzadi et al. [13] | 2SP | ✓ | Food | ✓ | | Profit | Weighted |
| Chen et al. [25] | IP | | | | ✓ | Time, cost (transp.) | Weighted |
| Fattahi et al. [15] | MSP | ✓ | Glass | ✓ | ✓ | Cost (opening, storage, prod., transp., stockout) | Weighted |
| Fahimnia et al. [26] | MIP | | Clothing | | ✓ | Cost (opening, transp., prod., lost sales) | Weighted |
| Ghavamifar et al. [27] | 2SP | ✓ | Automotive | | | Profit, lost sales, delivery time | Multiobjective |
| Behzadi et al. [11] | 2SP | ✓ | Food | | ✓ | Profit | Weighted |
| Fattahi et al. [18] | 2SP | ✓ | | | ✓ | Cost (opening, transp., storage, lost sales) | Weighted |
| Goldbeck et al. [28] | MSP | | | ✓ | ✓ | Cost (opening, prod., storage, penalties) | Weighted |
| Our paper | 2SP | ✓ | COVID safety, food | ✓ | ✓ | Cost (opening, transp., prod., outsource, lost sales, penalties) | Multiobjective |

*Note:* (IP: Integer programming, MIP: Mixed-integer programming, MILP: Mixed-integer linear programming, MINLP: Mixed-integer non-linear programming, 2SP: Two-stage stochastic programming, MSP: Multi-stage stochastic programming, PP: Probabilistic programming).

resilience metric and develop a recovery model for a supply chain when considering disruptions that arise when transporting goods. A recovery time window is allotted after the disruption to adjust the production and ordering schedule. A mixed-integer non-linear programming (MINLP) model is developed where the objective is to identify the optimal recovery schedule that minimizes the total recovery costs of the system. Fattahi et al. [18] create a new metric similar to the LPR by implementing a two-stage stochastic programming model that measures the expected value of the supply chain's cost increase due to a potential disruption during its recovery period.

Recovery level (RL) is also commonly used in supply chains as a resilience metric. It has various definitions, including customer service level, fill rate, and the percentage of demand that is unsatisfied [11]. Garcia-Herreros et al. [14] formulate a two-stage stochastic programming model where the objective is to minimize the investment and distribution costs during a finite time horizon. They penalize the unsatisfied demand via a linear function of volume. Fahimnia and Jabbarzadeh [24] use a stochastic fuzzy goal programming approach to analyse the trade-offs between a sustainable supply chain and a resilient supply chain using data from a clothing company. They implicitly incorporate the RL by analysing how much demand is satisfied when considering economic and environmental sustainability measures. Chen et al. [25] apply a resilience metric to a container transportation supply chain. They develop an integer programming (IP) model to minimize the cost of operations. They define their resilience metric as the ratio of the total satisfied container flow to the total container transport demand.

Employing resilience strategies is another common way to build supply chain resilience. A resilience strategy is generally defined as a reactive strategy that companies execute after the risk materializes [11]. Behzadi et al. [13] implement a two-stage stochastic programming model that incorporates three different resilience strategies applied to a kiwifruit supply chain. The first strategy is having a backup supplier, that is, obtaining supplies from one or more backup sources when the main source is not operating. The second one is by moderating the disruption probability by allocating resources to facilities in advance to lessen the chance of disruption. This is a proactive approach as opposed to a reactive one. The third strategy is to have multiple suppliers. The model maximizes the expected profit by choosing the best resilience strategies. Medal et al. [22] also implement proactive strategies to mitigate the level of disruption. Their strategy is facility hardening, which is a form of facility protection that allocates resources to a facility to make it less susceptible to failure. They implement a multi-objective MIP model that minimizes the post-disruption and pre-disruption radius.

Fattahi et al. [15] are one of the few to incorporate a resilience metric and a resilience strategy. They use the RL as the resilience metric and develop a multi-stage stochastic programming (MSP) model to address a multi-period supply chain network design for a glass company. The customer demand is dependent on how quickly the facilities can serve them, that is, their delivery lead times. Strengthening facilities and modifying sourcing decisions after disruptions are performed as mitigation and contingency strategies, respectively. Behzadi et al. [11] also consider both resilience metrics and a resilience strategy. They combine TTR, LPR, and RL while introducing a new metric, which considers the net present value of the lost profit

(NPV-LP) over the planning horizon. They develop a two-stage stochastic programming model to maximize expected profit, and as a resilience strategy, they incorporate backup suppliers. Goldbeck et al. [28] also include both a resilience metric and resilience strategies in their model. They develop an MSP model that minimizes pre-disruption and post-disruption costs. Their resilience strategies include the acquisition of resources to help in repairing the parts of the supply chain that were damaged and their allocation and transportation to those parts. They quantify the performance of a supply chain based on its ability to satisfy demand.

Pishvaee et al. [23] do not explicitly implement resilience metrics or strategies. They instead consider critical decisions such as the number, location, capacity, and technology of the required facilities in the supply chain to mitigate the risk. They develop a multi-objective possibilistic programming model for a sustainable medical supply chain that distributes needles and syringes. Ghavamifar et al. [27] also implicitly incorporate resilience by taking into account competition in the automotive parts industry. They use a bi-level multi-objective programming approach to design the network. The objective is to maximize total profit and minimize lost sales caused by competition.

Despite the abundance of studies on supply chain resilience, very few of these studies focus on empirical analysis with real-life scenarios [11]. That is, many of the quantitative studies that propose mathematical models for supply chain resilience do not include a specific industry and/or product category, and instead consider a general supply chain. They also seldom consider uncertainty in various model parameters including facility shutdowns and demand. Moreover, the various impacts of the recent COVID-19 pandemic stressed the importance of analysing how supply chains operate under a global crisis. Humanitarian logistics literature has largely studied the negative impacts of epidemics; however, there exists a research gap in analysing the impact of the pandemic on supply chains [20], particularly those of high demand, essential products. Our chapter aims to fill these research gaps in the domain of supply chain resilience.

## 13.3 PROBLEM DEFINITION AND MODEL FORMULATION

In this chapter, we are particularly interested in understanding how a supply chain can continue to meet demand when faced with disruptions. These disruptions might shut down facilities entirely so that they cannot produce, store, or ship any products. To establish a basis for this analysis, we examine a supply chain consisting of manufacturing plants, distribution centres, market zones, and outsourcing facilities. Manufacturing plants and distribution centres are subject to facility shutdowns. A set of scenarios represent situations where certain regions (e.g., cities and provinces) undergo a lockdown or facilities are forced to close and are unable to operate. Incorporation of outsourcing facilities can present a solution in such cases as they can be alternatives to satisfy the demand when the company cannot produce the products in-house. The proposed model aims to determine the network design decisions that minimize the overall supply chain cost while ensuring that demand is met.

We introduce the notations that will be used throughout the chapter in Table 13.2. The uncertain parameters, $a_{is}^1$ and $a_{js}^2$ indicate whether or not the manufacturing plants and distribution centres are operating in the given scenario, and $\delta_s$ represents the demand level in scenario $s$. We also assume that the probability distribution of the uncertain parameters is known to solve the problem via sample average approximation (SAA) [29].

The decisions are made in two stages. The first-stage decisions are independent of scenarios and include which manufacturing plants and distribution centres are to be used for production and transportation. The second stage decisions are scenario dependent. Specifically, they include determining quantities of products to be produced, outsourced, and transported. These decisions also reveal how much demand will not be met, resulting in lost sales.

**TABLE 13.2**
**Notation Used in the Two-Stage Stochastic Programming Model**

| | |
|---|---|
| **Sets:** | |
| $\mathcal{I}$: | Set of manufacturing plants, indexed by $i$ |
| $\mathcal{J}$: | Set of distribution centres, indexed by $j$ |
| $\mathcal{K}$: | Set of market zones, indexed by $k$ |
| $\mathcal{L}$: | Set of outsource facilities, indexed by $\ell$ |
| $\mathcal{M}$: | Set of product types, indexed by $m$ |
| $\mathcal{S}$: | Set of scenarios, indexed by $s$ |
| **Parameters:** | |
| *Deterministic:* | |
| $c_i^1$: | Capacity of manufacturing plant $i$ |
| $c_j^2$: | Capacity of distribution centre $j$ |
| $c_{\ell m}^3$ | Number of units of product $m$ outsource facility has available |
| $d_{km}$: | Nominal demand of product $m$ in market zone $k$ |
| $f_i^1$: | Fixed cost required to operate manufacturing plant $i$ |
| $f_j^2$: | Fixed cost required to operate distribution centre $j$ |
| $g_{im}$: | Variable cost of manufacturing product $m$ in manufacturing plant $i$ |
| $h_m$: | Volume of product $m$ |
| $N$: | Number of steps for step function |
| $u_{km}$ | Unit cost of lost sales for product $m$ at market zone $k$ |
| $w_{\ell m}^n$: | Unit cost of outsourcing product $m$ from outsource facility at cost level $n$ |
| $\eta_{km}$ : | Penalty associated with decision variable $\lambda_{kms}$ |
| $\gamma$: | Recovery level |
| $\Delta^n$: | Threshold for step function cost of purchasing from outsource facilities for $n \in \{1,2,\ldots,N\}$ |
| $\alpha_{ijm}^1$ : | Unit cost of shipping product $m$ from manufacturing plant $i$ to distribution centre $j$ |
| $\alpha_{jkm}^2$ | Unit cost of shipping product $m$ from distribution centre $j$ to market zone $k$ |
| $\alpha_{\ell km}^3$ | Unit cost of shipping product $m$ from outsource facility to market zone $k$ |

(*Continued*)

**TABLE 13.2**
**(Continued)**

| | |
|---|---|
| *Uncertain:* | |
| $a_{is}^{1}$: | Binary parameter indicating if manufacturing plant $i$ is operating in scenario $s$ |
| $a_{js}^{2}$: | Binary parameter indicating if distribution centre $j$ is operating in scenario $s$ |
| $\delta_s$: | Percentage demand multiplier for scenario $s$ |
| **Decision variables:** | |
| $q_{ims}$: | Quantity of product $m$ to produce at manufacturing plant $i$ in scenario $s$ |
| $r_{kms}$: | Quantity of lost sales for product $m$ at market zone $k$ in scenario $s$ |
| $t_{\ell kms}$: | Quantity of product $m$ to ship from outsource facility to market zone $k$ in scenario $s$ |
| $v_{\ell ms}^{n}$: | Quantity of product $m$ to outsource from outsource facility in scenario $s$ that falls below $\varepsilon^n$ |
| $x_i^1$: | A binary decision variable equal to 1 if manufacturing plant $i$ is opened |
| $x_j^2$: | A binary decision variable equal to 1 if distribution centre $j$ is operating |
| $y_{ijms}$: | Quantity of product $m$ to ship from manufacturing plant $i$ to distribution centre $j$ in scenario $s$ |
| $z_{jkms}$: | Quantity of product $m$ to ship from distribution centre $j$ to market zone $k$ in scenario $s$ |
| $\lambda_{kms}$ | Variable corresponding to difference of $\gamma$ and percentage of demand met for scenario $s$ |

The total supply chain cost under each scenario includes:

$$\text{Manufacturing plant costs} = MC = \sum_{i\in\mathcal{I}} f_i^1 x_i^1 \tag{1}$$

$$\text{Distribution centre costs} = DC = \sum_{j\in\mathcal{J}} f_j^2 x_j^2 \tag{2}$$

$$\text{Production costs} = PC_s = \sum_{i\in\mathcal{I}}\sum_{m\in\mathcal{M}} g_{im} q_{ims} \tag{3}$$

$$\text{In-house shipment costs} = ISC_s = \sum_{i\in\mathcal{I}}\sum_{j\in\mathcal{J}}\sum_{m\in\mathcal{M}} \alpha_{ijm}^1 y_{ijms} + \sum_{j\in\mathcal{J}}\sum_{k\in\mathcal{K}}\sum_{m\in\mathcal{M}} \alpha_{jkm}^2 z_{jkms} \tag{4}$$

$$\text{Outsource costs} = OC_s = \sum_{N}^{n=1}\sum_{\ell\in\mathcal{L}}\sum_{m\in\mathcal{M}} w_{\ell m}^n v_{\ell ms}^n \tag{5}$$

$$\text{Outsource shipment costs} = OSC_s = \sum_{\ell\in\mathcal{L}}\sum_{k\in\mathcal{K}}\sum_{m\in\mathcal{M}} \alpha_{\ell km}^3 t_{\ell kms} \tag{6}$$

$$\text{Cost of lost sales} = LC_s = \sum_{k\in\mathcal{K}}\sum_{m\in\mathcal{M}} u_{km} r_{kms} \tag{7}$$

$$\text{RL penalty} = RL_s = \sum_{m\in\mathcal{M}}\sum_{k\in\mathcal{K}} \eta_{km}\lambda_{kms} d_{km}\delta_s \tag{8}$$

Equations (1) and (2) refer to the fixed costs of operating manufacturing plants and distribution centres. Equation (3) corresponds to the total cost of producing, whereas Equation (4) corresponds to the total cost of shipping from the manufacturing plants to the distribution centres and from the distribution centres to the market zones. Equation (5) provides the total cost of outsourcing products, and Equation (6) provides the cost of shipping the outsourced products to the market zones. Note the assumption that outsourcing facilities directly ship the products to market zones. Equation (7) indicates the total cost of lost sales across all market zones and products. Finally, Equation (8) corresponds to the total cost of the additional penalties incurred when not meeting the demand at the given recovery level.

We formulate three separate objective functions to analyse the trade-offs between modelling objectives. Since we solve this problem with SAA, we use the scenario average equivalent in place of expected value terms.

$$f_1 = MC + DC + \frac{1}{|\mathcal{S}|}\sum_{s\in\mathcal{S}}\left(PC_s + ISC_s\right) \tag{9}$$

$$f_2 = \frac{1}{|\mathcal{S}|}\sum_{s\in\mathcal{S}}\left(OC_s + OSC_s\right) \tag{10}$$

$$f_3 = \frac{1}{|\mathcal{S}|}\sum_{s\in\mathcal{S}}\left(LC_s + RL_s\right) \tag{11}$$

In these objectives, $f_1$ refers to all the costs related to the in-house costs, including operating, producing, and shipping. The second objective, $f_2$, includes all the costs associated with obtaining products from the outsource facilities. This includes the costs to acquire and ship the products. In our step-cost function, the unit cost per product increases as more products are outsourced. Although it is common to offer a discount when products are bought in bulk, the particular products that we consider are typically scarce during the COVID-19 lockdown periods. Thus, the difficulty of providing such products results in an increased cost. The last objective, $f_3$, is the cost associated with not meeting demand and includes the total cost of lost sales and the additional recovery level penalties incurred. The resulting formulation for the multi-objective two-stage stochastic programming model is as follows:

$$\min\left(f_1, f_2, f_3\right) \tag{12a}$$

$$\text{s.t.} \quad q_{ims} \geq \sum_{j\in\mathcal{J}} y_{ijms} \quad \forall i \in \mathcal{I}, m \in \mathcal{M}, s \in \mathcal{S} \tag{12b}$$

$$\sum_{i\in\mathcal{I}} y_{ijms} \geq \sum_{k\in\mathcal{K}} z_{jkms} \quad \forall j \in \mathcal{J}, m \in \mathcal{M}, s \in \mathcal{S} \tag{12c}$$

$$\sum_{j\in\mathcal{J}} z_{jkms} + \sum_{\ell\in\mathcal{L}} t_{\ell kms} + r_{kms} \geq d_{km}\delta_s \quad \forall k \in \mathcal{K}, m \in \mathcal{M}, s \in \mathcal{S} \tag{12d}$$

$$\sum_{n=1}^{N} v_{\ell ms}^{n} \geq \sum_{k \in \mathcal{K}} t_{\ell kms} \quad \forall m \in \mathcal{M}, \ell \in \mathcal{L}, s \in \mathcal{S} \tag{12e}$$

$$\sum_{m \in M} h_m q_{ims} \leq a_{is}^{1} c_i^{1} x_i^{1} \quad \forall i \in \mathcal{I}, s \in \mathcal{S} \tag{12f}$$

$$\sum_{i \in \mathcal{I}} \sum_{m \in \mathcal{M}} h_m y_{ijms} \leq a_{js}^{2} c_j^{2} x_j^{2} \quad \forall j \in \mathcal{J}, s \in \mathcal{S} \tag{12g}$$

$$\lambda_{kms} \geq \gamma - \left(1 - \frac{r_{kms}}{d_{km}\delta_s}\right) \quad \forall k \in \mathcal{K}, m \in \mathcal{M}, s \in \mathcal{S} \tag{12h}$$

$$\sum_{n=1}^{N} v_{\ell ms}^{n} \leq c_{\ell m}^{3} \quad \forall \ell \in \mathcal{L}, m \in \mathcal{M}, s \in \mathcal{S} \tag{12i}$$

$$v_{\ell ms}^{n} \leq \Delta^{n} - \Delta^{n-1} \quad \forall n \leq N, \forall \ell \in \mathcal{L}, m \in \mathcal{M}, s \in \mathcal{S} \tag{12j}$$

$$x_i^{1}, x_j^{2} \in \{0,1\} \qquad \forall i \in \mathcal{I}, j \in \mathcal{J} \tag{12k}$$

$$q_{ims}, y_{ijms}, z_{jkms}, v_{\ell ms}, t_{\ell kms},$$

$$r_{kms}, \lambda_{kms} \geq 0 \quad \forall i \in \mathcal{I}, m \in \mathcal{M}, s \in \mathcal{S}, \ell \in \mathcal{L}, k \in \mathcal{K} \tag{12l}$$

Constraints (12b), (12c), (12d), and (12e) are network flow constraints. Constraint (12b) ensures that the quantity of products produced in the manufacturing plants equals the total amount shipped to the distribution centres. Constraint (12c) ensures that the total amount shipped to the distribution centres from the manufacturing plants equals the total amount shipped to the market zones from the distribution centres. Constraint (12d) indicates that the total amount of products shipped to the market zones from both the distribution centres and outsourcing facilities, and the amount in lost sales must be greater than or equal to the demand for the given scenario. Constraint (12e) ensures that the total number of units outsourced are shipped to the market zones. Constraints (12f) and (12g) enforce capacity limitations for manufacturing plants and distribution centres. Constraint (12h) enforces that $\lambda_{kms}$ is greater than or equal to the difference between the recovery level and the amount of demand met per product per market zone in each scenario. Constraint (12i) states that the quantity of outsourced products must be less than or equal to the available product quantity. Constraint (12j) enforces an upper bound on the quantity of units outsourced at each step in the step-cost function. Finally, constraints (12k) and (12l) define the logical constraints on the decision variables.

Our formulation is based on the assumption that the manufacturing plants and distribution centres are the only facilities subject to shutdowns. When they are forced to shut down, they cannot operate at any capacity. We also assume that the outsourcing facilities are always open; however, there is a limit on the number of products that can be outsourced.

## 13.4 CASE STUDY

The COVID-19 pandemic has led to a myriad of lockdowns to slow the spread of the virus. This placed a significant burden on supply chains, particularly those producing high-demand, essential products [5]. For instance, some of the largest companies in the world were severely impacted, as their facilities were located in regions where lockdowns took place [5]. The given circumstances made it exceedingly difficult to continue the supply chain operations, as some parts of the supply chain stopped entirely [30]. While the majority of manufacturing firms across different industries were affected by COVID-19, the particular impact of the effects varied depending on whether or not the demand for the products was high or low. For instance, the demand for some products such as toilet paper, soap, sanitizers, and medicines increased considerably. In contrast, the demand for other products, such as clothing and sports items, fell drastically [31]. Aside from health and safety products seeing an immense increase in demand, food-away-from-home purchases declined rapidly. Grocery stores and other food retail outlets experienced an increased demand for food items. This was a direct result of the temporary shutdown of restaurants and other food service outlets under lockdown rules. The effect of the shift in the demand for food to retail outlets was further exacerbated by panic buying and stockpiling behaviours among consumers, which lead to short-run stockouts [32]. This motivated us to use relatively high-demand products as our basis for the case study.

### 13.4.1 Dataset Specifications

In our analysis of the proposed stochastic programming model, we consider two different instances that involve two distinct product categories. The first one falls into the health and safety category and includes three products, namely, masks, disinfectants, and hand sanitizers. The second category falls under food, specifically meat products. We use three products again, including chicken, beef, and pork.

Both instances have a set of manufacturing plants, distribution centres, market zones, and outsourcing facilities. We use the Greater Toronto Area (GTA) in Canada as the location for the distribution centres and market zones. The manufacturing plants are spread across Canada, and the outsourcing facilities are in either Canada or the United States. The two instances share the same distribution centres and market zones since they are based on the same company but have different manufacturing plants and outsourcing facilities that are specific to the product categories.

Table 13.3 provides information on facility locations. The second column indicates the facilities, where MP stands for a manufacturing plant, and DC stands for a distribution centre. The third column (Location) provides the province where each facility is located. The last two columns indicate how far the facilities are from downtown Toronto, Canada, regarding the geodesic distance and how they are positioned, respectively. Note that the distribution centres are common for both instances.

Manufacturing plants and distribution centres can either be shut down or functioning. We assume outsourced facilities are always open. However, they have a limit to their product availability. Each instance has three possible facilities that the company can outsource if it cannot meet demand on its own. The facilities from which they outsource are product specific. We note that the ability to outsource meat

**TABLE 13.3**
**Facility Locations and Distances to Downtown Toronto (direction is relative to downtown Toronto)**

| Instance | Facility | Location | Dist. to Downtown (km) | Direction |
|---|---|---|---|---|
| COVID safety | MP1 | Manitoba | 1527 | NW |
| | MP2 | Manitoba | 1523 | NW |
| | MP3 | Ontario | 23 | NW |
| | MP4 | Ontario | 18 | NE |
| | MP5 | Quebec | 722 | NE |
| | MP6 | Quebec | 724 | NE |
| Food | MP1 | British Columbia | 3300 | NW |
| | MP2 | Alberta | 2712 | NW |
| | MP3 | Saskatchewan | 2120 | NW |
| | MP4 | Manitoba | 1650 | NW |
| | MP5 | Ontario | 124 | SW |
| | MP6 | Quebec | 445 | NE |
| Both | DC1 | Ontario | 79 | SW |
| | DC2 | Ontario | 26 | NW |
| | DC3 | Ontario | 34 | NE |
| | DC4 | Ontario | 38 | NE |

products is more difficult, given the many bottlenecks that affected the ability to transport animals from feedlots [10, 33]. Thus we limit the availability of outsourcing for the meat products instance, whereas, for the first instance, the company can outsource the majority of its demand if necessary.

### 13.4.2 Parameter Estimation

We have a fixed nominal demand that can increase or decrease by a scalar value, $\delta_s$, for each scenario $s$, which follows a discrete uniform distribution. That is, we assume that the demand for all products changes between 25% lower and higher than the nominal value; hence $\delta_s \in [0.75, 1.25]$. We discretize $\delta_s$ by increments of 0.005 so that there are 101 total possible demand realizations of equal probability. Then, the demand for the scenario is $\delta_s d_{km}$, where $d_{km}$ represents the nominal demand for product $m$ at market zone $k$.

To determine how disruptions occur, we adopt the following logic. If a province contains more than one manufacturing plant and the province goes into a lockdown, then all manufacturing plants within that province must be shut down. Distribution centre shutdowns are independent of each other. Both instances have six manufacturing plants and four distribution centres. For the first instance, the six manufacturing plants are evenly spread across three provinces in Canada. Hence, there are a total of $2^3 \times 2^4 = 128$ ways that the supply chain can be disrupted. For the second instance, all manufacturing plants are dispersed across six different provinces in Canada. As a result, there are $2^6 \times 2^4 = 1024$ ways that the supply chain can be disrupted.

We assume disruptions follow a binomial distribution, and we use 0.1 as the probability of a single disruption. This probability was also used by Fattahi et al. [18] and considered to be a moderate level, that is, not too high or too low. Thus, the sum of probabilities over all possible disruption scenarios sums to 1. Combining the possible demand realizations for each scenario, and the aforementioned possible disruptions, we have a total of $128 \times 101 = 12,928$ scenarios for the first instance and $1024 \times 101 = 103,424$ for the second instance. Accordingly, we sample a set of scenarios from the entire scenario set to solve our two-stage stochastic programming model.

The geographic data of the manufacturing plants, distribution centres, market zones, and outsource facilities was obtained from Google Maps [34]. We estimate the cost of transporting products by making them proportional to the geodesic distance between two points. The population data of the GTA was obtained from Statistics Canada [35] and was used to estimate the nominal customer demand for each product per market zone.

## 13.5 NUMERICAL RESULTS

In this section, we provide results from our detailed numerical study with two problem instances. We first explain sample average approximation, an algorithm that we use to determine an appropriate amount of scenarios to consider in the stochastic programming model. We then discuss the value of incorporating stochasticity in our supply chain resilience problem. Next, we present the Pareto optimal solutions calculated via the epsilon-constraint method to analyse trade-offs between the different objective function values. We then provide a sensitivity analysis for the important model parameters. Finally, for easier interpretation of the model solutions, we present detailed visualizations of the four-echelon network. All the mathematical models were implemented using Gurobi 9.0 and Python 3.7. Experiments were performed on a PC with Intel i7–8700 CPU and 32GB RAM. Model run times for the base case analysis (e.g., for 200 scenarios) with two-stage stochastic programming models took 2 minutes to 10 minutes, depending on the parameter settings.

### 13.5.1 Sample Average Approximation Analysis

We perform SAA for each instance to determine an appropriate number of scenarios for the stochastic programming model. Although the use of a large number of scenarios would result in a more accurate estimation of the objective function value, this is not always possible because of the underlying computational complexity. We implement a sampling-based method explained in Algorithm 1 to obtain a suitable number of scenarios for our problem [29]. In these experiments, we use $\tau = (64n)_{n\in[6]}$, where $\tau_1 = 64$, $\tau_6 = 384$, $\alpha = 0.025$, and $N_{LB} = 30$ for both instances, $N_{UB} = 1,000$ for the first instance and $N_{UB} = 10,000$ for the second instance. $Q(\cdot)$ refers to the scenario-dependent costs and $fx$ refers to the facility operating costs. The SAA analysis leads to the results in Table 13.4, where column $|S|$ refers to the number of scenarios, and mean ($F^{LB}$) and mean ($F^{UB}$) are the averages of the lower bound and upper bound on the objective values, respectively. Similarly, min $LB_s$ is the lower bound on the confidence interval for mean ($F^{LB}$), and max $UB_s$ is the upper bound on the confidence interval for mean ($F^{UB}$).

The gap values in Table 13.4, expressed in percentage, can be calculated as:

$$Gap = \min_{i \in N_{LB}} \frac{\left(\max_t UB_{t_i} - \min_t LB_t\right)}{\min_t LB_t} \times 100$$

For the instance regarding COVID safety products, starting at $|S| = 192$, the gap drops significantly and nearly always remains under 1%. For the second instance, also starting at $|S| = 192$, the gap drops significantly and remains under 2%. For both cases, fluctuations also fall within less than 1%. Thus, for the remaining numerical experiments, we use $|S| = 192$ as the default number of scenarios.

**Algorithm 1** SAA procedure

1: **procedure** SAA assessment $(S, N_{LB}, N_{UB}, \alpha)$

2: **for** $t \in [|\tau|]$ **do**

3: Generate $\left(\left(\xi^{ij}\right)_{j \in [\tau_t]}\right)_{i \in [N_{LB}]}$

4: $X^{LB} = \left(\arg\min_{x \in X}\left\{f^{\top}x + \frac{1}{\tau_t}\sum_{j \in [\tau_t]} Q\left(x, \xi^{ij}\right)\right\}\right)_{i \in [N_{LB}]}$

5: $F^{LB} = \left(\min_{x \in X}\left\{f^{\top}x + \frac{1}{\tau_t}\sum_{j \in [\tau_t]} Q\left(x, \xi^{ij}\right)\right\}\right)_{i \in [N_{LB}]}$

6: Generate $\left(\tilde{\xi}^{j}\right)_{j \in [N_{UB}]}$

7: $F^{UB} = \left(\left(f^{\top}x + Q\left(x, \tilde{\xi}^{j}\right)\right)_{j \in [N_{UB}]}\right)_{x \in X^{LB}}$

8: $LB_t = \text{Mean}\left(F^{LB}\right) \pm \frac{t_{1-\frac{\alpha}{2}, N_{LB}}}{\sqrt{N_{LB}}} \text{std}\left(F^{LB}, \text{ddof} = 1\right)$

$$9\colon UB_t = \left( \text{Mean}\left(F^{UB_i}\right) \pm \frac{t_{1-\frac{\alpha}{2}, N_{UB}}}{\sqrt{N_{UB}}} \text{std}\left(F^{UB_i}, \text{ddof}=1\right) \right)_{i\in[N_{LB}]}$$

10: **output**: $F^{LB}, F^{UB}, (LB_t)_{t\in[|\mathcal{T}|]}, (UB_t)_{t\in[|\mathcal{T}|]}$.

**TABLE 13.4**
**SAA Results per Instance**

| Instance | \|S\| | Mean ($F^{LB}$) | Mean ($F^{UB}$) | min$LB_s$ | max $UB_s$ | Gap ($) | Gap (%) |
|---|---|---|---|---|---|---|---|
| COVID safety | 64 | 18,172,722.0 | 18,092,861.5 | 17,795,081.5 | 18,155,552.0 | 360,470.6 | 2.0 |
| | 128 | 18,150,771.9 | 18,092,861.5 | 17,912,478.0 | 18,132,419.6 | 219,941.7 | 1.2 |
| | 192 | 18,580,983.1 | 18,388,823.4 | 18,395,294.2 | 18,419,648.8 | 24,354.6 | 0.1 |
| | 256 | 18,440,944.8 | 18,308,636.5 | 18,285,644.6 | 18,334,417.3 | 48,772.7 | 0.3 |
| | 320 | 18,421,424.8 | 18,483,934.1 | 18,297,993.5 | 18,504,424.5 | 206,430.9 | 1.1 |
| | 384 | 18,449,658.9 | 18,308,636.5 | 18,309,694.4 | 18,331,871.4 | 22,177.0 | 0.1 |
| Food | 64 | 18,299,598.8 | 18,879,316.5 | 18,019,661.4 | 18,894,012.0 | 874,350.6 | 4.9 |
| | 128 | 18,573,639.9 | 18,879,316.5 | 18,382,379.6 | 18,889,356.9 | 506,977.3 | 2.8 |
| | 192 | 18,753,761.1 | 18,879,316.5 | 18,596,890.9 | 18,887,551.5 | 290,660.6 | 1.6 |
| | 256 | 18,810,684.3 | 18,879,316.5 | 18,662,278.4 | 18,887,107.2 | 224,828.8 | 1.2 |
| | 320 | 18,742,354.3 | 18,879,316.5 | 18,615,574.6 | 18,885,971.9 | 270,397.3 | 1.5 |
| | 384 | 18,855,796.8 | 18,951,629.3 | 18,740,337.5 | 18,957,690.4 | 217,352.9 | 1.2 |

### 13.5.2 Value of Stochastic Solution

To measure the value of incorporating stochasticity, we compare the quality of the first-stage solutions from the stochastic model, that is, the decisions to choose which facilities to use for production and distribution, and the deterministic model by calculating the expected value of perfect information (EVPI) and expected value of stochastic solution (EVSS). The EVPI measures the maximum amount a decision-maker would be willing to pay in return for complete and accurate information about future uncertainty. The value of the stochastic solution (VSS) solves the mean-value problem to obtain a first-stage solution, fixes that solution, and then solves the stochastic programming model. However, the mean-value scenario does not work in this situation, as the scenario disruptions are represented by binary vectors indicating whether or not a facility is running. Thus we implement the EVSS [36]. Different from the standard method to obtain the value of stochasticity, the EVSS is applicable when there is no intuitive way to choose a representative scenario as the expected one. This was used by Mousavi et al. [36], as their scenarios were also represented by binary vectors. The EVSS is calculated by solving the deterministic problem for each scenario, taking those first-stage decisions that the model solved, and fixing them to solve the original stochastic programming model. These costs are then averaged, and we subtract the optimal value of the stochastic solution to obtain the EVSS. We calculate both the EVPI and EVSS at three different recovery levels: $\gamma \in \{0.50, 0.75, 0.95\}$.

Given a sample of scenarios, $\left\{\xi^s\right\}_{s \in \mathcal{S}}$, the EVPI is calculated as:

$$\text{EVPI} = \hat{\upsilon}(S) - \overline{\upsilon}(S) \tag{13}$$

where $\hat{\upsilon}(S)$ is the optimal objective value for the stochastic programming model and

$$\overline{\upsilon}(S) = \frac{1}{|\mathcal{S}|} \sum_{s \in \mathcal{S}} \hat{v}\left(\left\{\xi^s\right\}\right)$$

which is the average of objective values to the deterministic model for each scenario. Table 13.5 provides results for the EVPI calculations. We observe that the EVPI is quite large, starting at $\gamma = 0.50$, meaning that the value of perfect information about the future is considerable. A common trend we observe from the table is that, as the recovery level increases, the EVPI increases slightly. This implies that as the required level of demand to satisfy increases, the decision-maker would be willing to pay more for perfect information.

By defining the first-stage function as $F\left(x_1, x_2\right)$ and the second stage function as $Q(v, q, y, z, t, r, \lambda, \xi)$, the EVSS can be calculated as:

$$\text{EVSS} = \upsilon^{DET}(S) - \hat{v}(S) \tag{14}$$

where

$$\upsilon^{DET}(S) = \frac{1}{|\mathcal{S}|} \sum_{s \in \mathcal{S}} \upsilon^{DET}\left(\left\{\xi^s\right\}\right)$$

**TABLE 13.5**
**EVPI per Instance at Different $\gamma$ Values**

| Instance | $\gamma$ | $\overline{\upsilon}(S)$ | $\hat{\upsilon}(S)$ | EVPI ($) | EVPI (%) |
|---|---|---|---|---|---|
| COVID safety | 0.50 | 12,613,408.7 | 18,226,651.3 | 5,613,242.6 | 30.8 |
| | 0.75 | 12,650,274.2 | 18,461,727.8 | 5,811,453.6 | 31.5 |
| | 0.95 | 12,694,366.8 | 18,765,472.9 | 6,071,106.1 | 32.4 |
| Food | 0.50 | 12,545,978.7 | 18,536,959.3 | 5,990,980.6 | 32.3 |
| | 0.75 | 12,582,935.5 | 18,650,024.0 | 6,067,088.5 | 32.5 |
| | 0.95 | 12,644,805.9 | 18,856,486.5 | 6,211,680.6 | 32.9 |

and for each $s \in \mathcal{S}$

$$v^{DET}\left(\{\xi^s\}\right) = F\left(\hat{x}_1\left(\{\xi^s\}\right), \hat{x}_2\left(\{\xi^s\}\right)\right)$$
$$+\frac{1}{|\mathcal{S}|}\sum_{s' \in \mathcal{S}} Q\left(\hat{v}\left(\{\xi^s\}\right), \hat{q}\left(\{\xi^s\}\right), \hat{v}\left(\{\xi^s\}\right), \hat{z}\left(\{\xi^s\}\right), \hat{t}\left(\{\xi^s\}\right), \hat{r}\left(\{\xi^s\}\right), \hat{\lambda}\left(\{\xi^s\}\right), \xi^{s'}\right)$$

Table 13.6 shows that the stochastic solutions provide significant benefits over the deterministic ones. We observe the same trend as in EVPI; that is, as the recovery level increases, the EVSS increases slightly. Thus, as the required level of demand to satisfy increases, so does the value of incorporating stochasticity. This is clear in the second instance, regarding food, where the availability of outsourcing is much smaller. Therefore, without enough facilities available for production, the company is unable to produce many of the products in-house. With less opportunity to outsource, the model accumulates costs in lost sales and penalties, resulting in a high cost.

### 13.5.3 Pareto Frontier

Pareto optimal solutions in Figures 13.1 and 13.2 were calculated via the epsilon-constraint method [37]. The purpose of this implementation is to analyse the trade-offs that are present among the three objective functions: the in-house cost, the cost of outsourcing and shipping from the outsource facilities, and the costs incurred from lost sales and penalties, which are combined as a single objective component. The epsilon-constraint method is applied as follows. Let $f_1$ denote the primary objective function and $p$ denote the number of objectives; then we have the following model:

$$\begin{aligned} &\text{Minimize} \quad f_1(x) \\ &\text{Subject to} \quad f_i(x) \le \varepsilon_i \quad \forall i \in \{2, \ldots, p\} \\ &\qquad\qquad\quad x \in \mathcal{X} \end{aligned}$$

**TABLE 13.6**
**EVSS per Instance at Different $\gamma$ Values**

| Instance | $\gamma$ | $v^{DET}(S)$ | $\hat{v}(S)$ | EVSS ($) | EVSS (%) |
|---|---|---|---|---|---|
| COVID safety | 0.50 | 21,396,828.1 | 18,226,651.3 | 3,170,176.7 | 14.8 |
| | 0.75 | 21,874,757.1 | 18,461,727.8 | 3,413,029.3 | 15.6 |
| | 0.95 | 22,550,191.3 | 18,765,472.9 | 3,784,718.3 | 16.8 |
| Food | 0.50 | 28,259,486.1 | 18,536,959.3 | 9,722,526.9 | 34.4 |
| | 0.75 | 30,334,315.8 | 18,650,024.0 | 11,684,291.9 | 38.5 |
| | 0.95 | 33,276,662.6 | 18,856,486.5 | 14,420,176.1 | 43.3 |

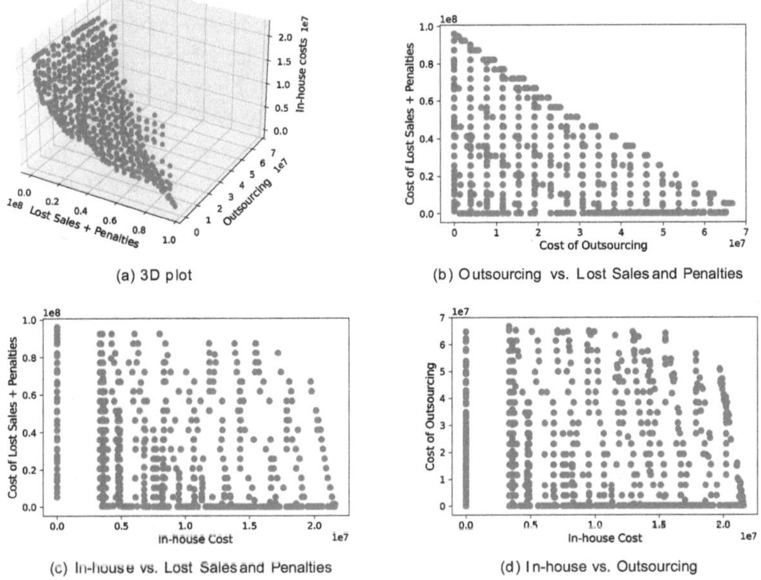


**FIGURE 13.1** 3D Pareto frontier and 2D plots for COVID safety instance. (a) 3D plot, (b) outsourcing vs. lost sales and penalties, (c) in-house vs. lost sales and penalties, (d) in-house vs. outsourcing.

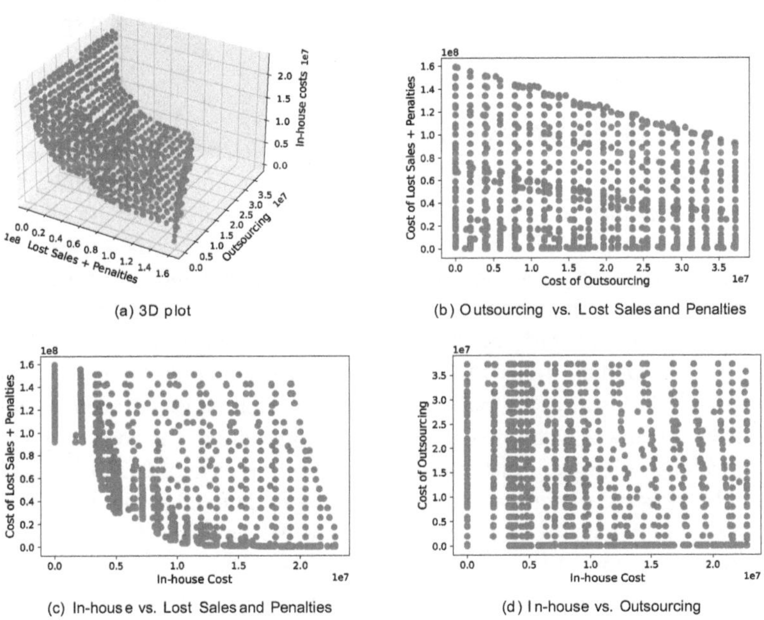


**FIGURE 13.2** 3D Pareto frontier and 2D plots for food instance: (a) 3D plot, (b) outsourcing vs. lost sales and penalties, (c) in-house vs. lost sales and penalties (d) in-house vs. outsourcing.

where $X$ is used to denote the feasible solution space. The $\varepsilon_i$s are upper-bound parameters for the objectives, excluding the primary one that is being minimized. To implement this method, we first generate a payoff table, which is a table containing the results from the individual optimization of the $p$ objective functions. This table contains the lower and upper bounds for each objective. We obtain the range of each one of the $p - 1$ objective functions by subtracting the lower bound from the upper bound. To use these ranges as constraints, each objective function's range is divided into $r_i$ equal intervals using $r_i - 1$ equally distant grid points, which leads to a total of $r_i + 1$ grid points ($\varepsilon_i$s). We use these $\varepsilon_i$s to bound the $i$th objective function. The total number of models run to obtain the Pareto frontier becomes $(r_2 +1) \times (r_3 +1) \times \ldots \times (r_p +1)$. By using the epsilon-constraint method, we are able to control how dense the representation of the efficient set is by carefully choosing values for $r_i$ [37]. That is, with more grid points, the representation of the efficient set would be more dense. However, this comes at the cost of higher computation time. We chose 20 grid points for each $r_i$, resulting in a total of 400 model runs for each objective.

In both instances, from Figures 13.1b and 13.2b, it is clear that a trade-off exists between the costs of outsourcing the products and lost sales and penalties. This means that the more demand outsourced, the less lost sales and penalty costs incurred. The trade-off is significantly more pronounced for the first instance. This is likely due to the availability of products to outsource, that is, the quantities of products to outsource were more limited in the second instance. Although the trade-offs are less significant, we observe from Figures 13.1c and 13.2c that the minimization of in-house costs results in a higher cost of lost sales and penalties. Conversely, from Figures 13.1d and 13.2d, minimizing both in-house and outsourcing costs are objectives that can be achieved simultaneously; that is, the costs incurred would be directly a result of lost sales and penalties.

### 13.5.4 Value of Incorporating Resilience

To measure the value of incorporating resilience, we compare the results from our model to a model that does not have the option to outsource. Specifically, we hold all parameters from our original model constant but remove the outsource facilities. Then, the only way for the company to satisfy demand is through in-house operations. We repeat this analysis for different recovery level parameter values, [3], and observe how the overall objective value and percentage of demand met differs between the two models. Table 13.7 shows that, when given the option to outsource, the objective value is always smaller (i.e., better). That is, under each $\gamma$ value, the overall costs are lower when provided with an alternative way to satisfy customer demand, which is observed for both instances. For the case of COVID safety products, we also observe that about half the time, more demand is met when given the option to outsource. The reason for the negative values in the final column is that the first-stage decisions differ, and the model without outsourcing chose to activate an extra facility to avoid significant costs in lost sales and penalties. Thus, in some cases, the level of demand met for the model without outsourcing is in fact greater than the model with outsourcing. In the second instance, while less significant, the demand met under the model that has the option to outsource is always greater than

**TABLE 13.7**
**Impact of Outsourcing for Different $\gamma$ Values**

| | $\gamma$ | Obj. with Outsourcing | Obj. without Outsourcing | Difference (%) | Demand Met with Outsourcing (%) | Demand met without Outsourcing (%) | Difference (%) |
|---|---|---|---|---|---|---|---|
| COVID safety | 0.50 | 18,226,651.3 | 18,833,546.3 | 3.2 | 94.1 | 91.1 | 3.0 |
| | 0.75 | 18,461,727.8 | 19,344,620.5 | 4.6 | 96.4 | 96.9 | −0.5 |
| | 0.95 | 18,765,472.9 | 19,922,836.7 | 5.8 | 99.0 | 96.9 | 2.1 |
| Food | 0.50 | 18,536,959.3 | 18,566,150.6 | 0.2 | 98.9 | 98.6 | 0.3 |
| | 0.75 | 18,650,024.0 | 18,847,962.3 | 1.1 | 99.1 | 98.6 | 0.6 |
| | 0.95 | 18,856,486.5 | 19,417,571.4 | 2.9 | 99.7 | 98.6 | 1.1 |

the model that does not. Thus, we can conclude that the strategy of outsourcing results in either cheaper overall supply chain costs or more demand satisfied or both.

### 13.5.5 Sensitivity Analysis

We perform a sensitivity analysis by varying the recovery level parameter $\gamma$. That is, we increase $\gamma \in [0.5, 0.95]$ by increments of 0.05, with 0.5 being the lowest level of demand that must be satisfied to avoid penalties and 0.95 being the highest. We only consider $\gamma$ for the sensitivity analysis since we are most concerned with how resilient the supply chain is with regard to satisfying demand. We analyse how the objectives, $f_1, f_2$, and $f_3$ change at different levels. We also observe how many facilities are opened up in the first stage, how much demand is met, and how much of that is outsourced.

From Table 13.8, we observe for both instances that as $\gamma$ increases, the in-house cost, $f_1$, is nearly stagnant, only fluctuating slightly. This would imply that the supply chain is producing and shipping around the same amount at each recovery level. On the other hand, the outsourcing cost, $f_2$, rises as the recovery level, $\gamma$, increases. This indicates more products are outsourced to avoid higher penalties. Therefore, when the company cannot produce in-house anymore, it must outsource the remaining customer demand. Finally, from the third objective, $f_3$, we observe a decrease in the lost sales and penalties as $\gamma$ increases, since more demand is being met.

The "Demand met" and "Demand outsourced" columns refer to the average total demand met and the percentage of demand that was outsourced, respectively. These values increase as $\gamma$ increases. The increase is smaller in the second instance, most likely due to the lower availability to outsource. In the last two columns, the symbol $\|\cdot\|$refers to the $_1$-norm. That is, $x^1_{i\ell 1}$ corresponds to the number of open manufacturing plants, and $x^2_{j\ell 1}$ corresponds to the number of open distribution centres. It is worth noting for both cases that the decisions on which facilities to use for production and distribution do not change as we increase $\gamma$. This would explain

**TABLE 13.8**
**Sensitivity Analysis Results for Different $\gamma$ Values**

| | $\gamma$ | Objective Value | $f_1$ | $f_2$ | $f_3$ | Demand Met (%) | Demand Outsourced (%) | $x^1_{i\ell 1}$ | $x^2_{j\ell 1}$ |
|---|---|---|---|---|---|---|---|---|---|
| COVID safety | 0.50 | 18,226,651.3 | 13,685,882.0 | 1,572,230.5 | 2,968,538.8 | 94.1 | 2.7 | 3 | 3 |
| | 0.55 | 18,262,208.0 | 13,685,964.4 | 1,810,649.0 | 2,765,594.6 | 94.5 | 3.1 | 3 | 3 |
| | 0.60 | 18,304,855.6 | 13,686,125.7 | 2,049,985.0 | 2,568,744.9 | 94.9 | 3.5 | 3 | 3 |
| | 0.65 | 18,352,685.5 | 13,686,274.1 | 2,316,852.6 | 2,349,558.7 | 95.4 | 4.0 | 3 | 3 |
| | 0.70 | 18,405,348.9 | 13,686,415.2 | 2,599,051.3 | 2,119,882.5 | 95.9 | 4.5 | 3 | 3 |
| | 0.75 | 18,461,727.8 | 13,686,481.5 | 2,904,063.8 | 1,871,182.6 | 96.4 | 5.0 | 3 | 3 |
| | 0.80 | 18,522,196.8 | 13,686,363.6 | 3,260,599.6 | 1,575,233.6 | 97.0 | 5.6 | 3 | 3 |
| | 0.85 | 18,589,356.2 | 13,686,257.6 | 3,662,605.7 | 1,240,493.0 | 97.6 | 6.2 | 3 | 3 |
| | 0.90 | 18,671,721.6 | 13,685,939.5 | 4,062,615.3 | 923,166.7 | 98.2 | 6.8 | 3 | 3 |
| | 0.95 | 18,765,472.9 | 13,685,661.5 | 4,501,526.5 | 578,285.0 | 99.0 | 7.6 | 3 | 3 |
| Food | 0.50 | 18,536,959.3 | 17,257,885.2 | 274,637.7 | 1,004,436.3 | 98.9 | 0.3 | 4 | 4 |
| | 0.55 | 18,554,380.4 | 17,257,749.3 | 301,981.2 | 994,649.9 | 98.9 | 0.3 | 4 | 4 |
| | 0.60 | 18,574,035.0 | 17,257,596.8 | 336,342.7 | 980,095.5 | 98.9 | 0.3 | 4 | 4 |
| | 0.65 | 18,595,787.2 | 17,257,575.9 | 406,934.1 | 931,277.2 | 99.0 | 0.4 | 4 | 4 |
| | 0.70 | 18,620,133.5 | 17,257,532.0 | 492,681.2 | 869,920.3 | 99.0 | 0.5 | 4 | 4 |
| | 0.75 | 18,650,024.0 | 17,257,296.4 | 611,543.0 | 781,184.6 | 99.1 | 0.6 | 4 | 4 |
| | 0.80 | 18,686,369.4 | 17,256,956.3 | 717,519.0 | 711,894.1 | 99.2 | 0.7 | 4 | 4 |
| | 0.85 | 18,730,115.6 | 17,256,670.9 | 855,472.2 | 617,972.4 | 99.4 | 0.8 | 4 | 4 |
| | 0.90 | 18,786,550.7 | 17,256,434.8 | 1,032,779.0 | 497,336.9 | 99.5 | 0.9 | 4 | 4 |
| | 0.95 | 18,856,486.5 | 17,256,085.7 | 1,306,394.5 | 294,006.2 | 99.7 | 1.2 | 4 | 4 |

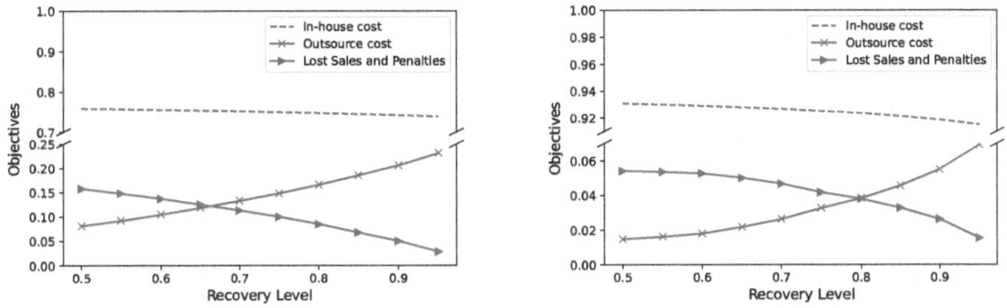


(a) Covid safety (b) Food

**FIGURE 13.3** Objective values vs. recovery level: (a) COVID safety; (b) food.

why the percentage of demand being out-sourced increases as the amount produced in-house hardly changes. These results indicate that model results are quite robust with regard to the changes in the $\gamma$ parameter, as the first-stage decisions do not change. The observed results are easy to conceptualize. For instance, it would make sense that as the required level of demand to satisfy increases, the company has to outsource more. To visualize how the different objective values change with $\gamma$, we plot each objective's percentage of the total cost over the ten $\gamma$ levels, as seen in Figure 13.3.

### 13.5.6 Visuals of Four-Echelon Network

We present visualizations with a four-echelon network in Figures 13.4 and 13.5 to analyse the results on a finer level. The thickness of each edge represents the proportion of transported products among all edges of the corresponding stage. To construct such a plot, we first build the sets of nodes $\mathcal{I}, \mathcal{J}, \mathcal{K}$, and $\mathcal{L}$. The frame of the plot is as follows: nodes $i \in \mathcal{I}$ were placed from the bottom to the top of the left side of the plot, followed by the nodes in $\mathcal{J}, \mathcal{K}$ and $\mathcal{L}$ that were placed the same way in the middle left, middle right, and far right, respectively. Let $(P_i)_{i\in[3]} = \{\mathcal{I}\times\mathcal{J}, \mathcal{J}\times\mathcal{K}, \mathcal{K}\times\mathcal{L}\}$ represent the sequence of stages. The final stage can happen simultaneously with the other two since the products are coming from outsourced facilities. To build three bipartite graphs, we define $\Pi_{P_i} \in \mathcal{Q}$ such that the summation of the width of the edges belonging to $P_i$ is equal to $\Pi_{P_i}$ for all $i \in [3]$ with the following formula:

$$\text{Width}_{(x,y)} = \Pi_{P_i} \frac{\text{Number of goods transported from } x \text{ to } y}{\sum\limits_{(x,y)\in P_i} \text{Number of goods transported from } x \text{ to } y}$$

With this formula, the visuals show that the thicker the edge, the more products are being sent through it. The visualizations are specific to a single scenario and we pick different recovery levels for that scenario. Figure 13.4 is based on the supply chain for COVID safety products, and Figure 13.5 is based on the supply chain for food products. The visuals provide a clear picture of the manufacturing plants and distribution centres that were either operational in the first stage and disrupted in the

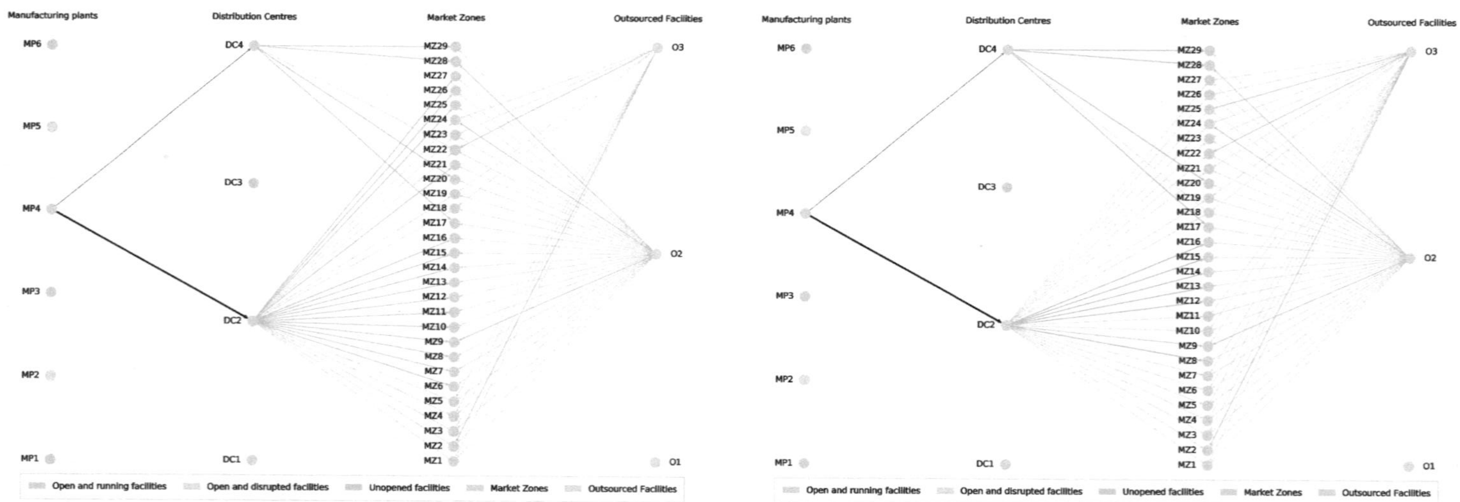


**FIGURE 13.4** Four-stage network visualization at two different γ values for COVID safety products: (a) network visualization at γ = 0.5 and (b) network visualization at γ = 0.95.

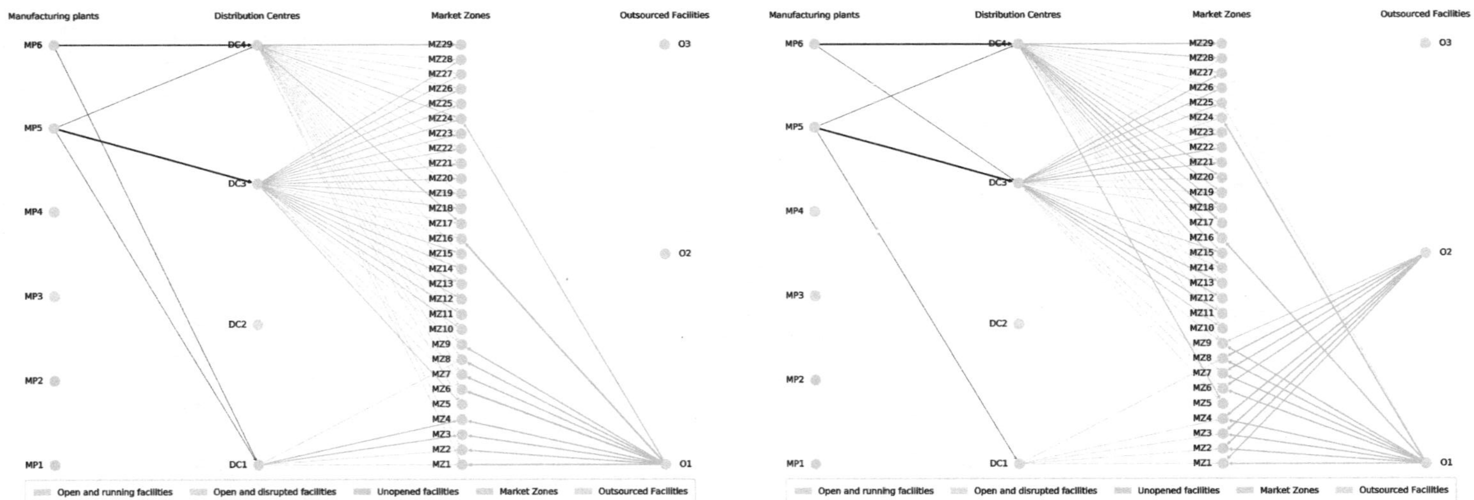


**FIGURE 13.5** Four-stage network visualization at two different $\gamma$ values for food products: (a) network visualization at $\gamma = 0.5$ and (b) network visualization at $\gamma = 0.95$.

scenario (blue), operational in the first stage and operating in the scenario (pink), or not used at all (green). For example, in Figure 13.4, although MP2 and MP5 were set to be operational in the first stage, those plants were disrupted under the given scenario. Thus, the manufacturing plants could not be used to produce products. This figure also gives insight into the usefulness of having outsourcing facilities. Although few manufacturing plants were used, two out of three outsource facilities (O2 and O3) were utilized to deliver products to the market zones. The outsource facilities, O1, O2, and O3, are specific to a single product, which means the company did not outsource one of the products at all. This could be because the product is too expensive to outsource or transport, or the company chooses to produce more of that product in-house. In Figure 13.5, there are two open manufacturing plants and three open distribution centres, leading to less outsourcing compared to the COVID-safety product case. For both instances, while less evident in Figure 13.4, we can see that at the higher [3] level, more outsourcing is done.

Table 13.9 is a numerical representation of the visuals with each row corresponding to Figures 13.4 and 13.5, respectively. We include the percentage of demand that was met under the given scenario, including the percentage met using in-house production and the percentage outsourced. The last three columns describe the average value of the degree of each node, $i \in \mathcal{I}$, $j \in \mathcal{J}$, and $\ell \in \mathcal{L}$. For the first instance (COVID safety), there is a significant increase in the percentage of demand met, for example, when $\gamma$ increases from 0.5 to 0.95, the overall demand met goes from 66.7% to 96.7%. This is because more penalty costs are incurred when demand is not satisfied at the given recovery level. Thus, to avoid significant penalty charges, it is important to meet the demand. For the second instance, there is a jump in the percentage of demand met, however not as significant. There were more operating facilities to begin with, so it was easier to satisfy the demand. Since the manufacturing plant operating decisions do not change at each level of $\gamma$, the demand that is satisfied by in-house production hardly changes. Thus, the percentage change in the overall demand satisfied is directly related to the percentage of demand outsourced, that is, the company outsources more products at higher levels of $\gamma$. This is consistent with what we observed when we averaged the demand met and outsourced over all scenarios in Table 13.8.

**TABLE 13.9**
**Summary Statistics for a Given Scenario Used for Visualizations**

| Instance | $\gamma$ | Demand Met (%) | Demand from In-House (%) | Demand Outsourced (%) | $avg(\delta(\mathcal{I},\mathcal{J}))$ | $avg(\delta(\mathcal{J},\mathcal{K}))$ | $avg(\delta(\mathcal{L},\mathcal{K}))$ |
|---|---|---|---|---|---|---|---|
| COVID | 0.50 | 66.7 | 44.5 | 22.2 | 0.33 | 8.0 | 13.0 |
| safety | 0.95 | 96.7 | 44.4 | 52.3 | 0.33 | 8.5 | 16.0 |
| Food | 0.50 | 83.4 | 77.9 | 5.5 | 0.83 | 13.75 | 3.33 |
| | 0.95 | 96.7 | 77.4 | 19.3 | 0.83 | 12.0 | 6.33 |

## 13.6 CONCLUSION AND FUTURE WORK

While supply chain resilience is widely discussed in the literature, quantitative studies to analyse the effects of pandemics on supply chains, particularly those of essential products, are limited. Many of the mathematical models implemented to analyse supply chain resilience do not consider uncertainty among parameters. Moreover, effective integration of both metrics and strategies to quantify and build resilience is seldom implemented. This chapter contributes to bridging these important gaps by analysing the results of a two-stage stochastic programming model for a supply chain that produces COVID-safety and meat products. We incorporate a resilience metric and investigate the usefulness of a resilience strategy under a pandemic setting that forces facilities into a lockdown. The metric enforces a level of demand that must be satisfied to avoid a penalty charge. The strategy offers an alternative to meeting demand when disruptions occur. The supply chain network consists of manufacturing plants, distribution centres, market zones, and outsourcing facilities. When the manufacturing plants and distribution centres are subject to facility closures, the company can outsource the demand. Given that an analysis of all possible scenarios can become a challenging task as the number of scenarios increases, we estimated an appropriate number of scenarios to be considered in the model through SAA analysis. We measured the importance of stochasticity by computing both the EVSS and the EVPI. We observed trade-offs among the different objective functions via the epsilon-constraint method and presented the Pareto optimal solutions. We performed a sensitivity analysis on the recovery level parameter and observed how the solutions varied at different levels. Finally, we visualized the results for select scenarios by showing how the products are transported from one point to another in the supply chain network.

Through our SAA analysis, we find that despite a large number of scenarios in our finite set, the value of the gap converges quickly and a small number of scenarios are needed to be representative of the entire scenario set. From our EVSS calculations, we observe that the stochastic solutions provide significant benefits over the deterministic ones and that the value of incorporating stochasticity increases as the recovery level increases. Our analysis with multi-objective optimization techniques shows the trade-offs between the cost of outsourcing and the cost of lost sales and penalties. The trade-offs between the other objective functions seem to be much less significant, demonstrating that the minimization of certain costs can be achieved at the same time. The sensitivity analysis demonstrates a clear trend that as we increase the recovery level, there is an increase in the overall demand satisfied and the percentage of demand outsourced, while the in-house decisions hardly change.

Although the methodology and models presented provide useful insights, there are certain limitations to our chapter. For instance, we assumed facilities could either be fully operating or not operating at all. It would be beneficial to analyse how the results differ when facilities operate at a percentage of their regular capacity. Also, although scenario-based approaches are necessary to capture the supply chain uncertainties stemming from demand and disruption events, detecting such scenarios and estimating their probabilities are challenging tasks. Further analysis of real-world situations to better estimate this uncertainty would provide valuable insights, which we leave for future research.

## Disclosure statement

No potential conflict of interest was reported by the authors.

## Data Availability Statement

Data is available upon request.

# REFERENCES


[1] Christopher, M., Peck, H., 2004. Building the resilient supply chain. *The International Journal of Logistics Management* 15, 1–13.

[2] Chen, J., Sohal, A. S., Prajogo, D. I., 2013. Supply chain operational risk mitigation: A collaborative approach. *International Journal of Production Research* 51, 2186–2199.

[3] Mandal, S., 2012. An empirical investigation into supply chain resilience. *IUP Journal of Supply Chain Management* 9, 46–61.

[4] Ivanov, D., 2020a. Predicting the impacts of epidemic outbreaks on global supply chains: A simulation-based analysis on the coronavirus outbreak (COVID-19/SARS-CoV-2) case. *Transportation Research Part E: Logistics and Transportation Review* 136, 101922.

[5] Paul, S. K., Chowdhury, P., 2020. A production recovery plan in manufacturing supply chains for a high-demand item during covid-19. *International Journal of Physical Distribution and Logistics Management* 51, 104–125.

[6] Handfield, R. B., Graham, G., Burns, L., 2020. Corona virus, tariffs, trade wars and supply chain evolutionary design. *International Journal of Operations and Production Management* 40, 1649–1660.

[7] Modgil, S., Gupta, S., Stekelorum, R., Laguir, I., 2021. AI technologies and their impact on supply chain resilience during COVID-19. *International Journal of Physical Distribution and Logistics Management.* Ahead-of-print. DOI: 10.1108/ijpdlm-12–2020–0434.

[8] Berardi, A., Perinelli, D. R., Merchant, H. A., Bisharat, L., Basheti, I. A., Bonacucina, G., Cespi, M., Palmieri, G. F., 2020. Hand sanitisers amid covid-19: A critical review of alcohol-based products on the market and formulation approaches to respond to increasing demand. *International Journal of Pharmaceutics* 584, 119431.

[9] Ivanov, D., 2021. Lean resilience: AURA (active usage of resilience assets) framework for post-COVID-19 supply chain management. *The International Journal of Logistics Management.* Ahead-of-print. DOI: 10.1108/ijlm-11-2020-0448.

[10] Tonsor, G. T., Lusk, J. L., Tonsor, S. L., 2021. Meat demand monitor during covid-19. *Animals* 11, 1040.

[11] Behzadi, G., O'Sullivan, M. J., Olsen, T. L., 2020. On metrics for supply chain resilience. *European Journal of Operational Research* 287, 145–158.

[12] Ivanov, D., Pavlov, A., Dolgui, A., Pavlov, D., Sokolov, B., 2016. Disruption-driven supply chain (re)-planning and performance impact assessment with consideration of pro-active and recovery policies. *Transportation Research Part E: Logistics and Transportation Review* 90, 7–24.

[13] Behzadi, G., O'Sullivan, M. J., Olsen, T. L., Scrimgeour, F., Zhang, A., 2017. Robust and resilient strategies for managing supply disruptions in an agribusiness supply chain. *International Journal of Production Economics* 191, 207–220.

[14] Garcia-Herreros, P., Wassick, J. M., Grossmann, I. E., 2014. Design of resilient supply chains with risk of facility disruptions. *Industrial and Engineering Chemistry Research* 53, 17240–17251.

[15] Fattahi, M., Govindan, K., Keyvanshokooh, E., 2017. Responsive and resilient supply chain network design under operational and disruption risks with delivery lead-time sensitive customers. *Transportation Research Part E: Logistics and Transportation Review* 101, 176–200.
[16] Losada, C., Scaparra, M. P., O'Hanley, J. R., 2012. Optimizing system resilience: A facility protection model with recovery time. *European Journal of Operational Research* 217, 519–530.
[17] Hishamuddin, H., Sarker, R. A., Essam, D., 2013. A recovery model for a two-echelon serial supply chain with consideration of transportation disruption. *Computers and Industrial Engineering* 64, 552–561.
[18] Fattahi, M., Govindan, K., Maihami, R., 2020. Stochastic optimization of disruption-driven supply chain network design with a new resilience metric. *International Journal of Production Economics* 230, 107755.
[19] Kouvelis, P., Kurawarwala, A. A., Gutierrez, G. J., 1992. Algorithms for robust single and multiple period layout planning for manufacturing systems. *European Journal of Operational Research* 63, 287–303.
[20] Queiroz, M. M., Ivanov, D., Dolgui, A., Wamba, S. F., 2020. Impacts of epidemic outbreaks on supply chains: Mapping a research agenda amid the covid-19 pandemic through a structured literature review. *Annals of Operations Research* 319, 1159–1196.
[21] Peng, P., Snyder, L. V., Lim, A., Liu, Z., 2011. Reliable logistics networks design with facility disruptions. *Transportation Research Part B: Methodological* 45, 1190–1211.
[22] Medal, H. R., Pohl, E. A., Rossetti, M. D., 2014. A multi-objective integrated facility location-hardening model: Analyzing the pre-and post-disruption tradeoff. *European Journal of Operational Research* 237, 257–270.
[23] Pishvaee, M. S., Razmi, J., Torabi, S. A., 2014. An accelerated benders decomposition algorithm for sustainable supply chain network design under uncertainty: A case study of medical needle and syringe supply chain. *Transportation Research Part E: Logistics and Transportation Review* 67, 14–38.
[24] Fahimnia, B., Jabbarzadeh, A., 2016. Marrying supply chain sustainability and resilience: A match made in heaven. *Transportation Research Part E: Logistics and Transportation Review* 91, 306–324.
[25] Chen, H., Cullinane, K., Liu, N., 2017. Developing a model for measuring the resilience of a port-hinterland container transportation network. *Transportation Research Part E: Logistics and Transportation Review* 97, 282–301.
[26] Fahimnia, B., Jabbarzadeh, A., Sarkis, J., 2018. Greening versus resilience: A supply chain design perspective. *Transportation Research Part E: Logistics and Transportation Review* 119, 129–148.
[27] Ghavamifar, A., Makui, A., Taleizadeh, A. A., 2018. Designing a resilient competitive supply chain network under disruption risks: A real-world application. *Transportation Research Part E: Logistics and Transportation Review* 115, 87–109.
[28] Goldbeck, N., Angeloudis, P., Ochieng, W., 2020. Optimal supply chain resilience with consideration of failure propagation and repair logistics. *Transportation Research Part E: Logistics and Transportation Review* 133, 101830.
[29] Kleywegt, A. J., Shapiro, A., Homem-de Mello, T., 2002. The sample average approximation method for stochastic discrete optimization. *SIAM Journal on Optimization* 12, 479–502.
[30] Ivanov, D., 2020b. Viable supply chain model: Integrating agility, resilience and sustainability perspectives—Lessons from and thinking beyond the COVID-19 pandemic. *Annals of Operations Research.* DOI: 10.1007/s10479-020-03640-6.
[31] Haren, P., Simchi-Levi, D., 2020. How coronavirus could impact the global supply chain by mid-march. *Harvard Business Review* 28.

[32] Hobbs, J. E., 2021. The covid-19 pandemic and meat supply chains. *Meat Science* 181, 108459.
[33] Rude, J., 2020. COVID-19 and the Canadian cattle/beef sector: Some preliminary analysis. *Canadian Journal of Agricultural Economics/Revue canadienne d'agroeconomie* 68, 207–213.
[34] Google Maps, 2021. URL: www.google.ca/maps.
[35] Statistics Canada, 2021. URL: www.statcan.gc.ca/eng/start.
[36] Mousavi, K., Bodur, M., Roorda, M. J., 2021. Stochastic last-mile delivery with crowd-shipping and mobile depots. URL: www.optimization-online.org/DB_HTML/2020/02/7619.html [Accessed: Sept. 18th, 2021].
[37] Mavrotas, G., 2009. Effective implementation of the *e*-constraint method in multi-objective mathematical programming problems. *Applied Mathematics and Computation* 213, 455–465.

# *Section 4*

## *Digital Innovation, Transformation, and Sustainability Industry Cases*

# 14 Profitability of Shipping and the Role of Fleet Ownership—North European Company Case Study

*Ulla Tapaninen and Olli-Pekka Hilmola*

## 14.1 INTRODUCTION

Sustaining profitability in shipping is difficult to achieve in the long term, and its drivers are a major question in the shipping business. Branch and Robarts (2014) identify fleet management as one of the main areas in successful shipping management. In fleet management strategies, there are various issues to be considered, such as markets to serve, fleet capacity, cargoes, revenues, sailing programmes, quality management, surveys and repairs, crew management, ports of call, IT, classification and choice of flag state.

There has been interest in the academic literature concerning whether to own or to charter the tonnage that a shipping company is operating (Meng and Wang, 2011; Park et al., 2018; Tapaninen, 2020; Hiekata et al., 2022). Some companies use a strategy of owning all the vessels they operate, while others only use chartered vessels. In addition, there are shipping companies that do both, owning their basic tonnage and chartering vessels for additional needs. However, the success of the ownership strategy is highly dependent on the level of charter prices and when the vessels are needed (Meng and Wang, 2011; Park et al., 2018).

These maritime phenomena are referred to as maritime fleet size and mix problems (Pantuso et al., 2014; Hiekata et al., 2022). First, we conducted a literature review of the methods. It was found that the specific sub-problem, whether to own or charter the vessels, has received limited attention in the academic literature, in particular when the ownership strategy is related to the financial performance of shipping companies. We also examined key findings of the case of Hanjin Shipping, which previously has been well documented (Shin et al., 2019).

The question of owning or chartering the vessel is very important for shipping companies. When the company owns a vessel, it has more power in developing the operating procedures of the vessels. In chartered vessels, this link is more distant,

DOI: 10.1201/9781003438748-18

as everything must also be agreed with the ship owner. As is shown in the literature review in Section 14.2, there exist theoretical studies of best possible ownership versus chartering strategies, but relevant literature lacks empirical case studies. In the real world, all managerial decisions are more complex than can be modelled with simulation or optimisation methods. Therefore, our empirical long-term case study (1994–2022) can give some insights that cannot be achieved in modelling studies.

The problem of the size of the personnel has received less interest in academic literature (see discussion, for example, in Branch and Robarts (2014) and Tapaninen (2020)). However, lately, as there has taken place a scientific breakthrough in autonomous shipping, the effect of minimising or eliminating crew costs has also received more attention. Furthermore, we will look at the potential of personnel costs in the profitability of a shipping company.

The methodology of the study is based on case study research with triangulation. The case study methodology is well suited for this type of study, where there does not exist enough data for pure statistical analysis. Yin (2009, p. 18) defines a case study as an empirical study investigating a phenomenon in its real-life context. Eisenhard (1989) states that a case study is well suited to studying phenomena that is new or where there does not exist sufficient statistical data, in particular when creating novel theory.

Triangulation means that multiple methods are used to study the research question (Patton, 1999). Triangulation is mainly used in qualitative research, but it's also commonly applied also in quantitative research. In this study we use mixed methods research, where we first study the phenomenon using both qualitative and quantitative data and only thereafter make some statistical analysis to possibly prove our observations. As background we have a literature review supporting the findings.

In this study, the following research questions are studied:

- How did the case company's sales, profit and return on investment (ROI, %) develop in 1994–2022?
- How did financial performance develop as fleet ownership strategy changed?
- What was the role of the amount of ship crew in this process?

According to principles of case study research, we look at the subject from various angles. First, we examine what has been stated in the literature review; then we explain the financial situation where the company has been operating during the time period under study. Then we examine the longitudinal development of major financial key performance indicators (KPIs, e.g., profit and sales). In addition to have more insight in the company, we perform a regression analysis of possible drivers of ROI and its sub-components (profit margin and asset turns). Finally, we draw conclusions and discuss findings and their validity.

In Section 14.2, a literature review is completed of both fleet size and the mix problem as well as the importance of crew costs within shipping. Thereafter, in Section 14.3, we introduce the operating environment of Finnish foreign trade and logistics (the main operating country of the case study shipping company). The examination is longitudinal, and it concerns overall cargo analysis but in addition sub-cargo groups such as ro-ro and ro-pax. Section 14.4 presents a case study from a roro (roll-on/roll-off) shipping company operating Finnish, Swedish and German routes (pseudonymously called Upsilon in the following). The company is one of the largest shipping companies operating short-sea shipping with ro-ro vessels between Finland and central Europe. The timeframe of this study is 1994–2022, a time when the company's geographical focus and operating model (type of vessels and customers) remained unchanged, but the profitability of the company altered significantly. We look at the reasons for the change and its financial outcomes. Together with these, we try to dive deeper into understanding the drivers of profitability in the observation period within this company by building regression models in Section 14.5. At the end of the research in Section 14.6, we discuss the findings and limitations of the study. In addition, we point out directions for future research.

## 14.2 LITERATURE REVIEW

A successful shipping business is very much about how to manage a company in changing shipping cycles. Stopford (2009) describes these cycles in three groups: long-term cycles that are driven by changes in world industry; short-term cycles due to changes in the world economy; and seasonal cycles, such as those driven by agriculture. Tapaninen (2020) mentions that the correct timing of ordering, selling and scrapping ships and buying and selling used tonnage is often considered more important than operational decisions (like selling cargo space at the right price). Often companies do not only acquire new ships but also focus on transactions in the second-hand market—sale and purchase activity and processes are an important part of profitability of shipping (Park et al., 2018).

Shipping companies have several ways to deal with demand cycles and other strategic issues. Dulebenets et al. (2021) define the problems a liner shipping company faces by three groups: strategic, tactical and operational. Strategic-level problems are fleet size and mix, alliance strategy and network design. Tactical-level problems are service frequency determination, fleet deployment, sailing speed optimisation and vessel scheduling. Operational-level decisions are cargo booking, cargo rooting and vessel rescheduling. Santos et al. (2022) have developed a methodology for short-sea-shipping service design within intermodal transport chains.

Tapaninen (2020) mentions that the size of the crew has an effect on shipping profitability. In principle, a bigger crew translates to a higher cost level for the shipping company. Consequently, there has been a constant tendency to decrease the size of the crew compared to the volumes transported, such as by increasing the size of the vessels. In addition, the number of crew members also depends on how watch keeping is organised, work and free time, the ship's area of operation, type of operation of the vessel, cargo, passengers, maintenance and servicing and safety aspects. Also, the flag state has a big effect on crew costs. Karvonen and Jousilahti (2020) calculated that in the Finnish ro-ro and ro-pax business, where the case study company of this research work operates, the manning costs were 5–8% of total costs.

During recent years, studies of autonomous vessels have been focusing on the importance of the costs of the crew. It has been estimated in various studies that autonomous shipping would decrease the crewing costs. Ghaderi (2019) and Kretschmann et al. (2017) found that short-sea shipping operators can benefit from cost reductions by implementing autonomous technologies, while Hannaford and Hassel

(2021) found that reducing crews and increasing shipboard automation have potentially negative effects on the deck officers, including sensor over-reliance, decreased situational awareness and increased complacency, while providing no reduction in onboard duties or fatigue. However, it should be noted that these studies look at only on the number of crew on board and not the whole personnel of the company.

The objective of maritime fleet size and mix problem (MFSMP) is typically to minimise the total cost of setting up and operating a fleet of a shipping company. The problem has been widely addressed by operations research models. For extensive review of MFSMP problems and algorithms see, for example, Pantuso et al. (2014) and Hoff et al. (2010).

Already half a century ago, Everett et al. (1972) presented the problem of composing the US merchant marine fleet to get the best ship designs and sizes for a fleet of tankers and bulkers supposed to carry 15% of the US foreign trade. In the 1980s, Davies (1983) studied the strategic-level problem of the link between fleet size and profitability, coming to the conclusion that profits and unit costs in liner shipping are highly sensitive to the degree of capacity utilisation. Later, Alizadeh and Nomikos (2007) studied the performance of trading strategies in the sale and purchase market of dry bulk ships. Fusillo (2004) showed that liner shipping fixed capacity brings problems as demand fluctuates.

Hiekata et al. (2022) states that research related to the MFSMP mainly focuses on the design of the overall fleet size and mix of vessel type for a static shipping demand and does not consider fleet change to meet a dynamic demand over time. Once the fleet composition has been determined, it must be updated in response to subsequent changes in demand. The fleet operator needs to make decisions such as buying, chartering and scrapping of vessels.

As Meng and Wang (2011) illustrate, multi-year liner fleet planning is a dynamic activity, and typically, for example, in growing transportation markets, smaller ships are being sold after some usage time, and fewer larger ones are acquired. Based on their study, in the short term, chartering of ships could be reasonable and even cheap, but in the long term, its benefits can be questionable, as the sale of ships often leads to sizeable profits.

Cariou and Wolff (2013) note that even though there are multiple strategies for the MFSMP problem, there are fewer studies on the decision of whether to own or to charter vessels. Liner operators always own a certain number of vessels to meet the minimum requirements of their contract engagements, but they also charter extra tonnage to meet specific market requirements (see also Plomaritou and Papadopoulos, 2019).

D'Agostini et al. (2019) mention chartering provides a more flexible solution, as ships are almost immediately available, and the capital required is usually lower except in low-demand markets, when vessels are sold almost at scrap prices. According to their study, large container shipping lines have very different chartering strategies. Hapag-Lloyd had the lowest share of chartered ships with 37.5% and ZIM the highest with 93.9% of ships on charter.

Cariou (2008) suggests that chartering instead of ordering new vessels provides more flexibility and reduces the initial capital requirement and the delivery time but is more costly in the long run. In addition, Cariou and Wolff (2013) name several advantages of chartering vessels instead of owning them: First, it reduces the debt-leverage ratio. Second, chartering offers the possibility of allocation of resources to other activities, like door-to-door services. Third, it secures future slot-costs that are pre-established in the charter party. Fourth, chartering provides indirect access to ship funds, leasing or partnership structures. Finally, it is a way for a company to benchmark its owned vessel performance with those of chartered vessels.

Bhonsle (2022) states even more advantages and disadvantages when it is more appropriate to charter or to own the fleet a company is operating. The advantages of owning are, for example, control of vessel long-term utilisation and maintenance, lower operating expenses and better cash flow as well as better long-term planning. The disadvantages of owning are higher capital need and possible interest burden and risk of high fixed costs when there is a lack of cargo. On the contrary, Bhonsle (2022) states that the advantages of chartering are minimal capital expenditures, vessels can be deployed into service very rapidly after demand increases, avoiding long construction time and having an option to return vessels in case of a downturn in the market.

However, the disadvantages of chartering are higher operative cash outflows and potential problems regarding maintenance, and, in case of high demand, it will be difficult to find appropriate tonnage to charter, leaving carriers unable to deploy adequate capacity, resulting in revenue losses and missing the chance to take advantage of cyclical upturns.

Finally, Cariou and Wolff (2013) mention that there is a lack of studies that connect shipping companies' chartering decisions with financial performance. The issue is, of course, very complicated, as there are multiple choices that have an influence on the financial performance of a shipping company.

In shipping, there have been decades of success as well as expansion, and then decades of declining markets. When markets are on a favourable path, new capacity is needed, and even fewer efficient shipping companies produce sufficient profits. However, in declining markets, capacity needs to meet demand, and it should be as efficient as possible. For example, in 1970s liquid bulk shipping (oil) was booming and offered shipping capacity was high. However, it took more than two decades that oil tanker fleet was at same level as it was in the year 1980 (Lun et al., 2013). So, a lot of fleets were simply removed from the markets (by scrapping).

Similar problems persisted in shipping after the global financial crisis (GFC) in 2008 and in the decade of 2010–2019. For example, container shipping companies ordered ever larger ships prior to the GFC, when demand was continuously increasing. These orders were not cancelled, and ship size and efficiency had major importance in container shipping strategy in 2010–2020. Smaller ships, still totally functional, were just removed from markets and scrapped as they were not economically competitive. In container shipping markets, growth of supply (ships) was larger than growth in demand during the years 2012–15 as well as in 2018–20 (United Nations, 2021). Only in 2010, as the global market recovered from the GFC, was demand clearly larger than ship supply growth.

An example of this container shipping crisis was Hanjin Shipping. The company went bankrupt in the autumn of 2016 (Shin et al., 2019). Hanjin operated mostly in the container shipping segment and was one of the top 10 companies in this industry (in 2015 Hanjin was the eighth largest with market share of 3.1%; United Nations,

2016). After the GFC it decided to charter larger vessels to keep its position and market share. In the depressed freight rates, this was financially less beneficial, as charter rates were too high for a low-cost strategy. As declining and low-cost emphasis markets in container shipping persisted longer than a few years, it was evident that this charter strategy led to financial difficulties. Eventually, bankruptcy was caused by an insufficient amount of working capital.

## 14.3 FOREIGN TRADE AND ITS LOGISTICS IN FINLAND

Finland is almost entirely dependent on maritime transport. In 2022, 92.5% of the Finnish foreign trade of goods was carried out by sea (Finnish Customs, 2024a). In principle, only Russia's and some Commonwealth of Independent States (CIS) countries' goods trade was also carried out by road or rail. The Russian military invasion of Ukraine in February 2022 changed this, and land-based modes have significantly lost volume. Not only has pure Russian trade serving land-based modes been hurt, but indirect effects have also taken place. During the COVID-19 pandemic, Chinese container trains connecting Finland were growing substantially but have now completely disappeared (e.g., Hilmola & Li, 2023).

Before the global financial crisis of 2008–2009 there was an almost constant predictable increase in Finnish maritime transport, which can be seen in Figure 14.1 (with some notable decreases, like in 1975, 1995 and 2005). After the GFC, Finnish total maritime transport has not been constantly growing any more but has rather varied between 82 and 104 million tons. The highest volume could be detected in the year 2018, and thereafter the country has faced a continuous decline in sea port

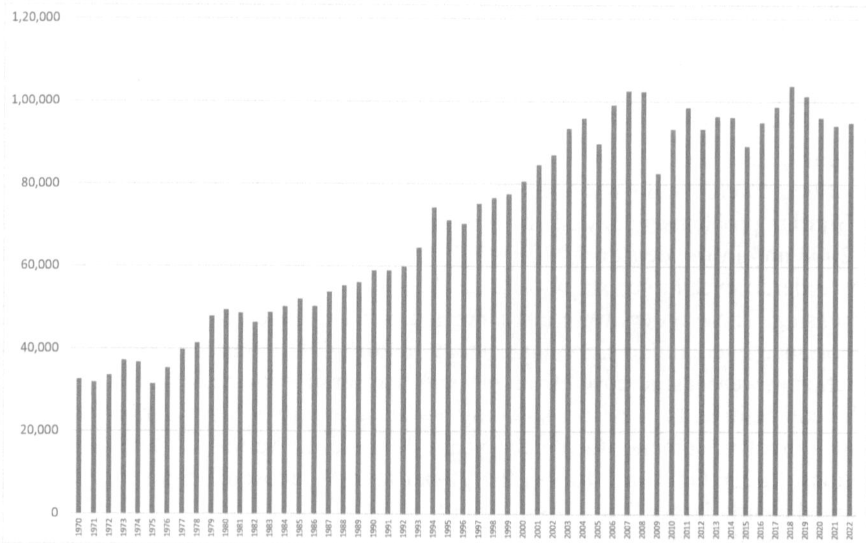


**FIGURE 14.1** Overall sea port handling of Finland during 1970–2022 ('000 tons).

**Source (data): Statistics Finland (2024a).**

handling (only to recover slightly in 2022). Overall, this means that Finnish tied shipping market has clearly matured after GFC. In this environment, revenue growth has been difficult to obtain. This also has its implications for the general profitability of actors of this sector. Both partly answer our research question 1 of the study.

Finland has been going through major transformations in the recent decades. Lower-value production exports still hold significance, but their volumes are not growing any longer. In higher-valued goods, changes have been rather significant and constant—these goods are not seen in large amounts (calculated in tons) in transport statistics. Moreover, the share of service export has reached to 27% of total exports in value, and cargo export is only 63% in value—hence, seaborne exports are only 55% of total exports from Finland in value. The country is also part of the European Union, and the debt crises of last decade have also affected its economy and consumption development.

In 2021, 11.4 million tons (12%) of Finnish foreign trade was carried in containers. Trucks and trailers (that are the main cargo groups of the case study company of this study)—are carried on ro-ro or ro-pax vessels or passenger-car ferries, in total 16.6 million tons (18% of Finnish foreign trade). In the recent two decades, trucks and semi-trailers (see Figure 14.2) in shorter maritime routes, particularly between Estonia and Finland, have shown consistent growth.

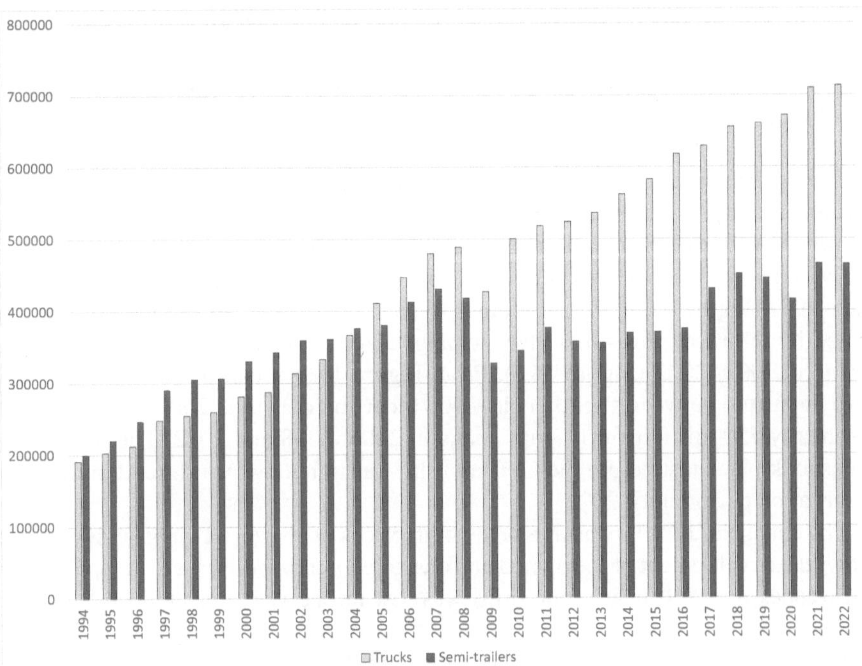


**FIGURE 14.2** Number of trucks with semi-trailers ("trucks") and solely semi-trailers handled through Finnish sea ports during period of 1994–2021.

**Source (data): Statistics Finland (2024b) and own research.**

## 14.4 CASE STUDY: LONGITUDINAL FINANCIAL RESULTS OF A SHIPPING COMPANY

In this study, we focus on a North European shipping company, pseudonymously named Upsilon. The company has been (during the years under study) in a stable situation in terms of geographical focus and trade volumes, but there have been significant changes in the ownership of the vessels and the number of personnel. In this study, we look at the effect of these two factors on the profitability of the company.

Upsilon has been one of the most successful shipping companies in Finland. It has operated in the last 30 years practically in the same geographical area: Finland–Germany, Finland–Sweden and some lines to North Sea (Finland–UK, The Netherlands and Belgium), the same vessel types (ro-ro/ro-pax vessels) and with the same customer base. The customer base has been Finnish export of break bulk (mainly forestry products) and export and import of trucks and trailers with consumer goods and industrial raw materials. The company has not been operating in fast-growing Finland-Estonia truck traffic enabled by a short sea route. Therefore, Upsilon has been positioned in mature markets, where further growth has been difficult to obtain after the GFC (the basis for research question 1 of the study).

The company was established in 1947 mainly to serve the Finnish export of forest products to the United States, UK and Western Europe. From the late 1960s onwards, the shipping company concentrated on ro-ro vessels. The paper industry and its export remained the main focus. In imports, it transported trailers and lorries carrying consumer goods and industrial products.

In 1986 a new daughter company was founded to carry cargoes between Finland and Sweden. In 1989 the company decided to separate their cargo-carrying operations from passenger operations.

During 2007 a larger international shipping company became the largest owner of Upsilon and expressed interest in purchasing the entire company. However, this took a number of years to materialise: In late 2015 the full acquisition was completed.

Even though the operations and net turnover of Upsilon have been quite stable since 1997 (Figure 14.3), the company's financial performance and profitability presented by return on investment have changed tremendously during the observation period of 1994–2022 (Figure 14.4). From Figure 14.3, it can be seen that net sales have fluctuated between slightly below 500 (years before 1998, 2009, 2016 and 2020) and nearly 750 million euros, with peaks in 2005, 2008 and 2022. The highest revenues took place just before the GFC and show a slight downward trend in the entire observation period until 2022.

The change of ROI-% was due to the high prices of chartered tonnage; Figure 14.5 shows the development of the owned and chartered number of vessels. A rapid decline of revenues in 2009 (nearly one third was lost in one year) resulted in declines of ROI-% (Figure 14.4).

Even if the turn in 2009 was rather rapid and deep, it could be said that ROI was in a declining state from 2006 onwards, when the ROI level was 9.9 %, which eventually declined to −1.7 % in 2009. The decline for the worse actually started earlier than 2008, when the ROI was already as low as 2.9 %. It could be estimated that the ROI had a clear cycle from a high level to lows, starting in the late 1990s and ending in 2009.

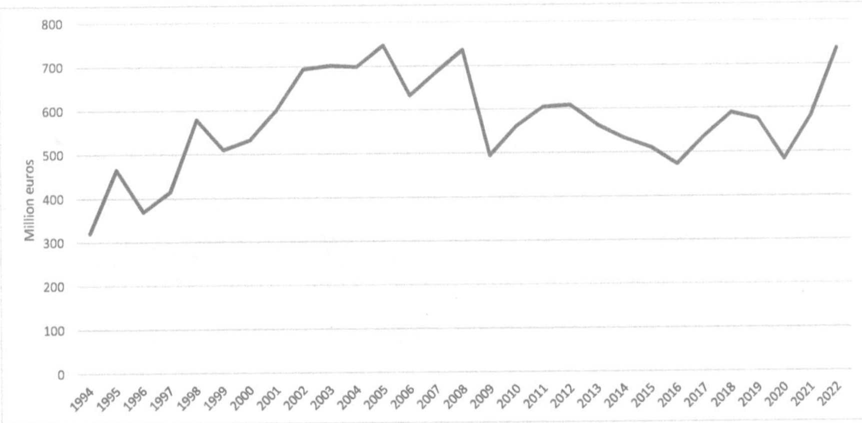


**FIGURE 14.3** Annual sales (net turnover) of the case study company during years 1994–2022.

**Source (data): annual reports.**

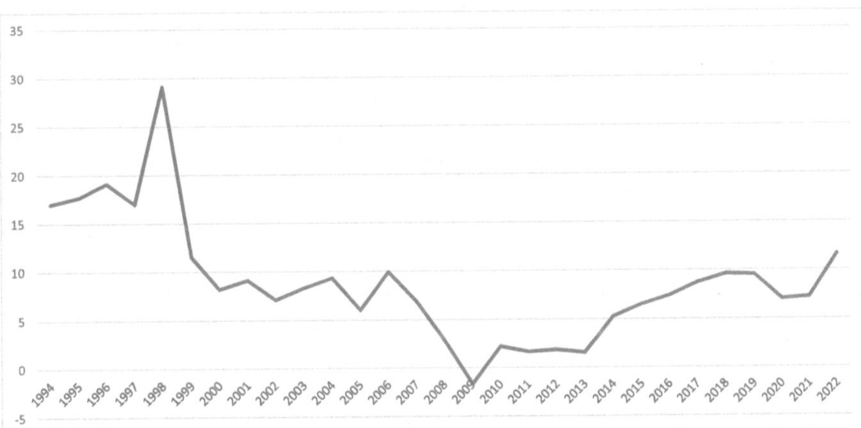


**FIGURE 14.4** Return on investment development of the case study company during 1994–2022.

**Source (data): annual reports.**

In 2007, the case study company changed its strategy from a mixture of ship chartering and ownership to just pure ship ownership. Figure 14.5 shows that starting in 2009, the number of chartered vessels decreased tremendously. However, it took five years before the ROI increased to 5% or higher (Figure 14.4), but of course some improvement was seen from 2010 onwards. This illustrates well that, with a delay, profitability started to improve in maturing markets due to fleet ownership strategy change. Earlier in the era of growth (prior to the GFC and especially in the 1990s), fleet ownership was based on both owning vessels but also on large-scale chartering and resulted in high profitability then. This partly answers our research question 2.

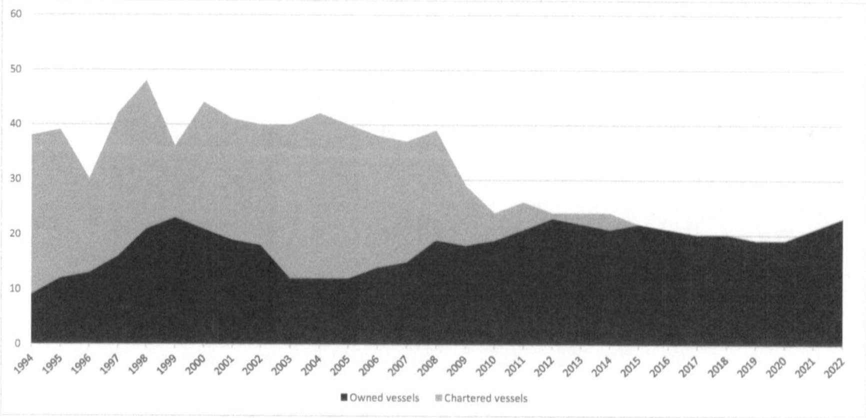


**FIGURE 14.5** Number and ownership of vessels in the case study company's fleet during years 1994–2022 (ro-ro/ro-pax vessels only in this chart).

**Source (data): annual reports.**

The change in vessel ownership strategy was significant: In 2000 the share of chartered vessels was somewhat above half (see Figure 14.5). Starting from 2015, there were no chartered vessels at all. The case study company's (from 2015) annual report highlighted the strategic shift, which required a decade to be completed and needed an investment of 1 billion EUR. The company stressed that they want to own their vessels and charters are not used at all. The charter market was described in the annual report as volatile, and these vessels did not always serve the purpose and needs of customers on a cost-efficient basis. This was also in line with the owner's strategy of maximising vessel transport capacity to achieve economies of scale and scope and the maximum environmental benefits at the same time (Haralambides, 2013).

The strategy change had implications for two components of the ROI formula—asset turns, referring to a company's sales relative to the value of its assets (Figure 14.6), and profit-loss margin (%), referring to profit or loss compared with sales (Figure 14.7). As a company concentrates more in the ownership of its used assets, typically relative margins improve, and asset turns decline. As Figure 14.6 illustrates, the asset turns of Upsilon were at the level of 0.6–0.8 at the beginning of observation period and in the heights of chartering. With the new strategy (favouring ownership), asset turns declined to a level of 0.4–0.5.

This shipping company has today more of its own capital than external debt in its assets. This has an upside. Profit margins (%) have improved a lot (see Figure 14.7)—in recent years, they have reached the level of 15%, which is substantially higher than what the situation was at the beginning of the century. Ownership of vessels saves costs in the profit and loss statement, as excessive premiums are not paid from charters to other organisations. This has been especially beneficial in times after the GFC, as interest rates were very low until 2022. Similarly, with profitability, improvements in profit margins needed time after the fleet ownership strategy changed (Figure 14.7). There is clearly a time delay factor with the fleet decision being made and its effect on profit margin (note research question 2).

**FIGURE 14.6** Asset turns of the case study company (sales/assets) during years 1994–2022.

**Source (data): annual reports.**

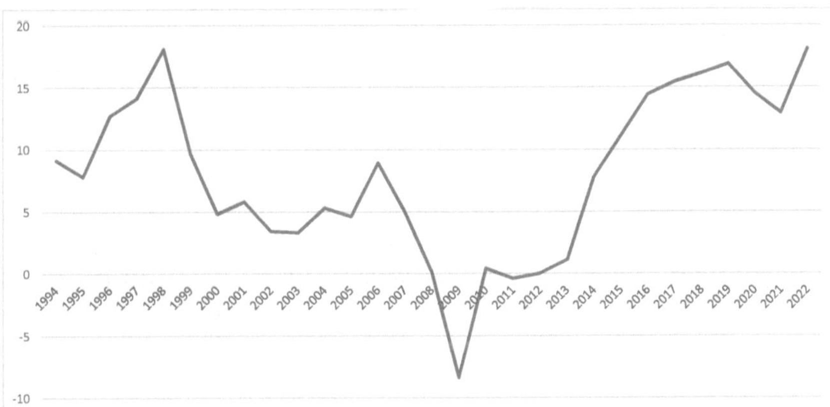


**FIGURE 14.7** Net profit-loss (%) of the case study company during years 1994–2022.

**Source (data): annual reports**

What is often overlooked in a vessel owning strategy within the literature is the fit, efficiency and productivity of the fleet in overall operations. When analysing the number of employees (entire shipping company) and in addition sea personnel, it is clear that a new strategy involves a decreased number of sea personnel (see Figure 14.8), and revenues per employee are higher. This is mainly due to decreasing the number of vessels and increasing vessel sizes.

The total number of Upsilon's employees increased up to the GFC and has thereafter clearly been on a declining path. The number of sea personnel in turn increased up to 2012, and after that it has been in a slight decline. In the five-year period of 2005–2009, revenues per employee (all, not only sea personnel) were around 0.29 million EUR, while within the five-year period of 2017–2021, the revenue per

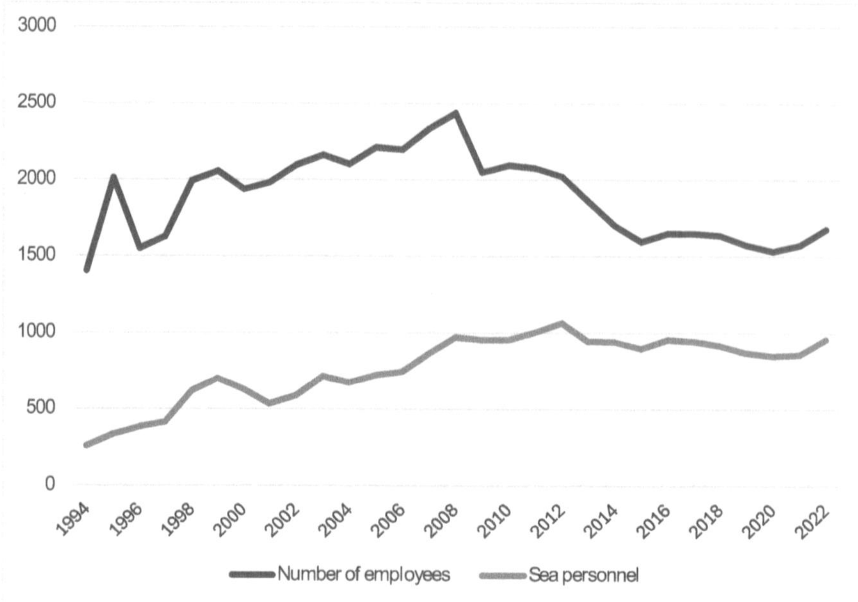


**FIGURE 14.8** Number of employees and sea personnel of the case study company during 1994–2022.

**Source (data): annual reports.**

employee increased to 0.35 million EUR (18.5% increase). The figures are not totally justifiable, as it is not known whether the chartered vessels were bareboat or voyage charters. It should be noted that as the number of vessels has decreased, their size has considerably grown, allowing fewer crew members per cargo ton transported (note research question 3 of the study).

## 14.5 REGRESSION MODELLING

According to case study research methodology and triangulation requirement, we looked at this case from various angles and with several methods. In this section we look at statistical analysis of the case study company, based on the data presented previously.

We developed regression models with the data of 1994–2022 to analyse what the drivers of profitability (ROI-%) were. In Appendix A, we analyse first regressors for ROI-%, then for two main components of ROI, profit-loss (%) margin and asset turns.

As dependent and independent variables are so interconnected, it should be noted that the used regression models contain possible multicollinearity issues. For example, chartered vessel data has a positive correlation with the number of employees (0.473) and a negative correlation with the amount of owned vessels (−0.73). The number of sea personnel is positively correlated with owned vessels (0.657). In some research works, the correlation coefficient threshold for multicollinearity of independent variables is 0.7 (Sy et al., 2017; Derbali and Jamel, 2018), while in others it has been 0.8 (Shrestha, 2020). So, multicollinearity could be an issue in these presented regression models, as it is in the area of the threshold.

In these models, we found that statistically significant drivers for higher ROI-% were both number of employees (the lower the number, the better the ROI) and number of chartered vessels (the higher the amount, the better the ROI). Owned vessels were rather close to statistical significance ($p = 0.056$ vs. 0.05), also indicating that a higher number of own vessels would result in a higher ROI. In general, one vessel adds around 0.6% units to ROI (chartered co-efficient higher than owned). This is a rather high ROI increase and illustrates well the critical decision of the amount of fleet in a shipping company in particular demand circumstances. However, it should be noted that adding 35 employees (co-efficient −0.0175) more will have the same-sized effect as adding a new vessel. Thus, this is negative. So, it is vital that larger-scale operations be efficient and even more efficient than what the already existing fleet offers.

In profit-loss (%) margin, number of employees was the single factor with statistical significance (the lower the amount, the better the margins; co-efficient −0.019). In asset turns, chartered vessels were analysed to be drivers (positive but small co-efficient, 0.007), where the number of sea personnel was negative (very small co-efficient, −0.0003).

These results highlight two issues: the importance of employee productivity (not only sea personnel but the entire shipping company) and the role of chartered vessels in building high profitability performance. This brings a very important finding to our discussion. ROI-% was at high levels with chartered strategy during the years 1994–2007. This is mostly because the era could be characterised as a high economic boom in shipping (in the markets of Upsilon). However, due to the GFC the situation became very difficult in 2008–2009, and strategic change was needed to tackle the crisis as overall markets matured.

After 2015, with the entire ownership of vessels, ROI-% is again at high levels in 2015–2022, as it was in the latter part of high-performance period (not nearly as high as in the 1990s, which is in a class of its own). However, at this time, the high ROI-% is based on lower workforce and therefore higher employee productivity. In the time of a transition out of charters, it could be said that the case study company was in a mixed state, where chartered vessels were still used, and the number of employees was in decline. Thus, this did not prove to produce high ROI-%. However, the role of personnel should be seen in a larger context, and an examination should concern the entire shipping company. In Upsilon, all personnel costs have been in recent years (2018–2022) somewhat above 15%, as they were in 2009–2010 at the level of 20% or higher. As said, labour productivity is a critical part of a shipping company's profitability. The amount of labour was affecting all three regression models. Our research question 3 was tied to the number of sea personnel and their role—it seems that it has a direct effect on asset turns, but number of employees should rather be seen on a total basis concerning the entire shipping company (it directly affects profitability).

Our research question 2 concerned the role of fleet ownership change and its impact on profitability. Regression modelling results indicate that the amount of owned vessels possibly has a direct link to profitability (nearly statistically significant regressor). A regression model also had a statistically significant driver of chartered vessels—this overall indicates that ownership strategy is linked to profitability and should be fit for a particular business environment. After the GFC, revenue growth has been more difficult to achieve.

## 14.6 CONCLUDING DISCUSSION

In this research, we have examined how the profitability of a roro shipping company serving North European markets has changed in roughly three decades of time and compared its strategy of owning vs. chartering vessels. We have relied on the information given by the annual reports; no inside knowledge or information has been available.

While answering research questions, first we notice that the company's sales (Figure 14.3) have varied during the years 1994–2022 between 321 million euros and 746 million euros. However, the difference in profit (Figure 14.7) was between −8.4% and 18.1% and the ROI-% between −1.7% and 29.1%. This comparison shows us that the variation in profit and losses is not only dependent on sales of marine traffic but also on other factors.

The next research question is based on the first one. If and when the profit and ROI-% are not mainly dependent on sales, what are the other factors that influence profit? Figure 14.5 shows that before the difficult years starting in 2007, the number of chartered vessels was high. Unfortunately, the full data of vessel hires was not publicly available for researchers. However, it could be seen that when the ownership strategy of the company changed from chartering to fully owning the vessels, the financial situation also improved considerably (but with a time delay). Owning vessels seems to be associated with mature markets.

As vessel hires are only one of the factors affecting the financial performance of shipping companies, we have also included a study of sea personnel (Figure 14.8). This shows that the number of employees in the company has remained at a constant level since 2008. It should be noted that these are only the employees under the company's own payroll. The seafarers who worked in the chartered vessels were not always included in this figure—depending on the terms of chartering. This information was not available.

To have further understanding of the changes, we also developed regression models with the data of 1994–2022 to analyse what the drivers of profitability (ROI-%) were. In these models, we found that statistically significant drivers for higher ROI-% were both the number of employees (the lower the number, the better the ROI) and the number of chartered vessels (the higher the amount, the better the ROI). The market position and turnover of the case study company have been quite stable during the years. However, the profitability of the company disappeared in 2008–09 due to high vessel hires and decreasing volumes. After that the company made a total change in its strategy to own vessels, the profitability of the case study company has risen significantly. In other words, the strategy of the company relying on chartered tonnage was not robust in turbulent financial and market conditions. Therefore, it is not surprising to find that the number of owned vessels was a nearly statistically significant positive driver of profitability too.

The regression model argues that decisive action concerning chartered and owned vessels would lead to much higher ROI-% than what was achieved. This supports the idea that vessel ownership strategy needs to be clear—it could either be solely owned vessels or in a model where charters play a significant role. However, this argument would require more analysis with several companies to have statistical value.

There are some problems in this study that should be taken into account when examining results. First of all, the data is acquired from the annual reports, and no other sources were used. This results in limitations in the understanding of the company situation. In addition, the company has been also operating in some feeder-container transport, small-tonnage transport and port operations in addition to ro-ro/ro-pax operations. The financial impact of these operations is in the financial figures of the study, but their effect has not been analysed. The port operations business was around 15% at the start of the century, and during the years it decreased to the level of 4% in 2020. Finally, the share of ro-pax business has been growing slightly, so there is a need for more personnel to serve passengers, and that increases the crew headcount. Therefore, the results of this study should be seen as indicative. Deeper analysis of these operations and their impact on company profitability would bring more value to this study.

This case adds value to the present state of literature: how the company owned/charter decision can be a significant issue in profitability, not just arising from the selling of old fleets to second-hand markets (Meng and Wang, 2011; Park et al., 2018). This study supports the findings of Ghaderi (2019) and Kretschmann et al. (2017) that short-sea shipping operators can benefit from cost reductions by reducing personnel if the vessels are under the control of the company itself. This chapter can be used as teaching material in studies of maritime economics and logistics, showing students the effect of major strategic decisions made by shipping company executives. In particular, we suggest using this research when executives of companies have a background in other business areas and start working in shipping and do not have experience in the cyclic nature and capital intensiveness of the shipping business.

Similarly, this chapter gives insight for policy makers about the fundamentals of the shipping business. It shows how dependent shipping companies are on investment decisions made at the right time. In particular, this is important now, when shipping companies are facing difficult times, when tightening environmental regulations change the business environment and force companies to make new types of investment decisions (see, e.g., Tapaninen and Palu, 2022; Laasma et al., 2022).

Finally, the research methodology of this research was a case study, which adds novelty to the academic literature in areas that cannot be studied by other methods. However, this brings the major weakness of the method: The results cannot be directly applied to all application areas. Therefore, this study should be continued by collecting similar data from several shipping companies and analysing how well these results can be applied to the shipping sector.

As further studies, we would recommend carrying out similar studies with more cases to see the effect of the charter/own decision as well as size of the crew on long-term profitability.

## REFERENCES

Alizadeh, A. H. and Nomikos, N. K. (2007). Investment Timing and Trading Strategies in the Sale and Purchase Market for Ships. *Transportation Research Part B: Methodological*, Vol. 41, No. 1, pp. 126–143. https://doi.org/10.1016/j.trb.2006.04.002.

Bhonsle, J. (2022). Chartering Vs Owning Vessels: Carrier Strategies, Rationale, and Advantages and Disadvantages. *Marine Insight*, 18 May 2022. www.marineinsight.com/maritime-law/chartering-vs-owning-vessels/ (accessed 1 June 2022).

Branch, A. E. and Robarts, M. (2014). *Branch's Elements of Shipping*. Routledge, Milton Park, Abingdon (UK).

Cariou, P. (2008). Liner Shipping Strategies: An Overview. *International Journal of Ocean Systems Management*, Vol. 1, No. 1, pp. 2–13. https://doi.org/10.1504/IJOSM.2008.017778.

Cariou, P. and Wolff, F.-C. (2013). Chartering Practices in Liner Shipping. *Maritime Policy and Management*, Vol. 40, No. 4, pp. 323–338. https://doi.org/10.1080/03088839.2013.781280.

D'Agostini, E., Nam, H.-S. and Kang, S.-H. (2019). Gaining Competitive Advantage at Sea: An Overview of Shipping Lines' Strategic Decisions. *International Journal of Transportation Engineering and Technology*, Vol. 5, No. 4, pp. 74–81. https://doi.org/10.11648/j.ijtet.20190504.12.

Davies, J. E. (1983). An Analysis of Cost and Supply Conditions in the Liner Shipping Industry. *The Journal of Industrial Economics*, Vol. 31, No. 4, pp. 417–435. https://doi.org/10.2307/2098339.

Derbali, A. and Jamel, L. (2018). Determinants of Performance of Tunisia Insurance Companies: Case of Life Insurance. *International Journal of Productivity and Quality Management*, Vol. 24, No. 4, pp. 531–542. https://dx.doi.org/10.1504/IJPQM.2018.093452.

Dulebenets, M. A., Pasha, J., Abioye, O. F. and Kavoosi, M. (2021). Vessel Scheduling in Liner Shipping: A Critical Literature Review and Future Research Needs. *Flexible Servicing and Manufacturing Journal*, Vol. 33, pp. 43–106. https://doi.org/10.1007/s10696-019-09367-2.

Eisenhardt, K. M. (1989). Building Theories From Case Study Research. *The Academy of Management Review*, Vol. 14, No. 4, pp. 532–550. https://doi.org/10.2307/258557.

Everett, J. L., Hax, A. C., Lewinson, V. A. and Nudds, D. (1972). Optimization of a Fleet of Large Tankers and Bulkers: A Linear Programming Approach. *Marine Technology and SNAME News*, Vol. 9, No. 4, pp. 430–438. https://doi.org/10.5957/mt1.1972.9.4.430.

Finnish Customs. (2024a). *Foreign Trade Transports*. Finnish Customs, Helsinki, Finland. https://tulli.fi/tilastot/taulukot/logistiikkatilastoja (accessed 15 January 2024).

Fusillo, M. (2004). Is Liner Shipping Supply Fixed? *Maritime Economics & Logistics*, Vol. 6, pp. 220–235. https://doi.org/10.1057/palgrave.mel.9100110.

Ghaderi, H. (2019). Autonomous Technologies in Short Sea Shipping: Trends, Feasibility and Implications. *Transport Reviews*, Vol. 39, No. 1, pp. 152–173. https://doi.org/10.1080/01441647.2018.1502834.

Hannaford, E. and Hassel, E. V. (2021). Risks and Benefits of Crew Reduction and/or Removal With Increased Automation on the Ship Operator: A Licensed Deck Officer's Perspective. *Applied Sciences*, Vol. 11, No. 8, p. 3569. https://doi.org/10.3390/app11083569.

Haralambides, H. E. (2013). Editorial. *Maritime Economics and Logistics*, Vol. 15, No. 3, pp. 289–291. https://doi.org/10.1057/mel.2013.11.

Hiekata, K., Wanaka, S. and Okubo, Y. (2022). Mining Rules of Decision-Making for Fleet Composition Under Market Uncertainty Using a Genetic Algorithm. *Journal of Marine Science and Technology*, Vol. 27, No. 1, pp. 730–739. https://doi.org/10.1007/s00773-021-00864-4.

Hilmola, O.-P. and Li, W. (2023). Drivers of Railway Container Transports Between China and Finland. *Journal of Shipping and Trade*, Vol. 8, No. 22. https://doi.org/10.1186/s41072-023-00151-y.

Hoff, A., Andersson, H., Christiansen, M., Hasle, G. and Løkketangen, A. (2010). Industrial Aspects and Literature Survey: Fleet Composition and Routing. *Computers and Operations Research*, Vol. 37, pp. 2041–2061. https://doi.org/10.1016/j.cor.2010.03.015.

Karvonen, T. and Jousilahti, J.-P. (2020). *Alusliikenteen yksikkökustannukset 2018 (in Finnish, Title in English: The Unit Costs of Vessel Traffic 2018)*. Finnish Transport Infrastructure Agency, Helsinki, Finland, 49. https://urn.fi/URN:ISBN:978-952-317-807-6.

Kretschmann, L., Burmeister, H.-C. and Jahn, C. (2017). Analyzing the Economic Benefit of Unmanned Autonomous Ships: An Exploratory Cost-Comparison Between an Autonomous and a Conventional Bulk Carrier. *Research in Transportation Business and Management*, Vol. 25, pp. 76–86. https://doi.org/10.1016/j.rtbm.2017.06.002.

Laasma, A., Otsason, R., Tapaninen, U. and Hilmola, O.-P. (2022). Evaluation of Alternative Fuels for Coastal Ferries. *Sustainability*, Vol. 14, No. 24, p. 16841. https://doi.org/10.3390/su142416841.

Lun, Y. H. V., Hilmola, O.-P., Goulielmos, A. M., Lai, K.-H. and Cheng, T. C. E. (2013). *Oil Transport Management*. Springer, London. https://doi.org/10.1007/978-1-4471-2921-9.

Meng, Q. and Wang, T. (2011). A Scenario-Based Dynamic Programming Model for Multi-Period Liner Ship Fleet Planning. *Transportation Research Part E*, Vol. 47, No. 4, pp. 401–413. https://doi.org/10.1016/j.tre.2010.12.005.

Pantuso, G., Fagerholt, K. and Hvattum, L. M. (2014). A Survey on Maritime Fleet Size and Mix Problems. *European Journal of Operational Research*, Vol. 235, No. 2, pp. 341–349. https://doi.org/10.1016/j.ejor.2013.04.058.

Park, K.-S., Seo, Y.-J., Kim, A. R. and Ha, M.-H. (2018). Ship Acquisition of Shipping Companies by Sale and Purchase Activities for Sustainable Growth: Exploratory Fuzzy-AHP Application. *Sustainability*, Vol. 10, No. 6, p. 1763. https://doi.org/10.3390/su10061763.

Patton, M. Q. (1999). Enhancing the Quality and Credibility of Qualitative Analysis. *Health Services Research*, Vol. 34, No. 5 Pt 2, pp. 1189–1208.

Plomaritou, E. and Papadopoulos, A. (2019). *Shipbroking and Chartering Practice*. Routledge, Milton Park, Abingdon (UK).

Santos, T. A., Santos, G. L. D., Martins, P. and Soares, C. G. (2022). A Methodology for Short-Sea-Shipping Service Design Within Intermodal Transport Chains. *Maritime Economics and Logistics*, Vol. 24, No. 1, pp. 138–167. https://doi.org/10.1057/s41278-021-00193-8.

Shin, S.-H., Lee, P. T.-W. and Lee, S.-W. (2019). Lessons From Bankruptcy of Hanjin Shipping Company in Chartering. *Maritime Policy and Management*, Vol. 46, No. 2, pp. 136–155. https://doi.org/10.1080/03088839.2018.1543909.

Shrestha, N. (2020). Detecting Multicollinearity in Regression Analysis. *American Journal of Applied Mathematics and Statistics*, Vol. 8, No. 2, pp. 39–42. https://doi.org/10.12691/ajams-8-2-1.

Statistics Finland. (2024a). *Foreign Sea Transportation Volumes by Sea Ports and Product Groups*. Statistics Finland, Helsinki, Finland. https://statfin.stat.fi/PxWeb/pxweb/fi/StatFin/StatFin__uvliik/ (accessed 15 January 2024).

Statistics Finland. (2024b). *Carriage of Vehicles and Other Transport Equipment by Sea by Port*. Statistics Finland, Helsinki, Finland. https://pxweb2.stat.fi/PXWeb/pxweb/fi/StatFin/StatFin__uvliik/statfin_uvliik_pxt_12ix.px (accessed 15 January 2024).

Stopford, M. (2009). *Maritime Economics*, 3rd edition. Routledge, London and New York.

Sy, A., Jamel, L. and Derbali, A. (2017). Do Ownership Structure and Quality of Financial Information Affect the Cost of Debt of Tunisian Listing Firms? *International Journal of Critical Accounting*, Vol. 9, No. 2, pp. 140–153. https://dx.doi.org/10.1504/IJCA.2017.084891.

Tapaninen, U. (2020). *Maritime Transport*. Kogan Page, London.

Tapaninen, U. and Palu, R. (2022). Recovery of ro-pax Ferry Traffic From COVID-19 Under Tightening Environmental Regulations: Case Helsinki-Tallinn. *Journal of Shipping and Trade*, Vol. 7, No. 10. https://doi.org/10.1186/s41072-022-00112-x.

United Nations. (2016). *Review of Maritime Transport*. United Nations Conference on Trade and Development, New York.

United Nations. (2021). *Review of Maritime Transport*. United Nations Conference on Trade and Development, New York.

Yin, R. (2009). *Case Study Research: Design and Methods*, 4th edition. Sage, Thousand Oaks.

# APPENDIX A

## Regression Models from Possible Drivers of Profitability (ROI, %), Profit-Loss Margin (%), and Asset Turns

### ROI (%)

| Regression Statistics | |
|---|---|
| Multiple *R* | 0.7728 |
| *R* Square | 0.5972 |
| Adjusted *R* Square | 0.5489 |
| Standard Error | 4.2715 |
| Observations | 29 |

### ANOVA

| | df | SS | MS | F | Significance F |
|---|---|---|---|---|---|
| Regression | 3 | 676.3876 | 225.4625 | 12.3568 | 3.76137E-05 |
| Residual | 25 | 456.1503 | 18.2460 | | |
| Total | 28 | 1132.5379 | | | |

| | Coefficients | Standard Error | t Stat | P-value |
|---|---|---|---|---|
| Intercept | 22.6788 | 7.5005 | 3.0237 | 0.006 |
| Number of employees | –0.0175 | 0.0035 | –4.9395 | 4.36E-05 |
| Owned vessels | 0.6311 | 0.3149 | 2.0041 | 0.056 |
| Chartered vessels | 0.5916 | 0.1201 | 4.9245 | 4.53E-05 |

### Profit-loss margin (%)

| Regression Statistics | |
|---|---|
| Multiple *R* | 0.6866 |
| *R* Square | 0.4714 |
| Adjusted *R* Square | 0.4080 |
| Standard Error | 5.1248 |
| Observations | 29 |

## ANOVA

| | df | SS | MS | F | Significance F |
|---|---|---|---|---|---|
| Regression | 3 | 585.5928 | 195.1976 | 7.4322 | 0.0010 |
| Residual | 25 | 656.5948 | 26.2638 | | |
| Total | 28 | 1242.1876 | | | |

| | Coefficients | Standard Error | t Stat | P-value |
|---|---|---|---|---|
| Intercept | 33.1863 | 8.9988 | 3.6879 | 0.0011 |
| Number of employees | –0.0190 | 0.0043 | –4.4788 | 0.0001 |
| Owned vessels | 0.4488 | 0.3778 | 1.1880 | 0.2460 |
| Chartered vessels | 0.2035 | 0.1441 | 1.4119 | 0.1703 |

## Asset turns

| Regression Statistics | |
|---|---|
| Multiple $R$ | 0.8899 |
| $R$ Square | 0.7919 |
| Adjusted $R$ Square | 0.7758 |
| Standard Error | 0.0733 |
| Observations | 29 |

## ANOVA

| | df | SS | MS | F | Significance F |
|---|---|---|---|---|---|
| Regression | 2 | 0.5311 | 0.2656 | 49.4569 | 1.37631E-09 |
| Residual | 26 | 0.1396 | 0.0054 | | |
| Total | 28 | 0.6707 | | | |

| | Coefficients | Standard Error | t Stat | P-value |
|---|---|---|---|---|
| Intercept | 0.6902 | 0.0907 | 7.6137 | 4.42 E-08 |
| Chartered vessels | 0.0070 | 0.0017 | 4.0463 | 0.0004 |
| Sea personnel | –0.0003 | 0.0001 | –3.1523 | 0.0041 |

# 15 A Case Study
## *Lean Transformation at Sultan Qaboos University Hospital in Oman*

*Yousuf Al Khamisi*

## 15.1 INTRODUCTION

Sultan Qaboos University Hospital (SQUH) is an educational hospital that is affiliated with a medical school in the university. It started its operations in 1990 and has over 600 beds. The hospital was awarded its first international standards certification award, ISO 9001:2000, in 2005 and has been re-certified twice in 2008 and 2011. The journey continued with the ISO accreditation and went further with the Canadian accreditation in 2014. Also, it was re-certified in 2017 and 2022.During the COVID-19 pandemic, the university administration realized the urgency of revising the current processes, re-structuring them, and removing the non–value-added activities. After a discussion, it was decided that the Lean concept would be implemented in order to achieve the changes.

## 15.2 LEAN AT HEALTHCARE SERVICES

Lean healthcare concentrates on efficiency and patient satisfaction by clearly defining value-added procedures, and it is relatively new. Healthcare institutions adopted the values of the lean system from manufacturing in which much attention is focused on work processes, quality, and efficiency.

Implementing Lean in healthcare and the broader public sector is often studied by researchers. Laursen et al. (2003) stated that Lean started in the 1940s, but it was launched in the healthcare system in 2002. Bridges (2006) proposed that Lean can be used not only to remove waste from the healthcare environment but also to produce real motivations for innovation and value creation. Al-Balushi et al. (2014) found seven important categories of readiness factors for a successful launch of Lean in healthcare: strong leadership support, initiating Lean with the strategic program, understanding what value and customer groups are present in healthcare, undertaking the end-to-end process view, staff training and participation in Lean principles and methods, measurement and reward systems aligned to Lean objectives, and equivalent demand and capacity levels to increase flow. According to Bhandari et al.


DOI: 10.1201/9781003438748-19

(2021), Lean Six Sigma helped the healthcare sector to be better prepared during the time of a global pandemic.

Going backward, in 2007, Buheji was the first author to discuss Lean in healthcare in Oman at the 1st International Conference on Quality Management in Healthcare at Sultan Qaboos University. In 2010, the Petroleum Development Oman (PDO) Company started the Lean implementation journey. According to its managing director, "Lean that was initiated in the PDO in the last 10 years has saved 0.5 billion Omani Rial (1.3 billion U.S.D) in terms of value added work and reducing cost". In 2014, Al Farsi et al. conducted two Lean pilot improvement projects in a selected Omani hospital.

In 2015, Sultan Qaboos University Hospital trained over 100 staff members in Lean Six Sigma principles at yellow and green belt levels. However, improvement projects have yet to be conducted as part of the training.

In 2017, Al Khamisi et al. evaluated Lean Six Sigma implementation in both SQUH and Royal Hospital. The study suggested that Lean Six Sigma teams must pay more attention during the mapping of value streams (VSM) in the new projects to increase the effectiveness of Lean Six Sigma implementation. Furthermore, Al Khamisi et al. (2019) developed a knowledge-based system to implement a Lean Six Sigma hybrid system for a healthcare environment. The proposed system as part of doctoral dissertation has been validated in actual implementation at three major hospitals: Sultan Qaboos University Hospital, Royal Hospital, and Khoula Hospital. They concluded that Lean Six Sigma team in SQUH is 70.53% below the benchmark to fulfill the requirement of full Lean Six Sigma implementation. The hybrid system analysis concluded that SQUH needs to improve the mapping value streams dimension in the pre-implementing of Lean Six Sigma.

In Royal Hospital, Al Khamisi et al. concluded that the key critical point identified by the system was pre-implementing Lean Six Sigma in which a notable gap was found in the dimension of mapping value streams. The system also identified that Royal Hospital administration didn't track communication with action, precisely by focusing on Lean Six Sigma issues in direct reports relating to the whole organization. On the other hand, Khoula Hospital did not implement Lean principles during the system implementation. According to Al-Farsi and Al-Balushi (2018), leadership must take Lean philosophy as a strategy and consider the holistic scope of Lean implementation.

Al-Balushi and Al-Mandhari (2018) implemented Lean management techniques at a radiation oncology department at Royal Hospital. Later, Royal Hospital published the Lean implementation story in 2019. In 2021, Sohal et al. assessed the readiness for Lean thinking in healthcare settings: the case for Oman.

## 15.3 THE SQUH STORY

Sultan Qaboos University realized the importance of implementing Lean philosophy. The decision was taken to start implementing the journey in the hospital. As a result of that decision, the decree of the Lean Operations Committee formulation was signed by the university vice-chancellor. The LOC consists of 14 representatives from 14 different areas.

## 15.4 THE IMPLEMENTATION FRAMEWORK

The implementation framework was developed based on literature reviews that discussed enablers and constraints (Al-Balushi et al., 2014; D'Andreamatteo et al., 2015; Ruiz and Ortiz, 2016; Yu, 2020). It was verified and validated based on experts' opinions. The refined version consists of four pillars, as seen in Figure 15.1. The first pillar is top management **commitment**, followed by the **involvement** of multi-functional teams to enable **change management** and build the desired **culture.** The last pillar **communicated** the quick wins and positive results to all stockholders to sustain the momentum among all hospital layers.

Figure 15.2 illustrates a sub-framework that was developed to fill the knowledge gap. It consists of inductions to Lean sessions in Arabic and English, extensive workshops in Lean implementation and monitoring tools and continuous coaching and follow up. The final goal at the end of this sub-framework is to achieve Lean culture.

## 15.5 TRAINING

a. **Awareness Virtual Sessions:**
After the formulation of the LOC, a Lean awareness video was created to enhance the awareness of Lean philosophy and how it works. Examples of processes can be targeted in the hospital followed by a series of online lectures targeting both Arabic- and English-speaking staff.

**FIGURE 15.1** SQUH Lean transformation framework.

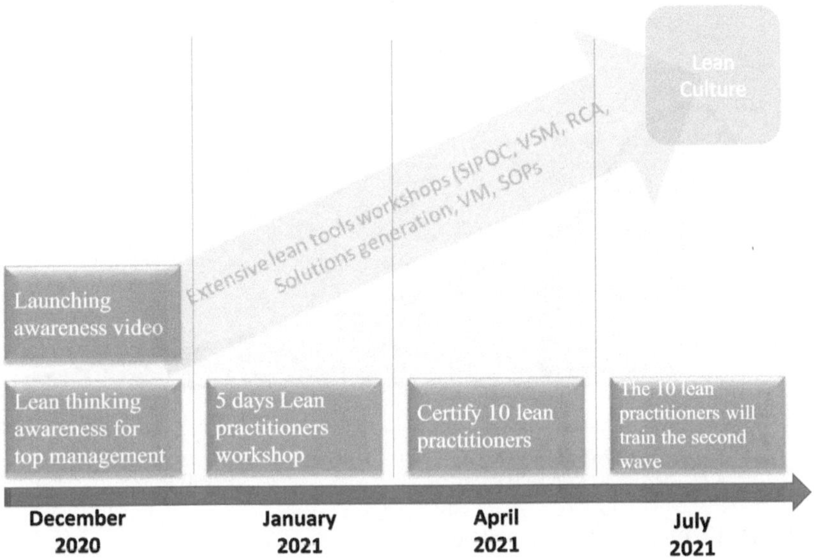


**FIGURE 15.2** Lean training and coaching timeframe for SQUH

b. **Call for Proposals:**
The LOC announced to all hospital wards/units and departments that they must identify two processes in their areas that consist of one or more of the following waste types:

1. Transport (T)
2. Inventory (I)
3. Motion (M)
4. Waiting (W)
5. Overprocessing ($O^a$)
6. Overproduction ($O^b$)
7. Defects or errors (D)

As discussed in the previous point, these waste types were detailed with examples for all staff through the induction sessions. The departments used an electronic link to apply.

c. **Filtering the Proposals:**
A total of 61 proposals were submitted to the LOC and were filtered based on their doability and impact, as Figure 15.3 shows.
The LOC selected 23 problems in the green area as part of a change management strategy that focuses on showing positive results. These positive results will inspire others to join the journey of Lean transformation.

d. The Shortlist:
The projects listed in Table 15.1 were selected for wave 1 of Lean implementation. Candidates who were nominated by their superiors were advised to attend the extensive workshop.

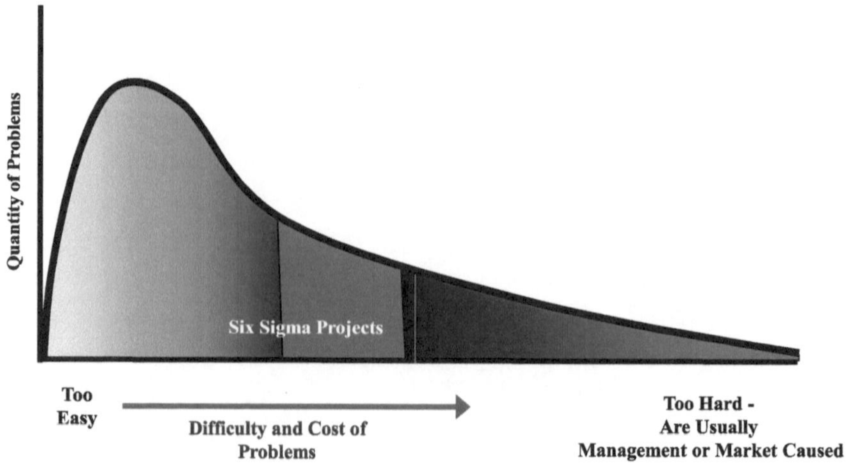


FIGURE 15.3 Project selection method.

## TABLE 15.1
## List of Lean Projects Proposed for the First Wave of Implementation at SQUH

| | Department | Proposed Projects | Waste Targeted ($TIMWO^aO^bD$) |
|---|---|---|---|
| 1 | Clinical Physiology Department | Reducing cycle time of reporting electromyography (EMG) and nerve conduction study (NCS) routine procedure process in clinical physiology (CPH) | W<br>$O^a$ |
| 2 | Directorate of Development and Quality | Reducing the number of missed critical incidents received via IRS | D<br>$O^a$ |
| 3 | Obstetrics and Gynecology Department | Reducing cycle time for elective cervical cerclage at SQUH | W<br>T<br>$O^a$ |
| 4 | Pathology Department | Optimizing the process of preparing cell blocks from body fluid samples in the cytology lab at SQUH | W<br>D<br>$O^a$<br>$O^b$ |
| 5 | Biochemistry Department | Improving reagent management system in biochemistry lab | W<br>I<br>D<br>$O^a$ |
| 6 | Child Health Department | Reducing unnecessary delays in transferring admitted pediatric patients from the emergency department to the general ward | W<br>T<br>$O^a$ |

*(Continued)*

**TABLE 15.1**
**(Continued)**

| | Department | Proposed Projects | Waste Targeted (TIMWO$^a$O$^b$D) |
|---|---|---|---|
| 7 | Procurement Directorate | Reducing the procurement cycle time of purchasing routine microbiology lab items at SQUH | W<br>O$^a$ |
| 8 | Ophthalmology Department | Reducing waiting time at ophthalmology adult OPD | W<br>O$^a$ |
| 9 | Genetics Department | Improving target mutation test process in the genetics lab | W<br>D<br>O$^a$ |
| 10 | Family and Community Medicine Department | Reducing the number of unnecessary repeated blood tests in patients with chronic disease (FMPH) | D<br>O$^a$<br>O$^b$ |
| 11 | Directorate of Development and Quality | Improving the revision process of departmental documents at SQUH | W<br>D<br>O$^a$ |
| 12 | Child Health Department | Increasing the percentage of general pediatric patients discharged by noon | W<br>O$^a$ |
| 13 | Pharmacy Department | Improving SQUH pharmacy dispensing process | W<br>O$^a$ |
| 14 | Hematology Department | Optimization of lab services for busulfan pharmacokinetic assay | D<br>O$^a$<br>O$^b$ |
| 15 | Directorate of Development and Quality | Decreasing the number of non-updated blood samples collected in the emergency department and processed in the biochemistry lab | W<br>D<br>O$^a$ |
| 16 | Medicine Department | Accelerating discharging of acute medical patients in medicine department at SQUH | W<br>O$^a$ |
| 17 | Radiology Department | Reducing the number of unnecessary (low back pain) MRI procedures | W<br>D<br>O$^a$ |
| 18 | Hospital Information System Directorate | Optimizing the process of generating new user accounts in the Hospital Information System Directorate | W<br>D<br>O$^a$ |
| 19 | Medical Record Department | Accelerating the external medical referral process to SQUH | W<br>O$^a$ |
| 20 | Nursing Directorate | Reducing the infection rate in the ICU | D |
| 21 | Behavioral Medicine Department | Reducing unimportant investigations at the Behavioral Medicine department | D<br>O$^a$ |
| 22 | Pharmacy Department | Accelerating the number of items from Family Medicine Department Pharmacy (minor pharmacy in the hospital) to the main pharmacy of the hospital | W<br>O$^a$ |
| 23 | Finance Directorate | Reducing the cycle time of the payment process | W<br>O$^a$ |

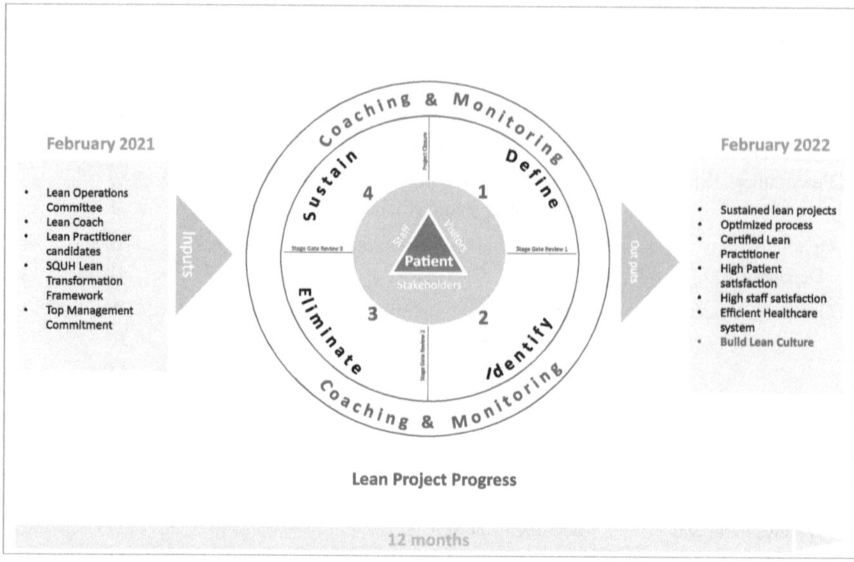


**FIGURE 15.4** Framework used to monitor Lean project progress.

It can be seen in Table 15.1 that most of the projects were concentrated on waiting waste. This is reasonable and understood, as Womack and Jones (1996) have suggested that 90% of waste removal can be achieved by reducing the lead time of an organization's processes.

e. **Lean Practitioner Workshop for 23 Candidates:**
Twenty-three candidates attended the four days of extensive workshops. The aim of the workshops was to enable participants to understand Lean and know how to implement sustainable and continuous improvement in their workplace

## 15.6 MONITORING THE PROGRESS OF THE PROJECTS

After completing the workshop, candidates returned to their areas to start their proposed processes for improvement. They were supervised by the Lean management representative in the LOC. Figure 15.4 was used to follow the progress of every project. The process of following Lean project progress consists of four phases, starting with **Define** and going through the **Identify** phase. The next phase is **Eliminate**, which enables the team to remove the waste from the targeted process. The last phase is **Sustain**, which will identify tools to maintain the improvement. After each stage, there is a review stage gate that will give the project's sponsor a brief about each phase and enables him/her to give guidance or inputs to move to the next stage.

The previous framework has been implemented in real situations, as detailed in Table 15.2.

**TABLE 15.2**
**Main Tasks Implemented during Lean Project Progress**

| | Stage | Main Tasks | |
|---|---|---|---|
| 1 | **Define** | 1.1 Create Charter | Update A3 (Problem solving form) |
| | | 1.2 Estimate Benefits | |
| | | 1.3 SIPOC | |
| 2 | **Identify** | 2.1 Gemba Walk | |
| | | 2.1 Current Value Stream Mapping (VSM) | |
| 3 | **Eliminate** | 3.1 Root Cause Analysis | |
| | | 3.2 Solution Generation | |
| | | 3.3 Future Value Stream Mapping | |
| | | 3.4 Implementation Action Plan | |
| | | 3.4 Pilot Study | |
| 4 | **Sustain** | 4.1 Generate and Communicate Standard Operating Procedure (SOP) | |
| | | 4.2 Create Visual Management Board | |
| | | 4.3 Create Leader Standard Work (LSW) Form | |

Like any other project, a Lean project needs a charter that states the current problem, proposed project scope, goal, team involved, and estimated time frame. Then, benefits from reviewing systems and processes beyond the basic expectations are estimated. These benefits can be defined according to several criteria summarized in the three levels illustrated in Table 15.2. The Define stage will end by detailing the SIPOC method to identify and specify suppliers, inputs, processes, outputs, and customers of the targeted process.

The Identify stage consist of two main steps. The first is a Gemba walk to see the actual place and understand the real problem; the second one is value stream mapping to detail the current process.

During the Eliminate stage, root cause analysis workshops are conducted to propose suitable solutions for each root. The proposed solutions are implemented via an action plan to start the pilot study and validate their effectiveness in the real practice. Finally, the Future value stream is mapped after removing the non-value-added tasks.

In the sustain stage, an SOP is generated and communicated, a visual management board is designed, and a Leader Standard Work Form is created to sustain all results and improvements made.

These benefits can be seen in terms of better information, improved decision-making process effectiveness, organizational flexibility and transparency, and better corporate image.

**TABLE 15.3**
**Level of Benefits That Can Be Obtained from the Lean Project**

| Level of Benefits | Description | Sub-Level | Description |
|---|---|---|---|
| 1 | Direct impact on profit and loss | 1.1 | Revenue improvement |
| | | 1.2 | Cost reduction |
| 2 | Avoidance of potential opportunities for future savings/revenues increase, but impact on the bottom line is uncertain | 2.1 | Delivery of process efficiency |
| | | 2.2 | Cost avoidance opportunities |
| | | 2.3 | Hours saved with no budget impact |
| | | 2.4 | Full-time equivalent (FTE) capacity, freeing people's time to work on key priorities |
| 3 | Non-quantifiable (in terms of monetary value) | 3.1 | Metrics improvements |
| | | 3.2 | Increase in customer or staff satisfaction |

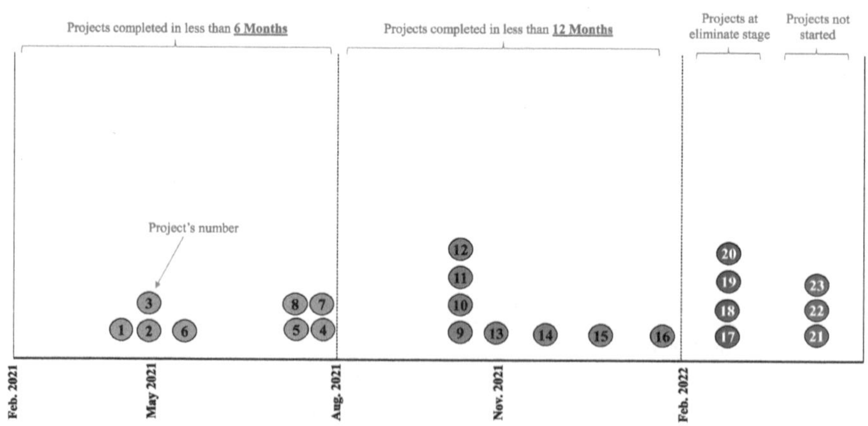


**FIGURE 15.5** Project completion time.

Figure 15.5 shows the duration of the projects' executions. It can be seen that eight projects were concluded within 6 months and another eight projects completed in 12 months. It also shows four projects still waiting in the Eliminate stage, while three did not start yet.

## 15.7 RESULTS

Table 15.4 summarizes main results and the benefits achieved of each project. One hundred percent of these projects have shown benefits from cost-saving and revenue opportunities, and 18% of the projects have shown a direct impact on hospital profit.

**TABLE 15.4**
**Results of the Completed Projects (to February 2022)**

| | Lean Project | Benefits Achieved (L1, L2 or L3) | Main Results |
|---|---|---|---|
| 1 | Reducing the cycle time of reporting EMG and NCS routine procedure process in CPH | L2.1<br>L2.3<br>L2.4<br>L3.1<br>L3.2 | The reporting cycle time of EMG and NCS reduced by 47% (from 17 days to 8 days). |
| 2 | Reducing the number of missed critical incidents received via IRS | L2.4<br>L3.1<br>L3.2 | No incident (0%) has been missed or overlocked. |
| 3 | Reducing cycle time for elective cervical cerclage at SQUH | L2.1<br>L2.2<br>L2.4<br>L3.1<br>L3.2 | The cycle time reduced from 52 hours to 11 hours. |
| 4 | Optimizing the process of preparing cell blocks from body fluid samples in cytology lab at SQUH | L2.1<br>L2.2<br>L2.4<br>L3.1<br>L3.2 | The cell block numbers from body fluid samples reduced by 48%. |
| 5 | Improving reagent management system in biochemistry lab | L2.1<br>L2.2<br>L2.4<br>L3.1<br>L3.2 | The cycle time for conducting monthly manual inventory was reduced from seven to five hours from March to May 2021.<br>Also, the number of errors was reduced from 10% to 1.2% after implementing the 5S tool partially. |
| 6 | Reducing unnecessary delays in transferring admitted pediatric patients from the emergency department to the general ward | L2.1<br>L2.3<br>L3.1<br>L3.2 | There was a significant improvement with a cycle time reduction of 74.3%. |
| 7 | Reducing the procurement cycle time of purchasing routine microbiology lab items at SQUH | L2.1<br>L2.3<br>L2.4<br>L3.1<br>L3.2 | The cycle time for purchasing the medical microbiology lab items was reduced from 47 days to 31 days (34% efficiency). |
| 8 | Reducing waiting time at ophthalmology adult OPD | L2.1<br>L2.3<br>L3.1<br>L3.2 | The follow-up client waiting time in ophthalmology OPD was reduced by 18%, starting from the registration step until being seen by the consultant. |
| 9 | Improving target mutation test process in the genetics lab | L2.1<br>L2.3<br>L2.4<br>L3.1<br>L3.2 | The target mutation testing process time was reduced by 19%, 21%, and 23% when 4, 10, and 30 samples were analyzed. |

*(Continued)*

**TABLE 15.4**
**(Continued)**

| | Lean Project | Benefits Achieved (L1, L2 or L3) | Main Results |
|---|---|---|---|
| 10 | Reducing the number of repeated blood tests in patients with chronic disease (FMPH) | L2.1<br>L2.2<br>L3.1<br>L3.2 | Reducing unnecessary testing by 12.5% resulted in one month of saving OMR 603 for nine patients. |
| 11 | Improving the revision process of departmental documents at SQUH | L2.1<br>L2.3<br>L2.4<br>L3.1<br>L3.2 | A pilot with two departments (seven documents) noted a significant improvement in time to submit the revised document and reduced errors. |
| 12 | Increasing the percentage of general pediatric patients discharged by noon | L1.2<br>L2.1<br>L2.2<br>L2.4<br>L3.1<br>L3.2 | Increased the percentage of general pediatric patients discharged by noon to 43%. |
| 13 | Improving SQUH pharmacy dispensing process | L2.1<br>L2.3<br>L2.4<br>L3.1<br>L3.2 | The results showed a median reduction of 48.5% (35 minutes to 18 minutes). The new system has shown a 90% staff and patient satisfaction. |
| 14 | Optimization of lab services for busulfan pharmacokinetic assay | L1.2<br>L2.1<br>L2.2<br>L3.1<br>L3.2 | One hundred percent reduction in lab BMS overtime (that will save over 2500 OMR per year for an average of the 20 days of overtime). Furthermore, the results showed that the busulfan pharmacokinetic test was activated in the HIS system; results are in the patient files, and test statistics are available at Trakcare with zero phone calls. |
| 15 | Decreasing the number of non-updated blood samples collected in the emergency department and processed in the biochemistry lab | L2.3<br>L2.4<br>L3.1<br>L3.2 | The number of non-updated blood samples was reduced by 71%. |
| 16 | Accelerating discharging of acute medical patients in the medicine department at SQUH | L1.2<br>L2.1<br>L2.2<br>L2.4<br>L3.1<br>L3.2 | The decision to discharge was accelerated by three hours. Also, the time of discharge summary paper submission was accelerated by two hours, and the time of patients leaving both wards was accelerated by two hours. |

## 15.8 CONCLUSIONS

The successful implementation of the 16 Lean projects at SQUH happened because of the commitment and support from the university vice-chancellor and his team. He emphasized that SQUH has to move toward making its processes Lean. The other factor that enabled the implementation was the quality culture among the hospital's staff. Over 20 years, SQUH has implemented different quality tools to improve and enhance patient experience. These tools have contributed to sharpen the skills and culture within the organization. Furthermore, having a full-time Lean coach who spent enough time with each project, attending each and every workshop and answering all queries and clarifications, contributed positively to accelerate the successful implementation.

On the other hand, some projects did not work properly according to the plan. Having protected time was a big challenge to bring Lean practitioners onboard. They could not balance their routine work and the project's requirements. The COVID-19 pandemic situation contributed to delaying these projects, as some healthcare practitioners were pulled to other services to fill the human resources at that time. Moreover, some Lean champions were not convinced to accept the change of having such projects.

## ACKNOWLEDGMENTS

Acknowledgments go to the Lean Operations Committee chairperson and all SQUH Lean practitioners for their involvement and participation in this success.

## REFERENCES

Al-Balushi, M. M. and Al-Mandhari, Z., 2018. Implementing lean management techniques at a radiation oncology department. *Sultan Qaboos University Medical Journal, 18*(3), p. e362.

Al-Balushi, S., Sohal, A. S., Singh, P. J., Al Hajri, A., Al Farsi, Y. M. and Al Abri, R., 2014. Readiness factors for lean implementation in healthcare settings–A literature review. *Journal of Health Organization and Management, 28*(2), pp. 135–153.

Al Farsi, Y., Al Abri, R., Al Hajri, A. and Al Balushi, S., 2014. The need for lean thinking in the Omani health care sector. *Oman Medical Journal, 29*(4), p. 248.

Al-Farsi, Y. M. and Al-Balushi, S. M., 2018. Go lean, get leaner: The application of lean management in Omani healthcare. *Sultan Qaboos University Medical Journal, 18*(4), p. e431.

Al Khamisi, Y. N. A., Khan, M. K. and Munive-Hernandez, J. E., 2019. The design of a knowledge-based system for quality management in healthcare: Case study. *International Journal of Advanced Operations Management, 11*(4), pp. 257–274.

Bhandari, P., Badar, M. A. and Childress, V., 2021, March. COVID-19 Surge planning in response to global pandemic in a healthcare setting: A Lean Six Sigma approach. In *Proceedings of the 11th International Conference on Industrial Engineering and Operations Management*, Singapore—Virtual.

Bridges, J. F., 2006. Lean systems approaches to health technology assessment. *Pharmacoeconomics, 24*(2), pp. 101–109.

D'Andreamatteo, A., Ianni, L., Lega, F. and Sargiacomo, M., 2015. Lean in healthcare: A comprehensive review. *Health Policy*, *119*(9), pp. 1197–1209.

Laursen, M. L., Gertsen, F. and Johansen, J., 2003. Applying lean thinking in hospitals-exploring implementation difficulties. *Udgivet På*. http://www.hctm.net/publications/publications.html.

Ruiz, E. and Ortiz, N., 2016. Lean healthcare: Barriers and enablers in the Colombian context. *World Academy of Science, Engineering, and Technology, International Journal of Social, Behavioral, Educational, Economic, Business and Industrial Engineering*, *10*(5), pp. 1508–1514.

Womack, J. P. and Jones, D. T., 1996. *Lean Thinking*. New York, NY: Simon and Schuster.

Yu, T., 2020. Lean transformation frameworks for hospital departments (Doctoral dissertation, Concordia University).

# 16 Tata Motors
## *Driving into the Future with Sustainability and Innovation*

*Vijaya Kittu Manda, Anuradha Yadav, and Radha Pavitra Yalamarti*

## 16.1 INTRODUCTION

The global automobile industry faces several challenges, including climate change, rising fuel prices, and increasing competition from emerging markets. The COVID pandemic has brought the industry to a standstill (Kakkar & Singh, 2023). Additionally, the shortage of semiconductor chips led to backlogs and delayed manufacturing (Wu et al., 2021). The declining sales numbers have shown that the Indian automobile sector's slowdown started significantly before 2020 (Manda & Beatrice, 2020), only to worsen during COVID-19 and then to rebound and move from resilience to resurgence (Kakkar & Singh, 2023). Only a few months earlier, several players in the industry began investing in new technologies, such as electric vehicles (EV) and autonomous driving. These innovative technologies are expected to make the automobile industry more sustainable and efficient. The importance of sustainable transportation practices has grown globally. Many new players entered the EV business and launched new models in the two-wheeler, passenger, and commercial vehicle segments (Manda et al., 2024). Much of the industry growth was attributed to customers' positive acceptance and subsidies as part of governmental EV policies.

The automobile industry is one of the most important contributors to India's economy. India is predominantly a two-wheeler market. The young and decently earning population, with their improved lifestyles and spending habits, have increased the penetration of cars over the years. India is home to some of the most influential players in the automobile industry, including the Tatas, the Mahindra, and the Maruti. Besides domestic players, India accommodates foreign players who import or make cars within India and sell them in the Indian market. Having a global presence alone does not confer a global competitive advantage, though some companies were able to buck the trend, be globally present, and stay globally competitive (Anil & Vijay, 2007). Most automobile manufacturers restored to upgradation, digitization, and automation to stay competitive (Liu & Racherla, 2019). The elite Indian class even directly imports cars. India also occupies a prominent position worldwide as a manufacturer and marketplace (IBEF, 2023).

DOI: 10.1201/9781003438748-20

Tata Motors Limited (fondly called TaMo) is a 78-year-old Indian multinational automotive company. It has been desperately working for a turnaround for several years now. The disruption of the industry by moving away from the traditional internal combustion engine to the electric engine is a turnaround opportunity for the company. However, with the recent motivation and support from its parent—the Tata Group—Tata Motors was inspired. It carried out a makeover, which became a successful turnaround in management circles (Vasudha, 2022). The group chairman, N Chandrasekaran, confirmed their strategy is to pick up anything digital. The group company has placed big bets on electric vehicles, among other things, and reinforced their commitment to developing clean energy cars.

TaMo quickly implemented the ideas and launched a series of successful products that helped it become a leading domestic EV car maker. The launch of the Tigor EV (June 2019), Nexon EV (January 2020), and the updated Tigor EV with Xpress T (2021) are some significant milestones in its long EV drive. The sales trends reflected EV's increasing adoption and acceptance in India (Autocar India, 2022). The 50k vehicle sales milestone was hit in about nine months (Tata Motors, 2023d). By mid-August 2023, the company reported 1 lakh EV sales. The company aims to meet the evolving needs of EV consumers by offering different body styles at various accessible price points. Not just that, TaMo has made some aggressive plans—to derive 50% of sales from its EV alone. Table 16.1 shows Tata Motors' three-phase EV strategy, known as "Go Beyond," that helps it expand in the EV market.

Tata Motors revitalized its global subsidiary—Jaguar Land Rover (JLR)—and made clear plans to turn it into an EV company. JLR had initially planned to have all its models be electric by 2025. However, the company has since pushed back that target to 2030. The high cost of developing and producing EVs is one of the main reasons for the delay. EVs also require a charging infrastructure different from that of traditional gasoline-powered cars. JLR also faces challenges in developing EVs

**TABLE 16.1**
**Tata Motors Three-Phase EV Plans**

| Phase | Description |
|---|---|
| First phase | Launch of Nexon EV and Tigor EV models based on a converted IC engine platform. The market has received these cars well and helped establish Tata Motors as a leader in the Indian EV space. The Nexon EV offers brisk acceleration and a reasonable range, making it one of the top-selling EVs in India. |
| Second phase | This phase will see the introduction of a bespoke EV platform, which will be used for cars such as the Altroz EV and Curvv EV. These cars will be more advanced than the Nexon EV and Tigor EV and will offer a more extended range and better performance. |
| Third phase | This phase will see the introduction of a skateboard platform, which will be used for smaller and more affordable EVs. This platform will be modular and scalable, allowing Tata Motors to develop various EVs for different segments. |

*Source:* Author compilation with data from multiple sources.

that meet its performance and luxury standards. The company's customers are used to high-performance cars with a luxurious interior. It is not easy to replicate this experience in an EV.

The case examines the strategies of TaMo, especially from the dimensions of product innovation and sustainability, that helped it to rise from a mere market player to a market leader. Much of the transformation has happened during and post-COVID times. The company showcased not just its strategic skills but also its financial expertise. The company was in deep losses and could not have an EV rebirth. However, prudent financial management helped it launch EVs as expected, even though new launches require enormous investments, especially considering the evolving technology. TaMo is not a cash-rich company and has suffered consecutive years of financial losses between 2018 and 2022. With the successful EV launch, the company is back in the green and reported its March 2023 financial numbers in green.

## 16.2 SUSTAINABILITY PLANS

The International Energy Agency (IEA), a Paris-based autonomous intergovernmental organization, said that "transport is responsible for around 23% of energy-related global greenhouse gas emissions, and these emissions grew faster than any other energy end-use sector from 2010 to 2019" (IEA, 2023). Much of the global carbon emissions are from burning fossil fuels, which are not sustainable forms of energy. Unfortunately, the automobile industry is the world's largest and most polluting industry. Cars and transport vehicles are responsible for significant greenhouse gas emissions, accelerating climate change. As concerns about climate change and environmental degradation grow, the automotive industry is under increasing pressure to become more sustainable.

There are several ways that the automotive industry can become more sustainable. One way is to develop and manufacture more fuel-efficient vehicles. Another way is to invest in renewable energy sources, such as solar and wind power, to power its factories and dealerships. The industry began exploring ways to reduce the amount of waste it produces and recycle and reuse materials to the extent possible.

### 16.2.1 Sustainability Practices at Tata Motors

Being the largest automaker in India and having several globally renowned car brands, Tata Motors knows the importance of using sustainable practices. Therefore, it is imperative to have a robust sustainability policy that places the environment at the top of its list of priorities. TaMo's sustainability strategy is still within the bounds of the law and concentrates on lifecycle thinking throughout many verticals. TaMo follows sustainability governance that spans functions strategically aligned around environmental, social, and governance (ESG) themes. The three ambitions of the company of the company are:

1. Driving Net Zero
2. Pioneering Circular Economy
3. Preserving Nature and Biodiversity

#### 16.2.1.1 Driving Net Zero

The company made a clear strategy that they will not just be manufacturing electric vehicles alone. However, they will use every new technology to drive this change to deliver zero carbon emissions. Tata Sons Chairman N Chandrasekaran also approved of the idea. Tata Motors has set an ambitious net zero goal for passenger vehicles by 2040 and commercial vehicles by 2045. The timeline difference for the two types of vehicles is mainly due to the operational and life cycle variations in both segments. India, as a country, has officially set 2070 as the target year for achieving net zero (Satish, 2023) and reaffirmed this commitment at the Conference of Parties 26 (COP26) summit in Glasgow. However, the company began investing $13 trillion to reach the target by 2050! (Bloomberg, 2023) It plans to slash the carbon dioxide emissions by a staggering 1 billion tons by the close of the current decade. The commitment is in line with the ambitions of the country, too. India committed to the EV30@30 campaign that aspires to make at least 30% of all vehicle sales in the country green vehicles by 2030.

To achieve this, the company is working on a twin strategy—a diverse technology mix in its product portfolio and focusing on renewable energy for its supply chain operations. Aikyam is a collaborative platform was created to bring all its vendors together to help in innovation, knowledge sharing, and co-creating. The company's supply chain sustainability roadmap aligns with circular economy and net zero ambitions. Building modular platforms was only possible because of the proper management of top leadership-level employees and strategic cooperation from partner firms (Malagihal, 2021). The company unveiled the Magic EV, an electrified version of its last-mile passenger transport vehicle, taking the EV trend beyond cars.

In the long-distance commercial vehicle segment, it understood the need for heavy-duty vehicles and launched appropriate commercial trucking and buses. Tata Motors has presented four clean propulsion technologies integrated into the Prima range for long-haul trucking. These technologies are aimed at providing sustainable alternatives for the transportation sector. The four propulsion technologies showcased are:

1. **Hydrogen ICE**: This technology uses hydrogen as a fuel in an internal combustion engine (ICE). Hydrogen is burned in the engine to generate power, producing only water vapor as a byproduct.
2. **Hydrogen Fuel Cell EV**: This technology utilizes hydrogen fuel cells to power an electric motor. Hydrogen is converted into electricity through a chemical reaction in the fuel cell, which powers the electric motor.
3. **Battery EV**: This technology uses rechargeable batteries to power an electric motor. The batteries store electrical energy that is used to propel the vehicle. The cost of batteries is a significant cost contributor for an EV, and Tata Motors feels that switching to pure EV models is not viable yet; hence, there is a need to continue with both internal combustion engine and electric models until the prices of batteries come down (Singh, 2023).
4. **Liquefied Natural Gas (LNG)**: This technology uses liquefied natural gas as a fuel source for the truck's engine. LNG is a cleaner-burning fuel compared to traditional diesel, resulting in reduced emissions.

#### 16.2.1.2 Pioneering Circular Economy

"Reduce, Reuse, and Recycle" is the core mantra of the circular economy (Geissdoerfer et al., 2017). The principles of circular economy come as an alternative to the traditional linear economy with "Make, Use, Dispose of" as its central theme. A circular economy emphasizes keeping resources in use for as long as possible, using fewer resources, extracting the maximum value from them, and then recovering and regenerating products and materials at the end of their life (Kirchherr et al., 2017).

By embracing circular economy principles, Tata Motors is committed to a sustainable and environmentally friendly future for the automotive industry. The slogan "value from waste" is prominently visible everywhere in the plant premises. The company took the first steps towards a circular economy in 2008, during which it launched the "Go Green" initiative. Initially aimed to reduce the environmental impact of its operations, the company, having tasted success, expanded the thought process to include several other sustainability goals. These include developing and commercializing electric and hybrid vehicles, Reducing the carbon footprint, improving water efficiency, increasing the use of recycled materials, and promoting sustainable manufacturing practices.

Later, in 2017, Tata Motors launched its sustainability policy, which outlines the company's commitment to sustainable development. The company policy is aligned with the UN Sustainable Development Goals and sets ambitious targets for the company to achieve by 2030. When company policies are painted with green words, it goes a long way.

Reva Electric Car Company is the first Indian domestic car company to launch EVs (Maini, 2005). The Reva was one of the pioneering electric cars in India and played a significant role in promoting electric mobility in the country. It laid the foundation for the growth of the electric vehicle industry in India. However, the company was acquired by Mahindra & Mahindra in 2010 and could never get the scale necessary to make a substantial impact.

Tata Motors is the second company to enter electric vehicle manufacturing and was able to build size and scale so that it rose to be the segment leader very swiftly. The company felt it could reduce carbon emissions and dependence on fossil fuels by introducing EVs, contributing to a more sustainable transportation system.

A critical step towards putting the words into practice is to recognize the importance of reducing waste and minimizing the environmental impact of its operations and products. Tata Motors has also focused on recycling and reusing materials and proper waste management practices in its manufacturing processes. It explored using recycled plastics, metals, and other materials in vehicle production, reducing the need for virgin resources. Tata Motors has taken steps to improve the end-of-life management of its vehicles. The company has established authorized recycling centers and initiatives to properly dispose of and recycle old vehicles. This helps to recover valuable materials and reduce the environmental impact of vehicle disposal.

Tata Motors Prolife offers re-manufactured engines, clutch, and pressure plates or cabins designed to provide long-lasting performance, reliability, and warranty. The company first ensures that the parts have a life of up to 80% of their original life, making them an economical choice for customers looking to extend the lifespan of their commercial vehicles.

Typically, in case of an engine failure, the customer goes for a repair, which means fixing by replacing or adjusting only the part of the component or engine that seems broken. The person repairing determines the quality. The engine is not even dismounted. The second option is reconditioning, which involves fixing and replacing all parts that seem to be broken or worn. There is no defined work process, and recommended tools may also not be available. Typically, the engines are assembled in an open, dusty environment. The technician makes a judgment as to which parts are acceptable for reuse. The engine assembly may use non-genuine, locally available, or even spurious parts of inferior quality. Here, the engine is dismounted, and minor machining, such as block boring or crank polishing, is done.

Tata Motors Prolife engineers classify vehicle engines into three categories. The green stage is where the engine has no change in performance. Its engine is new, within product warranty, and is tuned for optimum performance. The engine performance has come down in the yellow stage, but there is no breakdown. It becomes clear that there is a reduction in vehicle mileage, low compression, and high blow-by. The third phase is the red stage, whereby the vehicle performance reduces drastically, and the engine is very close to a breakdown. Performance has degraded, leading to a potential or even a complete breakdown. Products covered under the Prolife service are 2/4/6-cylinder engines, clutch and pressure plates, and cabin.

Tata Motors has collaborated with various organizations and stakeholders to drive circular economy practices in the automotive industry. These partnerships aim to develop innovative solutions, technologies, and business models that maximize resource efficiency and minimize waste generation. It initiated various training programs and education initiatives to make its workforce understand sustainability as a part of its organizational culture and align with the broader circular economy goals. Considering this, we can say that Tata Motors took a holistic view by involving all stakeholders in the circular economy process.

Tata Motors is a member of the Resource Efficiency and Circular Economy Industry Coalition (RECEIC) of the Group of 20 (G20), in which 39 major corporations from 11 countries participated. The RECEIC is an industry-driven forum with the governments only giving guidance and policy push. It is also a member of the RE100 campaign, an initiative led by Climate Group (Krishna Chaitanya, 2023, p. 27).

#### 16.2.1.3 Preserving Nature and Biodiversity

Tata Motors is working to reduce its carbon footprint. It pledged to conserve water by harvesting rainwater, recycling, recharging water, and other water conservation measures. The company also reduced its water consumption by 20% between 2015 and 2023.

Tata Motors did tree plantation drives at its manufacturing sites and other locations. Planting trees helps restore habitats, sequester carbon dioxide, and maintain biodiversity. It has set up several biodiversity parks and wetlands within the premises of its manufacturing facilities. These parks nurtured native flora and fauna. The company is also working with local communities to protect endangered species of plants and animals. It has set up a 245-acre wetland at its Pimpri plant in Pune, India. The wetland is home to various plants and animals, including the Indian marsh harrier, the Eurasian hobby, and the star tortoise.

It collaborates with environmental NGOs and other organizations on various conservation projects, ensuring a more holistic approach to biodiversity conservation. The company has also conducted workshops, training programs, and campaigns to raise awareness about the importance of biodiversity.

Tata Motors is sourcing 100% renewable energy for its operations in the UK. The company is also working to increase the use of renewable energy in its operations in India. Tata Sons group company Tata Power has worked closely with Tata Motors in implementing several renewable projects at Tata Motors. In August 2023, Tata Motors and Tata Power signed a key power purchase agreement (PPA). Tata Power Renewable, a subsidiary of Tata Power, has signed a PPA for a 9-megawatt solar plant at Tata Motors Pantnagar plant in Uttarakhand, India. Earlier, Tata Power Renewable Energy and Tata Motors agreed to establish a 7-megawatt solar project at the Pantnagar manufacturing facility. After the recent facility addition, the combined solar capacity will be 16 megawatts, generating 224 lakh units annually, meeting nearly 60% of their annual requirement (EquityPandit, 2023). In April 2024, TaMo and Infineum were recognized for leading the Indian lubricants market in developing a historic first heavy-duty engine lubricant designed to reduce vehicle emissions and improve air quality (Sustainable Transportation, 2024).

### 16.2.2 Electrification

Tata Motors has invested heavily in electric vehicle technology, which is essential to sustainable development. Their electric vehicle lineup, Tata Electric Mobility, includes the Tata Nexon EV, Tigor EV, Punch EV, Harrier EV, Curvv EV, and its concept EV codenamed Aniya.

Tata Motors plans to lead electrification in India's automobile industry and export its electric vehicles to global markets. The company is enhancing vertical integration and exploring battery and cell localization plans. Tata plans to localize 50% of its components, a shift from importing most parts to in-house production. The lower adoption of electric vehicles in India is mainly due to costs and limited charging infrastructure. Tata Motors' Tiago EV offers two battery options with varying ranges, but the company is working on expanding the charging infrastructure, considering the current 3,000 public charging points. The company has partnered strategically with Hyundai Motor India Limited to build an EV charging network (Erick, 2022).

Ziptron Technology is a core revolutionary electric vehicle in Tata EV vehicles. The technology complements other advanced features from Tata Motors' electric vehicle technology, such as a battery built on over 1 billion km of testing experience, an eight-year warranty, and a chilled liquid cool battery. The permanent magnet synchronous motor offers unmatched electric performance with instant torque and a dedicated cooling circuit. The water-cooled power electronics architecture ensures consistent performance and long-lasting durability, optimizing operations across various operating temperatures (Tata Motors, 2023b).

Tata Motors dominates the Indian electric vehicle market with its Nexon EV and Tigor EV, which cater to a niche audience due to their high prices. However, Tata has introduced the cheapest electric car in India, the Tiago EV, at a starting price of Rs

8.49 lakh. Tata's Nexon EV SUV was the most-sold electric car in H1 2022, followed by the Tigor EV and MG ZS EV. Introducing a budget EV would help consolidate Tata's market position. Launching the Tiago EV is significant, as IC-engine cars are becoming expensive, and petrol and diesel prices are already high. Tata Motors, a market leader in the passenger EV market, is now diversified with the Tiago EV, which is expected to be about a seventh of the gasoline version. This "compelling proposition" could open new opportunities and markets for Tata in smaller towns and cities.

Increased EVs on roads will mean there is an increased EV charging infrastructure. As a country, India aims to establish approximately 5 million public charging points by 2030. In July 2023, the company announced plans to introduce a master app that would provide its customers with information on the location and availability of electric vehicle charging stations. This was an effort to accelerate the adoption of EVs and help EV car users.

Paperback statistics show that Tata Power has installed more than 1,275 chargers in 175 societies. Each community charging facility has six to seven chargers and can cater to about 30 cars. Statistics as of July 2023 show that India has about 6,500 public charging stations. While the numbers are low compared to the demand/need, several companies are coming forward to install charging infrastructure. Bharat Petroleum Corporation Limited (BPCL), a leading Indian oil company, has started setting up charging points at their fuel pumps. BPCL targets establishing 7,000 charging points by 2025. India has set an ambitious target of 30% EV adoption by 2030. Research finds 11 significant challenges of EV charging infrastructure development in India have been identified. Seven challenges are identified through the comparative analysis of the literature review and four from the focused group study (Kore & Koul, 2022).

### 16.2.3 Manufacturing Processes

Sustainable practices play a pivotal role in the manufacturing sector of the automobile industry. With increasing environmental concerns, adopting sustainable practices becomes crucial for minimizing the industry's ecological footprint. By implementing eco-friendly manufacturing processes, such as using renewable energy sources and optimizing resource utilization, industry can reduce carbon emissions, conserve natural resources, and mitigate environmental degradation. Embracing sustainable practices enhances the industry's reputation and fosters long-term viability, ensuring a greener and more sustainable future for the automobile manufacturing sector.

Tata Motors is committed to optimizing its manufacturing processes for better efficiency and less waste. This includes reducing water use in their factories and moving to energy-efficient machinery and practices.

Tata Motors focuses on world-class manufacturing, ensuring high-quality standards throughout the process. With India's largest engine development facility and climate tests, the company is a leader in the commercial vehicle market. Tata's overseas plants also have state-of-the-art manufacturing and assembly lines. Tata Motors is known for its passion for excellence and quality talent, focusing on vehicle design, safety, and fuel efficiency. With world-class engineering and research centers, they offer complex vehicle designs through 3D visualization and hybrid technologies. Their latest offerings, Tigor and Nexon, and commercial vehicles Ultra and Signa, demonstrate their unmatched lead in auto technologies, securing their leadership in the automobile industry. Tata Motors' vehicle designers prioritize safety, maintenance, user comfort, and life cycle costs while conveying a trendy, trustworthy message (Tata Motors, 2023a).

Tata Sons has committed $5.1 billion (£400 million) to build an electric car battery factory that will come up at Hams Hall, near Birmingham, UK. The plan significantly boosts the development of the UK domestic car manufacturing industry. The industry welcomed the news because it will help create about 40,000 jobs, boost the economy, and produce 40 gigawatt hours (GWh) of batteries per year, reducing the UK's reliance on imported energy. However, some concerns are expressed about the potential to sweep up government support and any likely dent it can cause in the UK's future in the race to zero. The Tata EV battery plant is vital for the UK's needs, considering that Germany will have two giga-battery factories by the end of the decade, while France will have four!

Further funding is needed to enable innovative businesses to scale the battery ecosystem (Dimitris, 2023). The group wishes to use a mix of equity and debt sources and equipment financing without overly leveraging and paining the balance sheet.

In order to build its battery manufacturing capabilities, Tata Group began reaching out to multiple startups across the entire battery production spectrum—from cell chemistry and manufacturing to industrialization. It also entered talks at various value chain stages, including research and development, manufacturing innovation, and refining. Once the plant becomes operational, Jaguar Land Rover and Tata Motors will be the key customers. The plant is likely to produce 40 gigawatt hours of batteries starting in 2026 (ManufacturingToday, 2023).

The plant will be built by a group subsidiary company, Tata AutoComp Systems Limited. It will manufacture batteries using a mix of lithium iron phosphate (LFP) and nickel manganese cobalt (NMC). Renewable energy sources will power the plant. The plant is expected to be certified to the highest environmental standards. Mercedes-Benz Group AG, Stellantis NV, and Nissan Motor Co. plan to set up their EV battery plants and are preparing to revolutionize European battery supply (Nucleus_AI, 2023).

Tata Motors is focusing on restarting its Sanand plant, acquired from Ford India, to meet increasing vehicle demand. The plant, which can produce 3 lakh cars annually, is expected to be operational within the desired timelines. The company acquired a state-of-the-art facility in Sanand, Gujarat, adjacent to its existing plant, with a scalable capacity of 4.2 lakh units per annum. The company focuses on operationalizing the acquired plant in the current financial year. TML showcased 12 new passenger vehicles at Auto Expo 2023, gaining market shares against Maruti Suzuki, Hyundai, Kia, and Mahindra. With the unveiling of the Curvv and Avinya concepts, the already significant player in the EV segment has only increased its edge over the competition. TaMo has agreed to secure $1 billion in EV funding from TPG Rise for a stake of 11–15%. In FY23, TML became the third OEM in India to produce 5 lakh units, with EV sales exceeding 50,000 units (Nevin, 2023).

#### 16.2.3.1 Government Incentives

Tata Motors has obtained certification from the Automotive Research Association of India (ARAI) regarding its domestic value addition (DVA) level. The DVA is a critical metric for eligibility in India's government's production linked incentive (PLI) scheme, which is aimed at bolstering domestic manufacturing capabilities for advanced automotive technologies. Valued at Rs. 25,938 crores, the PLI program provides financial incentives to automakers contingent upon meeting predefined localization thresholds. Through achieving ARAI validation of its DVA percentage, Tata Motors has taken an important step toward qualifying to receive subsidies under this extensive national initiative seeking to catalyze India's automotive component ecosystem and boost competitiveness in electric and hydrogen fuel cell vehicles. The certification acknowledges Tata's ongoing efforts to incorporate locally sourced parts and develop supply chain integration within the country.

#### 16.2.3.2 Recycling and End-of-Life Vehicle Management

Tata Motors has initiatives for the recycling of vehicles and their components, with a focus on ensuring minimal environmental harm. They also aim to manage end-of-life vehicles effectively. "Returnable packaging" was another concept TaMo introduced in a similar vein. The broader plan is to reach a 99% level for its end-of-life vehicles by 2030.

#### 16.2.3.3 Re.Wi.Re Facility

The company used the Indian National Vehicle Scrappage Policy to promote a circular economy. The Indian government announced a plan to ensure that all commercial vehicles over 15 years old and all passenger vehicles that are 20 years old will go for mandatory fitness tests at automated testing stations (ATSs) in the coming years. Unfit vehicles will have the option to produce fitness certificates for a limited period. However, if the vehicle owner fails to do so, the vehicle will be treated as an end-of-life vehicle. It would be deregistered on the Vahan Portal with the help of the Registered Vehicle Scrapping Facility (RVSF). Vahan Portal allows the de-registration of vehicles located anywhere in India. Vehicle scrapping is a win-win situation, with the owner getting a 4% to 6% discount when buying a new vehicle. Certain other benefits accrue too—free of the registration charge and a road tax reduction of up to 25% for both commercial and non-commercial cars. The government wants to make India the South Asian hub for vehicle scrapping.

The Indian government estimates that 51 lakh light motor vehicles are older than 20 years, and another 34 lakh are older than 15 years. Around 17 lakh medium and heavy commercial vehicles (M&HCVs) are older than 15 years without a valid fitness certificate. Government estimates suggest that the vehicle scrappage ecosystem is expected to attract additional investments of around Rs 10,000 crore and 35,000 job opportunities (Autocar India, 2023).

Tata Motors began working on the idea of vehicle scrapping back in 2021. It contacted several European vendors to frame a holistic approach for a full-fledged vehicle scrappage facility model. Finally, the program became a reality after it inaugurated its first RVSF named Re.Wi.Re (for Recycle With Respect) on February 28, 2023. The first of these facilities was set up in Jaipur, Rajasthan, and has a capacity of 15,000 vehicles annually (Tata Motors, 2023c). The world-class facility follows eco-friendly processes for safe and sustainable end-of-life (EOL) vehicle dismantling. The facility is run by Tata Motor's partner Ganganagar Vaahan Udyog Pvt. Ltd. and will scrap end-of-life passenger and commercial vehicles of all brands to ensure de-polluting. The facility has dedicated processes and stations for dismantling tires, batteries, fuel, oils, liquids, and gasses. The second such facility was established in July 2023 in Bhubaneswar, Odisha, which can handle 10,000 vehicles a year. It chose Empreo Premium to run the facility.

After successful de-registration of the vehicle on the Vahan Portal, a certificate of deposit would be generated and given to the vehicle owner. A vehicle scrappage certificate will be issued to the vehicle owner after the complete dismantling and disposal of the vehicle. The company provides a free and fair online/offline assessment and gives the best price.

The success of these facilities will be the key to bringing in more service partners and starting similar facilities across the country. The company is exploring franchise routes to attract partners. Meanwhile, competitors such as Mahindra & Mahindra, Maruti Suzuki, and Toyotsu are running or planning to run scrapping hubs in India.

#### 16.2.3.4 Use of Renewable Energy

The company also uses renewable energy sources in its operations. The company has installed solar rooftop projects at its Pune, Pantnagar, Jamshedpur, and Dharwad manufacturing plants. It has also signed PPAs for wind and solar energy.

Their Pune plant in India has solar power capabilities, which help to reduce their carbon footprint. The company has installed solar panels on the roofs of its manufacturing plants, offices, and other facilities. These panels generate electricity from the sun's energy, reducing the reliance on traditional grid power and decreasing carbon emissions. The Sanand, Gujarat, facility is a significant solar energy installation.

Tata Motors has also explored wind power as a renewable energy source. The company has invested in wind farms and has signed PPAs with wind energy companies to procure electricity generated from wind turbines. It made changes to optimize manufacturing processes, upgraded equipment and machinery to more energy-efficient models, and implemented energy management systems to monitor and reduce energy consumption. It renovated existing ones to meet green building standards. These structures incorporate sustainable design principles to minimize energy consumption, maximize natural lighting, and improve energy efficiency.

#### 16.2.3.5 Solar Charging Stations

Tata Motors has collaborated with Tata Power, another entity within the Tata Group, to set up EV charging stations. While this is more in line with electric vehicle infrastructure, Tata Power's focus on renewable energy ensures that renewable sources can power many of these charging stations directly or indirectly.

In FY22, across all its plants in India, Tata Motors generated 92.39 million kWh of renewable electricity for its manufacturing operations, which is 19.4% of the total power consumption. This led to an avoidance of 72,992 metric tons of carbon dioxide equivalent and financial savings of Rs. 27.37 crores.

In 2021, the company commissioned a 7 MW solar rooftop project at its Pantnagar manufacturing facility. In 2022, Tata Motors and Tata Power entered a power purchase agreement (PPA) to develop a 7.25 MWp onsite solar project at Tata Motors' Jamshedpur commercial vehicle manufacturing facility. The project is important because it increased the onsite solar plant capacity at the Jamshedpur plant to reach 14 MWp.

#### 16.2.3.6 Reducing Emissions

Tata Motors leads the Indian EV market and has introduced several models, such as the Tata Nexon EV, Tata Tigor EV, and Altroz EV. Many more model launches, including SUVs and sedans, are planned for the next few years. By promoting EVs, Tata Motors aims to reduce emissions associated with traditional internal combustion engines. They have developed advanced powertrain technologies, lightweight materials, and aerodynamic designs to enhance fuel efficiency and reduce carbon dioxide ($CO_2$) emissions.

Tata Motors has been exploring and promoting alternative fuels such as compressed natural gas (CNG), biofuels, and hydrogen. Alternative fuels help lower emissions, thereby reducing dependence on traditional fossil fuels. The Indian automobile industry predominantly manufactures internal combustion engine vehicles (ICEVs), the emissions of which are the primary source of urban pollution and greenhouse gas (GHG). Much pressure falls on automakers to move to zero-emission vehicles such as EVs (Kumar et al., 2021). Tata Motors has been developing technology for their vehicles to reduce harmful emissions. They are also committed to monitoring and managing $CO_2$ emissions from their operations. Tata Motors has upgraded its vehicles with advanced emission control technologies to meet stringent emission norms like Bharat Stage VI (equivalent to Euro VI) in India.

#### 16.2.3.7 Sustainability in the Supply Chain

For automotive companies, sustainable supply chain management is crucial, given the environmental impacts of vehicle production and use. Car manufacturers have implemented several strategies to promote sustainability across their extensive networks of tiered suppliers. They conduct audits of suppliers' factories to evaluate compliance with emissions standards and monitor working conditions. Collaboration with materials suppliers focuses on developing lighter weight, recycled, and recyclable content for components to improve fuel efficiency. Partnerships support suppliers' transition to renewable energy sourcing to lower emissions footprint across the supply base. Transparency through reporting highlights progress on performance metrics like decreased water use and waste generation. Careful routing and optimization of parts deliveries reduces transport-related carbon emissions. Integrating circular economy principles, vehicle designers specify parts for disassembly and ensure recycled content targets are set to encourage closed-loop material flows. These collective efforts help automakers strengthen sustainability credentials to appeal to environmentally conscious consumers.

Tata Motors is committed to ensuring its suppliers and partners abide by sustainable practices, creating a holistic approach to sustainability. The company prioritizes long-term value creation through sustainability performance, focusing on the environmental, social, and governance aspects. They adopt a multiple-capitalism approach, focusing on reducing energy consumption, circular economy, water sourcing, social engagement, employee health and safety, and community empowerment. They also ensure compliance through supportive policies and monitoring systems. Jaguar Land Rover focuses on a circular economy, reducing waste and promoting environmental innovation. They develop technology for good, addressing social challenges and creating value beyond boundaries. They support global economies and a sustainable supply chain. Effective governance ensures that JLR's core values are delivered, ensuring sustainable profitability and a strong foundation for their business (Tata Motors, 2020).

Not just that, the company also asked its suppliers to adopt sustainable practices, leading to a greener supply chain. This includes using materials that are sustainable and can be recycled. The company is also developing new materials that are more recyclable and less environmentally harmful.

#### 16.2.3.8 Sustainable Mobility Solutions

Tata Motors focuses on providing clean, safe, and accessible mobility solutions. They are investing in research and development of fuel-efficient engines, alternative fuel options, and other technologies to make transportation more sustainable. Tata Motors is designing its products with sustainability in mind. The company is using lighter materials, more efficient engines, and other features that can help reduce its products' environmental impact.

Researchers studied numerous areas contributing to sustainability and innovation in the automotive sector using detailed structural content analysis (Vaz et al., 2017). The areas range from the role of gas emission not only to correct climate change but also to address various dangerous diseases in humankind (Resta et al., 2009; Van Den Hoed, 2007), life cycle assessment as technological product and process innovation, reverse logistics as innovative and eco-friendly initiative (Vaz et al., 2013), development of renewable energy alternatives along with creating technology-driven energy solutions (Bonilla et al., 2010), collectively known as eco-innovation. The automotive industry witnesses incremental innovation more commonly and practically. In addition to the factors contributing to eco-innovation (Van den Hoed, 2007), radical innovation is used much less frequently but is urgently needed (Tushman & Anderson, 1986; Utterback & Abernathy, 1975).

#### 16.2.3.9 Diversity

In the automobile industry, diversity in the workplace is essential for fostering innovation, understanding global markets, and driving competitiveness. Diverse teams bring varied perspectives, experiences, and problem-solving approaches, leading to creative vehicle designs and features that cater to a broader clientele. As the industry pivots towards electric vehicles, connectivity, and autonomous technologies, tapping into a rich tapestry of backgrounds can speed up adaptation and ideation processes. Furthermore, a diverse workforce better mirrors the global customer base, ensuring the needs of different demographics are met. Ultimately, embracing diversity is ethically sound and a strategic imperative in the fast-evolving automobile landscape.

Tata Motors understands the importance of diversity in the workplace and has formulated a clear roadmap and vision to bring in diversity and be a role model in the auto industry. The company has set up a dedicated position—chief diversity officer (CDO)—emphasizing inclusive leadership and ergonomically inclusive workspaces as catalysts for fostering diversity and inclusion (D&I) in organizations. The company has a diversity council, a development program called Gear-Up, conducts CXO Workshops for leaders, Power of Inclusive Management (POIM) workshops for middle management, a Reach Out platform for senior women leaders, and a Going Extra Miles (GEMS) Project, which is a shared platform to showcase projects (Mamta, 2023).

The company has set up an all-women assembly line at their PV plant in Pune. The company is using a comprehensive approach to meet the D&I objectives with the following:

1. Strategic Workforce Planning
2. Culture Transformation
3. Business Integration
4. Partnership and External Impact

## 16.3 INNOVATION PLANS

> Carrying out innovations is the only function which is fundamental in history.
>
> —Joseph Alois Schumpeter, 1943

While "innovation" has several definitions, the OSLO Manual (2005) definition is of special mention. "Innovation" is defined as "the implementation of a new or significantly improved product, good or service, or a process, or a new marketing method, or a new organizational method in business practices, in local organizations' work or external relations" (OECD & Statistical Office of the European Communities, 1995).

The automobile industry is traditionally one of the most innovative industries (Inna & Anastasiia, 2023). Production innovation in the industry can be examined in the dimensions of design, technology, and quality (Mamasioulas et al., 2020). Patent filings and design filings, safety awards, and computer technologies such as blockchain, metaverse, and NFTs have become benchmarks for benchmarking in automobile companies. Tata Motors stands in 16th place in the Global Innovation Index for India's most R&D-intensive firm.

### 16.3.1 Patent Filings

Patent filing is considered an essential way of measuring the innovation of an automobile company. Applying for patents and achieving successful issuance are distinct measures of innovative activity within an organization. Though submitting patent applications indicates efforts toward protecting intellectual property, only granted patents demonstrate validated novel concepts and ideas. A firm that can increase both the rate of patent filings and the proportion of approvals over time exhibits a sustained commitment to the research, development, and commercialization of new technologies. Demonstrating trajectory gains in pursuing and attaining patent rights reflects the ongoing prioritization of innovation within the company's strategic objectives and culture.

Tata Motors filed 158 patents during 2022–23. This is the highest number of patents filed by an Indian OEM manufacturing commercial vehicles and passenger cars. Some prominent patent filings include electrification, sustainability, safety, and connectivity. Figure 16.1 shows the number of patent filings and patents granted to Tata Motors between 2000 and 2022.

### 16.3.2 Design

Increasing filings and grants of designs speak to the company's commitment to innovative product designs. The company also won the prestigious IP Excellence Awards 2024 for Top Patent-Driven Industry by ASSOCHAM. Figure 16.2 shows an increasing trend regarding design filings and designs granted to the company.

### 16.3.3 Safety

After the US, China, Japan, and South Korea, India is the fifth country with a dedicated car crash safety program (Mainak, 2023). Besides country-specific safety tests, car manufacturers must also meet continent-specific safety tests such as ASEAN NCAP, Latin NCAP, Euro NCAP, and ANCAP (Mainak, 2023).

Tata Motors vehicles come with various safety features such as airbags, seat belts, an anti-lock braking system (ABS), electronic stability control (ESC), child safety seats, a lane departure warning system, a forward collision warning system,

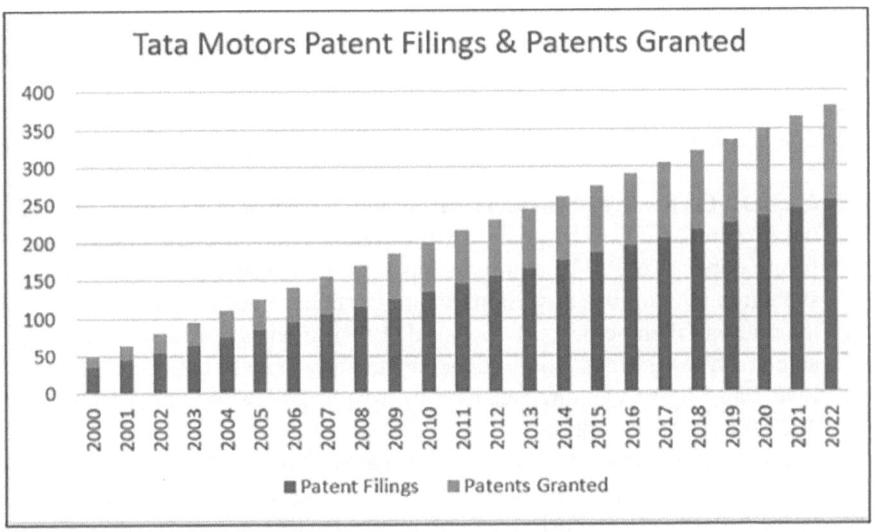


**FIGURE 16.1** Tata Motors—patent filings and patents granted, 2000–2022.

**Data Source: Author compilation from various sources.**

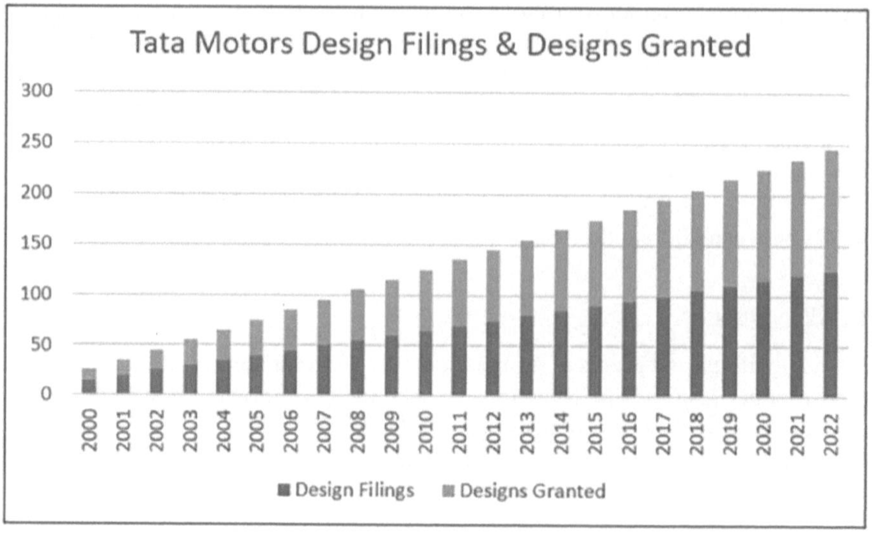


**FIGURE 16.2** Tata Motors—design filings and designs granted, 2000–2022.

**Data Source: Author compilation from various sources.**

and automatic emergency braking (AEB). Tata Motors focuses on building vehicles with solid, robust structures that can withstand crash forces and protect occupants. They utilize high-strength steel and advanced engineering techniques to enhance the structural integrity of their vehicles.

Some of Tata Motors' vehicles are equipped with connected car technology, which includes features like emergency braking notifications and real-time vehicle diagnostics, enhancing overall safety and security. Recognizing the importance of child safety, Tata Motors equips its cars with features like ISOFIX child seat mounts in many of its models.

The car marker is the recipient of various awards for its safety features, including:

1. **Global NCAP's five-star safety rating for the Nexon**: The Nexon became the first Indian car to be given the prestigious five-star adult safety rating from Global NCAP. The company has set a new benchmark in car safety standards by winning this. Significant efforts have been made to enhance the safety of their vehicles, and several of their models have achieved favorable crash test ratings from independent safety assessment organizations. Five Tata Motors cars are included in the top 20 list of Global NCAP in 2023 (Team Acko, 2023).
2. **SaferCarsForIndia's five-star safety rating for the Tiago and Tigor**: The Tiago and Tigor are the first Indian cars to receive a five-star safety rating from SaferCarsForIndia (Global NCAP, 2023).
3. **IIHS's Top Safety Pick+ award for the Harrier**: The Harrier became the first Indian car to receive a Top Safety Pick+ award from the Insurance Institute for Highway Safety (IIHS).

### 16.3.4 R&D Spending

Reasonable R&D spending is necessary for the present automobile industry to embrace technology and stay in the race against competitors. However, a point of cost in the short run, meaningful R&D spending will contribute to profits in the long run (Mansuri, 2019). The advances made by the automobile industry will result in technological development and value creation (Klink et al., 2014).

Over the years, the amount spent on research and development at Tata Motors has increased. One of the key focus areas of the R&D efforts is adopting renewable energy sources. This reduces their dependency on fossil fuels, contributing to a circular approach. It collaborated with academia, startups, and other industries to innovate and promote circular economy practices.

### 16.3.5 Engine and Transmission

Innovations and inventions in automobile technology have contributed immensely to the development of automobiles—right from the days of the James Watt steam engine of 1775 to date to technologies such as connected cars, regenerative braking, BladeScan technology headlights, and the like (Christopher, 2023).

Tata Motors has offered various engine options, including petrol, diesel, and electric. For instance, their Revotron and Revotorq series of engines balance performance and efficiency. They have introduced turbocharged petrol engines for better performance in the Nexon and Altroz models. Their transmissions, both manual and automated manual transmission (AMT), have evolved to offer smoother driving experiences.

The Automotive Sustainability Assessment Model (A-SAM) is a novel model for measuring the sustainability of the automotive industry. It provides 26 sustainability evaluation criteria for evaluating the effectiveness of the lifecycle of the cars. A-SAM has been demonstrated to be a valuable tool for making quick, high-quality judgments and can gauge, quantify, and translate the impact of various internal and external elements. The model assists design engineers in gaining a greater understanding of the effects of their actions, products, processes, and materials on the environment, resources, and society while maintaining cost-effectiveness in decision-making (Jasiński et al., 2021).

#### 16.3.5.1 Handling and Ride Quality

Tata Motors strongly emphasizes building cars with good ride quality and responsive handling. Even their entry-level vehicles are engineered to absorb bumps and vibrations caused by poor road conditions. The suspension systems are tuned to isolate passengers from impacts while maintaining stable control. The steering provides communicative feedback to drivers regarding traction and road placement. More premium Tata cars like the Tiago and Nexon feature advanced suspension designs with anti-roll bars and gas-filled shock absorbers. This enhances cornering ability and ensures a smooth, comfortable travel experience over diverse terrains. Overall, Tata aims to offer balanced dynamics that provide an enjoyable driving experience across diverse driving scenarios.

Modern Tata cars are often lauded for their balanced ride and handling. They are tuned to handle Indian road conditions effectively, offering comfort and stability. Car chassis and suspension setups like the Harrier and Nexon provide high-speed stability and confidence-inspiring handling dynamics.

### 16.3.6 Braking and Safety

Tata cars have essential safety features like ABS and electronic brakeforce distribution (EBD) that improve braking performance. The structural integrity of many of their cars has been rated highly in safety tests, which indirectly also speaks to their performance in crash scenarios.

### 16.3.7 Off-Road Capabilities

While Tata Motors primarily focuses on city and highway vehicles, they have offerings like the Safari and Harrier, which offer decent off-road capabilities. They are equipped with terrain response modes that adjust the car dynamics based on the driving surface.

### 16.3.8 Advancements and Innovations

Tata Motors has invested in research and development to incorporate advanced technologies, improving overall car performance. This includes better aerodynamics, weight reduction techniques, and more efficient powertrains.

### 16.3.9 Feedback and Reviews

Tata Motors has implemented innovative methods to continuously manage customer feedback and reviews to improve its products and services. The company uses online platforms and social media to collect customer reviews and insights actively. It employs advanced data analytics techniques like text and sentiment analysis on the feedback to identify key consumer pain points. Based on these, Tata Motors formulates solutions and action plans across product design, quality, service, and sales. The company also frequently interacts with customers through online discussion forums and surveys to understand changing needs. With its digital-first approach, robust analytics, and proactive customer engagement, Tata Motors leverages feedback optimally. Over the years, customer and expert reviews have highlighted the improved performance of Tata cars, noting their competitive edge in the Indian market.

### 16.3.10 Comfort and User Experience

Tata Motors is devoted to enhancing its customers' driving experience. The automaker focuses intently on comfort and intuitive user experience, ensuring all its vehicles provide a serene and stress-free journey. Spacious interiors with plush seating allow occupants to relax during any drive. The intelligently designed utility spaces give more storage than competitors, enhancing the travel experience.

Thoughtful ergonomic designs place all controls within easy reach. Smooth transmissions and suspensions glide over roads to minimize fatigue. Advanced infotainment systems offer engaging content to pass the time. The Juke-Car application allows users to create a virtual network and share a list of available songs on the device. Optional luxury features cater to discriminating tastes. Through constant refinements, Tata strives to satisfy every traveler with a soothing, convenient user experience inside each automobile it manufactures. Reports show that its bold design and focus on user experience contributed to a 66% increase in electric car sales (Vishal, 2023).

### 16.3.11 Connectivity

Connectivity in modern cars enhances safety, convenience, and user experience. It facilitates real-time navigation, traffic updates, and emergency assistance. Drivers access remote diagnostics, over-the-air software updates, and infotainment through connected features. Moreover, as autonomous vehicles evolve, seamless connectivity will be pivotal for vehicle-to-vehicle and vehicle-to-infrastructure communication, ensuring safer and more efficient roadways. Tata Motors has been proactive in embedding advanced connectivity features into their vehicles. These include:

1. Tata Fleetman: This is a telematics solution that provides fleet owners with real-time tracking of their vehicles, as well as information on fuel consumption, driver behavior, and other metrics. Currently, Tata Fleetman is offered as a part of fleet management services for medium- and heavy-range commercial vehicles. It has features such as real-time vehicle tracking, geofencing—sites/zones/point of interest, fuel consumption monitoring, engine idling monitoring, harsh acceleration, braking, and cornering detection, among others (Tata Fleetman, 2023).
2. Tata Connect: Customers can use this mobile app to connect with their vehicles remotely. The app helps check the vehicle's status, book service appointments, and even remotely start or stop the engine.
3. Tata iRA: This intelligent assistant, currently installed in the Tata Nexon EV range, allows voice commands to control various vehicle features. The features have applications in the infotainment system, climate control, and navigation.
4. Tata Altroz iTurbo Connected Edition: The hatchback comes with a range of connected features, a 7-inch touchscreen infotainment system with Android Auto and Apple CarPlay, a JBL sound system, and a rear parking camera.
5. Tata Curvv: An upcoming electric SUV, the Curvv is expected to have a host of connected features such as over-the-air updates, remote diagnostics, and geofencing.
6. ZConnect Mobile App in Tata Nexon EV: The ZConnect mobile app with this EV provides 35 different connectivity features, including remote commands, vehicle tracking, driving behavior analytics, and more.
7. Over-the-Air (OTA) Updates: Just like software updates for smartphones, Tata Motors has been working to enable OTA updates for its connected

vehicles, allowing for seamless updates of infotainment systems, performance enhancements, and new feature additions.

8. Remote Diagnostics and Prognostics: Some of Tata Motors' connected vehicle platforms offer this feature, enabling preventive maintenance and efficient breakdown services.
9. Voice Assist: Advanced voice recognition systems have been introduced in some models to improve the infotainment experience and aid in hands-free operations.

Using the collaboration route with tech firms and investing in research and development, Tata Motors is trying to enhance vehicle connectivity further. Three prominent vehicle community technologies are currently in the advanced research stages.

1. **Vehicle-to-vehicle (V2V) communication**: This technology allows vehicles to communicate with each other and share information about their location, speed, and other data. Preventing accidents by giving drivers more time to react to hazards is a potential benefit.
2. **Vehicle-to-infrastructure (V2I) communication**: This technology allows vehicles to communicate with roadside infrastructure, such as traffic lights and smart parking meters. Traffic flow improves, and drivers can find parking slots easily because of this.
3. **Vehicle-to-everything (V2X) communication**: This technology allows vehicles to communicate with various devices, such as pedestrians, cyclists, and other vehicles. These technologies are in the advanced form and are vital in creating a safer and more efficient transportation system, especially in the context of smart cities.

### 16.3.12 Computer and Networking Technologies

Keeping pace with the developments in computers and information technology, TaMo upgraded itself in three key Industry 4.0 and Web 3 areas—blockchain, metaverse, and non-fungible tokens (NFT)s (Yadav & Sagi, 2024, p. 3).

#### 16.3.12.1 Blockchain Plans

Blockchain is a disruptive technology that stores data in a cryptographically encrypted format on a secured storage, decentralized, and immutable ledger format. Automobile makers benefit from blockchain implementations in several areas, specifically supply chain and finance.

Tata Motors created a network called Tata Motors AutoMobility Collaboration Network 2.0 (TACNet 2.0), allowing the company to collaborate with startups for innovative solutions. Some solutions it is considering are its core automotive area, parking marketplace, NLP native chatbot, and demand prediction algorithm (Sanchita, 2019). Back in October 2021, Jaguar Land Rover did trials of the world-first digital supply chain for leather using blockchain technology. Jaguar Land Rover partnered with Circulor, Bridge of Weir Leather Company, and the University of Nottingham on an industry-leading leather trial (Tata Motors, 2021).

#### 16.3.12.2 Metaverse Plans

Metaverse is the next-generation platform for marketers as it creates new ground for reaching customers and helps the company showcase its products more effectively than traditional media formats. Backed with virtual and augmented reality, the metaverse provides an immersive experience for its users. Metaverse allows the co-existence of physical and virtual realities.

The metaverse offers the automobile industry new avenues for brand engagement, virtual showrooms, and interactive vehicle design feedback. Through immersive experiences, consumers can virtually test-drive vehicles, explore customization options, or attend digital car launches. This digital realm reshapes traditional marketing, enhancing customer connections and refining product development.

Tata Consultancy Services (TCS), a group company of Tata Sons, worked for a long time on metaverse and related technologies. In May 2022, the Tata Group gave the mandate to bring about Web3 technology advantages for all the group companies, especially metaverse solutions, so that the group companies can build a competitive edge and dominate in the metaverse (Rohit, 2022). Accordingly, TCS is helping Tata Motors build its presence in the metaverse (Ashok & Parthasarathi, 2022). Another Tata group company—Tata Elxsi—has expertise in building metaverse solutions with a blend of design and technology (ET CIO, 2021). In April 2022, Jaguar Land Rover, a subsidiary of Tata Motors, announced the launch of its open innovation platform that will build a mobility ecosystem incorporating the metaverse.

Tata Motors is not the lone Indian company making bets on metaverse. Competitor Maruti Suzuki announced its metaverse platform for the Auto Expo 2023 in January 2023. The platform is titled Maruti Suzuki EXPOVerse. It made its 1100-member–strong NEXA and AERNA dealerships across India with virtual reality devices for customer experience. It created a fantasy land to enthrall visitors before entering the Expo Pavilion virtually (BT, 2023). Another competitor—Hyundai—launched its Hyundai Pavilion on metaverse space at the Auto Expo 2023. The Hyundai Mobility Adventure is an immersive virtual city on the Roblox platform (IANS, 2023). So, Metaverse platforms are a necessary marketing function that automobile companies cannot ignore.

#### 16.3.12.3 NFT Plans

NFTs are a form of digital tokens representing a store of value of an asset. The automobile industry has begun to explore the utility of non-fungible tokens as a novel approach to value creation and customer engagement. NFTs, unique digital tokens representing ownership of a specific item on the blockchain, offer automakers an avenue to monetize digital car collectibles, virtual reality test drives, or even certify vehicle authenticity. For instance, limited edition cars can be paired with NFT-based certificates of authenticity. Furthermore, NFTs can facilitate digital documentation, ensuring tamper-proof provenance for parts or vintage cars. As the metaverse and virtual platforms expand, auto brands to capitalize on NFTs, blurring the lines between physical vehicles and their digital counterparts.

Global car makers such as BMW, Porsche, Lamborghini, and Formula 1 have begun using NFTs. These car makers used various strategies to leverage the new technology concepts. In some cases, they collaborated with digital artists to create the NFTs; in others, they created virtual car games and tracks. The most common method, of course, is to offer collectibles such as NFT trading cards. Being limited edition items, they are much sought after by their fans.

Tata Motors is a laggard in the Indian NFT space. Rival and first entrant Mahindra & Mahindra has an edge because of the iconic Thar offering and the Mahindra Gallery, which its subsidiary company, Tech Mahindra, has built (PTI, 2022). Tata Motors will have to take the help of its brother Tata Elxsi to launch its own NFT offering. Tata Elxsi has partnered with Cultos Global to integrate an NFT token mechanism for the Driver Reward Program, which Tata Motors can leverage (FPJ Web Desk, 2023).

## 16.4 ROAD AHEAD

Competition is always high in the automobile industry in general, both at the global level and at the domestic level. Tata Motors has enjoyed the first mover advantage by being the first domestic car maker to switch to EV. It did enjoy the monopolistic opportunity. However, going forward, Tata Motors will face increased competition as domestic players slowly start giving their EV offerings. Competitors like Maruti Suzuki and Mahindra & Mahindra, existing players such as MG, and prospective incumbents such as Tesla will likely launch their EVs in the Indian market.

The automobile industry is a distinct industry that needs a distinct way of measuring sustainability. Some models, such as the Automotive Sustainability Assessment Model (Jasiński et al., 2021), are yet to be widely included in regular elevations and measurements by the industry. Further econometric or empirical research can be done once similar and regular numbers are released, perhaps by industry associations.

## 16.5 CONCLUSION

Tata Motors Limited has been at the forefront of sustainability and innovation in the Indian automobile industry. TaMo's strategic foray into the EV market is encapsulated in its three-phase plan called "Go Beyond." This strategy underscores the company's commitment to diversify its EV offerings, catering to a broad spectrum of consumers. TaMo aims to meet the evolving demands of the EV market by providing a range of body styles and price points. The company has set ambitious targets for itself. It aspires to derive 50% of its sales solely from EVs, signaling a profound shift toward sustainable mobility.

Moreover, Tata Motors aims to rejuvenate its global subsidiary, Jaguar Land Rover, by transitioning it into a full-fledged EV manufacturer. Tata Motors has established aggressive net-zero goals that align with the imperative for global sustainability. They aim to achieve net-zero carbon emissions for passenger vehicles by 2040 and extend this goal to commercial vehicles by 2045. This commitment aligns with their broader objective to reduce carbon dioxide emissions by a staggering 1 billion tons within the current decade. It partnered with the Leadership Group for Industry Transition (LeadIT) to accelerate its transition towards net-zero emissions (Tata Motors, 2024).

Additionally, Tata Motors is actively participating in the EV30@30 campaign, pledging to make at least 30% of all vehicle sales in India eco-friendly by 2030. Tata Motors' visionary approach to sustainability not only reinforces its leadership in the Indian automotive landscape but also underscores the critical role the industry plays in addressing global environmental challenges. Through innovation and steadfast commitment, Tata Motors drives positive change towards a more sustainable and greener future.

## REFERENCES

Anil, K. G., & Vijay, G. (2007). Converting global presence into global competitive advantage. In *International Human Resource Management* (1st ed., p. 12). Imprint Routledge.

Ashok, M., & Parthasarathi, V. (2022, July). *The Unbearable Lightness of the Metaverse.* www.tata.com/newsroom/careers/unbearable-lightness-metaverse-tcs

Autocar India. (2022, November 7). *Tata Motors Achieves 50,000 Electric Vehicle Production Milestone.* www.autocarindia.com/car-news/tata-motors-achieves-50000-electric-vehicle-production-milestone-426262

Autocar India. (2023, July 24). *Tata Motors Opens Its Second Vehicle Scrapping Facility.* www.autocarindia.com/industry/tata-motors-opens-its-second-vehicle-scrapping-facility-428839

Bloomberg. (2023, August 24). India needs to invest $13 trillion to reach net zero emissions by 2050, says report. *Hindustan Times.* www.hindustantimes.com/india-news/india-needs-to-invest-13-trillion-to-reach-net-zero-emissions-by-2050-says-report-101692858267983.html

Bonilla, S. H., Almeida, C. M. V. B., Giannetti, B. F., & Huisingh, D. (2010). The roles of cleaner production in the sustainable development of modern societies: An introduction to this special issue. *Journal of Cleaner Production, 18*(1), 1–5. https://doi.org/10.1016/j.jclepro.2009.09.001

BT. (2023, January 11). Auto Expo 2023: Maruti Suzuki's Metaverse platform. *Business Today.* www.businesstoday.in/visualstories/auto/auto-expo-2023-maruti-suzukis-metaverse-platform-17767-11-01-2023

Christopher, M. (2023, June 23). *20+ Greatest Innovations And Inventions of Automobile Engineering: From the First Engine to Today.* https://interestingengineering.com/lists/20-greatest-innovations-and-inventions-of-automobile-engineering-from-the-first-engine-to-today

Dimitris, M. (2023, July 18). Tata Motors to build UK gigafactory in Somerset. *Energy Live News.* www.energylivenews.com/2023/07/18/tata-motors-to-build-uk-gigafactory-in-somerset/

EquityPandit. (2023, August 22). *Tata Power Signs a Power Purchase Agreement with Tata Motors.* www.equitypandit.com/tata-power-signs-a-power-purchase-agreement-with-tata-motors/

Erick, M. (2022, October 6). How Tata Motors is acing electric vehicle market in India with one EV at a time. *Outlook India.* www.outlookindia.com/business/tata-tiago-ev-launch-how-tata-motors-is-acing-electric-vehicle-market-in-india-with-one-ev-at-a-time-news-228052

ET CIO. (2021, December 23). Tata Elxsi preps up for the Metaverse. *Economic Times.* https://cio.economictimes.indiatimes.com/news/next-gen-technologies/tata-elxsi-preps-up-for-the-metaverse/88444358

FPJ Web Desk. (2023, June 2). Tata Elxsi Cultos global partner to integrate a NFT token mechanism for driver reward program. *The Free Press Journal.* www.freepressjournal.in/business/tata-elxsi-cultos-global-partner-to-integrate-a-nft-token-mechanism-for-driver-reward-program

Geissdoerfer, M., Savaget, P., Bocken, N. M. P., & Hultink, E. J. (2017). The circular economy—a new sustainability paradigm? *Journal of Cleaner Production, 143*, 757–768. https://doi.org/10.1016/j.jclepro.2016.12.048

Global NCAP. (2023). *#SAFECARSFORINDIA*. www.globalncap.org/safercarsforindia

IANS. (2023, January 5). Hyundai Motor India launches Hyundai Pavilion on Metaverse space. *Business Standard*. www.business-standard.com/article/companies/hyundai-motor-india-launches-hyundai-pavilion-on-metaverse-space-123010501050_1.html

IBEF. (2023, May). *Automobile Industry in India*. www.ibef.org/industry/india-automobiles

IEA. (2023). *IEA Transport*. www.iea.org/energy-system/transport

Inna, S., & Anastasiia, P. (2023, July 16). Trends in R&D spending in the automotive industry. *International Economics and International Relations*. 4th International Scientific and Practical Conference, Ottawa, Canada.

Jasiński, D., Meredith, J., & Kirwan, K. (2021). Sustainable development model for measuring and managing sustainability in the automotive sector. *Sustainable Development, 29*(6), 1123–1137. https://doi.org/10.1002/sd.2207

Kakkar, V., & Singh, A. (2023). From resilience to resurgence: COVID-19 and the automobile industry. In K. Antosz, J. C. Sa, M. Jasiulewicz-Kaczmarek, & J. Machado (Eds.), *Advances in Logistics, Operations, and Management Science* (pp. 90–124). IGI Global. https://doi.org/10.4018/978-1-6684-5606-4.ch005

Kirchherr, J., Reike, D., & Hekkert, M. (2017). Conceptualizing the circular economy: An analysis of 114 definitions. *Resources, Conservation and Recycling, 127*, 221–232. https://doi.org/10.1016/j.resconrec.2017.09.005

Klink, G., Mathur, M., Kidambi, R., & Sen, K. (2014). Contribution of the automobile industry to technology and value creation. *Auto Tech Review, 3*(7), 18–23. https://doi.org/10.1365/s40112-014-0688-5

Kore, H. H., & Koul, S. (2022). Electric vehicle charging infrastructure: Positioning in India. *Management of Environmental Quality: An International Journal, 33*(3), 776–799. https://doi.org/10.1108/MEQ-10-2021-0234

Krishna Chaitanya, S. V. (2023, July 27). G20's circular economy coalition of 39 corporate biggies kicks off on July 27. *The New Indian Express*. www.newindianexpress.com/states/tamil-nadu/2023/jul/27/g20s-circular-economy-coalition-of-39-corporate-biggies-kicks-off-on-july-27–2598959.html

Kumar, R., Lamba, K., & Raman, A. (2021). Role of zero emission vehicles in sustainable transformation of the Indian automobile industry. *Research in Transportation Economics, 90*, 101064. https://doi.org/10.1016/j.retrec.2021.101064

Liu, K.-C., & Racherla, U. S. (Eds.). (2019). *Innovation, Economic Development, and Intellectual Property in India and China: Comparing Six Economic Sectors*. Springer Nature Singapore. https://doi.org/10.1007/978-981-13-8102-7

Mainak, D. (2023, August 21). India is the world's fifth country to get a dedicated car crash safety program. *HT Auto Desk/Hindustan Times*. https://auto.hindustantimes.com/auto/cars/india-is-the-fifth-country-in-the-world-to-get-a-dedicated-car-crash-safety-program-know-more-41692588851966.html

Maini, C. K. (2005). REVA Electric car: A case study of innovation at RECC. *International Journal of Technology Management, 32*(1/2), 199. https://doi.org/10.1504/IJTM.2005.006824

Malagihal, S. S. (2021). Strategic options for automobile OEMs of Indian origin to have sustained competitive advantage: A case of Tata Motors. *International Journal of Global Business and Competitiveness, 16*(2), 139–152. https://doi.org/10.1007/s42943-021-00029-5

Mamasioulas, A., Mourtzis, D., & Chryssolouris, G. (2020). A manufacturing innovation overview: Concepts, models and metrics. *International Journal of Computer Integrated Manufacturing, 33*(8), 769–791. https://doi.org/10.1080/0951192X.2020.1780317

Mamta, S. (2023, May 17). *Accelerating diversity and inclusion in auto industry: The Tata Motors' roadmap*. www.peoplematters.in/article/diversity/accelerating-diversity-and-inclusion-in-the-automotive-industry-the-tata-motors-roadmap-37864

Manda, V. K., & Beatrice, B. B. (2020). The Indian automobile sector slowdown. *SSRN Electronic Journal*. https://doi.org/10.2139/ssrn.3528891

Manda, V. K., Mampilly, A. J., Gupta, V., & Nammi, L. (2024). The rise of two-wheeler electric vehicles in India. In S. Munuhwa (Ed.), *Advances in Civil and Industrial Engineering* (pp. 503–533). IGI Global. https://doi.org/10.4018/979-8-3693-3755-4.ch019

Mansuri, S. I. (2019). A study on R&D spending and its impact on profitability of selected automobiles companies. *SEMCOM*, *6*(2), 127–132.

ManufacturingToday. (2023, August 21). *Tata Invites Collaborators for UK Battery Plant, Energising Jaguar Land Rover's EV Revolution*. www.manufacturingtodayindia.com/sectors/tatas-ambitious-uk-battery-plant-project-to-supply-jaguar-land-rover-evs-gains-momentum

Nevin, J. (2023, June 30). Tata Motors out of capacity; restarting Sanand plant bought from Ford top priority. *Fortune India*. www.fortuneindia.com/enterprise/tata-motors-out-of-capacity-restarting-sanand-plant-bought-from-ford-top-priority/113203

Nucleus_AI. (2023, August 21). *Tata Plans UK Battery Factory for Land Rover and Jaguar EVs*. https://yourstory.com/2023/08/tata-uk-ev-battery-plant-innovation

OECD & Statistical Office of the European Communities. (1995). Measurement of scientific and technological activities: Manual on the measurement of human resources devoted to S&T—Canberra manual. *OECD*. https://doi.org/10.1787/9789264065581-en

PTI. (2022, March 25). Mahindra & Mahindra to auction Thar NFTs. *Economic Times*. https://economictimes.indiatimes.com/markets/cryptocurrency/mahindra-mahindra-to-auction-thar-nfts-on-march-29/articleshow/90447171.cms

Resta, B., Gaiardelli, P., & Pezzotta, G. (2009). Sustainability in the auto repair industry: A life cycle assessment application. *International Journal of Product Lifecycle Management*, *4*(1/2/3), 146. https://doi.org/10.1504/IJPLM.2009.031671

Rohit, K. (2022, May 17). *Tata Wants to Dominate Metaverse! All Tata Companies Will Deploy Metaverse Solutions: But How?* https://trak.in/tags/business/2022/05/17/tata-wants-to-dominate-metaverse-all-tata-companies-will-deploy-metaverse-solutions-but-how/

Sanchita, D. (2019, September 18). Tata Motors is asking startups for blockchain enabled solutions for the automotive industry. *Business Insider*. www.businessinsider.in/tata-motors-wants-blockchain-solutions-from-startups/articleshow/71185385.cms

Satish, N. (2023, June 19). India outperforms US, China in LEED net zero certifications. *Hindustan Times*. www.hindustantimes.com/cities/mumbai-news/india-leads-in-leed-zero-green-building-projects-outperforming-us-and-china-usgbc-and-gbci-101687115927271.html

Singh, R. S. (2023, August 11). *Better to Make EVs in Current ICE Models: Tata Motors*. www.thehindubusinessline.com/news/better-to-make-to-evs-in-current-ice-models-tata-motors/article67184813.ece

Sustainable Transportation. (2024, April 15). *Infineum and Tata Motors Deliver Historic First for Indian Lubricants Market*. www.infineum.com/media-centre/news/infineum-and-tata-motors-deliver-historic-first-for-indian-lubricants-market/

Tata Fleetman. (2023). *Tata Fleetman*. www.tatafleetman.com/

Tata Motors. (2020). *Tata Motors Annual Report 2019–2020: Sustainability Focus*. https://investors.tatamotors.com/financials/75-ar-html/sustainability-focus.html

Tata Motors. (2021, October 13). *Jaguar Land Rover Trials World's First Digital Supply Chain for Leather Using- Bockchain Technology*. www.tatamotors.com/jlr-press-release/jaguar-land-rover-trials-world-first-digital-supply-chain-for-leather-using-blockchain-technology/

Tata Motors. (2023a). *Tata Motors Facilities*. www.tatamotors.com/about-us/facilities/

Tata Motors. (2023b). *Tata Passenger Electric Mobility*. https://ev.tatamotors.com/technology/

Tata Motors. (2023c, February 28). *Tata Motors Launches Re.Wi.Re, Its First Registered Vehicle Scrapping Facility*. www.tatamotors.com/press/tata-motors-launches-re-wi-re-its-first-registered-vehicle-scrapping-facility/

Tata Motors. (2023d, August 11). *Tata Motors EV Family Is Now 1 Lakh Strong*. www.tatamotors.com/press/tata-motors-ev-family-is-now-1-lakh-strong/

Tata Motors. (2024). *Tata Motors Partners with UN-Backed LeadIT Initiative to Accelerate Transition Towards Net-Zero Emissions* [Press Release]. www.tatamotors.com/press-releases/tata-motors-partners-with-un-backed-leadit-initiative-to-accelerate-transition-towards-net-zero-emissions/

Team Acko. (2023, July 12). *20 Safest Cars in India: List of Best & Top Safety Cars in India with Global NCAP Rating*. ACKO. www.acko.com/car-guide/safest-cars-in-india/

Tushman, M. L., & Anderson, P. (1986). Technological discontinuities and organizational environments. *Administrative Science Quarterly*, *31*(3), 439. https://doi.org/10.2307/2392832

Utterback, J. M., & Abernathy, W. J. (1975). A dynamic model of process and product innovation. *Omega*, *3*(6), 639–656. https://doi.org/10.1016/0305-0483(75)90068-7

Van Den Hoed, R. (2007). Sources of radical technological innovation: The emergence of fuel cell technology in the automotive industry. *Journal of Cleaner Production*, *15*(11–12), 1014–1021. https://doi.org/10.1016/j.jclepro.2006.05.032

Vasudha, M. (2022). Tata Motors turnaround in 2021. *Amity Research Centers*, 19.

Vaz, C. R., Grabot, B., Maldonado, M. U., & Selig, P. M. (2013). Some reasons to implement reverse logistics in companies. *International Journal of Environmental Technology and Management*, *16*(5/6), 467. https://doi.org/10.1504/IJETM.2013.059447

Vaz, C. R., Rauen, T., & Lezana, Á. (2017). Sustainability and innovation in the automotive sector: A structured content analysis. *Sustainability*, *9*(6), 880. https://doi.org/10.3390/su9060880

Vishal, M. (2023, June 19). Car design must be a reflection of where it is from: Tata Motor's Martin Uhlarik. *Hindustan Times*. www.hindustantimes.com/business/tata-motors-electric-vehicle-sales-rise-66-in-may-credits-success-to-bold-design-and-user-experience-101687170794307.html

Wu, X., Zhang, C., & Du, W. (2021). An analysis on the crisis of "chips shortage" in automobile industry—based on the double influence of COVID-19 and trade friction. *Journal of Physics: Conference Series*, *1971*(1), 012100. https://doi.org/10.1088/1742-6596/1971/1/012100

Yadav, A., & Sagi, S. (2024). Management cases studies and technical use cases on web 3. In D. Darwish (Ed.), *Advances in Web Technologies and Engineering* (pp. 267–287). IGI Global. https://doi.org/10.4018/979-8-3693-1532-3.ch013

# 17 Blockchain in Healthcare
## *A Critical Evaluation*

*Ruchika Gupta, Shiv Ranjan, and M. Affan Badar*

### 17.1 INTRODUCTION

The healthcare industry offers commodities and services for curative, preventive, rehabilitative, and palliative care. It involves creating and selling health-related goods and services. Due to technological advances and the growing need for medical services from an ageing population, the healthcare industry has grown and is likely to continue to rise. However, at the same time, hospitals in the rural areas in the USA are being closed [1] because of profitability and other issues, which has negatively impacted rural communities during COVID-19. Regardless, use of technology and manufacturing industry tools has been growing in healthcare [2–4]. Furthermore, the ever-increasing complexity of healthcare delivery, which includes evolving technologies, will require accuracy and timely availability with safety and security of data and information, especially during exchange of information [5].

Despite significant investments in healthcare information, there has been very little electronic information exchange between healthcare providers (i.e., doctors, nurses, and other medical team members), hospitals, and information technology (IT) providers in the last five years, despite the fact that it plays a critical role in improving healthcare costs, quality, and patient experience [6]. The healthcare sector comprises several domains like healthcare providers, hospitals, patients, pharmacy, insurance, clinical research, medicine and medical device manufacturers, and many others. Exchange of healthcare information electronically is critical and crucial. Therefore, integrating a blockchain system into existing IT systems in the healthcare sector poses several technical, operational, governance, and economic challenges. Today's healthcare environment requires data protection and ownership. Medical records are unprotected, resulting in major data breaches. The Ponemon Institute, in partnership with IBM Security, found that the average cost of a data breach in the US was $7.91 million, with the health sector having the highest per capita cost. As personalised medicine and wearables grow increasingly popular, patients don't have full control over their medical data. Both of these situations have important moral ramifications that must be addressed.

In addition, the healthcare information systems of today are very distributed. While a multitude of sources produce increasing volumes of personal healthcare data—from doctors to dispensaries, research labs to healthcare institutions, and even our connected devices—the information is almost always isolated within data silos—stratified and difficult to access from outside the boundaries of closed and

DOI: 10.1201/9781003438748-21

privately owned systems. This is not only inefficient, but it also hinders a healthcare provider's ability to work with the most complete collection of patient data that is available in real time in order to achieve optimal health outcomes.

Interoperability, privacy, and supply chain traceability were all major concerns in the healthcare sector prior to the rise of COVID-19. Another significant issue is that private electronic health record systems from hundreds of suppliers often do not communicate with one another. As the epidemic spreads, healthcare providers have faced additional hurdles, including as adjusting supply systems to distribute protective gear and developing treatments, diagnostics, and vaccinations at a quick pace. Meanwhile, they're trying to figure out how to manage permission and keep personal health data safe while they try to use health data to securely resume to business. The electronic infrastructure that supports the health services on which so many of us have depended received relatively little attention during 2020 and into 2021. And despite all of the development and innovation in the healthcare business, its supporting data systems and information technology have traditionally trailed behind.

Instead of having numerous digital records for the industry, improved healthcare outcomes necessitate a single worldwide system that can scale enormously and efficiently handle large amounts of data to serve as the primary repository of information. Technologies such as blockchain potentially overcome these problems and bring change in healthcare by placing patients at the core of the system and providing secured, secluded, and interoperable heath data with immutable and transparent system.

A blockchain is a decentralised transaction ledger. Each record is stored in blocks with cryptographic keys (hashes). The shared ledger stores these keys or signatures, which are linked together by a set of nodes with each other. By copying it, each node keeps a synchronised and updated whole chain of the transactions. It allows group members to securely communicate data with one another without the use of an intermediary, as well as keeping logs of when, where, and what was transferred. Rather than being stored on a breachable single computer, the information is spread over several systems, making it exceedingly difficult to alter or erase [7].

Per the National Institute of Standards and Technology (NIST), the procedure of blockchain, which guarantees any data entered into the block is tamper-proof and legitimated, increases trust among group members. Blockchain might revolutionise health data storage and transmission. The firm may benefit from blockchain technology's ultra-secure encrypted database and shared ledger, which makes communication easier.

With blockchain, there'd be no need for a centralised system of administration since smart cryptography would eradicate the need to do so. Moreover, each member has total control over their personal information and activities. Blockchain can efficiently reorganise medical records and facilitate safe exchange because data dealing with the healthcare sector is confidential and needs speedy access to information for the benefit of patients. Blockchain delivers access security, data privacy, and scalability in a single step. This would allow healthcare facilities to devote more time and money to clinical outcomes and research instead of just focusing on administrative duties [8].

Despite numerous benefits offered by blockchain, integrating a blockchain system into existing IT systems in the healthcare sector poses several technical, operational, governance, and economic challenges. Companies need to access their strategic requirements and prioritise the implementation of the blockchain system in certain use cases. Also, due to recent security breaches that used Bitcoin as ransom, blockchain technology has a negatively perceived image that could limit its acceptance for handling sensitive health data in the near future. With these, there are challenges related to the technical perspective of decentralised cryptosystems, and the lack of confirmed business prototypes further limits the large-scale acceptance of blockchain systems. Verification/validation with blockchain also requires a lot of computing power and cannot be scaled for more complex health data. Further slowness of the healthcare industry in the implementation of new technologies and, above all, the conflict of interest between incumbent who keep health data in guardianship are likely to constitute significant obstacles. Therefore, this chapter tries to evaluate the various opportunities offered and challenges posed by the technology and also to critically evaluate the various blockchain platforms available for healthcare sector focusing on their strengths and applicability. This work also discusses several blockchain applications cases in healthcare sector and analyses barriers in adoption of this technology for the benefits of healthcare industry.

## 17.2 WHAT IS THE MECHANISM BEHIND BLOCKCHAIN?

Three essential components make up blockchain, a decentralised peer-to-peer (P2P) technology:

- Decentralised network: The P2P distributed architecture is made up of nodes, each of which maintains a copy of the blockchain and may authenticate and validate network transactions.
- Shared ledger: Network members record all digital transactions in a shared ledger. After a majority of members accept, the proposed transaction is posted to the shared ledger.
- Digital transaction: Any information or digital data that may be recorded on a blockchain is a digital transaction. Every transaction is grouped into a "block", which contains a cryptographic hash to add transactions in a linear, chronological order. Every blockchain transaction is the same [9].

Any blockchain's core concept of functioning may be split down into four separate, continuous steps [10]:

- Every transaction is documented: The digital signatures of the persons who made the transaction are used to validate this record, which contains specific data about them.
- Every transaction is double-checked for accuracy: The computers linked to the network finish the verification process by independently checking to confirm that the trade is authentic. Since this is a distributed procedure, it requires agreement from every node in the network before it can be completed.

- Addition of every confirmed transaction into hashed blocks: The term "block" refers to a collection of transaction records, each of which is distinct. Every block contains a hash value (or hash digest) that both uniquely identifies it and indicates where it is in the blockchain. The hash also protects the data's integrity, demonstrating that it hasn't been changed since the block was created.
- When the block is finished, it is added to the end of the blockchain: The blockchain construction and certification procedure has now come to a close. Following the completion of one block, another will be completed shortly.

## 17.3 BLOCKCHAIN TECHNOLOGY APPLICATIONS IN HEALTHCARE

In the field of healthcare, blockchain offers several applications and uses. Different researchers have reviewed blockchain technology applications for healthcare [11–14]. The following are some of the major areas in which blockchains might be used in healthcare [11]:

- Secure Storage and Transfer of Patients' Medical Records: Blockchain's ability to maintain an immutable, decentralised, and complete record of patient medical history makes it ideal for security checks and applications. When accessible to peers, blockchain is also private, disguising any individual's identity using sophisticated and secure algorithms that can protect medical information. Due to its decentralised nature, the technology allows patients, doctors, and healthcare professionals to share information quickly and securely. Blockchain technology allows patient medical records to be accessed and shared. Blockchain might unify stakeholder data and let patients securely access their medical records. Medical data is generally dispersed among several healthcare organisations, making it difficult to access.
- Remote Patient Monitoring: Remote patient monitoring entails gathering medical history by using Internet of Things (IoT), sensors, and mobile devices in order to remotely monitor a patient's condition. Remote patient monitoring technologies can help all sorts of experts execute their jobs more successfully and efficiently—and, eventually, joyfully—by removing the stress and chaos from healthcare for a wide spectrum of patients. Using sensors that convey alerts to family members as well medical professionals, this type of tele-health can also give comfort and reassurance for relatives of patients who may be at increased risk of falling or accidents. Blockchain is to be used to store, distribute, and recover biological history that has been acquired remotely. By providing remote healthcare services in a decentralised, traceable, tamper-proof, transparent, trustworthy, and secured manner, blockchain technologies can enhance tele-health and telemedicine services. It allows medical practitioners to properly detect scams involving physician educational qualifications and medical testing kits widely used for home diagnosis.

- Patient Access to Medical Records: Another benefit of an open ledger is patient access to medical records. Healthcare data interoperability requires regulatory and legal requirements. To help patients get the latest treatments, a healthcare blockchain may be connected to medical research. The doctor and patient would discuss science-based alternative therapies. Blockchain technology improves healthcare data interoperability by creating secure and verifiable medical histories, linking, recording, and allowing patients to access and agree to stakeholder communication [15, 16].
- Medicine Management Supply Chain: Medical product authenticity is a major issue in the healthcare business and others. Supply chain stakeholders may track things from manufacture to delivery using blockchain technology. Drug ledgers indicate origins (i.e. a laboratory). The ledger will record who handled and where data goes until it reaches the client. The system tracks labour and waste emissions. In developing markets, where counterfeit prescription pharmaceuticals kill thousands, this is a huge challenge for the industry. Medical equipment is growing due to pandemic, attracting unscrupulous players [16].
- Unlocks Genetic Code: Genomics' promise to improve human health is now a scientific and commercial reality. In 2001, processing a human genome cost $1 billion. Due to incurable hereditary disorders, 23andMe and Ancestry.com's DNA testing are now in millions of families. Genomic data theft is a growing threat as more companies provide DNA sequencing to individuals. Blockchain can circumvent this and establish an open marketplace where scientists may buy genetic data for study. This may promote secure selling and eliminate intermediaries.
- Using Smart Contracts for Medical Insurance Claims: Because it can reveal medical events as they happen, the health sector blockchain technology is suitable for insurance claim processing. It prevents data manipulation for fraud. By sharing common digital contracts between beneficiary, distribution agency, and healthcare institutions on a blockchain ledger, users may considerably reduce billing refund claims for pharmaceuticals and other things. Pharmaceutical firms, distributors, insurance companies, and healthcare experts may confirm their identities, record contract information, check services given, and claim payment data utilising blockchain-based platforms from companies like Curisium and Chronicled [17].
- Verifying Medical Staff Credentials: Blockchain technology can verify medical items and monitor medical practitioners' knowledge and skill. Medical institutions and healthcare corporations may register staff credentials to speed up recruiting. ProCredEx, headquartered in the US, has developed a medical credential verification system using R3 Corda blockchain technology [17].

## 17.4 BLOCKCHAIN IN HEALTHCARE: ISSUES OR CHALLENGES

Healthcare includes healthcare providers (medical teams), hospitals, patients, pharmacies, insurance, medical research, medicine, and equipment manufacturers. Blockchain might affect all of these healthcare stakeholders. Blockchain is gaining appeal as a solution to healthcare industry issues including wasteful spending and higher rates for providers, insurers, and consumers. New transparent business models are being established as sceptical opponents join together to test the technology.

This technology might change care by connecting different systems, creating insights, and measuring value. A nationwide blockchain network for electronic medical information may improve efficiency and patient outcomes. Blockchain, like any other new technology, has its own set of restrictions. The following are some of the most significant problems with blockchain applications in healthcare:

- Lack of Standardisation: The blockchain's public ledger allows anybody in the network to access personally identifiable healthcare data, which is a big drawback compared to conventional data storage methods. Since this technology is still young, its flexibility is limited, which leads to a lack of industry standards, which makes it difficult to fully realise its potential.
- Scalability: Another drawback of the blockchain system is scalability; as the number of blocks grows, the system becomes slower and demands more processing power and energy to operate. This may be made considerably more difficult in an IoT context, where data must flow in real time.
- Network Infrastructure Security: Information recorded in the blockchain database must be protected and validated for usage using a cryptographic procedure that involves the use of private and public keys at all levels of the network infrastructure. Because the keys required by various cryptographic procedures are accessible for all members, access control is essential. Conventional access control concepts are ineffective since one key for all blocks implies that if the key is compromised, all of the blocks may be accessed; similarly, using a separate key for each block is inefficient because it raises the cost of storing and retrieving data.
- Participants' Identify Verification and Authentication: The way a person's online identity is formed and utilised determines how safe and secure access to their health information seems to be while more healthcare professionals convert to electronic health records (EHRs) and provide online services through portals. Comprehending how patient identification is verified can assist healthcare workers in better securing electronically stored health data, but it is difficult to do. Systems in healthcare must account for adequate degrees of identity assurance and strike a challenging balance between security and usability in order to serve a diverse set of users. When designing techniques and operational procedures for identity formation and continuing user authentication, this balance must take into consideration individual, social, and intergenerational variations.
- Uniformity in Authorisation Patterns to Access EHRs: There is no standardisation in the usage of authentication techniques, as one might anticipate given today's constantly developing computer environments and the merging of numerous ancient information systems with more contemporary ones. Most systems rely on the vendor's built-in authentication methods, and the lengths and forms of valid account login credentials are frequently inconsistent.

## 17.5 BLOCKCHAIN PLATFORMS AVAILABLE FOR HEALTHCARE: COMPARATIVE ANALYSIS

One of the most important tasks in starting to design and construct a real-world healthcare blockchain project is to choose the best underlying blockchain platform. Ethereum, Hyperledger, and MultiChain are the primary platforms used for healthcare. A comprehensive evaluation of these platforms in terms of their technological characteristics is important to addressing the problems of choosing a blockchain platform for healthcare or medical applications [12, 13, 17]. Kuo et al. (2019) [17] have compared ten platforms, including Ethereum, Hyperledger, and MultiChain. Table 17.1 presents a comparative analysis of five platforms that can help researchers and application developers of healthcare blockchain system to select the best available platform for their unique problem and issues.

- Ethereum: These days, Ethereum is the most popular blockchain platform. Users may develop new financial apps, decentralise markets, and create games and Bitcoin wallets, among other things. The major goal of this platform is to eliminate third-party access to data for the purpose of tracking financial instruments. Ethereum boasts more core protocol engineers, researchers, cypherpunks, and miners' groups than any other cryptocurrency. Rapid deployment is a benefit of this platform. It is complete and features smart contract functionality. It has the ability to tokenise data [19].

  For patient-managed health information sharing apps, MedRec and Patientory suggest using a blockchain based on the Ethereum platform. Nebula Genomics recommends using an Ethereum-based blockchain platform to distribute and analyse genetic data. For use in clinical applications, Ethereum has also suggested clinical research data exchange and automatic remote healthcare monitoring.
- Hyperledger: A modular architectural platform for app and solution development, consensus, and membership. It has several modular and flexible designs for industrial use. It allows network creation. It safeguards users' confidential contracts and transactions. Modular and low-latency, platform solidity and multilingual intelligent contracts are supported [18].

  The Hyperledger platform is proposed for use in a cancer clinical data exchange framework for patient care. Hyperledger is also used to provide a basic structure for enforcing medical industry rules and conduct. Hyperledger is also being considered for a patient monitoring application as well as medical record storage and access apps.
- MultiChain: MultiChain is an open-source variant of Bitcoin. It is simple to use and can be used to construct both private and public blockchains. It includes a carefully selected set of features and upgrades geared towards corporate and commercial users. Compatibility with local assets and the storage of larger quantities of random data both sound interesting. A different but unique option is consensus-based permit management for consortial blockchains. The chain may be modified by editing a simple text file generated by underlying blockchain; it should have been done before

running MultiChain for the first time, as variables cannot be changed once the network is fully created. On the blockchain, it consists of huge number of monitored and validated assets that may be utilised to conduct secure multi-asset and multi-party transactions. Data broadcasting, sharing data, timestamping, and secure archival are all possible with it. It is based on GitHub. The programming languages C, C++, Python, and JavaScript are all supported [18].

ModelChain was utilised to assist decentralised cross-institutional predictive modelling, which helped to speed up research and boost quality improvement programs. ModelChain's usage of blockchain allows diverse institutions to increase prediction capability by adding facts and figures to a collaboratively built system without having to send the data or the system to a central place. ModelChain's initial purpose was for predictive analytics in anaesthesia, and it employed a permissioned blockchain called MultiChain.

- Corda (R3 Corda): A popular open-source blockchain initiative for enterprises is Corda. Secure and private blockchain systems may be built using Corda. It is cutting-edge contracting. It facilitates corporate communication. Businesses may monitor transaction processes in a shared ledger, eliminating the need for participants to double-check their transactions after engaging. Health, finance, and construction utilise it most. It protects you. It's flexible. It allows collaborative design and development on a blockchain platform. Corda's innovative privacy architecture and smart contract capacity can provide an effective and reliable framework for patient—provider identification sharing and verification.

  The Corda Healthcare Community blockchain effort is a collaboration between HSBlox and R3, a blockchain enterprise organisation. The Community will run on R3's Corda platform and will serve as a hub for companies interested in adopting or researching blockchain-based healthcare solutions. [19]

**TABLE 17.1**
**Comparative Analysis of Major Blockchain Platforms for Healthcare [17, 21, 22]**

| Blockchain Platforms/Basis of Distinction | Network Permission | Special Hardware Requirements | Support for Smart Contracts | Scripting Language | Open Source |
|---|---|---|---|---|---|
| Ethereum | Permissionless/ permissioned | No | Yes | Solidity, Serpent, LLL | Yes |
| Hyperledger | Permissioned | Various | Yes | Bitcoin Script | Yes |
| MultiChain | Permissioned | No | No | Various | Yes |
| Corda | Permissionless | No | Yes | Kotlin | Yes |
| Neo | Permissionless/ Permissioned | No | Yes | Various like C#, Kotlin, Python | Yes |

- Neo: It supports its own currency and enables you construct digital assets and smart contracts on a distributed ledger technology platform. It resembles the US-based Ethereum blockchain network. Neo's purpose is to use smart contracts to automate digital asset management and create a distributed network-based smart economy system. The Neo network can process up to 10,000 transactions per second using a decentralised Byzantine fault tolerant (dBFT) consensus technique among centrally sanctioned nodes [20].

Digital identification allows crucial information about involved people, organisations, and institutions to be verified. Smart contracts allow parties to trade without a court or government. The network's computer code enables contract execution and transaction traceability, transparency, and irreversibility.

## 17.6 PROPOSED BLOCKCHAIN MODEL FOR HEALTHCARE

The chapter proposes a model (Figure 17.1) using blockchain and smart contract technology for providing emergency treatment. The model is based on the concept of integrating all the stakeholders of healthcare, taking into consideration how timely and reliable access to information by stakeholders can save lives of patients and reduce the chances of casualty as well.

However, the administration of patient medical records has always been linked to the problems of privacy and security. Many existing healthcare models keep the individual, who is the foundation of the system, out of the loop when it comes to freely accessing information. Another key consideration is the integrity of health-related data [1, 14]. A breach in data integrity might lead to erroneous conclusions, putting a person's life at jeopardy. One of the most significant concerns is multi-organisational data sharing, which requires medical data obtained by a healthcare provider to be openly accessible to other organisations, such as a doctor or research institution.

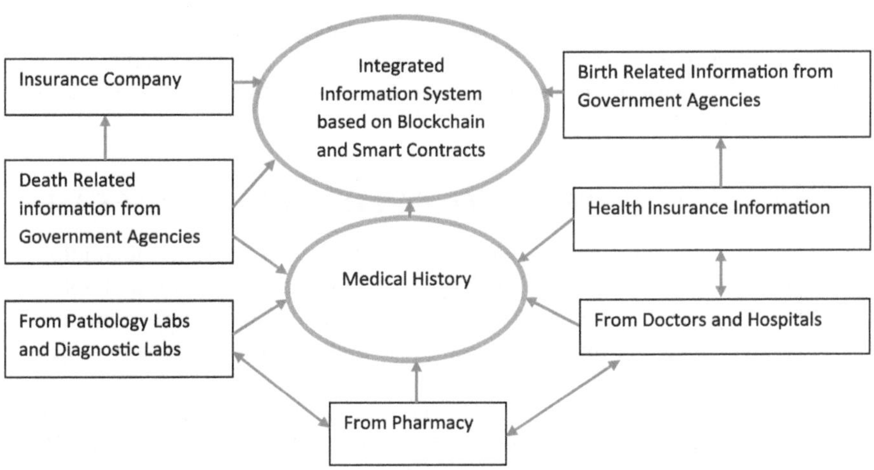


**FIGURE 17.1** Proposed model.

Older medical and healthcare systems only share resources internally and are not fully interoperable. However, evidence shows that connecting these networks for connected and enhanced healthcare offers various benefits, requiring connection across firms for health informatics researchers [2, 23]. Data-driven businesses must safeguard, store, trade, and integrate data. Blockchain technology may help healthcare providers solve these crucial issues [3, 4].

The model suggests how these stakeholders can work together and help in emergency treatment. It will also help achieve the best treatment, prevent fraud and false insurance claims, and, even in the worst case, help a deceased person's family get speedy insurance claims. Smart contracts record information in an irreversible manner, making it available to important stakeholders and avoiding conflicts resulting from the concealment or falsification of information in a more efficient and timely manner. That's the reason the technology has been chosen in the proposed model so that trust, reliability, and transparency in the system can be achieved.

It is expected that all stakeholders would benefit from this system. Decentralising health data would allow for complete interoperability, first across national health stakeholders and then globally. The adoption of smart contracts might blur the lines between different healthcare systems, allowing for more efficient information exchange across them.

The model is centred around the medical history of the patients, since it is of the utmost importance and critical due to the following reasons:

- The patients' medical history can help or hinder a diagnosis, since it is essential for avoiding the administration of drugs to which a patient is allergic.
- External actors can use patient health data for marketing reasons or even malevolent goals, which is why it's so important to keep it safe.

There is no method to ensure that data is not duplicated or transferred once it has been provided to an actor. Patients can, however, demand that actors and institutions sign a contract promising to delete personal data after the original reason for acquiring it has been met. This aspect is being taken care of by the proposed model since smart contracts integrating blockchain technology can help tracing back a series of transactions.

## 17.7 BLOCKCHAIN APPLICATIONS IN HEALTHCARE: USE CASES

BurstIQ: Blockchain improves medical data exchange. BurstIQ's technology helps hospitals securely handle huge patient data. Blockchain technology secures data storage, selling, sharing, and licencing under strict standards. BurstIQ's software might identify opioid and other prescription drug abuse [11].

MedRec: MedRec is a healthcare startup with a blockchain focus. MedRec uses a peer-to-peer ledger that makes it easy for providers to monitor files and information. In addition, the service enables for more direct connection with patients. Clinicians, facilities, and multiple systems are all connected on one platform, providing data transfer possible in most efficient way [11, 24].

Hashed Health: Anthem subsidiary national government services and insurer WellCare Health Plans are using blockchain to simplify the onerous process of validating professionals' credentials to practise in a clinical environment [24]. Hashed Health uses blockchain to provide a free and open forum for healthcare experts to discuss blockchain applications. The company also advises healthcare companies on how to integrate blockchain into their systems. Finally, the company has a blockchain lab to solve healthcare problems.

Dynostics: Another example is Dynostics. Individual users are catered to, and Dynostics assists them in determining their fitness based on current diagnostics data by giving rapid advice and provide one place for their data. Dynostics is concerned with security and privacy of data.

Chronicled: The Chronicled blockchain network ensures secure drug delivery and rigorous scrutiny. Chronicled builds ownership-proof blockchain networks. The networks help pharmaceutical companies deliver prescriptions on schedule and enable law enforcement to detect suspicious activity like drug trafficking. Chronicled launched the 2017 MediLedger Project, a distributed ledger solution for medical supply chain security, privacy, and efficiency. Chronicled claims that their blockchain-based MediLedger Project can make the pharmaceutical supply chain interoperable while satisfying data privacy regulations [11].

Medicalchain: Medicalchain's blockchain platform tracks patient identity and origin. Medicalchain's blockchain centralises and verifies medical data. Physicians, clinics, and labs may request traceable patient data to safeguard patient privacy. Medicalchain launched MyClinic.com in May 2018. MyClinic lets patients video chat with their doctors and pay using "MedTokens" [11].

EncrypGen: The blockchain-based EncrypGen Gene-Chain helps search, share, store, buy, and sell genetic data. Only other company members may buy genetic data using secure, traceable DNA tokens, protecting user privacy. Member businesses may utilise genetic data to strengthen their genetic knowledge and the industry. EncrypGen will add self-reported medical and behavioural data to user profiles [11, 16].

## 17.8 BARRIERS TO ADOPTION OF BLOCKCHAIN IN HEALTHCARE

Blockchains are susceptible to a 51% attack, often known as a double-spend attack [17]. When a miner or group of miners controls more than 50% of the network's mining hash rate (computing power), they may block or reverse transactions. The longest chain rule forces the network to switch to the attacker's chain since the malicious user has over 50% of the processing power and generates blocks quicker than the rest of the network (which stipulates that the longest blockchain is the legitimate chain). Despite 51% assaults, blockchain is one of the safest technologies since unbeaten attacks are rare.

Consider the cost of switching to blockchain. Transactions need payment for computation. Payments on Ethereum are done in "gas", or ETH, now worth $120 USD (EthereumPrice.org, 2019). Hospitals need many connections, which is costly. The issue is further exacerbated by the limited computing capacity of smart devices, which are not meant to do computations but rather to record and send data, which works well in a conventional database architecture but may cause problems in a blockchain network. Due to the overhead on the data flow that must be done, this may also increase the cost of the sensors.

Additionally, the technology has flaws. Blockchain's consensus method requires solving computationally hard problems. To overcome the cryptography problem, a miner must download all the data into the local machine, which may be gigabytes or petabytes. Furthermore, the number of transactions that may take place on a blockchain network in a unit of time is generally limited, which might limit the throughput of big chains. A malevolent actor can do significant harm to the network by withholding a portion of a mined block, reducing the motivation to mine and limiting the network's capacity to grow blocks. A double-spending attack may be carried out in a similar way by replicating a piece of digital information and so creating a clone of a network node.

Many doctors are still paper-bound. Thus, convincing them to switch from paper to blockchain-based EHR is difficult. Physicians, for instance, often prefer to leave certain questions unanswered in their notes, a practice that doesn't translate well to digital systems that require complete data entry. Technology isn't hard. However, change management is challenging. Changing behaviour is hard in any sector.

Healthcare providers and insurers handle records differently. Without a single payer, it would be impossible to convince all of these institutions to utilise blockchain. If some don't like it, the system's usefulness decreases.

## 17.9 POST-PANDEMIC MEASURES AND CONCLUDING INSIGHTS

The post-pandemic era has ushered in a period of profound transformation in the healthcare sector. This transformation has been driven by the need to adapt rapidly to the challenges presented by the COVID-19 pandemic and, more importantly, to create a healthcare ecosystem that is more resilient, patient-centric, and prepared for future crises.

In this chapter, we have examined several key post-pandemic measures undertaken by the healthcare sector to re-imagine its business practices:

1. Telehealth and Virtual Care Integration: Telemedicine and virtual care have become integral components of healthcare delivery, offering increased access to care and improved efficiency.
2. Data-Driven Decision-Making: Advanced data analytics and predictive modelling have enhanced healthcare organisations' agility, resource allocation, and response to emerging healthcare challenges.
3. Resilient Supply Chains: Strengthening supply chains and inventory management has become paramount to ensure the availability of essential medical supplies and medications during crises.

4. Patient-Centric Care Models: The healthcare sector has shifted towards patient-centred care models, emphasising shared decision-making, enhanced communication, and personalised treatment.
5. Workforce Well-Being: Prioritising the well-being of healthcare professionals, offering mental health support, and adapting staffing models have become critical to maintaining a motivated and resilient workforce.

In conclusion, the post-pandemic healthcare sector has demonstrated remarkable adaptability and innovation in re-imagining its business practices. However, to sustain and build upon these changes, several valuable suggestions are worth considering:

1. Invest in Digital Infrastructure: Continue to invest in robust digital infrastructure, ensuring that telehealth and data analytics capabilities are not only maintained but also enhanced.
2. Foster Collaboration: Strengthen public-private collaborations and global partnerships to facilitate rapid responses to emerging health crises and improve healthcare access for all.
3. Patient Empowerment: Continue to prioritise patient-centric care models that empower individuals to take an active role in their healthcare decisions.
4. Data Governance: Develop robust data governance frameworks that balance data privacy and security with the potential benefits of data sharing and interoperability.
5. Resilience Planning: Conduct comprehensive resilience planning that encompasses supply chain diversification, emergency preparedness, and workforce well-being as integral components.
6. Regulatory Flexibility: Advocate for regulatory flexibility that accommodates evolving healthcare practices and supports ongoing innovation.

As we move forward in this post-pandemic healthcare landscape, the lessons learned and the measures undertaken during this critical period will shape the future of healthcare. By embracing change, fostering innovation, and prioritising patient well-being, the healthcare sector can continue to evolve and provide more efficient, accessible, and patient-centred care to all.

To conclude, blockchain technology has great promise, but more research is required to understand, develop, and safely evaluate it. Blockchain might revolutionise healthcare by giving patients control over their medical records and data. The blockchain movement for patients is relatively young, but this change in responsibility might lead to patient-centred therapy.

## REFERENCES

1. Kissi, M., Walston, J. D., & Badar, M. A. (2021). Statistical Analysis of Rural Hospital Closures in the USA. *International Journal of Forensic Engineering and Management* 1(2): 209–224, https://doi.org/10.1504/IJFEM.2021.120186.

2. Bhandari, P., Badar, M. A., & Childress, V. (2020). On Socioeconomic Impacts of Technological Advancements in Healthcare. IEOM Proceed. of the 5th North American Conf. on Industrial Engr and Operations Mgmt, Detroit, MI, Aug 9–11, 2020, pp. 2921–2930, http://doi.org/10.46254/NA05.20200591.
3. Bhandari, P., Badar, M. A., & Childress, V. (2021). COVID-19 Surge Planning in Response to Global Pandemic in a Healthcare Setting: A Lean Six Sigma Approach. IEOM Proceed. of the 11th Int. Conf. on Industrial Engr and Operations Mgmt, Singapore—Virtual, March 7–11, 2021, pp. 315–326, http://doi.org/10.46254/AN11.20210055.
4. Hoseini, C., Badar, M. A., Shahhosseini, A. M., & Kluse, C. J. (2021). A Review of Machine Learning Methods Applicable to Quality Issues. IEOM Proceed. of the 11th Int. Conf. on Industrial Engr and Operations Mgmt, Singapore—Virtual, March 7–11, 2021, pp. 1225–1240, http://doi.org/10.46254/AN11.20210250.
5. Joynt Maddox, K. E., Bauchner, H., & Fontanarosa, P. B. (2019). US Health Policy—2020 and Beyond: Introducing a New JAMA Series. *JAMA* 321(17): 1670–1672, http://doi.org/10.1001/jama.2019.3451, last accessed 2021/08/17
6. Marchibroda, J. (2014). Interoperability. *Health Policy Brief*, www.healthaffairs.org/do/10.1377/hpb20140811.761828/full/, last accessed 2021/08/17
7. Shelby Livingston (2019). Will Blockchain Save the Healthcare System? www.modernhealthcare.com/article/20190209/TRANSFORMATION02/190209953/will-blockchain-save-the-healthcare-system, last accessed 2021/07/08
8. Abhinav Shashank (2018). 5 Benefits of Using Blockchain Technology in Healthcare, https://hitconsultant.net/2018/01/29/blockchain-technology-in-healthcare-benefits/#.YOcyGOgzbIU, last accessed 2021/07/08
9. Levis, D., Fontana, F., & Ughetto, E. (2021). A Look into the Future of Blockchain Technology. *PLoS ONE* 16(11): e0258995, https://doi.org/10.1371/journal.pone.0258995.
10. Daley, S. How Using Blockchain in Healthcare Is Reviving the Industry's Capabilities, https://builtin.com/blockchain/blockchain-healthcare-applications-companies, last accessed 2021/07/08
11. Agbo, C. C., Mahmoud, Q. H., & Eklund, J. M. (2019). Blockchain Technology in Healthcare: A Systematic Review. *Healthcare* 7(2): 56.
12. Ghosh, P. K., Chakraborty, A., Hasan, M., Rashid, K., & Siddique, A. H. (2023). Blockchain Application in Healthcare Systems: A Review. *Systems* 11(1): 38, https://doi.org/10.3390/systems11010038.
13. Rathore, H., Mohamed, A., & Guizani, M. (2020). Blockchain Applications for Healthcare, Chapter in *Energy Efficiency of Medical Devices and Healthcare Applications*, Editor: A. Mohamed, Academic Press, pp. 153–166, https://doi.org/10.1016/B978-0-12-819045-6.00008-X.
14. Haleem, A., Javaid, M., Singh, R. P., Suman, R., & Rab, S. (2021). Blockchain Technology Applications in Healthcare: An Overview. *International Journal of Intelligent Networks* 2: 130–139, https://doi.org/10.1016/j.ijin.2021.09.005.
15. Roehrs, A., da Costa, C. A., & da Rosa Righi, R. (2017). Omniphr: A Distributed Architecture Model to Integrate Personal Health Records. *Journal of Biomedical Informatics* 7(1): 70–81.
16. Bryan Weinberg (2019). 14 Major Real Use Cases of Blockchain in Healthcare, Blockchain Insights, 14 Major Real Use Cases of Blockchain in Healthcare | OpenLedger Insights, last accessed 2022/07/08
17. Kuo, T.-T., Rojas, H. Z., & Ohno-Machado, L. (2019). Comparison of Blockchain Platforms: A Systematic Review and Healthcare Examples. *Journal of the American Medical Informatics Association* 26(5): 462–478, https://doi.org/10.1093/jamia/ocy185.
18. Barley, M. R3 Partners with HSBlox to Launch Healthcare Community, www.ledger-insights.com/r3-corda-hsblox-healthcare-community/, last accessed 2021/07/08

19. Corbet, S., Lucey, B., Urquhart, A., & Yarovaya, L. (2019). Cryptocurrencies as a Financial Asset: A Systematic Analysis. *International Review of Financial Analysis* 62: 182–199, ISSN 1057–5219, https://doi.org/10.1016/j.irfa.2018.09.003
20. Lemieux, V. L. (2016). Trusting Records: Is Blockchain Technology the Answer? *Records Management Journal* 26: 110–139, https://doi.org/10.1108/RMJ-12-2015-0042.
21. Dinh, T. T. A., Wang, J., Chen, G., Liu, R., Ooi, B. C., & Tan, K.-L. (2017). BLOCKBENCH: A Framework for Analyzing Private Blockchains, SIGMOD '17: Proceedings of the 2017 ACM International Conference on Management of Data, pp. 1085–1100, https://doi.org/10.1145/3035918.3064033.
22. Yu, H., Sun, H., Wu, D., & Kuo, T. T. (2020 Mar 4). Comparison of Smart Contract Blockchains for Healthcare Applications. *AMIA Annual Symposium Proceedings* 2019: 1266–1275, PMID: 32308924, PMCID: PMC7153130.
23. Ekblaw, A., Azaria, A., Halamka, J. D., & Lippman, A. (2016). A Case Study for Blockchain in Healthcare: "MedRec" Prototype for Electronic Health Records and Medical Research Data. Whitepaper, www.media.mit.edu/publications/medrec-whitepaper, last accessed 2023/06/30
24. Yli-Huumo, J., Ko, D., Choi, S., Park, S., & Smolander, K. (2016). Where Is Current Research on Blockchain Technology? A Systematic Review. *PLoS ONE* 11(10): e0163477, http://doi.org/10.1371/journal.pone.0163477.

# 18 Digital Transformation in Advertising after COVID-19

## *Routing the New Landscape*

*Ganesha K. S. and K. G. Hemalatha*

## 18.1 INTRODUCTION

Advertising is the one industry in the world, which helps the firms to make their brand, identified as a separate one with the similar products and services by producing symbolic wrapping and creative contents. Advertising has been around for hundreds of years (Marketing Mind, n.d.). The evolution of advertising started in the 1600s with newspaper ads (Quick, 2020). Per studies from the OECD (2019), it is evident that the 1960s to 1980s was considered a "golden age" for advertising; during that time there was huge creativity, which peaked in advertising with the advent of televisions and radio devices in people's homes. After the 1990s the emergence of Web 1.0, or internet and cell phones, paved the way for digital advertising, the creation of websites, email ads, and pop-up ads. These led to Web 2.0, which changed the way customers search for information and make purchase decisions. With exposure to Web 3.0, consumers today use multiple information sources and take the help of influencers, bloggers, and user-generated content on Web 3.0 platforms (Ganesha & Ganapati, 2020).

The COVID-19 pandemic changed the lives of every individual and organization; also, it changed the way in which interactions happen between them. The pandemic and lockdowns, which forced us to employ social distancing, have changed the way the organizations promote products and services to end users. There were a lot of changes in social, behavioral, and cultural aspects in individuals, and it made organizations think of digital ecosystems for advertising.

Currently, most organizations use digital advertising as the main form of advertising, which was preceded by growth in the low-cost internet in countries like India. In addition, the growth of Artificial Intelligence (AI), Internet of Things (IoT), and data analytics made organizations rethink their strategy on traditional advertising and go for cost-effective digital advertising. With the emergence of Artificial Intelligence and Machine Learning (AI and ML), the metaverse, which is part of Web 4.0, led to the rise of "programmatic advertising", which uses the


DOI: 10.1201/9781003438748-22

complex algorithms with users data and makes the automated decisions regarding advertisements (Geradin & Katsifis, 2019). Immersive technologies like augmented reality (AR), virtual reality (VR), and mixed reality (MR) have also influenced the way firms advertise their products or services. Providing digital touch points through these immersive technologies has enhanced the way firms advertise their products and services.

## 18.2 CHANGE IN CONSUMER BEHAVIOR

Consumer behavior is a dynamic phenomenon, which constantly evolves, and changes its nature due to various factors. However, the pandemic has changed consumer behavior in a different way, as customers started more frequent online and omnichannel shopping, which also bought rapid changes to customer shopping practices, so these changes happened within a small time which otherwise might have taken years. Consumers during and after the pandemic got used to applications of new and innovative technologies. Zoom video services have become vital to connect and keep up with family, and many have learned to participate in Zoom meetings. Social media usage is prolific, including WhatsApp, WeChat, Facebook, YouTube, LinkedIn, and others. Therefore, consumers' lifestyle and behavior were affected by social media and other digital technologies (Sheth, 2020).

COVID-19 has changed the way consumers sense their security and affected their psychological responses, such as affective and behavioral responses (Campbell et al., 2020). Also, we are aware that consumer behavior is dynamic; it changes due to many factors. Factors such as growth in technology and COVID-19 have made consumers think swiftly and change the way they behave with products and services. However, to overcome the risk of product and psychological aspects, they employ strategies to reduce uncertainty and risk (Galoni et al., 2020; Kim, 2020; Sengupta et al., 2020; Giroux, 2021; Kim et al., 2020a, 2020b). So, as a part of an adaptive strategy, consumers started purchasing less risky products to get sense of control and to reduce risk, be it a low-priced stock preferred by a financial investor or FMCG products (Pena-Marin et al., 2022) Also, many researchers (Chalmers & Price, 2009; Zhang et al., 2014; Lee et al., 2019; Li et al., 2019; Park et al., 2022) proved that consumers, while avoiding risk, also check for the authenticity of messages, which in turn influences their perception and value towards advertising. Businesses that try to share a story of their purpose will have more perceived authenticity (Hamby et al., 2019; Taylor, 2020).

Since the beginning of the pandemic, firms' budgets towards marketing of their products and services have significantly decreased. Major costs associated with the mass advertisement are through traditional media channels such as outdoor advertising, magazines, and radio. However, advertisement expenses have not decreased but have increased in a few cases. During and after pandemic, many firms adopted a bottom-line approach to promotions, leveraging SMS marketing and social media to directly connect with their target audience, generate swift responses, and achieve cost savings compared to traditional methods.

## 18.3 ADVERTISING SETTING BEFORE COVID-19

Traditionally advertisers were relied on conventional media, such as print, radio, and television (Nichols, 2013). Nevertheless, even before COVID-19, organizations started using electronic and digital media for promotions and advertising (Valentine & Powers, 2013). In addition, social media marketing was significant for brand awareness and product reach even before COVID-19 (Ganesha & Ganapati, 2019). Advertisements traditionally envisioned reaching and motivating consumers to buy products or services from organizations (Ganesha, 2018). In addition, before COVID-19, the organizations were carrying out advertising and copywriting and all other related activities, and television networks and print media (Whitaker & Smith, 2009) were airing most of the advertisements. However, traditionally, organizations also used eye-tracking visuals with print ads, outdoor advertising, yellow pages advertising, and catalogs through which organizations tried to capture customer attention (Wedel & Pieters, 2017). Therefore, we can say that before COVID-19, advertisers used to focus on offline and online advertising, but usage of innovative advertisement methods such as AI–based and programmatic ads were limited. Advertisers turned to digital channels in order to effectively connect and engage with their target consumers because of physical store closings, postponed events, and social distancing policies

## 18.4 DISRUPTIVE CONSEQUENCES OF COVID-19 TOWARDS ADVERTISING

COVID-19 and the subsequent lockdowns have changed the perceptions of consumers towards brands, which in turn changed their brand loyalty, attractiveness, and switching behavior (Sirkeci, 2020). In addition, COVID-19 has disrupted the way the products are marketed and advertised. After the pandemic, organizations started focusing more on digital marketing, AI, IoT, and other innovative channels for advertising (Jaafar & Khan, 2022). Many industries, including pharma, started focusing on new and innovative marketing tactics during and after the pandemic (Khan & Basak, 2021). In addition, the pandemic led to digital revolution and transition in almost all sectors for advertising (Tien-Dung et al., 2022). However, these disruptions significantly influenced millennial customers and the way advertisers communicated with them (Rashi et al., 2021). So we can say that pandemic has not only disrupted many industries but also majorly disrupted advertising, and it has led to major digital transformation in the advertising industry. Now let us discuss those digital transformations.

## 18.5 DIGITAL TRANSITION IN ADVERTISING

The growth of the internet as a medium for advertising started during Web 1.0 with email marketing and website promotions. But after the 2000, it picked up growth with Web 2.0 and has shifted into social media marketing and other tools with Web 3.0. Now, during and after COVID-19, it has changed to Web 4.0. With the advent of AI, IoT, and immersive technologies, the digital transition has happened in most organizations.

Artificial intelligence technology is being applied to optimize reaching the targeted audience in internet-based advertising. The power of AI combined with machine learning techniques is very powerful, which optimizes digital advertisements through contextual or historical customer data, which gives digital advertising an edge over conventional methods (Chen et al., 2009). These advancements help improve user experience along with mitigating problems with the help of data.

Target identification: ML addresses the difficulty of effectively predicting a specific targeted audience in an ever-changing online setting (Chen et al., 2009) Targeted advertising strives to offer the most appropriate advertising messages to customers and to identify potential customers, get information, and segment the market through optimization and automation with the help of ML models. When we look at traditional segmentation, it used to use demographic and geographic data, but ML applications segment markets based on content shared and consumed by customers in the form of big data (Lo et al., 2015). User-generated content in social media platforms, which contains textual components, will help firms target audiences in a more sophisticated manner (Lo et al., 2015). Furthermore, this also helps increase return on investment, maximizing marketing efficiencies and improving customer relations through customization (Lo et al., 2015).

Behavioral targeting: Behavioral targeting (BT) is one method of reaching intended consumers with targeted adverts. BT uses user online behavior history and finds relevant ads based on this by recognizing "clicked links, pages visited, searches, and past purchases from the user's browsing history" (Chen et al., 2009). Searching content online and surfing websites for information are the most performed online activities, mostly due to search engines like Google and Yahoo. Web browsing behavior assists advertisers in drawing conclusions about users' interests and defining audience segmentation. Using the actual online behavior of users improves the way firms send prospective customers personalized and relevant advertising messages (Li et al., 2007). Furthermore, user searches aid in influencing the type of advertising that should be displayed to the user by matching them to the keywords of the advertiser.

Advertisers can also predict user click activity and improve the way they target advertising to users. For this, advertisers pay for advertisements only when a user clicks an ad, which acts as one of the most popular strategies for internet ads, called the cost-per-click model. This is measured by click-through rate, which predicts the probability of clicks an advertisement receive from its users (Ta, 2015), which in turn helps firms place ads, rank them, and filter and target them to the right audience (Tagami et al., 2013). In addition, when the ad keywords match a user's search query, it triggers the insertion of a textual ad known as a sponsored search result (Wang et al., 2018). When people search for keywords paid for by the advertiser, the appropriate advertisement is displayed. Increasing the efficiency of automatic selection of advertisements helps to improve user experience (Addis et al., 2009).

User profiling: Advertisers profile users based on their online interests, which they might have shared publicly, or through historical information and provide personalized ads through their preferences (Bilenko & Richardson, 2011). So this acts a pattern of recommendations to discover interesting and uninteresting aspects of an individual and profile them based on those (Addis et al., 2009). The capacity to discriminate between users through behavioral targeting is significant since message personalization is critical in improving user experience.

One type of important digital advertising is contextual advertising, where firms place an ad on a third-party web page like blogs or websites which match the content of the firm's products or customer interests (Zhang et al., 2008). Conveying relevant ads to consumers is very important, as "90% of consumers found tailored commercials appealing, and 95% of organizations utilizing customization tactics received three times ROI on their ad expenditure", according to *Forbes* (2020). Lot of backend works happen in real time to match keywords and offer relevant ads to web pages (Jiang et al., 2016), because if there is matching content on a webpage and ads placed, then there is a high chance of clicks and increased user experience and thereby high income for the firm (Tagami et al., 2013). Furthermore, display advertisements, video ads, blogs, web docs, and vehicle ads are additional sub-fields in content-based target advertising. Online display advertising is a type of behavioral retargeting that is based on the surfing habits of consumers (Bayer et al., 2020). Historically it was carried out through capturing information about gender, age, preferences, and locations (Gonzalvez-Cabañas & Mochón, 2016) with cookie data from browsers to establish a direct relationship between the relevance of ads and customer attitudes, which also leads to real-time bidding (Pepelyshev et al., 2015; Cai et al., 2017; Palos-Sanchez et al., 2019).

Click fraud not only inflates ad budgets through deceptive means but also undermines the integrity of the digital advertising industry as a whole. Click fraud happens when illegal clicks are made in which the user has no real interest in the advertisement in order to receive a dishonest reward (Taneja et al., 2015). False clicks on adverts can be replicated using either using a bot or automated clicking tools (Perera et al., 2013). Click fraud can drastically reduce firms' ROI on advertisement. Because publishers are frequently compensated via pay-per-click, the amount of clicks or impressions received by an advertisement may be inflated. According to Haider et al. (2018), click fraud accounts for 30% of ad revenue. As a result, it is critical to deploy a detection system for identifying click fraud in order to maintain confidence among publishers and advertisers, as well as to ensure the legitimacy of each click made (Almeida & Gondim, 2018). As a result, developing ML algorithms, which detect click fraud, is a critical tool for identifying and protecting advertisers from fake clicks (Perera et al., 2013).

Machine learning techniques for advertising: Machine learning can help advertisers come up with a content-centric and user-centric advertising approach to extract keywords from virtual communication contents (Li et al., 2007), and it also helps in detecting click fraud (Perera et al., 2013), as discussed earlier. These demonstrate how machine-learning technologies can be used to keep internet advertising viable. The use of programmatic advertising (PA) indicates that we can target customers in real time with online platforms by customized messaging via automated ad buys. It is being used as a suitable method in digital marketing, as it pools automation and personalization (Gonzalvez-Cabañas & Mochón, 2016), and because it targets and offers products based on the needs of users and getting real-time responses from them (Hoban & Bucklin, 2015). It also provides efficiency, reduces costs owing to automation (Miklosik et al., 2019), and allows you to reach clients on the go via their cellphones. Overall, reaching the proper audience with appealing ads helps advertisers to manage customers and their purchase journey in a better manner (Malthouse et al., 2019). Putting the customer at center stage and using their data and ML algorithms creates intelligent advertising, which could be the next stage in digital and programmed advertising (Li, 2019). Intelligent advertising introduces a new level of personalization by predicting customer requirements and desires in real-time context and recommending offerings using AI and ML technologies such as voice automation. Consumer views concerning these prescriptive strategies provide intriguing study opportunities. Ad relevance has an impact on more than just purchasing intentions (Kim & Huh, 2017; Hayes et al., 2020; Ciuchita et al., 2023).

Immersive technologies: Immersive technologies like AR, VR, and MR use cutting-edge technologies. Brands have an almost limitless opportunity to engage audiences through distinctive, memorable, and highly customized experiences thanks to immersive technologies. (Wang et al., 2022). Without needing to be moved to a page, customers may seamlessly explore a brand while viewing an ad unit online thanks to immersive ads. Another benefit of immersive advertising is that it allows consumers total power and flexibility to manage the goods and services because it uses rich media (Ambika et al., 2023; Ahn et al., 2023). Therefore, these immersive technologies will act as a bridge between the physical and virtual world for advertisers and customers, making the advertising experience a breathtaking one.

## 18.6 DIFFICULTIES AND ISSUES BECAUSE OF THE DIGITAL TRANSITION IN ADVERTISING

As all good things have challenges and difficulties, even the digital transition in advertising has many difficulties and issues. Let us discuss few of those issues, which are having a large impact on advertising. A major challenge for advertisers is dynamic consumer behavior in digital ecosystems (Ianenko et al., 2019). However, the difficulties affect small and medium organizations more than large organizations due to innovation and marketing capabilities (Teng et al., 2022; Sampe et al., 2022). In addition, lot of organizations might have a challenge attracting customers in a digital ecosystem because of the huge amount of information available to customers. In addition, there are instances of ad fraud, consumer privacy, and transparency issues.

Ad fraud: Ad fraud refers to defrauding of online advertisements to get more financial gain, mostly with the help of bots. Ad fraud is a major problem for advertisers in digital ecosystems (Richet, 2022). This fraud mostly happens in activities such as pay-per-click, impression management, and marketing spam. These things in ad fraud are a serious problem for advertisers and organizations, as they may spend more money thinking they have gotten good impressions or clicks, but in the end, they won't get that many customers in reality. Therefore, this fraud puts advertisers in a dream world, imagining more clicks and impressions, but the results will not reflect this.

Consumer privacy: It refers to managing and safeguarding consumer data and information. If the data and information of customers gets out due to a data breach or any other methods, it will be a great setback to the organizations, so to avoid this, many governments are creating laws to protect privacy in advertising (Johnson et al., 2022). In addition, organizations should also take the initiative to build privacy policies for their digital advertisements to protect consumer privacy (Hoffman et al., 2022).

Transparency in online advertising: Transparency in advertisements will have a different kind of impact on brands in the digital ecosystem (Ater & Rigbi, 2023). Transparency with respect to advertises and online platforms are both important. The Facebook ad library promises to improve accountability and transparency for online advertisements (Leerssen et al., 2023). However, maintaining transparency in advertising is challenging for any organization. Therefore, there is a need for organizations to adopt technologies and measures like blockchain technology to bring transparency (Tan & Saraniemi, 2023).

Changing regulatory environment: Many countries came up with data protection laws for digital ecosystems. The regulations were limited to traditional advertisements before, but after the pandemic almost all over the globe, most countries came up with data protection laws. Censorship and checking ethical factors in online advertisements, thereby regulating any unfair practices in online activities, also started in many countries. Therefore, current regulatory frameworks were also changed to adapt to the digital ecosystem. These changes also have an effect on advertisers to create online advertisements, which leads to innovations in advertisements to adapt to the changing ecosystem and regulations.

## 18.7 CASE STUDIES OF PRODUCTIVE DIGITAL TRANSFORMATION INITIATIVES IN ADVERTISING

### 18.7.1 Nike's AirMax 2023—Virtual Product Launch Success

#### 18.7.1.1 Background and Context

Nike, which is considered a global leader in the footwear and apparel industries, has a history of launching its products in an iconic way to capture the attention of enthusiasts and customers all over the globe. However, the pandemic has posed a lot of challenges that have forced brands to rethink and innovate their traditional advertising formats. So, in response to these challenges, Nike took the decision to embrace a virtual first approach to unveiling their AirMax 2023 line, combining cutting-edge technology with innovative storytelling.

#### 18.7.1.2 The Virtual Experience

Nike launched a first-of-its-kind gaming experience called Airphoria in Fortnite on June 20, 2023. "Airphoria represents a new, immersive experience for Nike as it amplifies its efforts in gaming and virtual products," says Ron Faris, VP/GM of Nike Virtual Studios.

> As the digital and physical worlds become more seamless across the products and experiences we offer, we're always seeking authentic ways to deepen our connection with fans, expand access points to Nike across our digital ecosystem, and inspire a love of sport and play. What's more, Nike is one of the first brands to use Epic Games' Unreal Editor for Fortnite to build Airphoria, paving the way for a continued partnership that will further unlock the future of gaming.

#### 18.7.1.3 Interactive AR Try-Ons

The incorporation of interactive AR try-ons was key to the virtual launch. Nike collaborated with augmented reality experts to create an app that allows customers to digitally "try on" multiple kinds of AirMax 2023 footwear using their cellphones or other devices. This feature not only provided a distinct and interesting experience, but it also addressed a long-standing issue in online footwear shopping: the inability to try on shoes physically.

#### 18.7.1.4 Livestreaming Event

The launch was livestreamed online, combining traditional product launch elements with the interaction activity of digital platforms. In the events, Nike designers demonstrated the design process and the innovations and inspirations that made AirMax 2023 a reality. The livestream was integrated with real-time question and answer sessions, allowing enthusiasts and customers to directly interact with designers and gain in-depth knowledge about Nike's products.

#### 18.7.1.5 Social Media Engagement

Nike used social media engagement to broaden the reach and effect of the virtual launch. The brand promoted the virtual launch of AirMax 2023 on platforms including Instagram, Twitter, and Pinterest, building anticipation and enthusiasm among their large online fanbase. Hashtags such as #AirMax2023 trended around the world, generating a sense of community and engagement.

#### 18.7.1.6 Results and Impact

Nike's virtual launch engaged millions globally through livestreams and interactive AR try-ons. This increased visibility, and the brand relationship showcased Nike's commitment to adapting to changing consumer habits. The digital launch also provided Nike with valuable data on user behavior and engagement patterns, enhancing its position as a forward-thinking brand. The virtual launch was more cost effective than traditional in-person events since it eliminated the need for venue rents, travel fees, and physical set-ups.

#### 18.7.1.7 Conclusion

Finally, Nike's AirMax 2023 virtual product introduction showcased innovation and versatility, integrating technology, interactive participation, and social media, highlighting the importance of digital transformation for thriving in a changing company world.

### 18.7.2 Coca-Cola's Adaptive Campaigns—Programmatic Ad Optimization

#### 18.7.2.1 Background and Context

Coca-Cola is widely recognized as one of the best-known brands globally, and it has earned a reputation for its exceptional marketing campaigns. In recent times, the company has embraced advertising to optimize its campaigns and effectively reach its target audience.

Programmatic advertising involves the automated buying and selling of advertising space. This approach allows advertisers to precisely target their ads and analyze their performance accurately.

As we are aware, Coca-Cola has been utilizing advertising for years now, and the results have been remarkable. In 2017, their programmatic marketing efforts resulted in a 20% increase in brand awareness and a 15% boost in sales. Most organizations are trying to use programmatic advertising as a tool after the pandemic, but Coca-Cola used it well before, and it shows how much Coca-Cola is ahead of its competitors.

#### 18.7.2.2 Programmatic Advertising at Coca-Cola

One of the factors contributing to Coca-Cola's success with advertising is its use of adaptable campaigns. These programmatic ad campaigns are designed to adjust themselves automatically in response to changing circumstances such as user behavior, inventory availability, or bidding dynamics.

Coca-Cola, for example, might utilize adaptive advertising to target those who are looking for information about its products or are interested in similar brands. The organizations can use adaptive campaigns with creativity based on the time of day or the weather in the region.

Coca-Cola's usage of programmatic advertising and adaptive campaigns is a perfect illustration of how brands may improve their marketing campaigns by leveraging technology. Brands may get more out of their advertising spend by more precisely targeting their ads and measuring their results.

#### 18.7.2.3 Optimizing Programmatic Campaigns

- **Select the best DSP**: Coca-Cola collaborated with a major DSP to obtain the data and insights required to optimize its ads.
- **Select the appropriate audience**: Coca-Cola reached its target audience by using a range of targeting variables, including demographics, hobbies, and online behavior.
- **Make good use of powerful creativity and landing pages**: Coca-Cola developed high-quality creative content that was relevant to its target audience and compelled consumers to act.
- **Gather and analyze data:** Coca-Cola gathered data on its programmatic initiatives in order to track their performance and make the required changes.
- **Improve creativity and locations**: There was an improvement in the creativity and placement of Coca-Cola ads based on the information received, ensuring that its advertisements were seen by the appropriate people at the right time.
- **Make use of a range of data sources:** In addition to demographic and interest data, brands may target their ads more accurately by using data on online behavior, purchase history, and other criteria.
- **A/B testing should be used:** A/B testing is a way of comparing the performance of different versions of an advertisement. Everything from the creative content to the targeting may be optimized using this method.

- **Keep a close eye on the findings:** It is critical to regularly monitor the success of programmatic initiatives so that adjustments can be made as needed. This will help to ensure that campaigns always get the required results.

Coca-Cola was able to achieve considerable results with its programmatic campaigns as a result of these actions. The company's brand awareness increased by 20%, while sales increased by 15%. This case of Coca-Cola demonstrates how programmatic advertising may be used to optimize marketing efforts and generate better outcomes. Brands can utilize programmatic advertising to reach their target demographic more effectively and drive more sales by following the procedures indicated previously. Programmatic advertising is a useful technique for optimizing marketing campaigns and achieving better outcomes. Brands may use programmatic advertising to reach their target demographic more effectively and drive more sales by following the suggestions suggested in this case study.

## 18.8 AMAZON X PRIME VIDEO ADVERTISING—SEAMLESS E-COMMERCE INTEGRATION

### 18.8.1 Background and Context

Amazon is the world's largest e-commerce company, and Prime Video is one of the most popular streaming services. Amazon has been trying in recent years to connect these two platforms in order to provide a more seamless experience for customers. Advertisers using this model are able to reach customers when they browse and discover new content with the Prime Video app. The different ad formats available to advertisers are sponsored channels, sponsored tiles, and feature rotators, which are curated and supported by Prime Video channel pages, and content detail pages to enrich the advertising model of advertisers. During and after the pandemic, the usage rate of OTT has been on an increasing curve, and this model of Amazon X Prime has led to changes in creativity and advertising types.

### 18.8.2 Seamless E-Commerce Integration

Amazon is accomplishing this in a number of ways, one of which is through Prime Video advertising. This program allows brands to put advertisements on Prime Video content and provides a variety of rewards, including:

**Targeted reach:** Brands can target their advertisements to specific audiences based on demographics, interests, and watching behavior.

**Brand safety:** Amazon ensures that all commercials on Prime Video are brand safe, so marketers can rest assured that the proper individuals will see their advertisements.

**Measurability:** Brands can track the performance of their advertising on Prime Video to see how they are increasing awareness, engagement, and revenue.

In addition to these advantages, Prime Video advertising provides a unique chance for marketers to link their commercials with e-commerce. This is accomplished through the use of a technology known as "shoppable ads." Shoppable ads allow viewers to click on an ad and be sent directly to the product page on Amazon. This makes it simple for viewers to buy the things they see in advertisements, which can lead to greater sales for firms.

Amazon has been attempting to make shoppable ads more effective by utilizing machine learning to customize the experience for each viewer. For example, Amazon may present a shoppable ad to a viewer for a product that they have previously viewed or that is comparable to things that they have purchased in the past.

Brands may reach their target audience and increase sales by combining Prime Video advertising with e-commerce. Brands may use this platform to reach a large and engaged audience, tailor their ads to the right individuals, and track the success of their campaigns.

### Advantages of Amazon X Prime Video advertising:

**Brand exposure is increased:** By running advertising on Prime Video, marketers may reach a large and engaged audience. This can aid in raising brand awareness and attention.

**Increase sales:** Shoppable ads make it simple for viewers to buy the things they see in ads. Brands may benefit from more significant sales as a result of this.

**Improved targeting**: Brands may now target their advertising on Prime Video based on demographics, interests, and viewing habits. This helps to ensure that advertisements are viewed by the intended audience.

**Measurable outcomes:** Brands can monitor the performance of their commercials on Prime Video to understand how they are influencing awareness, engagement, and purchases.

**Brand safety:** Amazon ensures that all advertisements on Prime Video are brand safe. This gives brands peace of mind, knowing that their advertisements are safe.

Overall, Amazon X Prime Video advertising is a powerful platform that may assist organizations in reaching their target audience, increasing sales, and improving marketing ROI.

### Issues while utilizing Amazon X Prime Video advertising:

**Cost:** Amazon Prime Video advertising can be costly, particularly for high-traffic terms.

**Complexity:** Setting up and managing the platform can be difficult.

**Competition:** Ad space on Prime Video is quite competitive.

Despite these difficulties, Amazon X Prime Video advertising is a significant tool for marketers looking to reach a large and engaged audience while driving sales. Brands can profit from properly assessing the platform's advantages and disadvantages.

## 18.9 IKEA PLACE APP—AUGMENTED REALITY SHOPPING

### 18.9.1 Introduction

The COVID-19 epidemic has had a significant impact on how people shop. With storefronts shuttered and customers staying at home, many retailers have had to shift their operations online. This has resulted in an increase in online shopping, and businesses are seeking innovative ways to engage customers and enhance the online buying experience.

Retailers are using augmented reality to accomplish this. AR is a technology that superimposes a computer-generated image on top of a user's real-world view. This can be used to make shopping more realistic and engaging.

### 18.9.2 IKEA Place App

IKEA is one of the retailers that is utilizing augmented reality to improve the shopping experience. IKEA Place, the company's app, allows customers to virtually install IKEA furniture in their homes. This allows users to see how the furniture will look in their space and make more informed purchasing decisions.

IKEA Place is available for both iOS and Android devices. To use the app, simply download it and scan your room with your phone's camera. The software will then provide a selection of IKEA furniture for users to digitally arrange in their rooms.

The app is extremely user friendly and simple to use. Users may effortlessly move and rotate the furniture to view how it appears from various perspectives. They can also zoom in and out for a better view.

### 18.9.3 The Benefits of AR Shopping

There are numerous advantages to using AR shopping. Here are a handful of the most significant:

AR shopping offers a more immersive and realistic shopping experience than traditional online shopping. This can assist users in making more informed purchasing decisions.

**Improved customer happiness:** By providing a more personalized and engaging purchasing experience, augmented reality shopping can help to boost customer contentment.

**Increased sales:** AR shopping can assist in increasing sales by allowing consumers to see how things will look in their homes.

**Reduced returns:** By allowing users to check out things before purchasing them, augmented reality shopping can assist in reducing returns.

**Cost savings:** AR shopping can help retailers save money on marketing and advertising.

### 18.9.4 The Challenges of AR Shopping

While there are numerous advantages to AR shopping, there are also drawbacks that companies should be aware of. Here are some of the most significant challenges:

Users must have a suitable device and an internet connection to shop with augmented reality. This can be difficult for certain users, particularly those in developing nations.

**Accuracy:** If the device is not correctly calibrated or if the surroundings are too dark or bright, AR shopping can be incorrect.

**Concerns about privacy:** Some consumers may be concerned about the privacy implications of augmented reality purchases. Retailers must be open about how they gather and use customer data.

**Adoption:** Because augmented reality shopping is still a relatively new technology, it has not yet been widely accepted by consumers. To boost awareness of AR shopping, retailers must spend on marketing and education.

### 18.9.5 Conclusion

AR shopping is an exciting new technology that has the potential to change the way people shop. However, there are some issues that retailers should be aware of. By addressing these issues, retailers can make augmented reality purchases more successful and widespread.

## 18.10 SPOTIFY'S PERSONALIZED ADS—DATA-DRIVEN PERSONALIZATION

### 18.10.1 Introduction

The COVID-19 pandemic has had a significant impact on the advertising sector. Advertisers have had to develop new ways to reach their target audiences as consumers spend more time at home.

Advertisers are accomplishing this in a variety of ways, including tailored advertising. Personalized advertising is advertising that is tailored to specific people or groups of people based on their interests, demographics, and other criteria.

Spotify is one of the companies that use targeted advertising to reach out to their customers. Personalized advertisements from the corporation are based on a range of characteristics, including the user's listening history, search history, and location.

### 18.10.2 Spotify's Personalized Ads

Personalized advertising on Spotify appears between songs or throughout podcast episodes. The ads are relevant and engaging since they are personalized to the user's interests.

Spotify personalizes its ads using a range of data sources, including:

Spotify tracks the songs and podcasts that the user listens to. This information is used to determine the user's preferences and interests.

The user's search history: Spotify keeps note of what songs and podcasts the user looks for. This information is also utilized to identify the user's interests and preferences.

The location of the user: Spotify tracks the user's location. This information is used to serve advertising to users based on their location.

Spotify also employs machine learning to customize its advertisements. Machine learning is a subset of artificial intelligence in which computers may learn and improve over time. Machine learning is used by Spotify to increase the accuracy of its tailored advertisements.

### 18.10.3 The Benefits of Data-Driven Personalization

There are numerous advantages to utilizing data-driven personalization in advertising. Here are a handful of the most significant:

Personalized ads are more relevant to the user than regular advertisements. This increases the likelihood that they will be clicked on and engaged with.

**Improved click-through rate:** The percentage of people who click on an ad is known as the click-through rate (CTR). Traditional advertisements have a lower CTR than personalized ads.

**Increased conversion rate:** The conversion rate is the proportion of people who take the intended action after clicking on an advertisement. Personalized advertisements convert at a higher rate than regular ads.

**Improved brand recognition:** By exposing users to appropriate ads more frequently, personalized ads can help to improve brand awareness.

**Reduced ad fatigue:** Ad fatigue is a phenomenon in which users become less likely to click on an ad after seeing it too many times. By giving people ads that are more relevant to them, personalized ads can assist in preventing ad fatigue.

### 18.10.4 The Future of Data-Driven Personalization

Personalization based on data is a rapidly expanding topic. Advertisers will be able to use more data to produce more tailored and engaging commercials as technology advances.

This will almost certainly result in a more effective and efficient advertising sector. Advertisers will be able to reach out to their target consumers more effectively and at a reduced cost.

### 18.10.5 Conclusion

Spotify exemplifies how data-driven personalization may be utilized to generate great advertising efforts. Personalized advertisements from the corporation are based on a range of characteristics, including the user's listening history, search history, and location.

This information is used to develop advertisements that are tailored to the user's interests and preferences. For example, if a user enjoys rock music, they may see an advertisement for a rock concert. If a person looks for information about a specific product, they may see an advertisement for that product.

Spotify's tailored ads are not just relevant but also interesting. The corporation employs a range of strategies to make its advertisements stand out, such as beautiful imagery and catchy copy.

Spotify's click-through and conversion rates have increased as a result of its tailored ads. The company's brand awareness has also increased.

## 18.11 L'ORÉAL'S MAKEUP TUTORIALS—VIDEO STORYTELLING AND LIVESTREAMING

### 18.11.1 Introduction

The COVID-19 pandemic has had a significant influence on the cosmetics sector. With people spending more time at home, in-store beauty expenditures have decreased. As a result, cosmetic companies have had to come up with new ways to communicate with customers.

Video tutorials are one way that cosmetic brands have done this. Video lessons are an excellent approach to teaching customers how to apply cosmetics and create different styles. They can also be used to tell stories and connect with customers.

### 18.11.2 L'Oréal's Makeup Tutorials

L'Oréal is one of the world's major beauty companies. For many years, the organization has used video tutorials to engage with customers. A team of expert makeup artists and hairstylists creates L'Oréal's beauty lessons. L'Oréal's website, YouTube channel, and social media accounts all have instructional content.

The cosmetics tutorials from L'Oréal are not only instructional but also interesting and amusing. The courses are frequently hosted by well-known influencers and celebrities. This contributes to a sense of authenticity and relatability among customers.

### 18.11.3 Video Storytelling and Livestreaming

L'Oréal is connecting with customers through video storytelling and live streaming, in addition to traditional video training. Video storytelling is a storytelling style that employs video to tell a tale. This can be accomplished in a variety of formats, including docuseries, vlogs, and short films.

Livestreaming is a method of watching and interacting with videos in real-time. This is an excellent approach to interacting with customers and addressing their questions.

L'Oréal has had significant success with video storytelling and livestreaming. For example, the company's docuseries "Beauty Secrets" has received over 100 million YouTube views.

### 18.11.4 The Benefits of L'Oréal's Makeup Tutorials

L'Oréal's makeup lessons provide numerous advantages. Here are a handful of the most significant:

**Educate customers:** L'Oréal's makeup tutorials can teach customers about various products and techniques. This can assist consumers in locating the best items for their needs and learning how to utilize them efficiently.

**Build relationships:** L'Oréal's makeup tutorials can aid in the development of consumer relationships. This is due to the fact that the tutorials are frequently offered by well-known influencers and celebrities. This contributes to a sense of authenticity and relatability among customers.

**Increase sales:** L'Oréal's cosmetics lessons can increase sales by raising awareness of the company's products. This is due to the fact that the tutorials frequently demonstrate to consumers how to apply the goods to achieve attractive beauty looks.

**Boost brand awareness:** Brand awareness can be increased by exposing consumers to L'Oréal's products and services through cosmetics tutorials. This is due to the fact that the tutorials are frequently shared on social media and other channels.

### 18.11.5 The Future of L'Oréal's Makeup Tutorials

L'Oréal is devoted to connecting with customers through video tutorials. The organization is always inventing and discovering new ways to share stories and establish relationships through video.

L'Oréal's cosmetic instructions are set to become more interactive and individualized in the future. The corporation will also most likely employ video to reach out to new groups, such as Generation Z and millennials.

### 18.11.6 Conclusion

L'Oréal's cosmetic tutorials are an excellent example of how marketers can use video to meaningfully interact with their customers. The tutorials provided by the company are instructive, interesting, and amusing. They also aid in the development of consumer relationships and the promotion of sales.

## 18.12 MODEL FOR DIGITAL TRANSITION IN ADVERTISING

The following model demonstrates how the advertising industry underwent a digital change and what process should be followed to successfully adopt digital transformation in advertising.

While referring to the model, it can be noted that the digital transition in advertising is dependent on several factors. The digital transition refers to the transition from a traditional way of advertising to a more sophisticated and dynamic digital ecosystem to promote products and services.

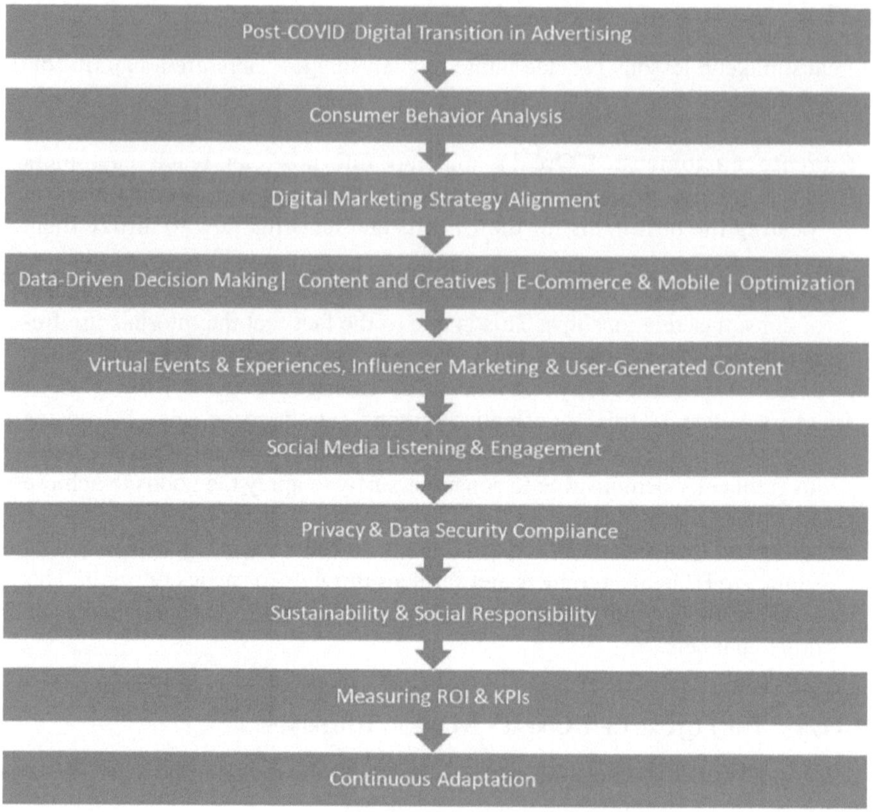


**FIGURE 18.1** Conceptual model for digital transition in advertising.

The first step showcases the organization's first need to have a plan for digital transition during a pandemic and what options are available for advertisements in the digital ecosystem.

The second step is consumer behavior analysis, where advertisers need to check the requirements and perceptions of consumers and then create digital advertisements based on this analysis.

Then firms need to align digital marketing strategies with the existing aspects of products and activities.

With the help of data the company has about the customers and its product, the firms need to carry out data-driven decision making, contents creation, E-commerce and mobile optimization, which helps firms manage their online advertisements.

Then the firms need to create virtual events and experiences for users and customers of its products and services. Firms also need to invest in influencer marketing and user-generated content to attract millennial customers.

However, there will be queries, complaints, and suggestions from customers. To manage this, firms need to be active on online platforms through social media, listening and engaging to be effective in the digital ecosystem.

In addition, with the data, there is the problem of privacy and data security compliance, so to manage this, firms need to invest in securing customer and user data and should also maintain the privacy of the data through effective methods.

Also, while advertising in the digital ecosystem, firms need to focus on sustainability and social responsibility. Ads should not be anti-social or hurt the feelings of any individual.

Once all the process is in place, firms need to take steps to measure the ROI and KPIs of advertising cost and check whether all the efforts the organization has made in digital advertisements have really worked.

Once the firms comes to know the advertising efforts are successful, they have to make it a continuous adoption for the firm. This step concludes the digital transformation and makes it a continuous process for the whole organization.

## 18.13 FUTURE OF ADVERTISING DUE TO DIGITAL TRANSFORMATION

With the digital transformation, firms should look for future trends in the field of advertising. As technology continues to evolve, the growth of augmented and virtual reality is poised to revolutionize the advertising landscape, with immersive 3D elements emerging as a key feature of future metaverse advertising campaigns (Eyada, 2023). Additionally, the growth of metaverse advertising over the last two years has provided the necessary space and confidence for marketers to adopt it as a viable future technology (Nie, 2023). So advertisers should understand the virtual world more clearly to make good advertising decisions in the metaverse (Samadet al., 2023). However, in a digital ecosystem, advertisements can be transparent with the help of blockchain technology and the concept of "trust exchange vs traditional exchange" (Tan & Saraniemi, 2023). This will bring transparency to the virtual world, with message content, social presence, and source credibility (Jayawardena et al., 2023). We can say that VR, AR, XR, and the evolving metaverse environment are bringing exciting innovations and challenges to the media and advertising industry (Ahn et al., 2023).

## 18.14 CONCLUSION

With all these considerations, we can say that the digital world is dynamic. The COVID-19 pandemic has expedited advertising's digital revolution. Consumers' reliance on digital platforms for information and entertainment has grown, and advertisers' methods have had to adjust correspondingly. As a result, a number of new trends in digital advertising have emerged, including:

**The rise of programmatic advertising:** Programmatic advertising is the practice of buying and selling advertising space using automated systems. This has made it possible for advertisers to target their adverts more accurately and better measure their results.

**The rise of social media advertising:** Because they allow advertisers to reach a big and engaged audience, social media platforms have become a key source of advertising revenue.

**The growing importance of data-driven advertising:** Advertisers are increasingly leveraging data to more effectively target their commercials. This information can be gathered from a variety of sources, including website traffic, social media activity, and purchase history.

New technologies, such as augmented reality and virtual reality, are being developed: These technologies are still in their infancy, but they have the potential to transform the way advertising is distributed. The digital change of advertising is still happening, and the future is unknown. One thing is certain: advertisers who fail to adapt to the digital landscape will fall behind. Aside from the patterns mentioned previously, the COVID-19 epidemic has created a slew of new hurdles for advertisers. Among these difficulties are:

**Traditional advertising channels are declining:** As consumers spend more time online, traditional advertising media such as television and print are declining in viewership and readership.

**The audience's increasing fragmentation:** Consumers have access to more media sources than ever before, making it more difficult for advertising to reach a large audience.

**Ad filters are on the rise:** Ad blockers are software tools that prohibit advertisements from appearing on websites. Advertisers may find it challenging to contact their target demographic as a result of this.

Despite these problems, the digital transformation of advertising provides advertisers with a variety of opportunities. Advertisers can reach their target audience more efficiently and accurately by embracing digital technology and data-driven marketing. They can also offer more tailored and engaging ad experiences that customers find appealing. Advertising's future is digital, and those that do not adapt will be left behind. Advertisers may position themselves for success in the new digital ecosystem by understanding the current trends and challenges.

## REFERENCES

Addis, A., Armano, G., & Vargiu, E. (2009). Profiling users to perform contextual advertising. In Proceedings of the 10th Workshop Dagli Oggetti agli Agenti (WOA).

Ahn, S. J., Kim, J., & Kim, J. (2023). The future of advertising research in virtual, augmented, and extended realities. International Journal of Advertising, 42(1), 162–170.

Almeida, P. S., & Gondim, J. J. (2018). Click fraud detection and prevention system for Ad Networks. Journal of Information Security and Cryptography, 5(1), 27–39.

Ambika, A., Shin, H., & Jain, V. (2023). Immersive technologies and consumer behavior: a systematic review of two decades of research. Australian Journal of Management, 03128962231181429.

Ater, I., & Rigbi, O. (2023). Price transparency, media, and informative advertising. American Economic Journal: Microeconomics, 15(1), 1–29.

Bayer, E., Srinivasan, S., Riedl, E. J., & Skiera, B. (2020). The impact of online display advertising and paid search advertising relative to offline advertising on firm performance and firm value. International Journal of Research in Marketing, 37(4), 789–804. doi: 10.1016/j.ijresmar.2020.02.002.

Bilenko, M., & Richardson, M. (2011). Predictive client-side profiles for personalized advertising. In Proceedings of the 17th ACM SIGKDD International Conference on Knowledge Discovery and Data Mining.

Cai, H., Ren, K., Zhang, W., Malialis, K., Wang, J., Yu, Y., & Guo, D. (2017). Real-time bidding by reinforcement learning in display advertising. In Proceedings of the Tenth ACM International Conference on Web Search and Data Mining (pp. 661–670).

Campbell, M. C., Inman, J. J., Kirmani, A., & Price, L. L. (2020). In times of trouble: A framework for understanding consumers' responses to threats. Journal of Consumer Research, 47(3), 311–326.

Chalmers, T. D., & Price, L. L. (2009). Perceptions of authenticity in advertisements: Negotiating the inauthentic. In McGill, A. L., & Shavitt, S. (Eds.), Advances in Consumer Research (Vol. 36, pp. 72–75). Association for Consumer Research.

Chen, Y., Kapralov, M., Canny, J., & Pavlov, D. Y. (2009). Factor modeling for advertisement targeting. Advances in Neural Information Processing Systems, 324–332.

Ciuchita, R., Gummerus, J. K., Holmlund, M., & Linhart, E. L. (2023). Programmatic advertising in online retailing: Consumer perceptions and future avenues. Journal of Service Management, 34(2), 231–255. doi: 10.1108/JOSM-06-2021-0238.

Eyada, B. (2023). Advertising in the metaverse: Opportunities and challenges. International Journal of Marketing Studies, 15(1).

Forbes, 50 Stats showing the power of personalization (2020). www.forbes.com/sites/blakemorgan/2020/02/18/50-stats-showing-thc-power-of personalization/#3fce20942a94.

Galoni, C., Carpenter, G. S., & Rao, H. (2020). Disgusted and afraid: Consumer choices under the threat of contagious disease. Journal of Consumer Research, 47(3), 373–392.

Ganesha, K. S. (2018). Effectiveness of advertisement on purchasing behavior of customers towards telecom product and services. BIMS International Journal of Social Science Research, 3(2), 73.

Ganesha, K. S., & Ganapati, B. S. (2020). Conceptual understanding of social media and tourist decision making. Shodh Sarita, 28(7), 29–34.

Ganesha, K. S., & Ganapati B. Sinnoor. (2019). Significance of social media marketing in brand awareness and product reach. International Journal of Management and Economics, 1(30), 201–205.

Geradin, D., & D. Katsifis (2019). 'Trust me, I'm fair': Analysing Google's latest practices in Ad Tech from the perspective of EU competition law, https://papers.ssrn.com/sol3/papers.cfm?abstract_id=3465780.

Giroux, H. A. (2021). The covid-19 pandemic Is exposing the plague of neoliberalism 1. In Collaborative Futures in Qualitative Inquiry (pp. 16-27). Routledge.

Gonzalvez-Cabañas, J. C., & Mochón, F. (2016). Operating an advertising programmatic buying platform: A case study. International Journal of Interactive Multimedia and Artificial Intelligence, 3(6), 6–15. doi: 10.9781/ijimai.2016.361.Haider, C. M. R., Iqbal, A., Rahman, A. H., & Rahman, M. S (2018). An ensemble learning based approach for impression fraud detection in mobile advertising. Journal of Network and Computer Applications, 112, 126–141.

Hamby, A., Brinberg, D., & Daniloski, K. (2019). It's about our values: How founder's stories influence brand authenticity. Psychology & Marketing, 36(11), 1014–1026.

Hayes, J. L., Golan, G., Britt, B., & Applequist, J. (2020). How advertising relevance and consumer–brand relationship strength limit disclosure effects of native ads on Twitter. International Journal of Advertising, 39(1), 131–165. doi: 10.1080/02650487.2019.1596446.

Hoban, P. R., & Bucklin, R. E. (2015). Effects of internet display advertising in the purchase funnel: Model-based insights from a randomized field experiment. Journal of Marketing Research, 52(3), 375–393. doi: 10.1509/jmr.13.0277.

Hoffman, D. L., Moreau, C. P., Stremersch, S., & Wedel, M. (2022). The rise of new technologies in marketing: a framework and outlook. Journal of Marketing, 86(1), 1–6.Ianenko, M., Ianenko, M., Huhlaev, D., & Martynenko, O. (2019, March). Digital transformation of trade: Problems and prospects of marketing activities. In IOP Conference Series: Materials Science and Engineering, 497(1), 012118. IOP Publishing.

Jaafar, N. S., & Khan, N. (2022). Impact of digital marketing innovation in competitive event industry during COVID-19: Evidence from Malaysia and the United States. International Journal of Interactive Mobile Technologies, 66(8).

Jayawardena, N. S., Thaichon, P., Quach, S., Razzaq, A., & Behl, A. (2023). The persuasion effects of virtual reality (VR) and augmented reality (AR) video advertisements: a conceptual review. Journal of Business Research, 160, 113739.

Jiang, Z., Gao, S., & Dai, W. (2016). Research on CTR prediction for contextual advertising based on deep architecture model. Journal of Control Engineering and Applied Informatics, 18(1), 11–19.

Johnson, G., Runge, J., & Seufert, E. (2022). Privacy-centric digital advertising: Implications for research. Customer Needs and Solutions, 9(1–2), 49–54.

Khan, M. M. R., & Basak, K. (2021). Shifts in pharma-marketing trends in post COVID-19 era. International Journal of Multidisciplinary: Applied Business and Education Research, 2(2), 108–114.

Kim, G. U., Kim, M. J., Ra, S. H., Lee, J., Bae, S., Jung, J., & Kim, S. H. (2020a). Clinical characteristics of asymptomatic and symptomatic patients with mild COVID-19. Clinical Microbiology and Infection, 26(7), 948-e1.

Kim, H., & Huh, J. (2017). Perceived relevance and privacy concern regarding online behavioral advertising (OBA) and their role in consumer responses. Journal of Current Issues and Research in Advertising, 38(1), 92–105. doi: 10.1080/10641734.2016.1233157.

Kim, J. (2020). Impact of the perceived threat of COVID-19 on variety-seeking. Australasian Marketing Journal, 28(3), 108–116.

Kim, J., Giroux, M., Gonzalez-Jimenez, H., Jang, S., Kim, S., Park, J., Kim, J. E., Lee, J. C., & Choi, Y. K. (2020b). Nudging to reduce the perceived threat of Coronavirus and stockpiling intention. Journal of Advertising, 49(5), 633–647.

Lee, S., Sung, B., Phau, I., & Lim, A. (2019). Communicating authenticity in packaging of Korean cosmetics. Journal of Retailing and Consumer Services, 48, 202–214.

Leerssen, P., Dobber, T., Helberger, N., & de Vreese, C. (2023). News from the ad archive: How journalists use the Facebook Ad Library to hold online advertising accountable. Information, Communication & Society, 26(7), 1381–1400.

Li, H. (2019). Special section introduction: Artificial intelligence and advertising. Journal of Advertising, 48(4), 333–337. doi: 10.1080/00913367.2019.1654947.

Li, H., Song, Y., & Xie, X. (2019). Altruistic or selfish? Responses when safety is threatened depend on childhood socioeconomic status. European Journal of Social Psychology, 50, 1001–1016.

Li, Y., Surendran, A. C., & Shen, D. (2007). Data mining and audience intelligence for advertising. ACM SIGKDD Explorations Newsletter, 9(2), 96–99. doi: 10.1145/1345448.1345470.

Lo, S. L., Chiong, R., & Cornforth, D. (2015). Using support vector machine ensembles for target audience classification on Twitter. PLoS One, 10(4).

Malthouse, E. C., Buoye, A., Line, N., El-Manstrly, D., Dogru, T., & Kandampully, J. (2019). Beyond reciprocal: The role of platforms in diffusing data value across multiple stakeholders. Journal of Service Management, 30(4), 507–518. doi: 10.1108/JOSM-12-2018-0381.

Marketing Mind (n.d.). Advertising through the ages, www.ashokcharan.com/Marketing-Analytics/~aw-advertising-throughages.php.

Miklosik, A., Kuchta, M., Evans, N., & Zak, S. (2019). Towards the adoption of machine learning-based analytical tools in digital marketing. IEEE Access, 7, 85705–85718. doi: 10.1109.

Nichols, W. (2013). Advertising analytics 2.0. Harvard Business Review, 91(3), 60–68.
Nie, H. (2023). Virtual augmented reality advertising based on metaverse platform. Information Systems and Economics, 4(4), 77–81.
OECD (2019). Online adverting: Trends, benefits and risks for consumers. doi: 10.1787/1f42c85d-en.
Palos-Sanchez, P., Saura, J. R., & Martin-Velicia, F. (2019). A study of the effects of programmatic advertising on users' concerns about privacy overtime. Journal of Business Research, 96, 61–72. doi: 10.1016/j.jbusres.2018.10.059.
Park, J., Kim, J., Lee, D. C., Kim, S. S., Voyer, B. G., Kim, C., Sung, B., Gonzalez-Jimenez, H., Fastoso, F., Choi, Y. K., & Yoon, S. (2022). The impact of COVID-19 on consumer evaluation of authentic advertising messages. Psychology & Marketing, 39(1), 76–89. doi: 10.1002/mar.21574. Epub 2021 Aug 10. PMID: 34539052; PMCID: PMC8441704.
Pena-Marin, J., Adaval, R., & Shen, L. (2022). Fear in the stock market: How COVID-19 affects preference for high-and low-priced stocks. Journal of the Association for Consumer Research, 7(1), 115–123.
Pepelyshev, A., Staroselskiy, Y., & Zhigljavsky, A. (2015). Adaptive targeting for online advertisement. In Proceedings of International Workshop on Machine Learning, Optimization and Big Data (pp. 240–251). Springer.
Perera, K. S., Neupane, B., Faisal, M. A., Aung, Z., & Woon, W. L. (2013). A novel ensemble learning-based approach for click fraud detection in mobile advertising. In Mining Intelligence and Knowledge Exploration (pp. 370–382). Springer.
Quick, T. (2020). Advertising evolution: How personalization has improved over time, https://instapage.com/blog/evolution-of-advertising.
Rashi, P., Bist, A. S., Asmawati, A., Budiarto, M., & Prihastiwi, W. Y. (2021). Influence of post covid change in consumer behaviour of millennials on advertising techniques and practices. Aptisi Transactions on Technopreneurship (ATT), 3(2), 201–208.
Richet, J. L. (2022). How cybercriminal communities grow and change: An investigation of ad-fraud communities. Technological Forecasting and Social Change, 174, 121282.
Samad, A., Izani, M., Razak, A., & Mustaffa, F. (2023, May). Understanding advertising in virtual worlds and best practices for metaverse advertising. In 2023 Zooming Innovation in Consumer Technologies Conference (ZINC) (pp. 45–50). IEEE.
Sampe, F., Yusuf, M., Pakiding, D. L., Haryono, A., & Sutrisno, S. (2022). Application of digital marketing in maintaining MSMEs during the COVID-19 pandemic. Jurnal Darma Agung, 30(2), 663–676.
Sengupta, V., Sengupta, S., Lazo, A., Woods, P., Nolan, A., & Bremer, N. (2020). Exosomes derived from bone marrow mesenchymal stem cells as treatment for severe COVID-19. Stem Cells and Development, 29(12), 747–754.
Sheth, J. (2020). Impact of COVID-19 on consumer behavior: Will the old habits return or die? Journal of Business Research, 117, 280–283. doi: 10.1016/j.jbusres.2020.05.059.
Sirkeci, I. (2020). Marketing and consumers in an era of disruption caused by COVID-19 pandemic. Transnational Marketing Journal (TMJ), 8(1), 1–6.
Ta, A.-P. (2015). Factorization machines with follow-the-regularized-leader for CTR prediction in display advertising. In Proceedings of 2015 IEEE International Conference on Big Data (Big Data) (pp. 2889–2891). IEEE.
Tagami, Y., Ono, S., Yamamoto, K., Tsukamoto, K., & Tajima, A. (2013). Prediction for contextual advertising: Learning-to-rank approach. In Proceedings of the Seventh International Workshop on Data Mining for Online Advertising, 2013 (pp. 1–8).
Tan, T. M., & Saraniemi, S. (2023). Trust in blockchain-enabled exchanges: Future directions in blockchain marketing. Journal of the Academy of marketing Science, 51(4), 914–939.
Taneja, M., Garg, K., Purwar, A., & Sharma, S. (2015). Prediction of click frauds in mobile advertising. In Proceedings of 2015 Eighth International Conference on Contemporary Computing (IC3), IEEE, pp. 162–166.

Taylor, C. R. (2020). Advertising and COVID-19. International Journal of Advertising, 39(5), 587–589.

Teng, X., Wu, Z., & Yang, F. (2022). Research on the relationship between digital transformation and performance of SMEs. Sustainability, 14(10), 6012.

Tien-Dung, P., Nguyen, X. D., & Das, S. (2022). Sustainable goal achievement by digital revolution during and after pandemic; how much one wins and losses: a bird's eye view for future planning. In Sustainable Development and Innovation of Digital Enterprises for Living with COVID-19 (pp. 133–147). Singapore: Springer Nature Singapore.

Valentine, D. B., & Powers, T. L. (2013). Generation Y values and lifestyle segments. Journal of Consumer Marketing, 30(7), 597–606.

Wang, Q., Liu, F., Xing, S., & Zhao, X. (2018). A new approach for advertising CTR prediction based on deep neural network via attention mechanism. Computational and Mathematical Methods in Medicine, 2018(1), 8056541.

Wang, R., Bush-Evans, R., Arden-Close, E., Bolat, E., McAlaney, J., Hodge, S., . . . Phalp, K. (2022). Transparency in persuasive technology, immersive technology, and online marketing: Facilitating users' informed decision-making and practical implications. Computers in Human Behavior, 107545.

Wedel, M., & Pieters, R. (2017). A review of eye-tracking research in marketing. Review of Marketing Research, 123–147.

Whitaker, W. R., & Smith, R. D. (2009). Mediawriting: Print, broadcast, and public relations. Routledge.

Zhang, H., Sun, J., Liu, F., & Knight, J. G. (2014). Be rational or be emotional: Advertising appeals, service types and consumer responses. European Journal of Marketing, 48(11/12), 2105–2126.

Zhang, Y., Surendran, A. C., Platt, J. C., & Narasimhan, M. (2008). Learning from multi-topic web documents for contextual advertisement. In Proceedings of the 14th ACM SIGKDD International Conference on Knowledge Discovery and Data Mining (pp. 1051–1059).

# 19 Identifying Predictors and Comparing Machine Learning Algorithms to Detect Erroneous Medicaid Adjudication in the USA

*Cyrus Hoseini, M. Affan Badar, A. Mehran Shahhosseini, and Christopher J. Kluse*

## 19.1 INTRODUCTION

In the USA, Medicaid is a subsidized healthcare program for low-income individuals or those with special needs (e.g., disabled and senior citizens). Federal and state governments collaboratively fund Medicaid to provide healthcare services to over 67 million individuals in 50 states and the District of Columbia (Medicaid.gov, 2018). Medicaid costs taxpayers over $570 billion a year (The Henry J. Kaiser Family Foundation, 2018) and is susceptible to improper payments estimated exceed $29 billion (*Testimony of Ann Maxwell Before the United States House of Representatives*, 2017). Medicaid programs involve two primary models: fee-for-service (FFS) and managed care organization (MCO). Both FFS and MCO models typically require a provider claim to be processed either by a health plan or a fiscal intermediary (FI). As a result, it is crucial to ensure all Medicaid claims are processed correctly and free from payment errors.

Due to the complexity of the systems and processes, the average nationwide improper payment for FFS Medicaid programs in 2015 was 10.6%. The federal share of Medicaid improper payments was over $20 billion (The Centers for Medicare and Medicaid Services, 2015). Traditional quality methods used by Medicaid programs to detect and prevent quality issues rely primarily on inferential statistics and sampling. This involves reviewing a relatively small sample of processed claims to detect errors and extrapolate the sample statistics to estimate population parameters. While this method provides broad estimations of quality issues, it falls short of identifying specific quality problems and their root causes.

DOI: 10.1201/9781003438748-23

The complexity inherent in Medicaid systems and processes contributes to improper payments and erroneous claim adjudication. This underscores the need for innovative approaches to improve accuracy and efficiency, including machine learning (ML). ML is a collective name for a group of computer algorithms and statistical methods that use computational learning and artificial intelligence principles to enable computers to learn a pattern or make a prediction without being explicitly programmed (Simon, 2013; Kohavi & Provost, 1998). ML algorithms are categorized into four major categories (Hoseini et al., 2021): (1) supervised learning algorithms in which historical data is used as a baseline (also called training data set) to produce an inferred function that predicts a pattern in the future datasets, commonly referred to as test data (Mohri et al., 2014); (2) unsupervised learning algorithms that do not require pre-labeled training data and analyze a set of unlabeled data to find and model hidden structures in the data set; (3) semi-supervised (or hybrid) learning algorithms; and (4) reinforcement learning algorithms,

Detecting quality issues is a form of classification problem and could be formulated as a supervised learning problem if labeled data is available. ML methods (particularly supervised learning) possess the theoretical capability to classify large volumes of claims accurately, provided they are properly set up and trained. ML has been used in fraud detection in electronic claim data in healthcare (Kose et al., 2015; Gallardo, 2017); fraud detection in capital markets (Yi et al., 2023); and detecting quality issues in manufacturing, food, and other industries, as described in the review of literature section. Thus, at least on an abstract level, an ML method can be used to detect all claim adjudication problems before FI sends the payment to a provider, translating to millions of dollars of savings for budget-constrained Medicaid programs. However, in practice, doing this requires careful formulating of the problem, selecting proper algorithms, and extensive data preparation. Many technological limitations may prevent us from using machine learning to solve this problem, including the processing power needed to handle massive computations. This chapter aims to offer a practical approach to using supervised learning to detect most claim adjudication issues in the entire population of claims processed by an FI. Examples of claim attributes (also known as features, predictors, or independent variables) could include International Classification of Diseases (ICD) codes, payment data, Current Procedural Terminology (CPT) codes, provider attributes, beneficiary attributes, claim adjudication team attributes, and policy attributes. In this work, 16 features for outpatient and 19 features for inpatient claim types were selected. Ten supervised learning algorithms were trained and tested on the labeled data: decision tree with two configurations—entropy and Gini, random forest with two configurations—entropy and Gini, naïve Bayes, K-nearest neighbor, logistic regression, neural network, discriminant analysis, and gradient boosting. This chapter aimed to find important features (or predictors) and effective supervised ML algorithms in detecting erroneously adjudicated Medicaid FFS inpatient or outpatient claims.

This work fits the scope of the book appropriately as it expounds on the use of digital innovation and sustainable operation. It focuses on business transformation from using traditional inferential statistics and sampling processes to applying ML algorithms for detecting erroneous Medicaid adjudication claims in order to improve business sustainability. The chapter is structured as follows. The next section summarizes a literature review related to ML algorithms and their applications in various industries. This is followed by a section on research design and procedure; a section on feature or attribute selection; and a section on building, training, and testing models. Then the results of hypothesis testing and answers to the research questions are explained. The paper concludes with a discussion and concluding remarks, the importance of the findings and practical implications, and recommendations for further research.

## 19.2 LITERATURE REVIEW

### 19.2.1 Application of ML to Solve Quality Issues in Different Industries

Machine learning techniques have been used to solve production control problems since the late 90s. As an example, Bowden and Bullington successfully applied an unsupervised learning method called the genetic algorithm rule discovery system (GARDS) to solve a flexible cellular manufacturing system (eMS) problem (Bowden & Bullington, 1996). However, leveraging machine learning to solve quality problems is a much newer trend. The literature has been reviewed extensively to find examples of the application of machine learning to solve quality problems. In most cases, machine learning methods are used to detect quality issues in a product, process, or service. Our literature review's focus was to summarize cases that are more relevant to this research's scope. This includes studies about leveraging ML to augment a quality control or quality assurance process or detect quality issues in service. The researchers have also included studies in which different machine learning algorithms' performance in solving a quality problem was compared.

Hoseini et al. (2021) and Hoseini (2020) have described the examples of ML application to detect quality issues in different industries like the service industry, product or manufacturing processes, food industry, software and information technology, healthcare, health insurance, and other industries. Table 19.1 summarizes the studies discussed in the literature review.

In addition to the industries listed in Table 19.1, ML has been used in fraud detection in capital markets (Yi et al., 2023). Hilal et al. (2022) reviewed anomaly detection techniques related to financial fraud. Kazemian and Shrestha (2023) applied ML to detect erroneous criminal identities.

### 19.2.2 Application of ML to Fraud in Medicaid and Its Challenges

Medicaid fraud analytics is a common application of analytics and ML in Medicaid. Generally, ML methods used for fraud detection in Medicaid are based on supervised methods. In supervised Medicaid fraud detection techniques, subject matter experts use prior information on class membership to select a set of training data (Travaille et al., 2011). This training data set is used by the algorithm to label each reviewed value as a suspicious or legal item. For example, subject matter experts review a set of paid and denied Medicaid claims and label each one legitimate or fraudulent. There are many supervised techniques listed as potential algorithms for

**TABLE 19.1**
**Summary of Literature Review of the Application of ML to Detect Quality Issues**

| Item | Description |
|---|---|
| Industry | Food |
| Quality problem | Detect the quality issues in salmon |
| Data type | Unstructured |
| Inputs | Images |
| ML algorithms applied | • Support vector machines (SVMs) with a nonlinear kernel function (RBF) [B][1]<br>• Nearest-neighbor classification |
| Note and reference | (Sture, 2015) |
| Industry | Food |
| Quality problem | QC test for olive oils |
| Data type | Structured and unstructured |
| Inputs | Chemical and visual measures |
| ML algorithms applied | • Naïve Bayesian [B]<br>• SVM (support vector machine)<br>• ANN (artificial neural network)<br>• k-nearest neighbor (kNN)<br>• Linear discriminant analysis (LDA)<br>• Decision tree |
| Note and reference | Researchers first used principal component analysis to reduce the number of predictors from 32 to 8 (Ordukaya & Karlik, 2017) |
| Industry | Food |
| Quality problem | Detect quality issues in French fries |
| Data type | Unstructured |
| Inputs | Images |
| ML algorithms applied | Vision processing techniques augmented by an ML method based on neural networks |
| Note and reference | (Lotfi et al., 2008) |
| Industry | Food |
| Quality problem | Measure the quality of mixed raisins |
| Data type | Unstructured |
| Inputs | Images |
| ML algorithms applied | • SVM [B]<br>• ANN |
| Note and reference | First used principal components analysis (PCA) method to reduce 146 predictors to an optimal level prediction (Karimi et al., 2017) |
| Industry | Food |
| Quality problem | Predict wine quality |

(*Continued*)

**TABLE 19.1**
**(Continued)**

| Item | Description |
|---|---|
| Data type | Structured |
| Inputs | Physicochemical characteristics |
| ML algorithms applied | • SVM [B]<br>• NN (neural network) |
| Note and reference | First used linear regression to measure the dependency of the target variable (wine quality) on predictors (Gupta, 2018) |
| Industry | Food |
| Quality problem | Predict the quality of the beer (defined by the intensity levels of sensory descriptors like perceived foam-related parameters and beer color) |
| Data type | Structured and unstructured |
| Inputs | Physicochemical characteristics |
| ML algorithms applied | • ANN<br>• Regression |
| Note and reference | Researchers first used principal component analysis to identify the most relevant predictors. The final resulted in an $R = 0.91$ correlation to predict the quality of the beer (Gonzalez Viejo et al., 2017) |
| Industry | Manufacturing |
| Quality problem | Detect rare quality defects in a high-conformance manufacturing environment |
| Data type | Structured |
| Inputs | Sensor data |
| ML algorithms applied | • Least absolute shrinkage and selection operator (LASSO)<br>• Logistic regression (LR) |
| Note and reference | Supervised learning (Escobar & Morales-Menendez, 2018) |
| Industry | Manufacturing |
| Quality problem | Predict potential quality failures of a process |
| Data type | Structured |
| Inputs | Process and sensor data |
| ML algorithms applied | N-dimensional Euclidean distance-based scoring algorithms |
| Note and reference | Resulting in a potential $300 million annual savings (RTInsights Team, 2017) |
| Industry | Manufacturing |
| Quality problem | Detect manufacturing quality issues |
| Data type | Structured |
| Inputs | QC data |
| ML algorithms applied | • Deep learning (deep restricted Boltzmann machine and the stacked autoencoder) [B]<br>• Shallow learning (feed-forward neural network with one hidden layer and the least-squares SVM with no hidden layers) |

(*Continued*)

**TABLE 19.1 (Continued)**

| Item | Description |
|---|---|
| Note and reference | Researchers proved deep learning methods outperform shallow learning methods. They also found the performance of deep learning algorithms is proportional to the sample size (Bai et al., 2017) |
| Industry | Manufacturing |
| Quality problem | Predict the quality of intermediate products in interlinked manufacturing processes |
| Data type | Structured |
| Inputs | Process and sensor data |
| ML algorithms applied | A hybrid model of unsupervised (k-means) and supervised (k-nearest neighbor [B] and SVM) algorithms |
| Note and reference | With the k-NN algorithm ($k = 11$), researchers achieved a 97% accuracy level in prediction (Lieber et al., 2013) |
| Industry | Manufacturing |
| Quality problem | Improve the product and process quality in the manufacturing process |
| Data type | Structured |
| Inputs | Process data |
| ML algorithms applied | The hybrid model consists of unsupervised (cluster analysis) and supervised ML algorithms (SVM) |
| Note and reference | (Irgens et al., 2012) |
| Industry | Manufacturing |
| Quality problem | Predict the quality of welding |
| Data type | Unstructured |
| Inputs | Arc sound and images |
| ML algorithms applied | • J48 [B]<br>• Random forest |
| Note and reference | 88.69% accuracy rate for J48 vs. 70.78% for random forest (Sumesh et al., 2015) |
| Industry | Manufacturing |
| Quality problem | Monitor quality issues in plastic injection molding |
| Data type | Structured |
| Inputs | Process data (cycle time, metering time, injection time, barrel temp., cushion, and injection velocity) |
| ML algorithms applied | SVMs (C-SVM and v-SVM) radial basis function NNs |
| Note and reference | Accuracy results ranged from 70.83% up to 99.16%. SVMs have a performance advantage over NNs (Ribeiro, 2005) |
| Industry | Call center |
| Quality problem | Detect quality issues (e.g., anger or delay) and categorize sentiment and emotions |

*(Continued)*

**TABLE 19.1**
**(Continued)**

| Item | Description |
|---|---|
| Data type | Unstructured |
| Inputs | Calls, text |
| ML algorithms applied | Rule-based natural language processing (NLP) |
| Note and reference | (Peddamuthu & Srivastava, 2014) |
| Industry | Call center |
| Quality problem | Predict the quality of service in a call center |
| Data type | Structured |
| Inputs | Call performance evaluation data |
| ML algorithms applied | • CART [B]<br>• Multi-layer perceptron (MLP)<br>• Linear NN<br>• Probabilistic NN<br>• SVM (with a third-degree polynomial kernel)<br>• Hybrid decision tree-ANN |
| Note and reference | Over 80% of customer service satisfaction prediction accuracy and close to 90% in predicting business need satisfaction (Paprzycki et al., 2004) |
| Industry | Hospitality |
| Quality problem | Measure service quality and customer satisfaction |
| Data type | Unstructured |
| Inputs | Text |
| ML algorithms applied | • Aspect sentiment analysis<br>• Decision trees |
| Note and reference | (Yussupova et al., 2016) |
| Industry | Utilities |
| Quality problem | Detect smart-grid power quality events (PQEs) |
| Data type | Structured |
| Inputs | Network data |
| ML algorithms applied | Discrete wavelet transform (DWT) |
| Note and reference | (Ucar et al., 2018) |
| Industry | Financial services |
| Quality problem | Detect inaccurate coding in financial contribution |
| Data type | Structured |
| Inputs | Financial data |
| ML algorithms applied | • Adaboost supported by ensemble classifier [B]<br>• SVM |
| Note and reference | Supervised learning (Blomquist & Möller, 2015) |
| Industry | Medical |

(*Continued*)

**TABLE 19.1**
**(Continued)**

| Item | Description |
|---|---|
| Quality problem | Improve the accuracy of quality assessment of MRI-derived data conducted by medical professionals |
| Data type | Unstructured |
| Inputs | Medical images |
| ML algorithms applied | • Kernelized SVM<br>• Gradient-boosted decision trees classifiers |
| Note and reference | Supervised learning using over 7,500 human-rated samples resulting in a human workload reduction by 30–70% (Petrov et al., 2017) |
| Industry | Natural resource |
| Quality problem | Measure water quality |
| Data type | Structured |
| Inputs | Lab data |
| ML algorithms applied | • ANN [B]<br>• SVM<br>• Classification and regression trees (CARTs)<br>• LR (logistic regression) |
| Note and reference | Very large sample size (10 years' worth of data from 20 reservoirs in Taiwan) (Chou et al., 2018) |
| Industry | Telecommunication |
| Quality problem | Estimate mobile phone provider service quality using social media |
| Data type | Unstructured |
| Inputs | Tweets from Twitter |
| ML algorithms applied | Naïve Bayes |
| Note and reference | Lack of proper validation method of test results (Calvin & Setiawan, 2014) |
| Industry | Telecommunication |
| Quality problem | Predict if a software module is fault-prone or not |
| Data type | Structured |
| Inputs | Software metrics |
| ML algorithms applied | • The SPRINT decision tree [B]<br>• CART decision tree |
| Note and reference | They found the way the SPRINT decision tree leverages a unique tree pruning technique based on the minimum description length (MDL) principle to improve the accuracy and stability of the model (Khoshgoftaar & Seliya, 2003) |
| Industry | Software |
| Quality problem | Predict which module in software has a higher chance of failure |
| Data type | Structured |

*(Continued)*

**TABLE 19.1**
**(Continued)**

| Item | Description |
|---|---|
| Inputs | Software metrics |
| ML algorithms applied | • SVM [B]<br>• LR<br>• KNN<br>• Multi-layer perceptrons<br>• Radial basis function<br>• Bayesian belief network (BBN)<br>• Naïve Bayes (NB)<br>• Random forest (RF)<br>• Decision tree (DT) |
| Note and reference | (Elish & Elish, 2008) |
| Industry | Healthcare |
| Quality problem | Detect fraud in outpatient clinic claims |
| Data type | Unstructured |
| Inputs | Hospital visit information |
| ML algorithms applied | • Unsupervised clustering methods:<br>• Decision tree |
| Note and reference | Used conditional probability distributions of the composite degree of anomaly (CDA) to adjust the risk score (Shin et al., 2012) |
| Industry | Healthcare |
| Quality problem | Detect fraud in electronic claim data |
| Data type | Unstructured |
| Inputs | Hospital visit information |
| ML algorithms applied | Non-parametric unsupervised method |
| Note and reference | AHP was used for weighting (Kose et al., 2015) |
| Industry | Healthcare |
| Quality problem | Detect fraud in electronic claim data |
| Data type | Unstructured |
| Inputs | Hospital visit information |
| ML algorithms applied | • Convolutional deep neural networks (CNNs)<br>• Recurrent neural networks (RNNs) |
| Note and reference | Patented proprietary algorithm (Gallardo, 2017) |

Medicaid fraud detection (Copeland et al., 2012; Li et al., 2008; Phua et al., 2005; Valdes & Skinner, 2000; Getchius, 2014), such as: multi-layer perceptron networks; artificial neural networks, including backpropagation (backward propagation of errors); support vector machine; decision trees; fuzzy logic; and Bayesian networks, including Bayesian belief networks (BBNs), probable graph models (PGMs), and linear Gaussian models.

Researchers also presented some innovative methods to evaluate supervised methods' performance by calculating their return on investment and costs versus benefits. At least two metrics are introduced in research to measure the performance of the anomaly detection method tradeoff between detection probability and false alarm ratio and the tradeoff between the false alarm ratio and detection delay (Siris & Papagalou, 2004). Phua and colleagues described how statistical methods should be classified, utilized, interpreted, and validated to detect healthcare fraud (2005). While supervised methods are more accurate in detecting previous fraud types, they also suffer from some shortcomings, primarily when used to detect Medicaid fraud (Bolton & Hand, 2001). These include changes in demographics, policies, and healthcare systems.

Unsupervised Medicaid fraud detection techniques do not assume prior class labels of legitimate or fraudulent behavior (Travaille et al., 2011). These algorithms mine the entire dataset and try to find a Medicaid data pattern and identify outliers or anomalies. All anomalies then need to be researched and validated. These algorithms' strengths are their sensitivity (detecting most true positives) and specificity (reducing false positives). Examples of unsupervised methods that could be used for fraud detection (including Medicaid fraud) are anomaly detection, K-means, Benford's law, linear regression, modified batch library method (MBLM), adaptive threshold algorithm, and cumulative sun (CUSUM) algorithms (Heino & Toivonen, 2003; Siris & Papagalou, 2004; Issa & Vasarhelyi, 2011; Tsung et al., 2007).

#### 19.2.2.1 Criteria for Selecting an ML Algorithm

There are quite a few supervised algorithms in the literature, and the list is evolving. Not all these methods are appropriate for the current work. The following criteria can be used to find some guidelines to help us select an initial list for the study, and then the results of the preliminary research can guide the final list.

- *Accuracy*: perhaps the most popular metric to select an appropriate machine learning algorithm is the model accuracy, which is defined as total true positives and true negatives of all results produced by an algorithm. As an example, Singh and Kaur (2016) compared the accuracy of J48 and REP Tree to predict the performance of computer science students and found J48 outperforms REP Tree in model accuracy 67.37% to 56.78%.
- *Stability*: an algorithm is defined as $\beta$-stable (or stable, in general) when its losses incurred by the corresponding hypotheses on two similar but different datasets are equal or less than $\beta$ (Mohri et al., 2014). There are specific limitations to how big $\beta$ could grow where a training algorithm can still converge.
- *Interpretability vs. prediction accuracy:* there is an inherited trade-off between the statistical models' interpretability and prediction accuracy (James et al., 2015). Many flexible and accurate ML algorithms (e.g., SVM and NN) lack interpretability and are like black-box approaches. When stakeholders demand the model logic be clearly explained, the choice is to compromise on flexibility and choose a more interpretable model (e.g., LASSO).

- *Bias-variance trade-off:* As a rule of thumb, as model flexibility and complexity increase, the bias will decrease, and variance will increase. So, complex models (e.g., NN and SVM) have an inherited tendency to over-fit and should be used with caution (James et al., 2015). Hypothesized variance can be expected to decrease as training set size increases, but no apparent effect of training set size on bias was observed (Brain & Webb, 1999).
- *The impact of data reduction*: one way to increase the efficiency of a ML model is to appropriately reduce the version of data to a lower number of attributes. El-Hasnony et al. (2015) compared the accuracy of nine classification algorithms using data reduction techniques like correlation feature selection (CFS), rough set attribute reduction (RSAR), fuzzy rough feature selection (FRFS), PCA, and gain ratio. These are the classification algorithms they tested in their research: C4.5, fuzzy rough nearest neighbor, multi-layer perceptron, nearest-neighbor-like algorithm using non-nested generalized exemplars (NNGEs), fuzzy nearest neighbor, sequential minimum optimization (SMO), classification via clustering, NB-tree, and naïve Bayes (El-Hasnony et al., 2015).
- *Model fitness*: Akaike's information criterion (AIC) is a statistical measure of the goodness of fit for a particular model. It maximizes the expression $-2(LL+ k)$, where $k$ is the number of features, and $LL$ is the maximized value of the log-likelihood function for the given model. The smaller the AIC, the better the model fits the data. Because of the $k$ term, a smaller number of model parameters is favored (Dean, 2014).

#### 19.2.2.2 Summary of Literature Review

In the literature review, details of several use cases were provided. These studies were related to ML application to solve quality problems and detect Medicaid fraud cases. Table 19.2 summarizes the appearance of various ML algorithms in these studies as a tested algorithm and, in some cases, as the best-performing one.

These are a few lessons learned from the research reviewed on the application of ML in detecting quality problems (in general) and Medicaid fraud. These are summarized here.

- The most common application of machine learning in the examined cases was to "detect quality issues through supervised learning". This problem could be formulated as a classification problem. The second most common application was to "predict quality" using supervised or semi-supervised learning. Depending on the nature of the quality measure (variable vs. attribute), the problem could be formulated as a regression or a classification problem.
- There is no magic bullet for the best algorithm, and each algorithm may perform better in solving a specific problem. The most popular choices among algorithms are decision tree (various configurations for both supervised and unsupervised learning), SVM (for supervised learning for binary classification problems), logistic regression (supervised), NN (for unsupervised and semi-supervised learnings), and naïve Bayes (supervised).

**TABLE 19.2**
**ML Algorithms Used in the Literature to Solve Quality Problems or Detect Medicaid Fraud**

| Algorithm | Frequency of Application | Times listed as the Best Performer |
|---|---|---|
| Decision tree | 9 | |
| Support vector machines | 9 | 2 |
| Logistic regression | 7 | |
| Neural network | 6 | 1 |
| Naïve Bayes | 4 | 1 |
| K-nearest neighbor | 3 | 1 |
| Random forest | 2 | |
| Discriminant analysis | 1 | |
| Gradient boosting | 1 | |
| Others | 6 | 1 |

- Complex (a.k.a. black box) algorithms like SVM and NN may perform better in many cases, but they are hard to explain. Simpler models like decision trees also have disadvantages, including low stability. Computation time for complex models tends to be higher. SVM requires massive computational power for mixed and massive datasets.
- The biggest challenge in leveraging supervised learning is the labeling process, usually a tedious and expensive process.
- Principal component analysis is frequently used to reduce the number of features without compromising the model performance.

#### 19.2.2.3 Research Gap, Questions, and Hypotheses

There exist several instances in the literature of applying big data analytics (BDA) and ML in detecting fraud in many industries, including Medicaid, and finding quality issues in several industries. However, the authors did not find any references in the literature (focusing on master's and PhD theses/dissertations and peer-reviewed journals) about applying ML to detect quality issues in adjudicating Medicaid FFS claims. Conventional methods of quality control, including statistical sampling methods and acceptance sampling (e.g., military standards), are still the most common method to control and manage the quality of operations in healthcare, from the claim adjudication process to pharmaceuticals (Fu et al., 2004; Borget et al., 2006).

Therefore, the following research questions are explored in the current chapter.

Research Question 1 (RQ1): What supervised machine learning algorithms can be used to determine Medicaid claim payment issues?

RQ2: What are the most critical measures to compare the performance of different machine learning algorithms (resulting from RQ1) for our problem?

RQ3: What are the claim attributes or features that could predict if a given FFS claim has been adjudicated correctly or erroneously (also known as predictors or independent variables)?

RQ4: Is there any statistically significant difference among predictors' predictability power in identifying erroneously processed Medicaid outpatient claims? This can be expressed as hypothesis 4.1: $H_{0_{4.1}}: \mu_{1.1} = \mu_{1.2} = \ldots = \mu_{1.n}$, where $\mu_{1i}$ is the feature importance score (Gini importance) of the $i$-th outpatient predictor.

RQ5: Same as RQ4 for Medicaid inpatient claims. $H_{0_{5.1}}: \mu_{2.1} = \mu_{2.2} = \ldots = \mu_{2.n}$ for inpatients.

RQ6: Is there any statistically significant difference among the selected supervised learning algorithms (the result of RQ1) in identifying erroneously adjudicated Medicaid outpatient claims (as measured by the result of the RQ2)? In the research design and procedures section, it was explained that ten algorithms were selected, and recall and F1-score were the measures to compare them: $H_{0_{6.1}}: \mu_{3.1} = \mu_{3.2} = \ldots = \mu_{3.10}$ ($\mu_{3.i}$: the mean recall of algorithm $i$) and $H_{0_{6.2}}: \mu_{41} = \mu_{42} = \ldots = \mu_{4.10}$ ($\mu_{4.i}$: the mean F1-score of algorithm $i$).

RQ7: Same as RQ6 for Medicaid inpatient claims. $H_{0_{7.1}}: \mu_{5.1} = \mu_{5.2} = \ldots = \mu_{5.10}$ ($\mu_{5.i}$: the mean recall of algorithm $i$) and $H_{0_{7.2}}: \mu_{6.1} = \mu_{6.2} = \ldots = \mu_{6.10}$ ($\mu_{6.i}$: the mean F1-score of algorithm $i$).

RQ8: Are the most powerful predictors different between outpatient and inpatient claims? Are the most accurate algorithms in detecting erroneously paid claims different between the two types of Medicaid claims studied?

## 19.3 RESEARCH DESIGN AND PROCEDURE

Table 19.3 describes the research procedures and statistical methods used to test the research hypotheses, answer research questions, and relate to each research methodology step.

### 19.3.1 Selecting ML Algorithms

There is no set of universally accepted "best" supervised learning algorithms, but SVM, naïve Bays, and neural network are among the algorithms most used in solving quality problems. Since SVM requires more computational power, especially for massive datasets, it was not selected for this work. The authors propose to use the following ten algorithms to build, test, and validate our supervised model:

1. Decision tree with two configurations: entropy (DTE) and Gini coefficient (DTG)
2. Random forest with two configurations: entropy (RFE) and Gini configuration (RFG)
3. Naïve Bayes
4. K nearest neighbor
5. Logistic regression

6. Neural network
7. Discriminant analysis (DA)
8. Gradient boosting (GB)

**TABLE 19.3**
**Research Procedures and Statistical Methods**

| Research Question | Goal | Method | Research Step |
|---|---|---|---|
| RQ1 | Selecting ML algorithms for this research | Literature review and select appropriate algorithms | Literature review and build, train, and test models |
| RQ2 | Selecting performance metrics to compare ML algorithm performance for the problem statement | Literature review and select appropriate metrics | Literature Review and build, train, and test models |
| RQ3 | Finding the most critical claim attributes to predict the outcome (classify the claim correctly) | Review the data and PCA | Manage data |
| RQ4 (H4.1) | Examine if there exist differences among predictors' predictability power in identifying erroneously processed Medicaid outpatient claims | Mean decrease in impurity (MDI) (Gini importance) and one-way ANOVA | Build, train, and test models |
| RQ5 (H5.1) | Examine if there exist differences among predictors' predictability power in identifying erroneously processed Medicaid inpatient claims | Mean decrease in impurity (MDI) (Gini importance) and one-way ANOVA | Build, train, and test models |
| RQ6 (H6.1, H6.2) | Examine if there are differences among the performance of selected algorithms to detect quality issues in outpatient claims | Classification algorithms (recall and F1-score), Gini measure, and one-way ANOVA | Evaluate models' performances |
| RQ7 (H7.1, H7.2) | Examine if there are differences among the performance of selected algorithms to detect quality issues in inpatient claims | Classification algorithms (recall and F1-score), Gini measure, and one-way ANOVA | Evaluate Models' Performances |
| RQ8 | Review the differences among the feature importance and algorithms performance between two selected claim types | Review the results | Evaluate Models' Performances |

### 19.3.2 Statistics for Model Selection

Our research problem uses both ratio and categorical data (as independent value) and predicts a categorical data type target. As a result, different algorithms' performances should be compared using proper statistics that work with categorical data. The most common statistics to compare supervised learning models' performance with categorical data are accuracy, recall, and precision extracted from the confusion matrix using the formula explained in Table 19.4 (Vihinen, 2012).

There are many metrics to select the optimal model to solve an ML problem with the massive dataset, including computation and tuning time, algorithm performance metric (e.g., accuracy, specificity, recall, balanced or weighted accuracy, F1 score, prediction power, etc.), explainability of the model and associated algorithm, stability, and scalability (Beeravalli, 2018), (Rácz et al., 2019). There are also many other metrics used to precisely evaluate the performance of supervised learning models. Each has its advantages and disadvantages based on the problem statement and the nature of the data. Researchers have offered various metrics for model validation and selecting the most appropriate model for each particular problem (Shah et al., 2016; Arsenault, 2017).

For our specific problem statement, recall is the most critical measure. This is because quality problems in a massive claim adjudication data are rare events, and it is paramount not to lose the chance to detect all of them. Precision has less importance; however, having many false positives will result in extra work by the quality team to manually research each false positive. To ensure our model considers the number of false positives as performance metrics, the researchers selected the F-measure (F1 score) as the second most essential metrics to compare model performance without losing sight of the recall. Receiver operating characteristic (ROC) separation is another popular metric; however, research shows F1-score's superiority to ROC separation in selecting the best performing model when dealing with imbalanced data (Yahya, 2018). The datasets used for this research are imbalanced datasets; thus, F1-score is selected as the second performance metric.

**TABLE 19.4**
**Confusion Matrix and Its Related Metrics**

| | | Actual | | Measure |
|---|---|---|---|---|
| | | + | – | |
| Predicted | + | True Positive (TP) | False Positive (FP) | Positive Predictive Value (PPV) = TP/(TP+FP) |
| | – | False Negative (FN) | True Negative (TN) | Negative Predictive Value (NPV) = TN/(FN+TN) |
| Measure | | Recall or Sensitivity = TP/(TP+FN) | Precision or Specificity = TN/(TP+TN) | Accuracy= (TP+TN)/(TP+FP+FN+TN)<br>F-measure (F1 Score) = 2. (Precision. Recall)/(Precision+Recall) |

### 19.3.3 Selecting Proper ML Toolset

Several machine learning software programs, including open source predictive modeling applications, were reviewed. Other researchers and data scientists were also contacted to gather their inputs about ML tools to solve problems with similar datasets. Python is a high-level and general-purpose programming language and very popular among data scientists as the top selection (Piatetsky, 2018). As a result, it has an excellent user community. It is easy to find examples of scripts that solve problems with Python. Its modular architecture and simple syntax improve its readability. Python is easy to install and performs well with complex datasets. Online documentation of Python features and libraries also is considerable and makes it easier to get the result. Therefore, Python was selected.

### 19.3.4 Collecting and Creating Research Dataset

To create a dataset for this research, several steps were taken to collect data from publicly available sources, protect the sensitive data, test our hypotheses, and examine our models' performance in a safe environment.

#### 19.3.4.1 Step 1: Select a Source for the Claim Data

This chapter focuses on finding quality issues in the Medicaid FFS claim data containing sensitive information. An alternative to using real-world data is to use publicly available PHI-free data. There are two principal methods to protect PHI data: de-identifying a real dataset and synthesizing a dataset that behaves like a real one. The Centers for Medicare and Medicaid Services (CMS) has created five massive datasets called Synthetic Public Use File (DE-SynPUF) (The Centers for Medicare and Medicaid Services, 2013b). To make such a dataset, CMS has selected five percent random samples of Medicare beneficiaries in 2008 and their claims from 2008 to 2010 and used them to generate millions of Medicare claim data samples for various claim types, as shown in Table 19.5.

A unique identification number is assigned to each synthetic beneficiary to link synthetic claims to a synthetic beneficiary. This beneficiary ID carries no information about the enrollee or any patient records and is provided solely for reference

**TABLE 19.5**
**Claim Data**

| File Name | Type of Data | Claim Type | Number of Records per Year | | |
|---|---|---|---|---|---|
| | | | 2008 | 2009 | 2010 |
| Beneficiary Summary DE-SynPUF | Beneficiary | N/A | 2,326,856 | 2,291,320 | 2,255,098 |
| Inpatient Claims DE-SynPUF | Claim | Inpatient | 547,800 | 504,941 | 280,081 |
| Outpatient Claims DE-SynPUF | Claim | Outpatient | 5,673,808 | 6,519,340 | 3,633,839 |
| Carrier Claims DE-SynPUF | Claim | Medical | 34,276,324 | 37,304,993 | 23,282,135 |
| Prescription Drug Events (PDE) DE-SynPUF | Claim | Pharmacy | 39,927,827 | 43,379,293 | 27,778,849 |

and data processing purposes. Table 19.6 shows the list of features or attributes (data elements) for outpatient claims with their data types (The Centers for Medicare and Medicaid Services, 2013a):

**TABLE 19.6**
**List of Features (Data Elements) for Outpatient Claims**

| # | Feature Name | Labels | Data Type |
|---|---|---|---|
| 1 | DESYNPUF_ID | DESYNPUF: Beneficiary Code | Categorical/Nominal |
| 2 | CLM_ID | DESYNPUF: Claim ID | Categorical/Nominal |
| 3 | SEGMENT | DESYNPUF: Claim Line Segment | Categorical/Nominal |
| 4 | CLM_FROM_DT | DESYNPUF: Claims start date | Ordinal (Date) |
| 5 | CLM_THRU_DT | DESYNPUF: Claims end date | Ordinal (Date) |
| 6 | PRVDR_NUM | DESYNPUF: Provider Institution | Categorical/Nominal |
| 7 | CLM_PMT_AMT | DESYNPUF: Claim Payment Amount | Ratio/Real number (Payment) |
| 8 | NCH_PRMRY_PYR_ CLM_PD_AMT | DESYNPUF: NCH Primary Payer Claim Paid Amount | Ratio/Real number (Payment) |
| 9 | AT_PHYSN_NPI | DESYNPUF: Attending Physician—National Provider Identifier Number | Categorical/Nominal |
| 10 | OP_PHYSN_NPI | DESYNPUF: Operating Physician—National Provider Identifier Number | Categorical/Nominal |
| 11 | OT_PHYSN_NPI | DESYNPUF: Other Physician—National Provider Identifier Number | Categorical/Nominal |
| 12 | NCH_BENE_BLOOD_ DDCTBL_LBLTY_AM | DESYNPUF: NCH Beneficiary Blood Deductible Liability Amount | Ratio/Real number (Payment) |
| 13–22 | ICD9_DGNS_CD_1—ICD9_DGNS_CD_10 | DESYNPUF: Claim Diagnosis Code 1—Claim Diagnosis Code 10 | Categorical/Nominal |
| 23–28 | ICD9_PRCDR_CD_1—ICD9_PRCDR_CD_6 | DESYNPUF: Claim Procedure Code 1—Claim Procedure Code 6 | Categorical/Nominal |
| 29 | NCH_BENE_PTB_ DDCTBL_AMT | DESYNPUF: NCH Beneficiary Part B Deductible Amount | Ratio/Real number (Payment) |
| 30 | NCH_BENE_PTB_ COINSRNC_AMT | DESYNPUF: NCH Beneficiary Part B Coinsurance Amount | Ratio/Real number (Payment) |
| 31 | ADMTNG_ICD9_ DGNS_CD | DESYNPUF: Claim Admitting Diagnosis Code | Categorical/Nominal |
| 32–76 | HCPCS_CD_1—HCPCS_CD_45 | DESYNPUF: Revenue Center HCFA Common Procedure Coding System 1—Revenue Center HCFA Common Procedure Coding System 45 | Categorical/Nominal |
| 77 | Year | Year | Ordinal (Date/year) |

Table 19.7 shows the list of features (data elements) for inpatient claims with their data types (The Centers for Medicare and Medicaid Services, 2013a).

**TABLE 19.7**
**List of Features (Data Elements) for Inpatient Claims**

| # | Variable Name | Labels | |
|---|---|---|---|
| 10 | *OP_PHYSN_NPI* | DESYNPUF: Operating Physician—National Provider Identifier Number | Categorical/Nominal |
| 11 | *OT_PHYSN_NPI* | DESYNPUF: Other Physician—National Provider Identifier Number | Categorical/Nominal |
| 12 | *CLM_ADMSN_DT* | DESYNPUF: Inpatient admission date | Ordinal (Date) |
| 13 | *ADMTNG_ICD9_DGNS_CD* | DESYNPUF: Claim Admitting Diagnosis Code | Categorical/Nominal |
| 14 | *CLM_PASS_THRU_PER_DIEM_AMT* | DESYNPUF: Claim Pass-Thru Per Diem Amount | Ratio/Real number (Payment) |
| 15 | *NCH_BENE_IP_DDCTBL_AMT* | DESYNPUF: NCH Beneficiary Inpatient Deductible Amount | Ratio/Real number (Payment) |
| 16 | *NCH_BENE_PTA_COINSRNC_LBLTY_AM* | DESYNPUF: NCH Beneficiary Part A Coinsurance Liability Amount | Ratio/Real number (Payment) |
| 17 | *NCH_BENE_BLOOD_DDCTBL_LBLTY_AM* | DESYNPUF: NCH Beneficiary Blood Deductible Liability Amount | Ratio/Real number (Payment) |
| 18 | *CLM_UTLZTN_DAY_CNT* | DESYNPUF: Claim Utilization Day Count | Categorical/Nominal |
| 19 | *NCH_BENE_DSCHRG_DT* | DESYNPUF: Inpatient discharged date | Ordinal (Date) |
| 20 | *CLM_DRG_CD* | DESYNPUF: Claim Diagnosis Related Group Code | Categorical/Nominal |
| 21–30 | *ICD9_DGNS_CD_1—ICD9_DGNS_CD_10* | DESYNPUF: Claim Diagnosis Code 1—Claim Diagnosis Code 10 | Categorical/Nominal |
| 31–36 | *ICD9_PRCDR_CD_1—ICD9_PRCDR_CD_6* | DESYNPUF: Claim Procedure Code 1—Claim Procedure Code 6 | Categorical/Nominal |
| 37–81 | *HCPCS_CD_1—HCPCS_CD_45* | DESYNPUF: Revenue Center HCFA Common Procedure Coding System 1—Revenue Center HCFA Common Procedure Coding System 45 | Categorical/Nominal |

#### 19.3.4.2 Step 2: Ingest the Claim Data

CMS's DE-SynPUF database contains five massive datasets and over 120 separate flat files. With a database developer's help, a data pipeline was created with a reporting layer to ingest all 120 flat files. The separate files were then merged using the data pipeline's capabilities, creating five massive tables: one for beneficiary data and four for each claim type (Inpatient, Outpatient, Medical, and Pharmacy claims). Our database is a platform-agnostic solution using a web server and a firebase database that consists of four components: webserver in Express.js, UI/UX design using React, Postgres for persistence database, and Firebase real-time database as the message broker. Researchers do not necessarily need to create a data pipeline to use CMS's DE-SynPUF datasets. The purpose of the data pipeline for this project was to ensure our simple random sampling of the synthesized claims is genuinely random. This requires the entire population of synthesized claims data to be available for random sampling and each member of the population to have the same chance of being selected. Appendix A in Hoseini (2020) provides the architecture designed for the data pipeline. A cluster sampling does not require ingestion and is less costly. This could be done quickly in Excel. The drawback is that cluster sampling makes the statistical analysis more complicated and increases the sampling error (Kerry & Bland, 1998). The 120 files available in the CMS's DE-SynPUF dataset are cultured so that claims for a particular beneficiary are in samples with the same number are in the same file as much as possible (The Centers for Medicare and Medicaid Services, 2019b). As a result, records within each file are similar (homogenous); however, the variation across clusters (files) is high, which means cluster sampling will increase error in the estimations.

#### 19.3.4.3 Step 3: Study the Selected Dataset to Verify It Works for the Medicaid Data

Medicaid Analytic eXtract Data (MAX) was used as the primary source of available public data to understand Medicaid inpatient and outpatient claims' typical coverage (Williams & Baugh, 2016). MAX is defined as a "research-ready data source for Medicaid and CHIP calendar year person-level data on eligibility, service utilization and payment information in the 50 states and the District of Columbia (DC)" (Williams & Baugh, 2016). The procedure codes (International Classification of Diseases and Current Procedural Terminology) on our claim dataset were compared against the Medicaid coverage of several states, including California, New Hampshire, Arizona, Nevada, and Hawaii, to verify the state level coverage of the Medicare-covered inpatient and outpatient FFS claims (Hawaii Department of Human Services & Med-QUEST, n.d.; NM Human Services Department (HSD), n.d.; Arizona Health Care Cost Containment System, n.d.; Nevada Department of Health and Human Services, n.d.). As appropriate, the researchers used the state level to add some adjustments and try to make the dataset as generic as possible.

#### 19.3.4.4 Step 4: Make Synthetic Claim Data Available for Sampling

The designed data pipeline was essential to ingest and merge split files into a coherent dataset. However, running queries in the pipeline is not fast, and it is costly. To overcome this challenge and draw random samples from various tables in the dataset quickly, the researchers with the help of a database admin, moved the datasets into

Amazon Web Services (AWS) cloud. Anyone interested can request access to this dataset for independent validation and verification of the data and study results.

#### 19.3.4.5 Step 5: Select a Random Sample from the Dataset

Our ingested dataset consists of over 1.3 million inpatient claims and close to 16 million outpatient claim payments. Two random samples were selected to build our supervised learning models, as illustrated in Table 19.8. SQL query or Excel function (RAND and RANDBETWEEN) can be used to get random samples. For example, the following SQL query can pull 5,000 random samples of inpatient claims for the year 2010:

```
SELECT * FROM “de_synpuf”.”inpatient_claims”
WHERE “de_synpuf”.”inpatient_claims”.”clm_from_dt” >= 20100101 and
“de_synpuf”.”inpatient_claims”.”clm_from_dt” <= 20101230
ORDER BY random() limit 5000;
```

#### 19.3.4.6 Step 6: Cleansing Sampled Dataset

Data cleansing is an extensive task in real data and mainly consists of finding and dealing with missing data and anomalies. Two random samples in Table 19.8 were selected from the ingested dataset. Many such issues are not prevalent in our synthesized dataset. However, following two steps were taken in our datasets: (1) eliminate negative payments: in the world of Medicaid FFS claims, a negative payment amount is usually an indication that the provider was overpaid in the past and had a negative balance. In those situations, the claim will result in a zero payment with a negative number on the claim explanation documents, also known as remittance advice details (RAD) Codes. Such incidents were removed from the sample to avoid unnecessary complications. (2) Rows with missing data in key fields (e.g., payment, NPIs) were eliminated.

#### 19.3.4.7 Step 7: Understand the Type and Percentage of Quality Issues in Medicaid FFS Claims Adjudication Using PERM Audits

The Payment Error Rate Measurement (PERM) program audit's goal is "to measure and report a national improper payment rate for Medicaid and the Children's Health Insurance Program (CHIP) to comply with the requirements of the Improper

**TABLE 19.8**
**Selected Sample Size**

| Claim Type | Category | 2008 | 2009 | 2010 | Total |
|---|---|---|---|---|---|
| Inpatient Claims | Total Population in DB | 547,800 | 504,941 | 280,081 | 1,332,822 |
| | Initial Sample Size | 5,478 | 5,049 | 2,801 | 13,328 |
| Outpatient Claims | Total Population in DB | 5,673,808 | 6,519,340 | 3,633,839 | 15,826,987 |
| | Initial Sample Size | 56,738 | 65,193 | 36,338 | 158,269 |

Payments Elimination and Recovery Improvement Act" (The Centers for Medicare & Medicaid Services, 2015). Typical Medicaid claim adjudication processing errors more or less follow the same pattern as edited criteria, including the following categories: eligibility, pricing, correct coding, third party liability-related issues, suspected duplicate, and frequency or utilization of services (*Iowa Medicaid Enterprise Performance Report*, 2006).

To get an overall sense of the percentage of the pricing and data entry errors, we used the PERM reports for 2019 and 2017, as shown in Table 19.9 (The Centers for Medicare and Medicaid Services, 2017, 2019b).

This chapter focuses on a subset of the improper payments reported in PERM audits related to claim adjudication, including claim pricing issues and data entry issues. The results of PERM audits show 1%–5% should be an overall acceptable interval estimate for data entry and claim pricing issues in a Medicaid FFS program.

#### 19.3.4.8 Step 8: Estimate the Overall Percentage of Claim Adjudication Errors (Population Proportion) in a Given Medicaid FFS Program

The PERM audit report provides a high-level understanding of the type of quality issues in a typical Medicaid FFS claim adjudication process. However, to label the target group in our synthesized dataset, detailed level information about the pattern of quality issues in each Medicaid program is needed. To achieve this goal, the researchers selected the California Medicaid Management Information System and submitted several formal Public Information Request (PRAs) to the California Department of Health Care Services to receive two quality reports: Monthly Quality Management Performance Report (MQMPR) and Weekly Payment Data Review (WPDR) report for three years (2016–2019). CA-MMIS was selected, as it is the most extensive Medicaid FFS program in the nation that processes over 200K claims per day.

**TABLE 19.9**
**Selected Error Rates in PERM Audit Reports for 2017 and 2019**

| Year[2] | Claim Type | Claims Sampled | Improper Payments | Improper Payment Rate | 95% CI |
|---|---|---|---|---|---|
| 2017 | Inpatient and Outpatient Hospital Services-Medical Review Errors | 3,672 | 38 | 1.3% | 0.9%–1.8% |
| 2017 | Inpatient and Outpatient Hospital Services-Data Processing Errors | 3,672 | 235 | 3.8% | 2.9%–4.8% |
| 2019 | Inpatient Hospital Services-Medical Review Errors | 2,651 | 9 | 0.69% | 0–1.47% |
| 2019 | Outpatient Hospital Services-Medical Review Errors | 1,670 | 29 | 1.24% | 0.39%–2.10% |
| 2019 | Inpatient Hospital Services-Data Processing Errors | 2,651 | 156 | 2.17 % | 1.30%–3.05% |
| 2019 | Outpatient Hospital Services-Data Processing Errors | 1,670 | 78 | 3.27% | 1.01%–5.52% |

**TABLE 19.10**
**Summary of Claim Adjudication Errors per CA-MMIS MQMPR for 2018**

| Month | Sample Size | Overpayments | Underpayments | Overpayment% | Underpayment% | Erroneous Payment% |
|---|---|---|---|---|---|---|
| Jan | 1,040 | 11 | 12 | 1.1% | 1.2% | 2.2% |
| Feb | 927 | 9 | 6 | 1.0% | 0.6% | 1.6% |
| Mar | 987 | 11 | 8 | 1.1% | 0.8% | 1.9% |
| Apr | 954 | 8 | 12 | 0.8% | 1.3% | 2.1% |
| May | 921 | 12 | 11 | 1.3% | 1.2% | 2.5% |
| Jun | 938 | 8 | 9 | 0.9% | 1.0% | 1.8% |
| Jul | 907 | 10 | 11 | 1.1% | 1.2% | 2.3% |
| Aug | 911 | 13 | 9 | 1.4% | 1.0% | 2.4% |
| Sep | 902 | 8 | 8 | 0.9% | 0.9% | 1.8% |
| Oct | 945 | 7 | 3 | 0.7% | 0.3% | 1.1% |
| Nov | 969 | 13 | 17 | 1.3% | 1.8% | 3.1% |
| Dec | 1,008 | 19 | 8 | 1.9% | 0.8% | 2.7% |
| Average | | | | | | 2.1% |

The researchers received the PHI-free version of these reports in which individually identifiable information (e.g., providers' names and NPIs) were redacted. MQMPR contains large samples (over 1,000 samples from all adjudicated claims per month), and WPDR contains statistically significant (over 300 per week) samples of high-dollar claims. MQMPR did not provide detailed information on each quality issue per claim type, so the researchers used it to estimate the population proportion of quality issues and study the pattern of data entry issues in general. Per the MQMPR report of 2018, the estimated overall error rate on the CA-MMIS claim adjudication process was 2.1%, as shown in Table 19.10.

#### 19.3.4.9 Step 9: Understand the Underlying Pattern of Quality Issues in Medicaid FFS Claims Adjudication Using Detailed Reports from a Selected Medicaid Program

CA-MMIS MQMPR reports do not provide detailed information on each quality issue per claim type. However, CA-MMIS WPDR reports include detail of the quality issues per claim type, so the researchers extracted the detail of the quality issues (erroneous payments) in the CA-MMIS WPDR 9/9/2016–5/1/2020. Table 19.11 summarizes the high-level statistics.

The researchers used the WPDR statistics to study and simulate the pattern of erroneous payments. It is important to note all reported findings in the CA-MMIS WPDR reports were caught and fixed before the providers' final payment, so they are not essentially erroneous "payments." However, they carry the same characteristics of erroneous payments in a typical Medicaid FFS. CA-MMIS WPDR targets high dollar value claims, and it does not use a simple random sample of all adjudicated

**TABLE 19.11**
**Claim Adjudication Errors Reported in CA-MMIS WPDR 9/9/2016–5/1/2020**

| Claim Types | Count of Error |
|---|---|
| Outpatient | 376 |
| Inpatient | 52 |
| Medical | 117 |
| Long Term Care | 6 |
| Pharmacy | 4 |
| Total | 555 |

claims. Therefore, while the findings of CA-MMIS WPDR reports are appropriate for data labeling, they do not provide a statistically valid interval estimate of the population proportion (estimated rate of erroneous payments in a typical Medicaid FFS). Therefore, PERM audit reports in step 8 were used to provide an interval estimate of erroneous payments in a typical Medicaid FFS program.

We simulated four types of erroneous payments to create labeled data for the target group: pricing errors, data entry error, erroneously paid, and erroneously denied claims. We divided them into two major categories:

- P0: Negatives (not targeted group, no issues). We did not change the payment amount for these groups and only applied feature engineering for data cleansing.
- P1: Positives (target group). These were simulated labels and were categorized into three subgroups:
  - P1.1: Erroneously denied (should have been paid)
  - P1.2: Erroneously paid (should have been denied)
  - P1.3: Pricing and data entry issues

Table 19.12 presents the number of labels created for the target category for both inpatient and outpatient datasets. A detailed explanation of this simulation process is provided in steps 9.1 to 9.3.

#### *19.3.4.9.1 Step 9.1: Simulating Pricing Errors and Data Entry Errors*

In the CA-MMIS WPDR findings between 9/9/2016 and 5/1/2020, other erroneously paid and documentation issues were removed because these were not related to pricing issues and simulated in step 9.2. The claim adjudication errors' underlying probability distribution functions (PDFs) for the absolute percentage difference between the paid and correct amounts for outpatient and inpatient claims were studied. Anderson-Darling (AD) tests (Upton & Cook, 2014) were conducted. Researchers have been using the AD test for estimating the distribution function of operational risk, which is very similar to how it is used in this research (Feuerverger, 2016). The smaller the AD statistics, the better the fit for the specific distribution. Minitab was used

**TABLE 19.12**
**Number of Labels Created for the Target Category**

| | | Outpatient | | Inpatient | |
|---|---|---|---|---|---|
| **Labels** | | **Count** | **Percentage** | **Count** | **Percentage** |
| P0 | | 112,335 | 98.65% | 12,833 | 97.07% |
| P1 | | 1,532 | 1.35% | 388 | 2.93% |
| | P1.1 | 800 | 0.70% | 200 | 0.18% |
| | P1.2 | 400 | 0.35% | 100 | 0.09% |
| | P1.3 | 332 | 0.29% | 88 | 0.08% |
| Total | | 113,867 | 100.00% | 13,221 | 100.00% |

because it tests 16 different distribution functions. However, since $p < 0.05$ for each distribution, all the null hypotheses that the outpatient claim data followed specific distributions were rejected. Readers may refer to Hoseini (2020) for details. Hence, no distribution was selected, even with the lowest AD. For the inpatient claim data, a similar conclusion was reached. As a result, during the labeling process, the exact pricing issue percentage differences are directly applied to the labeled data (instead of using a function to generate a dataset with the desired PDF) following these steps:

1. Select a random sample of 332 outpatient claims and 88 inpatient claims (numbers of datapoints matching to Table 19.12) with a paid amount greater than zero. This could be done in SQL or using the Excel RAND function.
2. Extract the absolute difference between paid and correct payment amounts from the CA-MMIS WPDR reports. The Excel ABS function could be used for this purpose.
3. Select a random percentage from step 2 and multiply it by the selected sample from step 1. Sampling without replacement should be used. When the number of samples is greater than the number of percentage differences (step 2), repeat the process from step 1. All selected samples were labeled P1.

*19.3.4.9.2 Step 9.2: Simulating Erroneously Denied Claims*

Using numbers shown in Table 19.12, 800 random samples of outpatient claims and 200 samples of inpatient claims were selected. This could be done in SQL, as described in step 6 of using the Excel RAND function. The total payment amount of these selected samples was changed to zero, and they were labeled as P1.

*19.3.4.9.3 Step 9.3: Simulating Erroneously Paid Claims*

Using numbers presented in Table 19.12, 400 random samples of outpatient claims with payment amount equal to zero, and 100 samples of inpatient claims with payment amount equal to zero were selected. For each claim, the researcher found the first ICD code used in the claim and queried claims with similar first ICD codes. A random sample of those claims was selected, and their claim paid amounts were used to replace the zero paid amount. This could be done in SQL, as described in step 6 of using the Excel RAND function. These were labeled as P1.

## 19.4 FEATURE SELECTION

This section discusses methods applied for feature selection to reduce the feature space dimension for our data. Feature selection aims to find and eliminate the independent variables (from all *N* features), which are less informative, making the computation difficult and even cause confusion. The goal is to create a simplified representation of the data using fewer dimensions (M) than the original number of features (N). We select the most important features—those along which the data shows the greatest variation or difference (PCA For Categorical Features?—Stack Overflow, n.d.). As discussed by many data scientists, there is no universal best method for feature selection, and the complexity of the problem, nature of the dataset and business, and computational power availability and limitation may dictate choosing one method over another (Shamsaei & Gao, 2016;(Brownlee, 2019). While there are several methods for feature selections, the most popular methods are principal component analysis, which is appropriate for continuous (ratio) data, and multiple correspondence analysis (MCA), which works with categorical data.

Our problem consists of a mix of ratio and categorical features (independent variables) and categorical target (dependent variable). We tried methods recommended by researchers to conduct PCA/MCA on mixed data (ration and categorical) but found they require extreme and cost-prohibitive processing power (Stackexchange, n.d.). Even if the processing capability is available to conduct complex PCA/MCA on mixed data to identify all variables' statistical significance, one should not assume the features selected by such methods are the best predictors of the target group. Relying solely on the statistical significance of the predictor in a similar situation may increase classification errors (Lo et al., 2015). For this research, we used a heuristic and hybrid method for feature selection consisting of three steps:

1. Eliminate data with no-low information value
2. Conduct PCA for continuous (ratio) features
3. Use business knowledge to select categorical features

Independent values with one (unique), irrelevant, identical, or near-identical values do not improve a machine learning model's prediction power. This is even more important when dealing with high dimensional data (datasets with over 15–20 features) and massive (hundreds and thousands of rows). Our original datasets each had over 70 elements and more than 10,000 rows. We reviewed datasets and found the following features to have unique or near-unique values, so they were removed from the model: Segment, QAREVIEW_RESULT, hcpcs_cd_45, and nch_bene_blood_ddctbl_lblty_am. We also removed feature with no information values, including identifiers, as they have no prediction power by nature. So the following features were removed: at_physn_npi, op_physn_npi, ot_physn_npi, prvdr_num, clm_from_dt, clm_thru_dt.

The PCA function of the Minitab (Stat > Multivariate > Principal Components) was used on the reduced dataset resulted from the previous step and selected five features (predictors) for the outpatient dataset PCA: clm_pmt_amt, nch_prmry_pyr_clm_pd_amt, nch_bene_ptb_ddctbl_amt, nch_bene_ptb_coinsrnc_amt, and clm_utlztn_day_cnt. Following the same process, six features were selected for the inpatient dataset PCA: clm_pmt_amt, nch_prmry_pyr_clm_pd_amt, clm_pass_thru_per_diem_amt, nch_bene_pta_coinsrnc_lblty_am, nch_bene_blood_ddctbl_lblty_am, and nch_bene_ip_ddctbl_amt.

The summary results of the PCA from Minitab are illustrated in Tables 19.13 and 19.14.

The first principal component for outpatient claims accounts for 22.2% of the total variance. To keep the most information in the selected features, a 90% cumulative cut-off threshold was set for the correlation matrix's Eigen analysis. This means all ratio features must be chosen to explain at least 90% of the variance in our dataset.

**TABLE 19.13**
**PCA for Five Ratio Features for the Outpatient Dataset**

| | | | | | |
|---|---|---|---|---|---|
| Eigen analysis of the correlation matrix | | | | | |
| Eigenvalue | 1.2556 | 1.1104 | 0.9846 | 0.8969 | 0.7525 |
| Proportion | 0.251 | 0.222 | 0.197 | 0.179 | 0.151 |
| Cumulative | 0.251 | 0.473 | 0.67 | 0.849 | 1 |
| 12794 cases used; 727 cases contain missing values | | | | | |
| Eigenvectors | | | | | |
| Variable | PC1 | PC2 | PC3 | PC4 | PC5 |
| clm_pmt_amt | 0.698 | −0.017 | 0.09 | −0.071 | 0.707 |
| nch_prmry_pyr_clm_pd_amt | 0.211 | 0.214 | −0.944 | 0.116 | −0.071 |
| clm_pass_thru_per_diem_amt | 0.1 | −0.683 | −0.204 | −0.677 | −0.158 |
| nch_bene_ip_ddctbl_amt | 0.047 | −0.691 | −0.059 | 0.718 | 0.017 |
| clm_utlztn_day_cnt | 0.676 | 0.1 | 0.236 | 0.088 | −0.685 |

**TABLE 19.14**
**PCA for Five Ratio Features for the Inpatient Dataset**

| | | | | | |
|---|---|---|---|---|---|
| Eigen analysis of the correlation matrix | | | | | |
| Eigenvalue | 1.1655 | 1.1056 | 1.0034 | 0.9832 | 0.8896 |
| Proportion | 0.194 | 0.184 | 0.167 | 0.164 | 0.148 |
| Cumulative | 0.194 | 0.379 | 0.546 | 0.71 | 0.858 |
| 12504 cases used; 717 cases contain missing values | | | | | |
| Eigenvectors | | | | | |
| Variable | PC1 | PC2 | PC3 | PC4 | PC5 |
| clm_pmt_amt | 0.658 | −0.209 | −0.057 | −0.007 | −0.075 |
| nch_prmry_pyr_clm_pd_amt | 0.298 | −0.351 | 0.338 | −0.738 | −0.019 |
| clm_pass_thru_per_diem_amt | 0.38 | 0.567 | −0.071 | −0.11 | 0.71 |
| nch_bene_pta_coinsrnc_lblty_am | 0.518 | −0.233 | 0.059 | 0.612 | −0.079 |
| nch_bene_blood_ddctbl_lblty_am | 0.071 | −0.174 | −0.935 | −0.219 | −0.059 |

As shown in Table 19.13, even if one ratio feature is removed, the remaining ratio features only account for up to 82.2% of the variation. As a result, all ratio features are kept in the outpatient model.

The first principal component for inpatient claims accounts for 19.4% of the total variance. To keep the most information in the selected features, a 90% cumulative cut-off threshold was set for the correlation matrix's Eigen analysis. As shown in Table 19.14, even if just one ratio feature is removed, the remaining ratio features only account for up to 85.8% of the variation. As a result, all ratio features are kept in the inpatient model.

### 19.4.1 Use Business Knowledge to Select Categorical Features

Medicaid claims data allows a researcher to use business knowledge and make a judgment call to eliminate categorical features with low predictability power. Two candidates for such heuristic feature selection are diagnosing and treating medical codes (CPT and ICD codes). A Medicaid FFS claim may contain several diagnoses and treatments, so there are up to 45 columns (features) to capture all codes related to one claim in the dataset. However, most claims contain less than a handful of CPT and

**TABLE 19.15**
**PCA for Five Ratio Features for the Inpatient and Outpatient Dataset**

| Inpatient | Outpatient |
|---|---|
| clm_pmt_amt | clm_pmt_amt |
| nch_prmry_pyr_clm_pd_amt | nch_prmry_pyr_clm_pd_amt |
| at_physn_npi | at_physn_npi |
| op_physn_npi | op_physn_npi |
| ot_physn_npi | ot_physn_npi |
| admtng_icd9_dgns_cd | icd9_dgns_cd_1 |
| clm_pass_thru_per_diem_amt | icd9_dgns_cd_2 |
| nch_bene_ip_ddctbl_amt | icd9_dgns_cd_3 |
| nch_bene_pta_coinsrnc_lblty_am | icd9_prcdr_cd_1 |
| nch_bene_blood_ddctbl_lblty_am | nch_bene_ptb_ddctbl_amt |
| clm_utlztn_day_cnt | nch_bene_ptb_coinsrnc_amt |
| clm_drg_cd | admtng_icd9_dgns_cd |
| icd9_dgns_cd_1 | hcpcs_cd_1 |
| icd9_dgns_cd_2 | hcpcs_cd_2 |
| icd9_dgns_cd_3 | hcpcs_cd_3 |
| hcpcs_cd_1 | YEAR |
| hcpcs_cd_2 | |
| hcpcs_cd_3 | |
| YEAR | |

ICD codes. A simple exploratory analysis of the data shows that only the first three CPT and ICD codes for each claim have considerable values. The rest are usually with no value. Table 19.15 listed all features selected at the end of the feature selection process to build our supervised learning models: 19 for inpatient and 16 for outpatient.

### 19.4.2 Build, Train, and Test Models

In this section, ML models are built and algorithms trained. First, the challenge of dealing with rare events and proposing event-based sampling to improve model performance will be discussed. An approach to oversampling, stratification strategy and the cross-validation method will be presented. Decision tree, random forest, and gradient boosting models are the first models to be trained. The outputs of these models are required to calculate the feature importance score for both dataset's features. Finally, the ML model for the rest of the chosen algorithms will be built and trained, and test results will be presented.

#### 19.4.2.1 Train and Test Decision Tree, Random Forest, and Gradient Boosting

Decision tree, random forest, and gradient boosting are the first three models to be trained and tested. We needed their output to calculate the feature importance score (Gini importance) for independent variables and test the first two hypotheses, H4.1 and H5.1. One crucial notion here is the difference between an "algorithm" and a predictive (or supervised learning) "model." An algorithm is a generic method that could apply to various datasets and could usually be tuned through some configuration features. For example, a decision tree is the name of a group of algorithms with similar approaches and have several configuration features like a maximum number of leaves. A model (in our definition) is a mathematical equation that resulted from running an algorithm with specific configuration features on a dataset. It is impractical to compare "algorithm performance" in the abstract due to the massive permutation among configuration features. The focus of this research is on the models derived from running selected algorithms. As a result, an algorithm with and one without oversampling creates two different predictive models. If common themes are found across various models, it shows that the algorithm performs similarly across different configurations and datasets.

To build and train the decision tree, we used the *DecisionTreeClassifier* function in Python Scikit-learn library with the following features:

- Decision tree with Gini classifier: *default, random_state=0, max_depth=10, max_leaf_nodes=115*
- Decision tree with entropy classifier: *criterion=entropy, random_state=0, max_depth=10, max_leaf_nodes=115*

DTE and DTG models were trained on both inpatient and outpatient datasets with and without oversampling. Random forest Gini and entropy and gradient boosting algorithms on inpatient and outpatient datasets were trained with and without applying the oversampling methods. *RandomForestClassifier* and *GradientBoostingClassifier* from the Python Scikit-learn library were utilized, and results were tested. Test results for all algorithms are presented together in Appendix H in Hoseini (2020).

#### 19.4.2.2 Review Feature Importance

Mean decrease in impurity (MDI) is a method in the Python Scikit-learn library to calculate the feature importance score (Gini importance). This method and associated codes were applied for the selected independent variables for inpatient and outpatient models. The results of these calculations are presented in Appendix G in Hoseini (2020) and discussed in detail here in the "Results of Testing Research Hypotheses and Answers to Research Questions" section.

#### 19.4.2.3 Train and Test the Rest of the Selected Models and Review the Results

The remaining models were trained and tested. The details can be found in the appendix in Hoseini (2020).

### 19.4.3 Results of Testing Research Hypotheses and Answers to Research Questions

The results of testing the hypotheses of our research are presented in this section. Here the assumptions, results of the primary hypothesis test, and post-hoc analysis performed are discussed.

#### 19.4.3.1 Research Hypothesis H4.1

Hypothesis 4.1 stated there is no significant difference among the predictors' feature importance in identifying erroneously processed Medicaid outpatient claims. This means all predictors (selected independent variables) are the same from the feature importance or mean decrease in impurity perspective. Detailed results of these calculations are included in Appendix G and of the one-way ANOVA test from Minitab in Appendix I in Hoseini (2020). Table 19.16 indicates that there was enough evidence at the $\alpha = 0.05$ significance level to reject the null hypothesis ($p = 0.000$). This means at least one of the feature importance scores (Gini importance) for outpatient claim independent variables performed significantly different from the rest of the features.

The Tukey pairwise comparisons using the Tukey method and a 95% confidence level in Table 19.17 show clm_pmt_amt has the highest mean *F*-value, significantly higher than the rest of the features.

**TABLE 19.16**
**One-Way ANOVA Results for Comparing Feature Importance Scores (Gini Importance) for Outpatient Claim Independent Variables**

| Source | DF | Adj SS | Adj MS | *F*-Value | *P*-Value |
|---|---|---|---|---|---|
| Factor | 15 | 0.6359 | 0.042394 | 8.02 | 0.000 |
| Error | 64 | 0.3383 | 0.005286 | | |
| Total | 79 | 0.9742 | | | |

The Fisher pairwise comparisons using the LSD method and a 95% confidence level show that clm_pmt_amt has the highest mean $F$-value, significantly higher than the rest of the features. The Fisher LSD test also presents a slight difference among other predictors' importance. In both tests, icd9_dgns_cd_1 and at_physn_npi have the second and third highest $F$-value for the feature importance, significantly lower than clm_pmt_amt.

#### 19.4.3.2 Research Hypothesis H5.1

Hypothesis 5.1 stated there is no significant difference among the predictors' feature importance in identifying erroneously processed Medicaid inpatient claims. Table 19.18 summarizes the results of the one-way ANOVA test from Minitab and indicates that there was enough evidence at the $\alpha = 0.05$ significance level to reject the null hypothesis ($p = 0.000$).

The Tukey pairwise comparisons using the Tukey method and a 95% confidence level in Table 19.19 show clm_pmt_amt and nch_prmry_pyr_clm_pd_amt have the highest mean $F$-values.

The Fisher pairwise comparisons using the Fisher LSD method and a 95% confidence level show that clm_pmt_amt has the highest $F$-measure, significantly higher than the rest of the predictors. nch_prmry_pyr_clm_pd_amt has the second-highest $F$-measure, significantly higher than the rest of the predictors. The Fisher LSD test

**TABLE 19.17**
**Tukey Pairwise Comparisons for Feature Importance Scores (Gini Importance) for Outpatient Claim Independent Variables**

| Factor | N | Mean | Grouping | |
|---|---|---|---|---|
| clm_pmt_amt | 5 | 0.389 | A | |
| icd9_dgns_cd_1 | 5 | 0.1100 | | B |
| at_physn_npi | 5 | 0.0983 | | B |
| hcpcs_cd_1 | 5 | 0.0539 | | B |
| icd9_dgns_cd_2 | 5 | 0.0521 | | B |
| hcpcs_cd_2 | 5 | 0.0484 | | B |
| nch_bene_ptb_coinsrnc_amt | 5 | 0.04511 | | B |
| ot_physn_npi | 5 | 0.04454 | | B |
| icd9_dgns_cd_3 | 5 | 0.04191 | | B |
| hcpcs_cd_3 | 5 | 0.03497 | | B |
| admtng_icd9_dgns_cd | 5 | 0.03077 | | B |
| op_physn_npi | 5 | 0.02340 | | B |
| YEAR | 5 | 0.01583 | | B |
| nch_bene_ptb_ddctbl_amt | 5 | 0.00787 | | B |
| nch_prmry_pyr_clm_pd_amt | 5 | 0.004190 | | B |
| icd9_prcdr_cd_1 | 5 | 0.000005 | | B |

*Source:* Means that do not share a letter are significantly different.

**TABLE 19.18**
**One-Way ANOVA Results for Comparing Feature Importance Scores (Gini Importance) for Inpatient Claim Independent Variables**

| Source | DF | Adj SS | Adj MS | *F*-Value | *P*-Value |
|---|---|---|---|---|---|
| Factor | 18 | 0.8838 | 0.049103 | 23.60 | 0.000 |
| Error | 76 | 0.1581 | 0.002080 | | |
| Total | 94 | 1.0419 | | | |

**TABLE 19.19**
**Tukey Pairwise Comparisons for Feature Importance Scores (Gini Importance) for Inpatient Claim Independent Variables**

| Factor | N | Mean | Grouping | |
|---|---|---|---|---|
| clm_pmt_amt | 5 | 0.3601 | A | |
| nch_prmry_pyr_clm_pd_amt | 5 | 0.2898 | A | |
| clm_drg_cd | 5 | 0.08992 | | B |
| at_physn_npi | 5 | 0.0441 | | B |
| icd9_dgns_cd_1 | 5 | 0.0352 | | B |
| icd9_dgns_cd_2 | 5 | 0.0334 | | B |
| icd9_dgns_cd_3 | 5 | 0.0332 | | B |
| admtng_icd9_dgns_cd | 5 | 0.0302 | | B |
| op_physn_npi | 5 | 0.02552 | | B |
| clm_utlztn_day_cnt | 5 | 0.01947 | | B |
| clm_pass_thru_per_diem_amt | 5 | 0.01288 | | B |
| YEAR | 5 | 0.00761 | | B |
| nch_bene_ip_ddctbl_amt | 5 | 0.00721 | | B |
| ot_physn_npi | 5 | 0.00694 | | B |
| nch_bene_pta_coinsrnc_lblty_am | 5 | 0.002416 | | B |
| nch_bene_blood_ddctbl_lblty_am | 5 | 0.002045 | | B |
| hcpcs_cd_3 | 5 | 0.000000 | | B |
| hcpcs_cd_2 | 5 | 0.000000 | | B |
| hcpcs_cd_1 | 5 | 0.000000 | | B |

*Source:* Means that do not share a letter are significantly different.

results are slightly different from the Tukey test results in the grouping, but the predictors' ranking is similar.

### 19.4.3.3 Research Hypothesis H6.1

Hypothesis 6.1 stated there is no significant difference among the average recall of algorithms in identifying Medicaid outpatient claim issues. We ran a one-way ANOVA based on fivefold cross-validation recall results across all selected algorithms to investigate this hypothesis. Table 19.20 shows that there was enough evidence at the $\alpha = 0.05$ significance level to reject the null hypothesis ($p = 0.000$).

The Tukey pairwise comparisons using the Tukey method and a 95% confidence level in Table 19.21 show the logistic regression model (with oversampling) has

**TABLE 19.20**
**One-Way ANOVA Results for Comparing Algorithms' Recall for Outpatient Claim Issues**

| Source | DF | Adj SS | Adj MS | *F*-Value | *P*-Value |
|---|---|---|---|---|---|
| Factor | 19 | 6.2851 | 0.33080 | 29.21 | 0.000 |
| Error | 80 | 0.9060 | 0.01132 | | |
| Total | 99 | 7.1911 | | | |

**TABLE 19.21**
**Tukey Pairwise Comparisons for Algorithms' Recall for Outpatient Claim Issues**

| Factor | N | Mean | Grouping | | | | | | |
|---|---|---|---|---|---|---|---|---|---|
| LR (OS) | 5 | 0.7825 | A | | | | | | |
| LR | 5 | 0.7639 | A | B | | | | | |
| NB | 5 | 0.5242 | | B | C | | | | |
| NB (OS) | 5 | 0.5229 | | B | C | | | | |
| DA (OS) | 5 | 0.48565 | | | C | D | | | |
| DTE (OS) | 5 | 0.4700 | | | C | D | | | |
| DTG (OS) | 5 | 0.4224 | | | C | D | | | |
| NN (OS) | 5 | 0.365 | | | C | D | E | | |
| GB (OS) | 5 | 0.25000 | | | | D | E | F | |
| kNN (OS) | 5 | 0.16383 | | | | | E | F | G |
| RFE (OS) | 5 | 0.11424 | | | | | | F | G |
| GB | 5 | 0.1129 | | | | | | F | G |
| RFG (OS) | 5 | 0.1103 | | | | | | F | G |
| DTG | 5 | 0.06266 | | | | | | F | G |
| DTE | 5 | 0.05680 | | | | | | F | G |
| RFG | 5 | 0.02741 | | | | | | F | G |
| RFE | 5 | 0.02611 | | | | | | F | G |
| NN | 5 | 0.00782 | | | | | | F | G |
| kNN | 5 | 0.00457 | | | | | | F | G |
| DA | 5 | 0.00196 | | | | | | | G |

*Source:* Means that do not share a letter are significantly different.

the highest $F$-value (this is not the F1-score used as a performance metrics) and significantly outperforms other algorithms in detecting erroneous payments in the research outpatient data. This will make the logistic regression (regardless of sampling approach) the favorite algorithm in finding true positives in outpatient claim adjudication. Logistic regression and naïve Bayes (with and without oversampling) are the next best performers and group together (group B). The Tukey test also shows the recall of some algorithms, including discriminant analysis, neural network, and decision trees, improves with oversampling. This means these algorithms may not perform well on unbalanced data. Oversampling improves the performance of algorithms like logistic regression, naïve Bayes, and gradient boosting. However, there is no statistical significance to this difference (Naïve Bayes slightly performed better without oversampling, which could be due to random effects in the cross-validation process). Tukey grouped algorithms from the recall point of view into several overlapping groups, whereas the groups that do not share a letter are significantly different.

The Fisher LSD pairwise comparisons with a 95% confidence level show the same ranking of the algorithms' recall means with a slightly different pairwise grouping.

#### 19.4.3.4 Research Hypothesis H6.2

Hypothesis 6.2 stated there is no significant difference among the average F1-score of selected supervised learning algorithms in identifying erroneously processed Medicaid outpatient claims. Table 19.22 presents one-way ANOVA result based on fivefold cross-validation F1-score results across all selected algorithms to investigate this hypothesis. There was enough evidence at the $\alpha = 0.05$ significance level to reject the null hypothesis ($p = 0.000$).

The Tukey pairwise comparisons using the Tukey method and a 95% confidence level in Table 19.23 show the naïve Bayes (without oversampling) and gradient boosting (with oversampling) outperformed the rest of the algorithms from the F1 score perspective. Gradient boosting (without oversampling) is the next performing algorithm on F1 score for outpatient claims. Oversampling significantly improves the F1-score of the gradient boosting algorithm on the outpatient data. Various configurations of decision tree and random forest (with entropy and Gini, with and without oversampling) are the next group in the rank.

The Fisher LSD pairwise comparisons with a 95% confidence level present the same ranking of the algorithms' F1 score means, with slight differences in the grouping of the low-

**TABLE 19.22**
**One-Way ANOVA Results for Comparing Algorithms' F1-Score for Outpatient Claim Issues**

| Source | DF | Adj SS | Adj MS | $F$-Value | $P$-Value |
|---|---|---|---|---|---|
| Factor | 19 | 0.31920 | 0.016800 | 126.77 | 0.000 |
| Error | 80 | 0.01060 | 0.000133 | | |
| Total | 99 | 0.32980 | | | |

**TABLE 19.23**
**Tukey Pairwise Comparisons for Algorithms' F1 Score for Outpatient Claim Issues**

| Factor | N | Mean | Grouping | | | | | | | | | |
|---|---|---|---|---|---|---|---|---|---|---|---|---|
| NB | 5 | 0.19742 | A | | | | | | | | | |
| GB (OS) | 5 | 0.19513 | A | | | | | | | | | |
| GB | 5 | 0.1400 | | B | | | | | | | | |
| DTG | 5 | 0.10924 | | | C | | | | | | | |
| DTE | 5 | 0.10178 | | | C | | | | | | | |
| RFE (OS) | 5 | 0.09074 | | | C | D | | | | | | |
| RFG (OS) | 5 | 0.08799 | | | C | D | | | | | | |
| DTE (OS) | 5 | 0.07190 | | | | D | E | | | | | |
| DTG (OS) | 5 | 0.06783 | | | | D | E | F | | | | |
| RFG | 5 | 0.05316 | | | | | E | F | G | | | |
| RFE | 5 | 0.05063 | | | | | E | F | G | H | | |
| NB (OS) | 5 | 0.042270 | | | | | | F | G | H | I | |
| DA (OS) | 5 | 0.027570 | | | | | | | G | H | I | J |
| LR | 5 | 0.027106 | | | | | | | G | H | I | J |
| LR (OS) | 5 | 0.026677 | | | | | | | G | H | I | J |
| kNN (OS) | 5 | 0.02572 | | | | | | | | H | I | J |
| NN (OS) | 5 | 0.02195 | | | | | | | | | I | J |
| kNN | 5 | 0.00702 | | | | | | | | | | J |
| NN | 5 | 0.00449 | | | | | | | | | | J |
| DA | 5 | 0.00390 | | | | | | | | | | J |

*Source:* Means that do not share a letter are significantly different.

### 19.4.3.5 Research Hypothesis H7.1

Hypothesis 7.1 stated there is no significant difference among the average recall of selected supervised learning algorithms in identifying erroneously processed Medicaid inpatient claims. Table 19.24 shows the one-way ANOVA and indicates that there was enough evidence at the $\alpha = 0.05$ significance level to reject the null hypothesis ($p = 0.000$).

The Tukey pairwise comparisons using the Tukey method and a 95% confidence level in Table 19.25 show the gradient boosting (with oversampling) outperformed the rest of the models in recalling quality issues of the inpatient claim data. However, the Tukey test does not suggest a statistical difference among the recall of the top 14 algorithms and models. This test shows k-nearest neighbor, neural network, discriminant analysis, and logistic regression (without sampling) have significantly lower recall than the rest of the algorithms. Oversampling has significantly improved the recall of the logistic regression model, which is a sign of the impact of imbalanced data on this algorithm's performance.

**TABLE 19.24**
**One-Way ANOVA Results for Comparing Algorithms' Recall for Inpatient Claim Issues**

| Source | DF | Adj SS | Adj MS | *F*-Value | *P*-Value |
|---|---|---|---|---|---|
| Factor | 19 | 6.0195 | 0.316815 | 33.38 | 0.000 |
| Error | 80 | 0.7594 | 0.009492 | | |
| Total | 99 | 6.7789 | | | |

**TABLE 19.25**
**Tukey Pairwise Comparisons for Algorithms' Recall for Inpatient Claim Issues**

| Factor | N | Mean | Grouping | | |
|---|---|---|---|---|---|
| GB (OS) | 5 | 0.6389 | A | | |
| DTG (OS) | 5 | 0.6055 | A | B | |
| GB | 5 | 0.5976 | A | B | |
| DTE (OS) | 5 | 0.5877 | A | B | |
| DTE | 5 | 0.5850 | A | B | |
| DTG | 5 | 0.5847 | A | B | |
| RFG | 5 | 0.5719 | A | B | |
| NB (OS) | 5 | 0.5594 | A | B | |
| RFE | 5 | 0.5541 | A | B | |
| NB | 5 | 0.5439 | A | B | |
| RFE (OS) | 5 | 0.5232 | A | B | |
| RFG (OS) | 5 | 0.4973 | A | B | |
| NN (OS) | 5 | 0.484 | A | B | |
| LR (OS) | 5 | 0.391 | | B | |
| kNN (OS) | 5 | 0.1365 | | | C |
| NN | 5 | 0.0205 | | | C |
| kNN | 5 | 0.01548 | | | C |
| LR | 5 | 0.01289 | | | C |
| DA (OS) | 5 | 0.000000 | | | C |
| DA | 5 | 0.000000 | | | C |

*Source:* Means that do not share a letter are significantly different.

The Fisher LSD pairwise comparisons with a 95% confidence level present the same ranking of the algorithms' recall means, with a slightly different pairwise grouping.

### 19.4.3.6 Research Hypothesis H7.2

Hypothesis 7.2 stated there is no significant difference among the average F1-score of selected supervised learning algorithms in identifying erroneously processed Medicaid inpatient claims. Table 19.26 presents one-way ANOVA results and indicates that there was enough evidence at the $\alpha = 0.05$ significance level to reject the null hypothesis ($\alpha = 0.000$).

The Tukey pairwise comparisons using the Tukey method and a 95% confidence level in Table 19.27 show gradient boosting (without and with oversampling), decision tree (using entropy and Gini) without oversampling, and random forest (with

**TABLE 19.26**
**One-Way ANOVA Results for Comparing Algorithms' F1-Score for Inpatient Claim Issues**

| Source | DF | Adj SS | Adj MS | *F*-Value | *P*-Value |
|---|---|---|---|---|---|
| Factor | 19 | 8.2962 | 0.436642 | 237.27 | 0.000 |
| Error | 80 | 0.1472 | 0.001840 | | |
| Total | 99 | 8.4434 | | | |

**TABLE 19.27**
**Tukey Pairwise Comparisons for Algorithms' Recall for Inpatient Claim Issues**

| Factor | N | Mean | Grouping | | | | | |
|---|---|---|---|---|---|---|---|---|
| GB | 5 | 0.7346 | A | | | | | |
| DTE | 5 | 0.7270 | A | | | | | |
| RFG | 5 | 0.7252 | A | | | | | |
| DTG | 5 | 0.7138 | A | | | | | |
| RFE | 5 | 0.7122 | A | | | | | |
| GB (OS) | 5 | 0.6633 | A | B | | | | |
| DA | 5 | 0.6002 | | B | C | | | |
| DTE (OS) | 5 | 0.5488 | | | C | | | |
| DTG (OS) | 5 | 0.5305 | | | C | | | |
| RFE (OS) | 5 | 0.5235 | | | C | | | |
| RFG (OS) | 5 | 0.5052 | | | C | | | |
| NB | 5 | 0.3825 | | | | D | | |
| NB (OS) | 5 | 0.2015 | | | | | E | |
| DA (OS) | 5 | 0.07102 | | | | | | F |
| NN (OS) | 5 | 0.05532 | | | | | | F |
| LR (OS) | 5 | 0.04701 | | | | | | F |
| kNN (OS) | 5 | 0.03712 | | | | | | F |
| kNN | 5 | 0.02125 | | | | | | F |
| LR | 5 | 0.01959 | | | | | | F |
| NN | 5 | 0.00766 | | | | | | F |

*Source:* Means that do not share a letter are significantly different.

entropy and Gini) without oversampling) outperformed the rest of the algorithms from the F1 score perspective on inpatient data. There is evidence to support a statistically significant difference between their performance and the rest of the algorithms. Tukey's ranking suggests that oversampling has reduced the F1-score (by decreasing the precision) for inpatient claims. This may result from oversampling and applying the SMOTE method on a very complex dataset that resulted in simulated samples not carrying the characteristics that could be easily modeled as a feature.

The Fisher LSD pairwise comparisons with a 95% confidence level present the same ranking of the algorithms' F1 score, with some differences in the algorithms' grouping.

#### 19.4.3.7 Summary of Results

Table 19.28 summarizes the research questions, hypotheses, and results.

## 19.5 DISCUSSION AND CONCLUDING REMARKS

The purpose of this chapter was to develop a supervised learning model to detect the Medicaid FFS claims with a high chance of being adjudicated erroneously. Eight research questions were articulated, six hypotheses were formulated, hypothesis testing and other statistical tests were performed, and the results were reviewed. Here crucial findings will be reviewed and discussed.

To answer the first research question, the researchers started reviewing the literature to understand what supervised machine learning algorithms can determine Medicaid claim payment issues. The researchers examined the literature extensively and found that among major classes of machine learning algorithms,

**TABLE 19.28**
**Summary of Research Questions and Results**

| Item | Description |
|---|---|
| Research Question 1 | What supervised machine learning algorithms can be used to determine Medicaid claim payment issues? |
| Result | The researchers selected ten algorithms for this study: decision tree with entropy configuration, decision tree with Gini coefficient configuration, random forest with entropy configuration, random forest with Gini configuration, naïve Bayes, K-nearest neighbor, logistic regression, neural network, discriminant analysis, and gradient boosting. |
| Research Question 2 | What are the most critical measures to compare the performance of different machine learning algorithms (resulting from RQ1) for our problem? |
| Result | Based on the nature of our problem at hand and the literature review, the researchers selected recall and F1-score as two metrics to compare our supervised learning models' performance. |
| Research Question 3 | What are the claim attributes that could predict if a given FFS claim has been adjudicated correctly or erroneously (also known as predictors or independent variables)? |

(*Continued*)

**TABLE 19.28**
**(Continued)**

| Item | Description |
|---|---|
| Result | The researchers used PCA on the ratio features and concluded all independent ratio variables should be kept in the model to account for over 90% of the variation. The researcher also used business knowledge and subject matter expertise to eliminate the low-information categorical features and selected 16 features for outpatient and 19 features for the inpatient dataset. |
| Research Question 4 | Is there any statistically significant difference among predictors' predictability power in identifying erroneously processed Medicaid outpatient claims? |
| Hypothesis | $H_{0_{4.1}}$: $\mu_{11} = \mu_{12} = \ldots = \mu_{116}$ , $\mu_{1i}$ = feature importance score (Gini importance) of the $i$-th outpatient predictor. |
| Result | clm_pmt_amt is the most powerful predictor. See the finding in the previous testing of hypotheses sub-section. |
| RQ5 | Same as RQ4 for inpatient claims. |
| Hypothesis | $H_{0_{5.1}}$: $\mu_{21} = \mu_{22} = \ldots = \mu_{2.19}$, $\mu_{1i}$ = importance of $i$-th inpatient predictor. |
| Result | clm_pmt_amt and nch_prmry_pyr_clm_pd_amt have the highest mean $F$-values in our inpatient claim dataset. See the previous sub-section. |
| Research Question 6 | Is there any statistically significant difference among the selected supervised learning algorithms in identifying erroneously adjudicated Medicaid outpatient claims? |
| Result | See the finding in the previous sub-section. |
| RQ7 | Same as RQ6 for inpatient claims. |
| Result | See the finding in the previous sub-section. |
| Research Question 8 | Are the most powerful predictors different between outpatient and inpatient claims? Are the most accurate algorithms in detecting erroneously paid claims different between the two types of Medicaid claims studied? |
| Result | 1. *clm_pmt_amt* is the most powerful predictor for both datasets.<br>2. For outpatient claims, *icd9_dgns_cd_1* and *at_physn_npi* have the next highest importance scores.<br>3. For inpatient claims, *nch_prmry_pyr_clm_pd_amt*, and *clm_drg_cdn* have the next highest importance scores.<br>4. NB and LR performed considerably better in outpatient data than inpatient data. This shows this algorithm requires more labeled data to train and perform well.<br>5. GB and DT (with various configurations and sampling strategies) outperform the rest of the algorithms in recall and F1-measure on both datasets. |

supervised learning was the most used method to detect quality issues, which is essentially a classification problem. While there is no set of universally accepted "best" supervised learning algorithms, SVM, naïve Bayes, and neural networks are among the algorithms most used in solving quality problems. SVM requires more computational power, especially for massive datasets. Hence, SVM was not selected in this chapter. The researchers also found principal component analysis is the most used dimensionality reduction technique allowing researchers to limit the number of features or predictors without compromising overall performance. Many factors should be considered in selecting proper algorithms, including their accuracy, detection power (recall), stability, and computation time. There are several trade-offs in selecting classification algorithms, including interpretability vs. prediction accuracy and bias-variance trade-offs. The researchers concluded this review by selecting ten supervised learning algorithms for the current problem.

The second research question was around finding the most critical measures to compare different machine learning algorithms' performance. While there are many metrics that could be used to compare machine learning algorithms' performance, the problem's nature dictates the superior metrics. Significant findings related to this research question could be summarized in three points:

1. The most famous performance metrics are around false and true positives and all derived from the confusion metrics, which are easy to calculate and communicate.
2. For the problem dealing with imbalanced data, accuracy will be inflated due to the high number of true negatives in the denominator. Recall and F1-score can measure the sensitivity and specificity of the model more realistically.
3. F1-score outperforms the ROC chart which measures the performance of an algorithm on imbalanced data.

Based on the nature of our problem at hand and the literature review, Recall and F1-score were selected as two metrics to compare the supervised learning models' performance.

The third research question was about finding the predictors to detect quality issues in FFS inpatient and outpatient claims. There were two massive datasets with over 100 features. In theory, all features could be used to build a supervised learning model. The researchers found thorough the literature review and some tests that this will increase the models' complexity without improving their performances. Dealing with mixed predictors (categorical and ratio) adds to the complexity of feature reduction. PCA on the ratio features showed all ratio features should be kept in our model to account for over 90% of the variation. Business knowledge and subject matter expertise were applied to eliminate the low-information categorical features. The researchers finally reduced the number of features by over 80% and selected 16 features or attributes for outpatient models and 19 features for inpatient models.

Research questions 4 and 5 were related to finding the most powerful predictors in detecting quality issues in our claim datasets for outpatient claims (RQ4) and inpatient claims (RQ5). Gini importance was calculated, and the hypothesis was tested with one-way ANOVA.

For both outpatient and inpatient claims, the Tukey pairwise comparisons using the Tukey method and a 95% confidence level show the feature *clm_pmt_amt* has the highest mean *F*-value. For outpatient claims, *icd9_dgns_cd_1* and *at_physn_npi* have the second and third highest values. For inpatient claims, *nch_prmry_pyr_clm_pd_amt*, and *clm_drg_cd* have the next highest mean *F*-value after *clm_pmt_amt*. The Fisher LSD method also presented a significant difference in *clm_pmt_amt*'s feature importance scores compared to other features. However, the Fisher test results are slightly different from the Tukey test results for inpatient claims. This test grouped *nch_prmry_pyr_clm_pd_amt* and *clm_drg_cdn* together, with a significant difference between these two features and the rest of the features. This is because the Tukey test is more conservative (Andy, 2015). Also, Fisher's LSD does not control the family-wise error rate (FWER) (Scheffe, 1953). Some researchers suggested using a more liberal method like Fisher's LSD to avoid being unnecessarily conservative (with weak statistical power) and detect a real difference when the goal is to prevent type II errors (Lee & Lee, 2018). In our case, it is preferred not to lose the opportunity to detect a real power of a feature in predicting the result, so both Tukey and Fisher LSD tests were conducted, and results were presented. From a practical point of view, proving that "Claim Payment Amount" (*clm_pmt_amt*) is our most crucial predictor was somehow expected. Differences in claim payment, especially for large under- or overpayments, make a payment an outlier in our dataset. The researchers applied the pattern of adjudication errors on the claim payments using the claim errors distribution function, without providing and adjusting relevant data elements, including:

- Who has processed the claim?
- What were applicable policies impacting the claim pricing methodology?
- What was the degree of difficulty in interpreting specific claim pricing policies?

For inpatient claims, the second most powerful predictor is a payment-related feature, too: "NCH Primary Payer Claim Paid Amount" (nch_prmry_pyr_clm_pd_amt). Its prediction power probably contributed to the fact that this amount provides a comparison point across various payments. The third predictor with a significantly higher feature important score is: "Claim Diagnosis Related Group Code" (clm_drg_cd). Its prediction power could be explained by the fact that for inpatient claims, pay-per-performance methods require providers like hospitals to bundle the related claims, thus providing another point of comparison for the algorithm to create a baseline and detect anomalies.

To answer RQ6 if there is a significant difference among the selected supervised learning algorithms in identifying quality issues in outpatient claims, the researchers conducted a one-way ANOVA test on two measures: recall and F1-score. The result showed the logistic regression model (with oversampling) has the best detection power (recall). The logistic regression model (without oversampling) and naïve Bayes (with and without oversampling) are the next best performers and group together (group B). Naïve Bayes (without oversampling) and gradient boosting (with oversampling) outperformed the rest of the algorithms from an F1 score perspective in outpatient data. Oversampling significantly improves the F1-score of the gradient boosting algorithm in outpatient data. The SMOTE algorithm was used for oversampling of the minority class and improved the gradient boosting performance. This is congruent with findings of other researchers working on imbalanced data (Cahyana et al., 2019).

RQ7 related to inpatient claims was like RQ6 for outpatient claims. Gradient boosting (with oversampling) outperformed the rest of the models in recalling quality issues of the inpatient claim data. k-nearest neighbor, neural network, discriminant analysis, and logistic regression (without sampling) have significantly lower recall than the rest of the algorithms on the inpatient dataset. Gradient boosting (without and with oversampling), decision tree (using entropy and Gini, and without oversampling), and random forest (with entropy and Gini, and without oversampling) have significantly higher F1 values on inpatient data. Oversampling significantly improved LR recall, a sign of the potential impact of imbalanced data on the performance of this algorithm.

The oversampling strategy applied in this research improved the recall and F1-score of models most of the time. In a few cases, the improvement done by oversampling was statistically significant (e.g., neural network recall improvement for inpatient claims, F1-score of the gradient boosting in outpatient data, etc.).

Research question 8 has two parts:

1. Are the most powerful predictors different between outpatient and inpatient claims? The result of our ANOVA tests shows *clm_pmt_amt* is the most powerful predictor for both datasets, which was expected. For outpatient claims, *icd9_dgns_cd_1* and *at_physn_npi* have the second-highest importance scores. For inpatient claims, *nch_prmry_pyr_clm_pd_amt* and *clm_drg_cdn* the have the second-highest importance scores.
2. Are the most accurate algorithms in detecting erroneously paid claims different between the two types of Medicaid claims studied?

This question is complex and requires comparing several measures. Table 19.29 illustrates the difference among various tested algorithms on both datasets. Tested algorithms are sorted based on their recall and F1-score from 1 to 20, where 1 shows the algorithm with the highest score for recall or F1-score. Then an average rank from recall and F1-score ranks is calculated, and algorithms are sorted based on their average rank.

Table 19.29 shows there are commonalities between the best and least effective algorithms across both datasets. Gradient boosting (with or without sampling), decision tree (both with entropy and Gini classifiers), and random forests have the best overall performance from both recall and F1-scores. There are some critical differences in the performance of some algorithms on two different datasets. Naïve Bayes is the best-performing algorithm on outpatient data on F1-score. This algorithm's precision, and consequently its F1-score, decreased with oversampling. This could have been because of overfitting of the data. Logistic regression has excellent recall on an outpatient dataset, but not so much for inpatient claims. Random forest models performed better on the inpatient dataset.

**TABLE 19.29**
**Ranking of ML Algorithms**

**Ranking of Algorithms on Outpatient Data**

| Algorithm | Recall Rank | F1-Score Rank | Average Rank |
|---|---|---|---|
| NB | 3 | 1 | 2.0 |
| GB (OS) | 9 | 2 | 5.5 |
| DTE (OS) | 6 | 8 | 7.0 |
| GB | 12 | 3 | 7.5 |
| DTG (OS) | 7 | 9 | 8.0 |
| LR | 2 | 14 | 8.0 |
| LR (OS) | 1 | 15 | 8.0 |
| NB (OS) | 4 | 12 | 8.0 |
| RFE (OS) | 11 | 6 | 8.5 |
| DA (OS) | 5 | 13 | 9.0 |
| DTG | 14 | 4 | 9.0 |
| DTE | 15 | 5 | 10.0 |
| RFG (OS) | 13 | 7 | 10.0 |
| NN (OS) | 8 | 17 | 12.5 |
| kNN (OS) | 10 | 16 | 13.0 |
| RFG | 16 | 10 | 13.0 |
| RFE | 17 | 11 | 14.0 |
| kNN | 19 | 18 | 18.5 |
| NN | 18 | 19 | 18.5 |
| DA | 20 | 20 | 20.0 |

**Ranking of Algorithms on Inpatient Data**

| Algorithm | Recall Rank | F1-Score Rank | Average Rank |
|---|---|---|---|
| GB | 4 | 1 | 2.5 |
| GB (OS) | 1 | 6 | 3.5 |
| DTE | 6 | 2 | 4.0 |
| DTG (OS) | 2 | 8 | 5.0 |
| DTG | 7 | 4 | 5.5 |
| RFG | 8 | 3 | 5.5 |
| DTE (OS) | 5 | 7 | 6.0 |
| RFE | 10 | 5 | 7.5 |
| DA (OS) | 3 | 13 | 8.0 |
| NB (OS) | 9 | 12 | 10.5 |
| RFE (OS) | 12 | 9 | 10.5 |
| NB | 11 | 11 | 11.0 |
| RFG (OS) | 13 | 10 | 11.5 |
| NN (OS) | 14 | 14 | 14.0 |
| LR (OS) | 15 | 15 | 15.0 |
| kNN (OS) | 16 | 16 | 16.0 |
| kNN | 18 | 17 | 17.5 |
| NN | 17 | 19 | 18.0 |
| LR | 19 | 18 | 18.5 |
| DA | 20 | 20 | 20.0 |

## 19.6 IMPORTANCE OF THE FINDINGS AND PRACTICAL IMPLICATIONS

Sampling and inferential statistics are the dominant quality tools utilized by Medicaid FFS programs. Such methods simply offer an interval estimate of the size of the problem in payments (e.g., estimated erroneous payments). They are not designed to detect potential erroneous payment at scale. This chapter aimed to apply supervised learning as a method to detect such possible erroneous payments at scale. The average accuracy and recall rate for the test algorithms are not as high as one would like to see. However, if this method is implemented successfully using operational data, it would be a game-changer. Let us provide a rough estimate of the potential saving that could be realized by implementing this method in a typical Medicaid FFS. Assume the following:

- The number of adjudicated claims a year 1,000,000 (many Medicaid FFS programs process a considerably higher volume).

- Percentage of erroneous payments 5% (a conservative estimate per PERM reports).
- Estimate the number of erroneously processed claims: 50,000.
- The average cost per erroneously adjudicated claim: $300. The average cost of a claim varies based on the claim type, but $300 is a conservative estimate (Substance Abuse and Mental Health Services Administration, 2015).
- Total potential saving: $15 million a year.
- Average manually sampled and review claims by the QM team at a typical Medicaid FFS: 50 a day, 10,000 reviewed claim a year.
- Potential saving through detecting potential erroneous payment (if they are detected pre-payment): 10,000 * 5% * $300 = $150K.
- Cost of each review: $5, so the total cost of review for 10,000 samples is $50 K.
- Actual saving for the Medicaid FFS program: $150k – $50k = $100K.

So, in this scenario, conventional sampling will save about $100K. If the same program can design and implement a model with a recall rate of 50% and precision of 10% (which equals an F1-score of 16.7%), the researchers calculate the potential saving using the assumptions listed in the previous scenario:

- Recall 50% means the model could potentially detect half of 50K erroneous payments or 25K claims.
- 10% precision means to detect 25K erroneous claims, 250K claims should be reviewed.
- Potential saving through detecting possible erroneous payment (if they are caught pre-payment): 25,000 * $300 = $7.5M
- The total cost of reviewing 250K claims: 250K * $10 = $2.5M
- Actual saving for the Medicaid FFS program: $7.5 – $2.5M = $5M

When real-world data is used, and the model is fine-tuned, then the accuracy rate will increase, and the cost savings would be even more substantial. One recommendation is to use several algorithms in tandem and create a scoring model to improve overall accuracy. A Medicaid FFS program can set up a multi-phase approach in which algorithms with higher recall used to collect as many true positives as possible and then use algorithms with higher accuracy to detect true positives.

## 19.7 RECOMMENDATIONS FOR FURTHER RESEARCH

The most prominent recommendation for further research based on this work's findings is to apply the same method to detect quality issues on other Medicaid FFS claim types, including medical (e.g., DME, vision, etc.), dental, and crossover claims. Medicare does not pay for most of the long-term care (LTS) claims. So, one should start with finding a set of Medicaid FFS LTC claims before applying the method recommended in this chapter. As shown in the result gathered from the California CA-MMIS project, quality issues in the Medicaid FFS claim adjudication process are rare events. As a result, collecting enough data to label target training datasets for specific claim types may require access to years' worth of data and be very time consuming. One way to overcome this challenge is to study the possibility of using text analytics and unsupervised learning methods to automate the data labeling process and then train the supervised algorithms built in this project.

Another aspect of this research that could be expanded is to apply the same process and include other algorithms. For example, SVM was not selected for testing because it requires massive computation power on the mix and massive datasets like datasets used for this research. However, it is recommended that researchers with more computational power at their disposal include SVM in their list of algorithms. As shown in the literature review, SVM has been successfully applied to solve many classification quality issues and frequently came on the top of this list from the accuracy and recall point of view.

As discussed in the managing data section, a synthesized dataset created by CMS was used for this research. The method explained in CMS's DE-SynPUF codebook user manual (The Centers for Medicare and Medicaid Services, 2013b) describes the data synthesizing process in detail. Then, to create labels, the general pattern of quality issues on a Medicaid FFS program was studied (e.g., population proportion, underlying distribution function, descriptive statistics, etc.). This approach eliminates the chance of any PHI disclosure. The biggest drawback of using a synthesized dataset is the limitation of multivariate modeling and model validation. As explained in the CMS's DE-SynPUF codebook, when the dynamic relationships between variables are altered (in this case, demographic, clinical, financial, and provider data), analyses from multivariate modeling should be interpreted with caution because the generated dataset may or may not inherit all the critical characteristic of the original dataset (The Centers for Medicare and Medicaid Services, 2013a). It may be appropriate for future research to use the claim and quality issues data from the same Medicaid program and start with the deidentification of sensitive data. In this chapter, beneficiary data, claim examiners' information, and information related to the pricing policies (e.g., error codes) were not included. A researcher can include them in their dataset, follow the steps explained in this chapter, and examine if they can find better predictors for quality issues in the Medicaid FFS claim adjudication data.

## NOTES

1 [B] best-performing algorithm among tested ones.
2 2018 PERM is not available on the CMS website.

## REFERENCES

Andy. (2015, August 31). *Tukey's Test or LSD Fisher—Cross Validated*. Stats.Stackexchange. https://stats.stackexchange.com/questions/169412/tukeys-test-or-lsd-fisher

Arizona Health Care Cost Containment System. (n.d.). *AHCCCS Fee-for-Service Fee Schedules Proposed Fee Schedules*. Retrieved June 17, 2018, from www.azahcccs.gov/PlansProviders/RatesAndBilling/FFS/

Arsenault, M.-O. (2017). KOLMOGOROV–SMIRNOV TEST. A Needed Tool in Your Data Science. *Towards Data Science*. https://towardsdatascience.com/kolmogorov-smirnov-test-84c92fb4158d

Bai, Y., Sun, Z., Deng, J., Li, L., Long, J., & Li, C. (2017). Manufacturing Quality Prediction Using Intelligent Learning Approaches: A Comparative Study. *Sustainability*, *10*, 85.

Beeravalli, V. (2018, September 30). Comparison of Machine Learning Classification Models for Credit Card Default Data. *Medium.* https://medium.com/@vijaya.beeravalli/comparison-of-machine-learning-classification-models-for-credit-card-default-data-c3cf-805c9a5a

Blomquist, H., & Möller, J. (2015). Anomaly Detection with Machine Learning—Quality Assurance of Statistical Data in the Aid Community. www.utn.uu.se/sts/cms/node/908

Bolton, R. J., & Hand, D. J. (2001). Unsupervised Profiling Methods for Fraud Detection. *Credit Scoring and Credit Control Conference VII*, Edinburgh, UK.

Borget, I., Laville, I., Paci, A., Michiels, S., Mercier, L., Desmaris, R.-P., & Bourget, P. (2006). Application of an Acceptance Sampling Plan for Post-Production Quality Control of Chemotherapeutic Batches in an Hospital Pharmacy. *European Journal of Pharmaceutics and Biopharmaceutics*, *64*(1), 92–98. https://doi.org/https://doi.org/10.1016/j.ejpb.2006.04.002

Bowden, R., & Bullington, S. F. (1996). Development of Manufacturing Control Strategies Using Unsupervised Machine Learning. *IIE Transactions*, *28*(4), 319–331.

Brain, D., & Webb, G. I. (1999). On the Effect of Data Set Size on Bias and Variance in Classification Learning. *Proceedings of the Fourth Australian Knowledge Acquisition Workshop (AKAW '99)*, 117–128.

Brownlee, J. (2019). *How to Choose a Feature Selection Method For Machine Learning.* https://machinelearningmastery.com/feature-selection-with-real-and-categorical-data/

Cahyana, N., Khomsah, S., & Aribowo, A. S. (2019). Improving Imbalanced Dataset Classification Using Oversampling and Gradient Boosting. *2019 5th International Conference on Science in Information Technology (ICSITech)*, 217–222. https://doi.org/10.1109/ICSITech46713.2019.8987499

Calvin, & Setiawan, J. (2014). Using Text Mining to Analyze Mobile Phone Provider Service Quality (Case Study: Social Media Twitter). *International Journal of Machine Learning and Computing*, *4*(1), 106–109. https://doi.org/10.7763/IJMLC.2014.V4.395

The Centers for Medicare and Medicaid Services. (2013a). *CMS Linkable 2008–2010 Medicare Data Entrepreneurs ' Synthetic Public Use File (DE-SynPUF): Codebook* (p. 125). www.cms.gov/Research-Statistics-Data-and-Systems/Downloadable-Public-Use-Files/SynPUFs/index.html

The Centers for Medicare and Medicaid Services. (2013b). *CMS Linkable 2008–2010 Medicare Data Entrepreneurs ' Synthetic Public Use File (DE-SynPUF): User Namual* (p. 34). www.cms.gov/Research-Statistics-Data-and-Systems/Downloadable-Public-Use-Files/SynPUFs/index.html

The Centers for Medicare and Medicaid Services. (2015). *Medicaid and CHIP 2015 Improper Payments Report.* www.cms.gov/Research-Statistics-Data-and-Systems/Monitoring-Programs/Medicaid-and-CHIP-Compliance/Downloads/2015MedicaidandCHIPImproperPaymentsReport.pdf

The Centers for Medicare and Medicaid Services. (2017). 2017 *Medicaid & CHIP Supplemental Improper Payment Data.* https://www.cms.gov/files/document/2017-medicaid-chip-supplemental-improper-payment-data.pdf

The Centers for Medicare and Medicaid Services. (2019a). *2019 Medicaid & CHIP Supplemental Improper Payment Data.* https://www.cms.gov/files/document/2019-medicaid-chip-supplemental-improper-payment-data.pdf-1

The Centers for Medicare and Medicaid Services. (2019b). *CMS 2008–2010 Data Entrepreneurs' Synthetic Public Use File (DE-SynPUF) | CMS.* www.cms.gov/Research-Statistics-Data-and-Systems/Downloadable-Public-Use-Files/SynPUFs/DE_Syn_PUF

Chou, J.-S., Ho, C.-C., & Hoang, H.-S. (2018). Determining Quality of Water in Reservoir Using Machine Learning. *Ecological Informatics*, *44*, 57–75. https://doi.org/10.1016/J.ECOINF.2018.01.005

Copeland, L., Edberg, D., Panorska, A. K., & Wendel, J. (2012). Applying Business Intelligence Concepts to Medicaid Claim Fraud Detection. *Journal of Information Systems Applied Research (JISAR)*, *5*(1), 51–61. www.aitp-edsig.org

Dean, J. (2014). Big Data, Data Mining, and Machine Learning: Value Creation for Business Leaders and Practitioners. In *Wiley & SAS Business Series*. Wiley.

El-Hasnony, I. M., Bakry, H. M. El, & Saleh, A. A. (2015). Comparative Study Among Data Reduction Techniques over Classification Accuracy. *International Journal of Computer Applications*, *122*(2), 8–15. https://doi.org/10.5120/21671-4752

Elish, K. O., & Elish, M. O. (2008). Predicting Defect-Prone Software Modules Using Support Vector Machines. *Journal of Systems and Software*. https://doi.org/10.1016/j.jss.2007.07.040

Escobar, C. A., & Morales-Menendez, R. (2018). Machine Learning Techniques for Quality Control in High Conformance Manufacturing Environment. *Advances in Mechanical Engineering*, *10*(2), 1–16. https://doi.org/10.1177/1687814018755519

Feuerverger, A. (2016). On Goodness of Fit for Operational Risk. *International Statistical Review*, *84*(3), 434–455. https://doi.org/10.1111/insr.12112

Fu, H.-H., Tsai, H.-T., Lin, C.-W., & Wei, D. (2004). Application of a Single Sampling Plan for Auditing Medical-Claim Payments Made by Taiwan National Health Insurance. *Health Policy*, *70*(2), 185–195. https://doi.org/10.1016/J.HEALTHPOL.2004.02.006

Gallardo, K. (2017). *Analytics Engine for Detecting Medical Fraud, Waste, and Abuse* (Patent No. US20170270435A1). https://patents.google.com/patent/US20170270435A1/en

Getchius, J. M. (2014). *Healthcare Fraud Detection with Machine Learning* (Patent No. US20140149128 A1).

Gonzalez Viejo, C., Fuentes, S., Torrico, D., Howell, K., & Dunshea, F. R. (2017). Assessment of Beer Quality Based on Foamability and Chemical Composition Using Computer Vision Algorithms, Near Infrared Spectroscopy and Machine Learning Algorithms. *Journal of the Science of Food and Agriculture*. https://doi.org/10.1002/jsfa.8506

Gupta, Y. (2018). Selection of Important Features and Predicting Wine Quality Using Machine Learning Techniques. *Procedia Computer Science*, *125*, 305–312. https://doi.org/10.1016/J.PROCS.2017.12.041

Hawaii Department of Human Services, & Med-QUEST. (n.d.). *Medicaid Fee-for-Service Eligibility*. Retrieved June 17, 2018, from http://humanservices.hawaii.gov/mqd/medicaid-fee-for-service/#Eligibility-0

Heino, J., & Toivonen, H. (2003). Automated Detection of Epidemics from the Usage Logs of a Physicians' Reference Database. *7th European Conference on Principles and Practice of Knowledge Discovery in Databases*, 180–191. https://doi.org/10.1007/978-3-540-39804-2_18

The Henry J. Kaiser Family Foundation. (2018). *Federal and State Share of Medicaid Spending*. www.kff.org/medicaid/state-indicator/federalstate-share-of-spending/?dataView=1&currentTimeframe=0&sortModel=%7B%22colId%22:%22Location%22,%22sort%22:%22asc%22%7D

Hilal, W., Gadsden, S. A., & Yawney, J. (2022). Financial Fraud: A Review of Anomaly Detection Techniques and Recent Advances. *Expert Systems with Applications*, *193*, 116429. https://doi.org/10.1016/j.eswa.2021.116429

Hoseini, C. (2020). Leveraging Machine Learning to Identify Quality Issues in the Medicaid Claim Adjudication Process. PhD Dissertation. Indiana State University.

Hoseini, C., Badar, M. A., Shahhosseini, A. M., & Kluse, C. J. (2021). A Review of Machine Learning Methods Applicable to Quality Issues. *IEOM Proceedings of the 11th International Conf. on Industrial Engineering and Operations Mgmt*, March 7–11, 2021, Singapore—Virtual, 1225–1240, www.ieomsociety.org/singapore2021/papers/250.pdf

Iowa Medicaid Enterprise Performance Report. (2006). www.legis.iowa.gov/docs/publications/IH/4906.pdf

Irgens, C., Wuest, T., & Thoben, K.-D. (2012). Product State Based View and Machine Learning: A Suitable Approach to Increase Quality? *IFAC Proceedings Volumes*, *45*(6), 1733–1738. https://doi.org/10.3182/20120523-3-RO-2023.00083

Issa, H., & Vasarhelyi, M. A. (2011). Application of Anomaly Detection Techniques to Identify Fraudulent Refunds. *Available at SSRN 1910468* (pp. 1–19). https://doi.org/10.2139/ssrn.1910468

James, G., Witten, D., Hastie, T., & Tibshirani, R. (2015). An Introduction to Statistical Learning. In *Springer Texts in Statistics*. https://doi.org/10.1016/j.peva.2007.06.006

Karimi, N., Ranjbarzadeh Kondrood, R., & Alizadeh, T. (2017). An Intelligent System for Quality Measurement of Golden Bleached Raisins Using Two Comparative Machine Learning Algorithms. *Measurement*, *107*, 68–76. https://doi.org/10.1016/J.MEASUREMENT.2017.05.009

Kazemian, H., & Shrestha, S. (2023, November 1). Comparisons of Machine Learning Techniques for Detecting Fraudulent Criminal Identities. *Expert Systems with Applications*, *229*(Part A), 120591. https://doi.org/10.1016/j.eswa.2023.120591

Kerry, S. M., & Bland, J. M. (1998). The Intracluster Correlation Coefficient in Cluster Randomisation. *BMJ*, *316*(7142), 1455–1460. https://doi.org/10.1136/bmj.316.7142.1455

Khoshgoftaar, T. M., & Seliya, N. (2003). Software Quality Classification Modeling Using the SPRINT Decision Tree Algorithm. *International Journal on Artificial Intelligence Tools*, *12*(03), 207–225. https://doi.org/10.1142/S0218213003001204

Kohavi, R., & Provost, F. (1998). *Glossary of Terms*. Machine Learning. http://ai.stanford.edu/~ronnyk/glossary.html

Kose, I., Gokturk, M., & Kilic, K. (2015). An Interactive Machine-Learning-Based Electronic Fraud and Abuse Detection System in Healthcare Insurance. *Applied Soft Computing Journal*. https://doi.org/10.1016/j.asoc.2015.07.018

Lee, S., & Lee, D. K. (2018). What Is the Proper Way to Apply the Multiple Comparison Test? *Korean Journal of Anesthesiology*, *71*(5), 353–360. https://doi.org/10.4097/kja.d.18.00242

Li, J., Huang, K.-Y., Jin, J., & Shi, J. (2008). A Survey on Statistical Methods for Health Care Fraud Detection. *Health Care Management Science*, *11*(3), 275–287. https://doi.org/10.1007/s10729-007-9045-4

Lieber, D., Stolpe, M., Konrad, B., Deuse, J., & Morik, K. (2013). Quality Prediction in Interlinked Manufacturing Processes Based on Supervised & Unsupervised Machine Learning. *Procedia CIRP*. https://doi.org/10.1016/j.procir.2013.05.033

Lo, A., Chernoff, H., Zheng, T., & Lo, S.-H. (2015). Why Significant Variables Aren't Automatically Good Predictors. *Proceedings of the National Academy of Sciences of the United States of America*, *112*(45), 13892–13897. https://doi.org/10.1073/pnas.1518285112

Lotfi, E., Yaghoobi, M., & Pourreza, H. R. (2008). A New Approach for Automatic Quality Control of Fried Potatoes Using Machine Learning. *2008 7th IEEE International Conference on Cybernetic Intelligent Systems*, 1–4. https://doi.org/10.1109/UKRICIS.2008.4798934

Medicaid.gov. (2018). *March 2018 Medicaid & CHIP Enrollment Data Highlights*. Medicaid.Gov. www.medicaid.gov/medicaid/program-information/medicaid-and-chip-enrollment-data/report-highlights/index.html

Mohri, M., Rostamizadeh, A., & Talwalkar, A. (2014). Foundations of Machine Learning. In *Adaptive Computation and Machine Learning Serries*. MIT Press.

Nevada Department of Health and Human Services Division of Health Care Financing and Policy. (n.d.). *NHHS Fee Schedules*. Retrieved June 17, 2018, from http://dhcfp.nv.gov/Resources/Rates/FeeSchedules/

NM Human Services Department (HSD). (n.d.). *NM Fee for Service Fee Schedules*. Retrieved June 17, 2018, from www.hsd.state.nm.us/providers/fee-for-service.aspx
Ordukaya, E., & Karlik, B. (2017). Quality Control of Olive Oils Using Machine Learning and Electronic Nose. *Journal of Food Quality*. https://doi.org/10.1155/2017/9272404
Paprzycki, M., Abraham, A., Guo, R., & Abraham, A. (2004). Analyzing Call Center Performance: A Data Mining Approach. *2.Softcomputing.Net, Section 3*. http://2.softcomputing.net/icfai-km.pdf%5Cnhttp://link.springer.com/chapter/10.1007/978-3-540-24677-0_112
*PCA for Categorical Features?—Stack Overflow*. (n.d.). 2020. Retrieved September 29, 2020, from https://stackoverflow.com/questions/40795141/pca-for-categorical-features
Peddamuthu, B., & Srivastava, S. (2014). *QualityAssurance in Real-Time (QART)*. Xerox Research Centre India. www.xrci.xerox.com/quality-assurance-in-real-time-qart
Petrov, D., Gutman, B. A., Shih-Hua, Yu, van Erp, T. G. M., Turner, J. A., Schmaal, L., Veltman, D., Wang, L., Alpert, K., Isaev, D., Zavaliangos-Petropulu, A., Ching, C. R. K., Calhoun, V., Glahn, D., Satterthwaite, T. D., Andreasen, O. A., Borgwardt, S., Howells, F., . . . Thompson, P. M. (2017). *Machine Learning for Large-Scale Quality Control of 3D Shape Models in Neuroimaging*. http://arxiv.org/abs/1707.06353
Phua, C., Lee, V., Smith-Miles, K., & Gayler, R. (2005). *A Comprehensive Survey of Data Mining-Based Fraud Detection Research*. School of Business Systems, Faculty of Information Technology, Monash University.
Piatetsky, G. (2018). *Python Eats Away at R: Top Software for Analytics, Data Science, Machine Learning in 2018: Trends and Analysis*. KDnuggets. www.kdnuggets.com/2018/05/poll-tools-analytics-data-science-machine-learning-results.html/2
Rácz, A., Bajusz, D., & Héberger, K. (2019). Multi-Level Comparison of Machine Learning Classifiers and Their Performance Metrics. *Molecules (Basel, Switzerland)*, *24*(15), 2811. https://doi.org/10.3390/molecules24152811
Ribeiro, B. (2005). Support Vector Machines for Quality Monitoring in a Plastic Injection Molding Process. *IEEE Transactions on Systems, Man and Cybernetics, Part C (Applications and Reviews)*, *35*(3), 401–410. https://doi.org/10.1109/TSMCC.2004.843228
RTInsights Team. (2017). *Predicting Quality Assurance Outcomes for Process Manufacturing*. www.rtinsights.com/predicting-quality-assurance-outcomes-for-process-manufacturing/
Scheffe, H. (1953). A Method for Judging Contracts in the Analysis of Variance. *Biometrika*, *40*(1–2), 87–110. https://doi.org/10.1093/biomet/40.1-2.87
Shah, N., Chatterjee, A., Bhargava, A., & Joshi, N. (2016). *Model Validation, KS Test and Lorenz Curve*. Kharagpur Data Analytics Group-KDAG. https://kgpdag.wordpress.com/2016/03/14/model-validation-ks-test-and-lorenz-curve/
Shamsaei, B., & Gao, C. (2016). Comparison of Some Machine Learning and Statistical Algorithms for Classification and Prediction of Human Cancer Type. *3rd IEEE EMBS International Conference on Biomedical and Health Informatics, BHI 2016*, 296–299. https://doi.org/10.1109/BHI.2016.7455893
Shin, H., Park, H., Lee, J., & Jhee, W. C. (2012). A Scoring Model to Detect Abusive Billing Patterns in Health Insurance Claims. *Expert Systems with Applications*. https://doi.org/10.1016/j.eswa.2012.01.105
Simon, P. (2013). *Too Big to Ignore: The Business Case for Big Data*. Wiley.
Singh, W., & Kaur, P. (2016). Comparative Analysis of Classification Techniques for Predicting Computer Engineering Students ' Academic Performance. *International Journal of Advanced Research in Computer Science Research*, *7*(6).
Siris, V. A., & Papagalou, F. (2004). Application of Anomaly Detection Algorithms for Detecting SYN Flooding Attacks. *Global Telecommunications Conference, 2004. GLOBECOM '04. IEEE*, *4*, 2050–2054. https://doi.org/10.1109/GLOCOM.2004.1378372
Stackexchange. (n.d.). Can Principal Component Analysis Be Applied to Datasets Containing a Mix of Continuous and Categorical Variables?—Cross Validated. Retrieved September 29, 2020, from https://stats.stackexchange.com/questions/5774/can-principal-component-analysis-be-applied-to-datasets-containing-a-mix-of-cont

Sture, Ø. (2015). Automatic Quality Control of Salmon—Using Machine Learning Algorithms Based on Input from a 3D Machine Vision System. *138*. https://brage.bibsys.no/xmlui/handle/11250/2352578#.WwSR2hm3SjU.mendeley

Substance Abuse and Mental Health Services Administration. (2015). Analyses of MAX Claims: SAMHSA Fee-for-Service Spending Estimates, Medicare-Medicaid Enrollee Analysis, and Managed Care Summary. Substance Abuse and Mental Health Services Administration. https://store.samhsa.gov/sites/default/files/sma15-4908.pdf

Sumesh, A., Rameshkumar, K., Mohandas, K., & Babu, R. S. (2015). Use of Machine Learning Algorithms for Weld Quality Monitoring Using Acoustic Signature. *Procedia Computer Science*. https://doi.org/10.1016/j.procs.2015.04.042

*Testimony of Ann Maxwell Before the United States House of Representatives*. (2017). http://docs.house.gov/meetings/IF/IF02/20170131/105493/HHRG-115-IF02-Wstate-ArchambaultJ-20170131.pdf

Travaille, P., Thornton, D., Müller, R. M., & van Hillegersberg, J. (2011). Electronic Fraud Detection in the U. S. Medicaid Healthcare Program: Lessons Learned from other Industries. *Seventeenth Americas Conference on Information Systems*, 1–10. http://doc.utwente.nl/78000/1/Travaille11electronic.pdf

Tsung, F., Zhou, Z., & Jiang, W. (2007). Applying Manufacturing Batch Techniques to Fraud Detection with Incomplete Customer Information. *IIE Transactions*, *39*(6), 671–680. https://doi.org/10.1080/07408170600897510

Ucar, F., Alcin, O., Dandil, B., & Ata, F. (2018). Power Quality Event Detection Using a Fast Extreme Learning Machine. *Energies*. https://doi.org/10.3390/en11010145

Upton, G., & Cook, I. (2014). Anderson–Darling Test. In *A Dictionary of Statistics* (3rd ed.). Oxford University Press. https://doi.org/10.1093/acref/9780199679188.001.0001

Valdes, A., & Skinner, K. (2000). Adaptive, Model-Based Monitoring for Cyber Attack Detection. In *Recent Advances in Intrusion Detection* (pp. 80–92). https://doi.org/10.1007/3-540-39945-3_6

Vihinen, M. (2012). How to Evaluate Performance of Prediction Methods? Measures and Their Interpretation in Variation Effect Analysis. *BMC Genomics*, *13*(Suppl 4), S2. https://doi.org/10.1186/1471-2164-13-S4-S2

Williams, S., & Baugh, D. (2016). *Medicaid Analytic eXtract Data (MAX) Providing Data to Researchers and Policy Analysts*. www.cms.gov/Research-Statistics-Data-and-Systems/Computer-Data-and-Systems/MedicaidDataSourcesGenInfo/Downloads/MAXintrotoData.zip

Yahya. (2018). *F1 Score vs ROC AUC*. Stack Overflow. https://stackoverflow.com/questions/44172162/f1-score-vs-roc-auc

Yi, Z., Cao, X., Pu, X., Wu, Y., Chen, Z., Khan, A. T., Francis, A., & Li, S. (2023, November 30). Fraud Detection in Capital Markets: A Novel Machine Learning Approach. *Expert Systems with Applications*, 231, 120760. https://doi.org/10.1016/j.eswa.2023.120760.

Yussupova, N., Kovács, G., Boyko, M., & Bogdanova, D. (2016). Models and Methods for Quality Management Based on Artificial Intelligence Applications. *Acta Polytechnica Hungarica*, 13(3), 45–60.

# Index

## A



## B



## C

## D



## E

## Q



## R

## U



## V



## W